# Cell Movements

## FROM MOLECULES TO MOTILITY

Second Edition

# Cell Movements

## FROM MOLECULES TO MOTILITY

Second Edition

DENNIS BRAY

Vice President: Denise Schanck
Editor: Matthew Day
Editorial Assistant: Kirsten Jenner
Production Editor: Emma Hunt
Production Assistant: Angela Bennett
Copy Editor: Len Cegielka
Illustrator: Nigel Orme
Indexer: Dorothy Jahoda



Visit the Garland Science web site at
http://www.garlandscience.com

**Library of Congress Catalogue-in-Publication Data**
Bray, Dennis.
Cell movements : from molecules to motility / Dennis Bray.--2nd ed.
p. cm.
Includes bibliographical references and index.
ISBN 0-8153-3282-3 (alk. paper)
1. Cells--Motility. I. Title.

QH647 .B73 2000
571.6'7--dc21

00-055159

Published by Garland Publishing, a member of the Taylor & Francis Group
29 West 35th Street, New York, NY 10001-2299

Printed in the United States of America
15 14 13 12 11 10 9 8 7 6 5 4 3 2 1

# Preface

*"The first thing to say about cell movements is that they are fascinating to watch. Even to someone forearmed with bookish knowledge, an* Amoeba *crawling over a microscope slide or a* Paramecium *swimming in a drop of water, is an arresting sight. Not only are these scraps of living matter, a few tenths of a millimeter in diameter, capable of rapid motion but they often move with the hesitations and vagaries we know from the behavior of higher animals.*

*Cell movements are even more remarkable when one considers how close they are to molecular dimensions. The smallest moving part of a cell, such as a cilium on its surface or a vesicle moving through its cytoplasm, is only one step away from molecules in size. Another factor of ten in magnification and the underlying pattern of protein molecules begins to emerge. Watch a cilium beat and you see a compact molecular automaton in action.*

*Much of the appeal of cell motility, in fact, probably arises from a type of closet vitalism. The primitive association of life with movement, which we still retain, makes it a continual source of wonder that cell movements can have a simple mechanistic explanation. Certainly for myself, even after years of research on the motile tips of growing nerve axons, I still find it hard to accept that such complex, integrated, seemingly sentient structures can arise from dumb molecules. The mystery is doubtless of my own making but it was the mainspring for writing this book." (Cell Movements, first edition, preface).*

Coming back to this book a decade on has been an enduring pleasure—a favourite tune sung to a richer harmony. It was a chance to put right what was wrong last time and to take into account many comments and criticisms of readers. This edition also benefited enormously from dedicated editorial help and the informed advice of experts.

It is also true that the subject of cell motility has matured over the past ten years. Major advances have been made in many directions, some expected others not. The seismic impact of recombinant DNA technology was felt in cell motility as in other fields, enormously expanding our knowledge of the protein molecules involved and our understanding of their function. Sequence studies revealed the plethora of molecular motors that fill the cytoplasm of eucaryotic cells, for example, and guided our analysis of their structures. The dazzling renaissance of light microscopy revealed to us the cytoskeleton with unprecedented clarity, giving real-time living images of molecular processes that previously could only be guessed at. Modern light microscopes also underpinned novel techniques to manipulate and measure single molecules, teaching us at first-hand how statistical mechanics underlies all cellular processes. Less predictable discoveries were sparked by simple curiosity. Who would have guessed that investigating the pathogenic action of Listeria bacteria would lead researchers finally to solve the puzzle of how fibroblasts extend their leading edge? And who could have anticipated the insights into the control of cell movements that would emerge from the study of development in worms and fruit flies?

There are certainly reasons to congratulate ourselves, collectively, on the outcome of this wonderful enterprise. Puzzles that have been with us since biologists first focused their microscopes onto cells that crawl, swim, divide or change shape seem finally within reach. It is true that we still do not yet have a complete understanding of how a muscle contracts, or a crawling cell extends a filopodium, or a dividing cell moves a chromosome. But the questions that remain to be answered, many of them highlighted in this book, are now at a higher level of enquiry. We see, dimly, that cell motility is moving from an era of data-collection to one of hypothesis-driven research, in which informed and increasingly quantitative theories can be proposed and tested.

This edition was not only completely rewritten to take account of the many major advances in the field, but also produced to a much higher standard. A major factor in this has been the dedicated professional support of the editorial staff at Garland—especially Matthew Day, Emma Hunt, Angela Bennett, Kirsten Jenner and Nigel Orme—and the critical comments of experts. Every chapter was reviewed by professional scientists actively working in cell motility and their recommendations and comments had a major influence on the content and accuracy of the final book. In this regard, I wish to thank Dick Macintosh, Joel Rosenbaum, John Tucker, Jeff Hardin, John Cooper, Tom Pollard, Judy Armitage, Dan Kiehart, Paul Janmey, Donald Ingber, Vikki Allan, Albert Harris, John Harris, Mary Porter, Birgit Lane, Andrew Somlyo, Vic Norris, Bob Bourret, Joe Howard, Eric Karsenti, and Fred Fay, for taking time out of their busy schedules to help in this way, and the many cell biologists who provided facts and advice, or who freely provided images to be used as illustrations. The enthusiastic cooperation of these fine scientists was a continual source of inspiration and encouragement during the writing of the book.

Dennis Bray
Cambridge, UK
Summer 2000

# Contents


| | | |
|---|---|---|
| **PART 1** | **Cell Movements in the Light Microscope** | |
| Chapter 1 | Cell Swimming | 3 |
| Chapter 2 | Migration of Cells Over Surfaces | 17 |
| Chapter 3 | Cell Behavior | 29 |
| Chapter 4 | The Cytoskeleton | 41 |
| **PART 2** | **Molecular Basis of Cell Movements** | |
| Chapter 5 | Actin Filaments | 63 |
| Chapter 6 | Actin and Membranes | 81 |
| Chapter 7 | Myosin | 103 |
| Chapter 8 | Fibroblast Locomotion | 119 |
| Chapter 9 | The Molecular Basis of Muscle Contraction | 137 |
| Chapter 10 | Muscle Development | 155 |
| Chapter 11 | Microtubules | 171 |
| Chapter 12 | Organelle Transport | 189 |
| Chapter 13 | Mitosis | 207 |
| Chapter 14 | Cilia | 225 |
| Chapter 15 | Centrioles and Basal Bodies | 243 |
| Chapter 16 | Bacterial Movements | 257 |
| **PART 3** | **Integration of Cell Movements** | |
| Chapter 17 | Intermediate Filaments | 277 |
| Chapter 18 | Cell Mechanics | 293 |
| Chapter 19 | Cell Shape | 315 |
| Chapter 20 | Cell Movements in Embryos | 333 |
| Glossary | | 353 |
| Index | | 363 |

# List of Headings


**PART 1 Cell Movements in the Light Microscope** 1

Chapter 1 Cell Swimming 3

A cell suspended in water moves passively by diffusion 3
Diffusion is ineffective over large distances 4
The movement of cells through water is governed by viscous forces 6
Swimming consumes only a small fraction of the cell's energy 7
Bacteria swim by means of flagella 8
Bacterial flagella are rigid helical structures that rotate 9
Flagellated bacteria come in a menagerie of different forms 9
Bacterial flagella and eucaryotic flagella differ in structure and mode of action 10
Eucaryotic flagella generate a wide variety of waveforms 11
Many protozoa move by the coordinated beating of cilia on their surface 12
Fields of cilia move adjacent layers of water 13
Modified cilia and flagella are used for purposes other than swimming 14

Chapter 2 Migration of Cells Over Surfaces 17

Cells close to a surface experience large viscous drag 17
Amoebae crawl by means of pseudopodia 18
A stream of fluid cytoplasm fills a growing pseudopodium 19
White blood cells migrate freely in the vertebrate body 20
Fibroblasts move within connective tissue 21
Cell migration is fundamental to vertebrate embryonic development 21
Even cells from differentiated tissues have an innate capacity for locomotion 22
Fibroblasts crawl by means of lamellipodia 23
Filopodia act like feelers for the cell 24
Growth cones crawl over surfaces 25
Myxobacteria, gregarines, and diatoms glide over surfaces 25
Cilia and flagella are also used to move over surfaces 26

Chapter 3 Cell Behavior 29

Bacteria and amoebae migrate toward distant sources of chemoattractant 29
Cells migrate over long distances in the body 31
Adaptation is an essential part of the chemotactic response 31
Cells are sensitive to weak electric fields 32
Paramecia avoid obstacles by reversing their cilia 33
Cells are guided by their substrata 34
Nerve cells have a highly developed capacity for pathfinding 35
Contacts between cells modify their migratory behavior 36
Slime mold amoebae congregate in response to chemical signals 36
Single cells respond to light 37
Protozoa carry out complex sequences of actions 38

Chapter 4 The Cytoskeleton 41

Shapes and movements are generated by the cytoskeleton 42
The cytoskeleton contains actin filaments, microtubules, and intermediate filaments 43
Most eucaryotic movements depend on actin filaments and microtubules 44
Accessory proteins hold the cytoskeleton together and link it to membranes 45
Why are protein filaments used in cell shape determination and movements? 46
Cytoskeletal filaments are produced by the polymerization of subunits 47
Polymerization of actin and tubulin is controlled by nucleotide hydrolysis 48
Polymerization is a form of movement 49
Changes in protein conformation can produce large-scale movements 50
Protein motors move along actin filaments and microtubules 50
Motor proteins are driven by ATP hydrolysis 51
Shape and movement are regulated by ions and small diffusible molecules 52
Cytoskeletal responses are coordinated by protein kinases 54
A motile cell is an 'intelligent' cell 54
The cytoskeleton was probably crucial for the evolution of eucaryotic cells 55
Bacteria have an organized cytoplasm 56
The cytoskeleton carries epigenetic information 56
Cytoskeletal proteins show overlapping redundancy in their functions 58

**PART 2 Molecular Basis of Cell Movements 61**

Chapter 5 Actin Filaments 63

Actin is abundant in most eucaryotic cells 64
Actin is a highly conserved protein 64
Vertebrates have a family of actin genes 64
Actin assembles spontaneously into filaments 66
Actin filaments grow at their free ends by a reversible association of actin molecules 67
The two ends of an actin filament grow at different rates 69
ATP hydrolysis changes the rate of growth of actin filaments 70
Specific toxins bind to actin and affect its polymerization 71
Thymosin sequesters unpolymerized actin 72
Profilin and cofilin regulate actin assembly into filaments 72
Pathogenic bacteria provide a simplified model of actin nucleation 73
VASP and Arp2/3 nucleate actin filaments 74
Actin-binding proteins have a modular construction 75
Actin filaments are an important site of protein synthesis 77

Chapter 6 Actin and Membranes 81

Filamin and α-actinin cross-link actin filaments 81
Actin-fragmenting proteins rapidly reorganize the cortex 83
Actin filaments attach, usually indirectly, to the plasma membrane 84
The red blood cell membrane is supported by a network of spectrin and actin 84
Ankyrin and band 4.1 link spectrin to the plasma membrane 86
How does the cell regulate assembly of its spectrin network? 87
Adherens junctions provide a stable anchorage for actin filaments 87
Cultured cells attach to their substratum at focal adhesions 89
Focal adhesions send and receive signals 90
Focal adhesions are triggered by both external and internal influences 91

Rho GTPases control actin's association with the membrane 90
Lamellipodia, filopodia, and microvilli form in response to external stimuli 94
Filopodia grow by controlled actin polymerization 94
Bundling proteins increase the rigidity of cell extensions 95
Intestinal microvilli are held together by fimbrin and villin 96
Stereocilia are specialized microvilli that perform a sensory function 97
Cells control the number, length, and position of actin filaments in their cortex 97

Chapter 7 Myosin 103

Muscle myosin is a two-headed, long-tailed molecule 104
Myosin heads bind to actin filaments and hydrolyze ATP 104
Motility assays probe the interaction between myosin and actin 105
Single myosin molecules can be measured at work 106
The tail of myosin II self-assembles into a bipolar filament 107
Myosin II molecules cause bundles of actin filaments to contract 108
Assembly of myosin into filaments is influenced by phosphorylation 108
Light chain phosphorylation can regulate myosin activity 109
Tropomyosin helps to stabilize actin filaments for contraction 109
Large animal cells divide by a contractile ring of actin filaments 110
Genetic analyses probe the function of myosins 111
Recombinant DNA techniques reveal a large family of unconventional myosins 112
Myosin I's associate with membranes 113
Myosin V attaches vesicles to actin filaments 114
Myosins are used in hearing and vision 115

Chapter 8 Fibroblast Locomotion 119

Crawling cells extend, adhere, contract, detach 119
Nematode sperm crawl without actin 120
The crawling of most cells is based on actin 122
A migrating cell advances by polymerizing actin 122
Arp2/3 and cofilin support treadmilling in the leading lamella 123
Receptors on the underbelly of the cell attach it to the substratum 125
Crawling cells must pull themselves over the surface 125
Detachment is an essential part of cell migration 125
Myosin II is needed for efficient, directed cell crawling 126
Unconventional myosins may be even more important 127
Cell crawling is based on a cyclical flow of cortical actin 127
There are interesting parallels between cell migration and cell division 128
Cytoplasmic streaming and hydrostatic pressure contribute to amoeboid locomotion 128
How does the membrane recycle? 130
Microtubules influence the migration of fibroblasts 131
Chemoattractants polarize the motile machinery 132
Networks of signals regulate the motile machinery 132

Chapter 9 The Molecular Basis of Muscle Contraction 137

Tension measurements define the performance of a muscle 137
Muscles can work aerobically or anaerobically 138
Creatine phosphate is an energy carrier for muscle contraction 139
Each muscle fiber is a large multinucleated cell 140
Myofibrils are the contractile units of a muscle cell 141
Internal membranes relay nerve stimulation to myofibrils 141
Muscle contraction is caused by the sliding of actin and myosin filaments 142
Actin and myosin filaments are held in precise positions in the myofibril 144
Cross-bridges made of myosin heads interact cyclically with actin filaments 145

A rigid α-helix in the myosin head acts like a lever 146
Troponin and tropomyosin make muscle contraction $Ca^{2+}$-sensitive 147
Troponin C belongs to a family of calcium-binding proteins 148
Accessory proteins maintain the precise architecture of a myofibril 149
Titin and nebulin act as molecular rulers for the sarcomere 150

Chapter 10 Muscle Development 155

Skeletal muscle forms by fusion of mononucleated myoblasts 156
Stem cells restore damaged muscle 156
Titin plays a crucial role in myofibril assembly 157
Microtubules align forming myoblasts 158
Skeletal muscle contains fast and slow fibers 158
Muscle type is influenced by innervation 159
Exercise affects muscle development and gene expression 160
Different genes are expressed at different stages of muscle development 160
Myofilaments can add and lose subunits in the cell 161
Muscle proteins are further diversified by alternative RNA splicing 161
The molecular lesion in muscular dystrophy is known 162
Cardiac muscle fibers consist of chains of mononucleated cells coupled together 163
Smooth muscle contains a network of myofibrils linked by dense bodies 164
Smooth-muscle contraction is regulated by multiple enzyme cascades 165
Insect flight muscles and molluscan catch muscles are highly specialized 166
Genetic analysis enables muscle development to be dissected 167

Chapter 11 Microtubules 171

Tubulin is present in most eucaryotic cells but is usually purified from brain 171
Tubulin is a GTPase distantly related to Ras 172
Vertebrates have multiple tubulin genes 173
Tubulin polypeptides need molecular chaperones to fold correctly 174
Specific drugs bind to tubulin and affect its polymerization 175
Purified tubulin assembles into microtubules 176
Microtubules are hollow tubes made of protofilaments 176
Microtubules are polarized structures 177
Dynamic instability is driven by hydrolysis of GTP 178
Growing and shrinking ends are structurally different 179
The centrosome is the major microtubule-organizing center in animal cells 180
Asymmetric microtubule arrays can be generated by dynamic instability 181
Microtubules undergo a slow maturation 182
Microtubules treadmill through the cytoplasm 183
Many proteins bind to microtubules and modify their function in the cell 184

Chapter 12 Organelle Transport 189

Membrane-bound organelles move rapidly along microtubules 190
Nerve cells have an exaggerated dependence on organelle transport 191
Axons and dendrites have different arrays of microtubules 192
MAP2 and tau stabilize neuronal microtubules 193
Kinesin carries organelles toward the plus end of microtubules 193
Cells contain a superfamily of kinesinlike proteins 194
Kinesin is a processive motor 196
Cytoplasmic dynein carries organelles toward the minus end of microtubules 197
Dynactin couples dynein to its cargo 198
Dynein can cause microtubules to slide 199
Motor proteins are responsible for neuronal polarity 199

Vesicles can carry more than one type of motor 200
Pigment granule movement may be regulated by phosphorylation 201
Microtubules form the basis of feeding tentacles of protozoa 202
Ribosomes, viruses, and nuclei also move along microtubules 203

Chapter 13 Mitosis 207

Bacteria use FtsZ, a distant relative of tubulin, to divide 208
Eucaryotic cells go through a fixed sequence of events as they divide 208
Phosphorylation and proteolysis control the eucaryotic cell cycle 209
Breakdown of the nuclear membrane commits the cell to mitosis 210
Microtubules assemble into a spindle 210
Chromosomes attach to the mitotic spindle 212
Chromosomes are both pushed and pulled by spindle microtubules 213
Molecular motors have multiple roles in mitosis 214
The kinetochore–microtubule link is stabilized by tension 214
Activation of a protease signals commencement of anaphase 215
Anaphase consists of two distinct mechanisms 216
Motor proteins in the kinetochore pull a chromosome to a spindle pole 216
The mitotic poles are pushed apart by microtubule sliding 217
Nuclei reform during telophase 218
The mitotic spindle dictates the plane of cytoplasmic cleavage 218
During the division of plant cells, new wall formation takes place within the spindle 219
Meiotic divisions in eggs proceed without a centrosome 219
Fusion of egg and sperm nuclei is orchestrated by microtubules 220
Bacteria can manipulate the mechanism of reproduction 221

Chapter 14 Cilia 225

Ciliary axonemes contain the machinery of wave propagation 226
The ciliary axoneme is built from a 9+2 array of microtubules 226
Ciliary microtubules are decorated with side-arms, spokes, and cross-links of protein 227
Tektin forms one of the protofilaments in ciliary doublet microtubules 228
Genetic analysis reveals the biochemical complexity of flagella 229
Ciliary bending depends on the sliding of adjacent microtubules 230
Microtubule sliding is produced by dynein arms 231
Dynein is a high-molecular-weight ATPase that binds to microtubules 232
Cilia contain a diversity of dynein molecules 233
Slide-to-bend conversion depends on radial spokes 233
Ciliary beating is influenced by $Ca^{2+}$ and cAMP 234
Inner dynein arms are regulated by phosphorylation 235
The left–right asymmetry of the body depends on ciliary beating 235
Many naturally occurring variants of sperm flagella exist 236
Large bundles of parallel microtubules bend through coordinated dynein action 237
Protein bridges can specify the geometry of microtubule bundles 238

Chapter 15 Centrioles and Basal Bodies 243

Flagella and cilia grow rapidly to a characteristic length 244
A pool of assembly-competent axonemal proteins exists in the cytoplasm 244
Cilia and flagella are nucleated by basal bodies 245
Axonemal parts preassemble near the basal body 246
Motor proteins move rafts of protein to the tip of the cilium 247
Many cells produce an apparently nonfunctional primary cilium 248
Contractile bundles control the position and orientation of flagella 248
Basal bodies can form centrioles, and vice versa 249
New centrioles usually form close to mature centrioles 250
Mother and daughter centrioles differ in structure and function 251

Parent/daughter differences also exist in basal bodies 252
Centrioles can appear *de novo* 252
Could centrioles have had a symbiotic origin? 253

Chapter 16 Bacterial Movements 257

Bacterial flagella are hollow cylinders made of flagellin 258
Flagellin molecules travel through the hollow core of the flagellum 258
The flagella motor is driven by a flux of protons 259
How does the flagellar motor work? 261
Transient reversals of the motor steer the bacterium 262
The direction of flagellar rotation is influenced by chemoattractants 262
Four categories of genes control chemotaxis 263
Chemotactic signals are relayed by a chain of coupled protein phosphorylations 264
Adaptation of the chemotactic response depends on protein methylation 265
Conformational changes are central to chemotaxis 266
Cooperative interactions between receptors could increase their sensitivity 266
Many variants of the chemotaxis system exist 267
Bacteria communicate 268
Bacteria use flagella to swarm over surfaces 269
Spiral bacteria migrate by screwing 269
Cyanobacteria glide by secreting mucilage 270
Disease-causing bacteria and viruses harness the actin cytoskeleton to enter cells 271
*Synechococcus* swims without flagella 272

**PART 3 Integration of Cell Movements 275**

Chapter 17 Intermediate Filaments 277

Intermediate filaments are tough and insoluble 278
The protein subunits of intermediate filament proteins are coiled-coil dimers 278
Intermediate filament assembly involves longitudinal annealing 280
Head and tail domains of the protein subunits enhance filament assembly 281
Intermediate filaments are not polar 282
Assembly of nuclear lamins is controlled by phosphorylation 282
Keratin filaments form a meshwork in epithelial cells 283
Keratin filaments are anchored in desmosomes 284
Vimentin filaments extend throughout the cytoplasm 284
Desmin filaments integrate muscle contraction 286
Neurofilaments give tensile strength to long axons and determine their diameters 287
Linker proteins connect intermediate filaments to the rest of the cell 287
Keratins are implicated in many genetic diseases 289
What is the function of cytoplasmic intermediate filaments? 289

Chapter 18 Cell Mechanics 293

Cytoskeletal filaments are flexible rods 294
Protein filaments have a similar elasticity to that of hard plastics 294
Bundles of filaments make rigid stiffeners for the cell 295
Bundle formation is a complex physicochemical process 295
Filament bundles can store strain energy 296
Systems of polymerizing filaments can exert a directed force 297
Networks of cytoskeletal filaments have viscoelastic properties 298
Cytoplasm is viscoelastic 299
Motor proteins make the cytoskeleton contract 300
Cytoskeletal filaments form a tensegrity structure 301
Myosin II makes a major contribution to cortical tension 302

What is the structure of cytoplasm? 303
Many 'soluble' proteins are associated with the cytoskeleton 303
Cytoplasm has a low microviscosity and a high macroviscosity 305
Proteins compete for water in the cytoplasm 305
Hydrostatic pressure may drive cell protrusions in plants and lower eucaryotes 306
How do membranes contribute to the mechanical properties of the cell? 308
Cells respond to sound, touch, gravity, and pressure 309

Chapter 19 Cell Shape 315

Largest and smallest cells 316
Bacteria, fungi, and plant cells have rigid cell walls 316
Plant morphogenesis depends on oriented cell division and expansion 316
Division and growth are controlled by subcortical microtubules 317
Yeast budding is controlled by small G proteins 318
Microtubules maintain the polarity of fission yeast 320
Cell wall synthesis in *Fucus* responds to external signals 321
In many tissues the cells are close-packed polyhedra 321
Cell function follows form 322
Localized contractions of the cortex provide a basis for changes in cell shape 323
Microtubules influence the actin cortex 324
A marginal band of microtubules determines the shape of blood cells 324
Systems of microtubules automatically find the cell center 325
Microtubule arrays assemble spontaneously 326
Actin filaments and microtubules act together to polarize the cytoskeleton 327
An intricate cytoskeletal framework transmits vibrations in the ear 328
The polarized secretion of extracellular matrix is a major source of cell asymmetry 329

Chapter 20 Cell Movements in Embryos 333

The egg cytoplasm carries positional information 334
mRNA molecules can be precisely positioned in the cell 334
Particles containing RNA and protein associate with the cytoskeleton 335
The insect egg is compartmentalized by the cortical cytoskeleton 335
Rotation of the cortex in amphibian eggs determines the axis of the future embryo 336
Spindle growth influences the plane of cleavage 337
The cell cortex can influence the plane of cell division 338
Dynein molecules in the cell cortex may pull on microtubules 339
Cell movements are widespread in vertebrate embryos 340
Morphogenetic movements are guided by cell adhesion molecules 340
Signals pass from one cell to the next in an embryo 341
Cell junctions stabilize the blastula 341
Major form-shaping movements are driven by epithelia 342
Cells flatten and intercalate during epiboly 343
Apical constrictions initiate blastopore formation 344
Embryonic invaginations are driven by actin-based contractions 345
Cell rearrangements support the elongation of an embryo 345
The nervous system begins as an epithelium rolls up 346
Cell migrations establish the anatomy of the brain 346
Growth cones are guided by both soluble and substrate-bound factors 348
Growth cones carry a complex 'on-board' navigational system 350

# Cell Movements in the Light Microscope

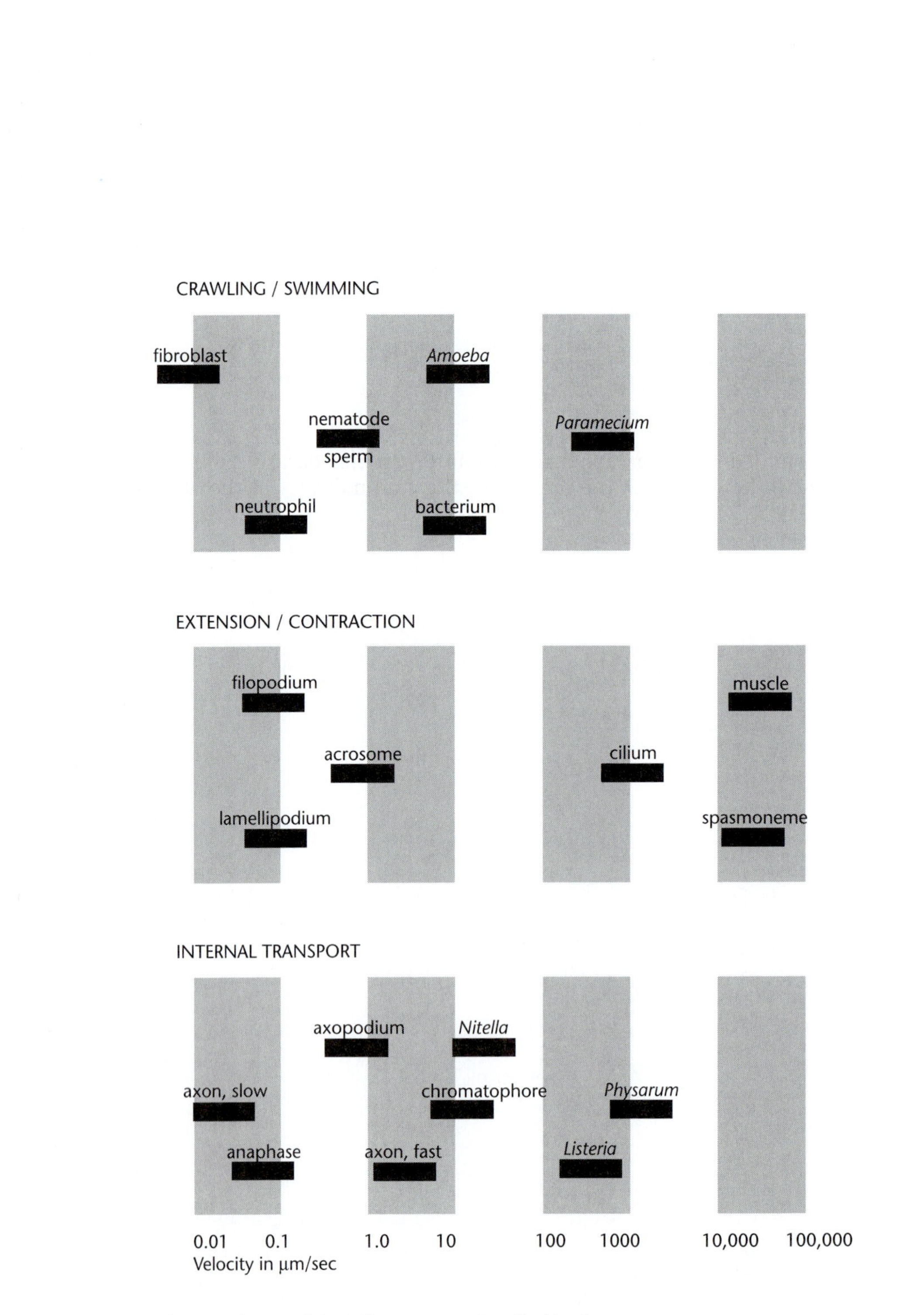


Velocities of some of the cell movements described in the text.

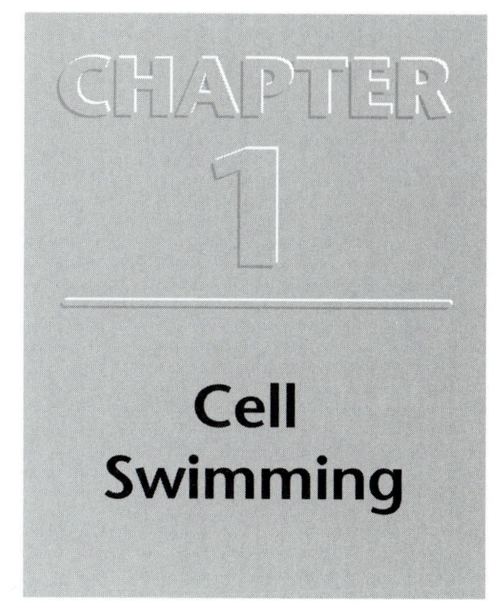


The first cells ever seen were swimming. When Anthony van Leeuwenhoek, in 1674, brought the glass bead that served him as a primitive microscope close to a drop of water taken from a pool close to his home in Delft, he was astounded to observe it full of minute particles darting here and there. Later he wrote: "... the motion of these animalcules in the water was so swift and various, upwards, downwards and round about, that 'twas wonderful to see ...." The organisms he saw were probably ciliated protozoa, each a single cell a few tenths of a millimeter in length, swimming by the agitated, but coordinated motion of thousands of hairlike cilia on their surface. The motion of such cells is so obviously directed and purposeful that van Leeuwenhoek knew it could only come from some living creature and not specks of inanimate matter.

Today, with the power and convenience of modern microscopes, we can visualize in rich detail the abundant variety of living cells, each capable of independent, directed motion through water. Sperm cells swim with a characteristic wriggling motion due to their long whiplike tails, or flagella. They often travel large distances to find an egg of the same species to fertilize, especially in aquatic species such as sea urchins. A huge menagerie of different kinds of protozoa exists including the ciliates seen by van Leeuwenhoek. Many of these swim with cilia and/or flagella, searching endlessly for food or for a mate. Even minute bacteria, less than 2 μm long, are capable of moving rapidly through water with a fishlike undulating motion.

How do cells swim, and why? These questions, in various forms and at different levels of enquiry, reoccur throughout this book. In this first chapter, we start by presenting information that is visible in the light microscope, surveying the types of cells that swim and their means of propulsion. We begin by discussing some of the special problems encountered by cells traveling through water because of their minute size.

## *A cell suspended in water moves passively by diffusion*

Because cells are so small, they encounter different physical problems when moving through water than fish or other aquatic animals. The progress of a cell through water is dominated by the viscosity of its environment; inertia, which carries a salmon or an otter forward, plays a negligible part. As for a human swimmer in a bath of honey, a cell moves through water by using a rowing or swimming action in which viscous resistance to the backward stroke is greater than viscous resistance to the forward stroke. It follows from the principle of action and reaction that swimming cells continually push backward against the surrounding water. Usually, this is accomplished by the repetitive beating of minute surface appendages. We will see below that these appendages, called *cilia* and *flagella*, encounter greater viscous resistance when they move backward than when they move forward.

Even without cilia and flagella, however, a cell suspended in water moves passively by diffusion. Like any other particle of similar size, a cell in water is pushed first one way and then the other by the thermal motion of surrounding molecules. Since chaotic thermal motion is an ever-present factor for all objects in solution, including objects within the cytoplasm as well as cells themselves, it is useful here to summarize its physical basis.

The speed of the cell undergoing such *Brownian movements* is directly related to its size. The average kinetic energy of any molecular or small particle suspended in liquid is given by $kT/2$, where $k$ is the Boltzmann constant and $T$ is the absolute temperature.[1] The instantaneous velocity of such a particle will change frequently as it bumps into another molecule or particle, but its average can be calculated by relating the average energy, as expressed above with temperature, and the equation for kinetic energy of any moving body: $MV^2/2$, where $M$ is the mass (kg) and $v$ its linear velocity (m/sec) in any direction. Equating these two expressions of energy, we find:

$$Mv^2/2 = kT/2 \qquad \text{or} \qquad v = (kT/M)^{1/2}$$

Using this simple equation we can calculate that at room temperature the instantaneous speed of a protein molecule will be about 10 m/sec, whereas for a cell the size of a bacterium it will be of the order of 1 mm/sec.

Neither a molecule nor a cell will travel at this speed very far before it bumps into a water molecule and changes both its direction and speed. The continual bombardment by surrounding molecules means that a cell undergoes a *random walk*, wandering over an ever longer and more tortuous path. In Figure 1-1, for example, the positions of a particle about 1 μm in diameter suspended in water is shown every 30 seconds. The trace of this particle is marked by a succession of straight lines, but if the measurements had been made at intervals of one second instead of 30 seconds, then each straight line segment in the figure would have to be replaced by a series of 30 smaller segments in a path just as tortuous as the one shown ... and so on, at smaller and smaller dimensions. In fact, it can be calculated that in order to measure the true steps of such a particle, one would have to take measurements at less than a microsecond and over distances less than a nanometer ($10^{-9}$ m).

Brownian movements are random and we cannot predict precisely where the cell will be at any time. But if many cells were to start out from the same point, then we would be able to say what their average distribution would be at different times (equivalently, we could calculate the probability that any one cell would be in a particular location after a given time).

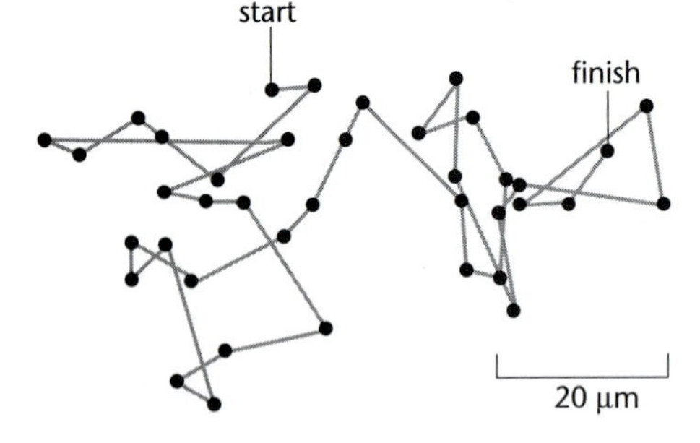


**Figure 1-1** Brownian motion. The paths followed by a minute particles as seen under a microscope. Positions of the particles were measured every 30 seconds and joined by straight lines. If the particles had been examined at higher magnification or measured at shorter intervals of time, the paths would have been far more complex than indicated here, since the true trajectory of a particle changes direction on a time scale of microseconds. (Based on Perrin, 1909.)

## *Diffusion is ineffective over large distances*

The distribution of diffusing particles (or cells, or molecules) in solution conforms to precisely defined spatial and temporal laws. These were first expressed mathematically in 1855 by the physiologist Adolf Fick, who adopted a set of differential equations already in use for the diffusive spread of heat in a solid. *Fick's first law of diffusion* says that the rate of diffusion of particles from one point to another is proportional to the difference in concentration of the particles between the two points. That is

$$J_x = -D\frac{dN}{dx}$$

**Figure 1-2** One-dimensional random walk. (a) At each step the cell has an equal chance of moving to the left or the right a distance $\alpha$. If the distance of the cell from its starting point after $n$ steps is $r_n$ then

$$r_n = r_{n-1} \pm \alpha$$

squaring this gives

$$r_n^2 = r_{n-1}^2 + 2\alpha r_{n-1} + \alpha^2$$

and

$$r_n^2 = r_{n-1}^2 - 2\alpha r_{n-1} + \alpha^2$$

If we average these two values of $r_n^2$ over a large population of similar cells, the terms $\pm\, 2\alpha r_{n-1}$ cancel (because steps to the left and right are equal in number), giving the mean square displacement

$$|r_n^2| = |r_{n-1}^2| + \alpha^2$$

Since all cells start at point zero,

$$|r_0^2| = 0; \quad |r_1^2| = \alpha^2; \quad |r_2^2| = 2\alpha^2; \quad |r_3^2| = 3\alpha^2$$

and so on. In other words, the mean square displacement $|r_n^2|$ increases linearly with the number of steps, $n$, and hence with the elapsed time. That is

$$|r_n^2| = \text{constant} \times \text{time}$$

(b) Spreading of a population of cells undergoing a one-dimensional walk. Probability distributions at three times after commencing the random walk described in the text. The root mean square displacement (white bars) increases as the square root of time.

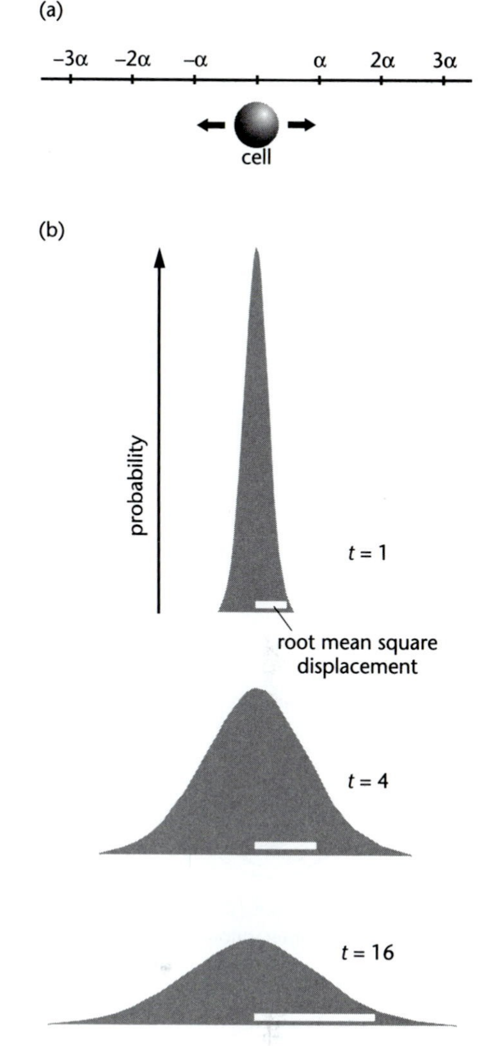


where $J_x$ is the *flux* of particles (the number passing through a window of unit area in unit time); $N$ is the number of particles per unit volume, $x$ is the distance, so $\frac{dN}{dx}$ is the concentration gradient. The constant $D$, known as the diffusion constant, depends on the size of the particle, its shape and other factors.

Fick's law says nothing about the origins of the force driving the molecules. There is, however, an alternative description of diffusion that is more closely tied to the physical reality of the situation. This description, given by Albert Einstein in 1905, attributes diffusion to the thermal (Brownian) motions of the particles that cause them to undergo the random walk described above.

To illustrate this second approach to diffusion, consider a population of particles constrained to move along a linear track (Figure 1-2a). All of the particles start at the same initial point and then take individually random steps either to the right or to the left. As time passes and more steps are taken, the particles will on average travel farther and farther from their starting point. However, the probability of a step to the right is equal to that of a step to the left, so on average the particles go nowhere! The result of many such steps by many particles is, on average, a bell-shaped distribution that becomes broader with time (Figure 1-2b).

A convenient measure of this spreading tendency is the mean square displacement of the particles $|r^2|$, which is always positive, a linear function of the duration of the random walk. That is

$$|r^2| = \text{constant} \times \text{time}$$

or, as it is usually written,

$$|r^2| = 2\,Dt$$

where $t$ is the time elapsed and $D$ is the diffusion constant already mentioned. In one dimension $|r^2| = 2\,Dt$ as shown; in two dimensions $|r^2| = 4\,Dt$ and in three dimensions $|r^2| = 6\,Dt$. Typical values of $D$, together with the average time taken to diffuse specific distances, are given in Table 1-1 for molecules, organelles and cells.

[1] Boltzmann's constant, $k$, has a value of $1.38 \times 10^{-16}$ erg/degree. $kT$ at room temperature is around $4 \times 10^{-14}$ erg, or $4 \times 10^{-14}$ g cm$^2$/sec$^2$. A typical protein molecule (with molecular weight 30,000) has a mass of $5 \times 10^{-20}$ g. A typical bacterium has a mass of $10^{-12}$ g.

**Table 1-1 Diffusion times in water**

| | Diffusion coefficient $cm^2$/sec | Typical time taken to diffuse indicated distance 1 μm | 10 μm | 1 mm |
|---|---|---|---|---|
| small molecule | $5 \times 10^{-6}$ | 1 msec | 0.1 sec | 15 min |
| protein molecule | $5 \times 10^{-7}$ | 10 msec | 1 sec | 3 hr |
| virus particle | $5 \times 10^{-8}$ | 0.1 sec | 10 sec | 1 day |
| bacterial cell | $5 \times 10^{-9}$ | 1 sec | 100 sec | 10 days |
| animal cell | $5 \times 10^{-10}$ | 10 sec | 20 min | 100 days |

Approximate values are given solely to indicate the magnitudes involved. The times are calculated for three-dimensional diffusion and represent the average, or root mean square displacement, of the population (see Figure 1-2). For sources see Hille (1992), Berg (1993), and Atkins (1994).

Note that according to the above equations—and also as shown in Table 1-1—the average distance traveled by diffusion is proportional to the *square root* of the time elapsed. Short distances are traveled rapidly; long distances much more slowly. It takes 100 times as long for a cell to wander 10 times farther.

Brownian movement and diffusion are relevant not only to the movement of individual cells suspended in water. We shall see in later chapters that they are also important in understanding the movements of molecules inside cells. On the molecular scale, the interior of a living cell is a place of ceaseless motion due to the agitated thermal motion of small and larger molecules.

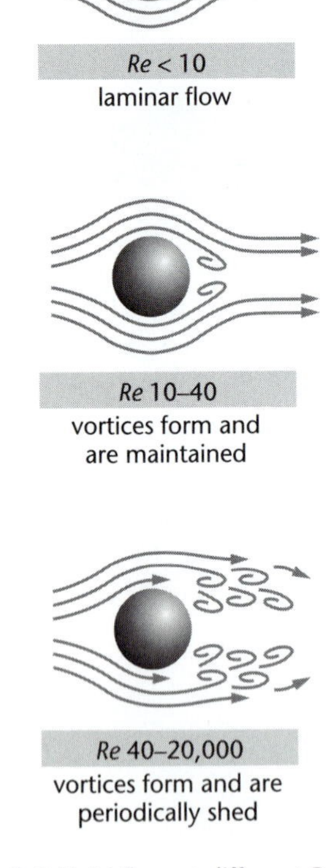


**Figure 1-3** Fluid flow at different Reynolds number (*Re*). At low *Re* the fluid flows smoothly, following the contours of the obstacle, but as *Re* increases small vortices develop. These are at first stationary with respect to the obstacle but at higher *Re* they are shed periodically. At very large *Re* fluid movements downstream of the obstacle become chaotic.

## *The movement of cells through water is governed by viscous forces*

Brownian movements are aimless and provide no net motion in a particular direction. They are consequently an inefficient way to travel toward a source of food or away from a noxious stimulus. Not surprisingly, therefore, many cells have evolved special structures that drive them purposefully through the surrounding water. As they swim, such cells encounter two kinds of resistance to forward motion. There is a *viscous drag* which is a frictional resistance, and there is an *inertial resistance* of the fluid that must be displaced, a function of the density of the fluid. The relative importance of inertial and viscous forces is expressed in a dimensionless constant, termed the Reynolds number (*Re*). As a rough guide, an *Re* greater than 100 signifies that resistance to movements is mainly inertial whereas an *Re* less than 0.001 means that movement is governed largely by viscous forces. We will see shortly that the *Re* for single cells is extremely small.

The Reynolds number was originally determined by studying the passage of water down a long pipe. At low rates of passage, the flow of water is 'laminar', meaning that streamlines do not cross or separate. As the flow rate increases, turbulent flow begins; the streamlines mix and separate, giving rise to eddies. The latter create extra resistance to flow, predominantly due to the resistance of the water to acceleration. Similarly, as a free body moves through water, it encounters either laminar or inertial flow, depending on its size, shape and velocity (Figure 1-3).

When used to describe a swimming organism, the Reynolds number depends on the size (average diameter) of the organism $L$ and its velocity $v$ as well as the density ρ and viscosity η of the liquid:

$$\mathrm{Re} = \frac{\text{initial force}}{\text{viscous force}} = \frac{v^2L^2\rho}{vL\eta} = \frac{vL\rho}{\eta}$$

This equation indicates that for a large body (large $L$), the Reynolds number is also likely to be large. The value of *Re* for a swimming mackerel, about $10^5$, indicates that the progression of this organism through water

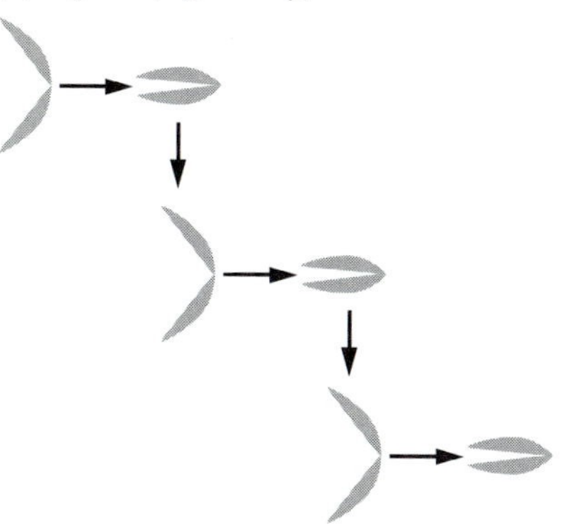


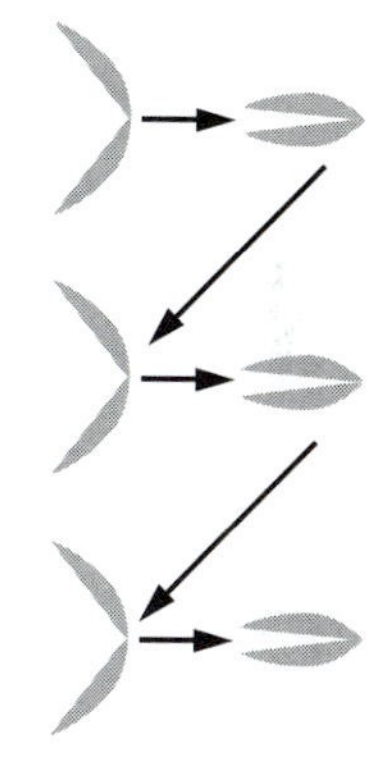


**Figure 1-4** Jet propulsion at low Reynolds number. (a) An organism such as a scallop that works at large Reynolds number can propel itself by slowly opening and then suddenly shutting the two halves of its shell. By periodically ejecting spurts of water, the organism makes net progress. (b) A hypothetical microscopic bivalve the size of a single cell that tried to operate at low Reynolds number could not swim by such a mechanism. Viscous drag would bring the organism back to its starting point after each slow opening.

is governed principally by inertia: a mackerel propels itself mainly by accelerating water. On the other hand, a very small body, like a single cell will have a very small Reynolds number: for a protozoan 100 μm long, swimming through water at top speed, $Re$ might be around $10^{-3}$ and for a 1 μm long bacterium around $10^{-5}$. In such cases, viscous forces dominate motion.

Because they move at low Reynolds number, cells find no advantage in having a streamlined shape. The smooth contours of an airplane or racing car are needed to reduce drag in an inertial flow situation by reducing turbulence. Cells do not create turbulence and need not be streamlined. Secondly, any thoughts of jet propulsion as a means of locomotion for cells can be rejected. A microorganism that tried to move like a scallop or jellyfish, by slowly filling a chamber with water and then rapidly expelling the water in a jet, would get nowhere (Figure 1-4). Jet propulsion works well in inertial situations but is extremely inefficient when the flow of liquid is entirely laminar.

In fact, most swimming cells are propelled through water by the repetitive movement of hairlike cilia or flagella, the movements of which are governed by viscosity. A typical cilium that is 0.2 μm in diameter and projects 10–20 μm from the surface of a cell will have no more tendency to continue motion when its driving force is removed than a bamboo stick in tar. Consequently a cilium can propel a cell only by altering its shape in a cyclical fashion, maximizing viscous resistance in one direction (during its 'power stroke') and minimizing viscous resistance in the other 'recovery stroke.' In other words, cilia propel cells by viscous shear. The same is true of flagella: if the flagella on the surface of a bacterium were instantaneously arrested, for example, inertia would carry the cell a distance of less than 0.01 nm—less than one-tenth the diameter of a hydrogen atom!

## *Swimming consumes only a small fraction of the cell's energy*

The power required to propel a cell in water is quite small. It may be calculated as the viscous drag multiplied by the velocity. For a spherical cell

$$\text{power} = (3\pi L \eta v) \times v$$

where $\eta$ is the viscosity of water ($10^{-2}$ g/(cm sec)), $L$ the length (or diameter) of cell, and $v$ its velocity. Thus, if the cell has a radius of 1 μm and is traveling at 10 μm/sec, the power consumed is $2 \times 10^{-11}$ erg/sec ($2 \times 10^{-18}$ J/sec).

What is this power requirement in terms of ATP molecules, the principal currency of energy in the cell? Hydrolysis of one gram-molecule of ATP (about 500 g) releases about 40 kJ of useful energy; hydrolysis of a single ATP molecule releases about $10^{-19}$ J. The cell in the above calculation therefore must split about 20 molecules of ATP every second in order to maintain its speed of 10 μm/sec. If we assume a plausibly low efficiency such as 2%, then the ATP hydrolysis required for swimming might be 1000 molecules per second. Although this sounds a large number at first hearing, in fact it is very small compared to the total expenditure of energy by the cell: the metabolic rate of a typical cell is around $10^7$ ATP molecules per second. Evidently, swimming per se is not a major energy cost—although building the apparatus that drives swimming could be expensive.

## *Bacteria swim by means of flagella*

A powerful advantage is gained by any single-cell organism that can move more rapidly than its competitors toward a source of food or away from a potentially harmful environment. Bacteria, with characteristic versatility, have evolved a rich variety of mechanisms for this purpose. Species of bacteria are known that are attracted, or in some cases repelled, by sugars, amino acids, metal ions, extremes of temperature, pH, oxygen, light, hyper- or hypo-osmotic solutions, and magnetic fields. In some cases movement is achieved by indirect or passive means, as in the myriad small organisms that float to the ocean surface by developing small gas vacuoles in their cytoplasm, or that glide slowly over surfaces by the secretion of viscous slime. But the most common and effective mode of transport is provided by bacterial flagella.

The swimming of a single bacterial cell of a species such as *Escherichia coli* (the common intestinal bacterium) or *Salmonella typhimurium* is impressively rapid. Although the cell itself is only about 2 μm long, it swims at speeds of 20–30 μm per second following an undulating path, like a small fish swimming against the current. Careful analysis of the motion reveals that a swimming bacterium alternates rapidly between two swimming modes: smooth swimming, during which it progresses in a roughly constant direction for a second or so, interrupted by brief periods of tumbling during which it changes it direction in a highly erratic manner (Figure 1-5).

The meandering path that results from this kind of swimming differs fundamentally from Brownian motion. First, the distance traveled in a given time is much greater than that traveled by a bacterium carried passively by thermal motion. This is because a swimming bacterium will travel in one direction for a second or more, whereas a bacterium undergoing thermal motion changes direction every few nanoseconds. Second, although the track of a swimming bacterial cell appears aimless, it can, unlike a cell exhibiting Brownian motion, be influenced by its local chemical or physical environment. As we shall describe in Chapter 3, a bacterium that senses it is approaching a source of nutrient is less likely to undergo a tumble, and hence is less likely to change its current direction of travel. This is enough to ensure that over the course of many runs and tumbles the bacterium moves toward the distant source.

The structures that drive a bacterium become visible if the cell is illuminated by a very intense light and viewed against a dark background. Fine threadlike flagella can then be seen streaming from the cell (Figure 1-6). The flagella are seen even more clearly if the viscosity of the medium is

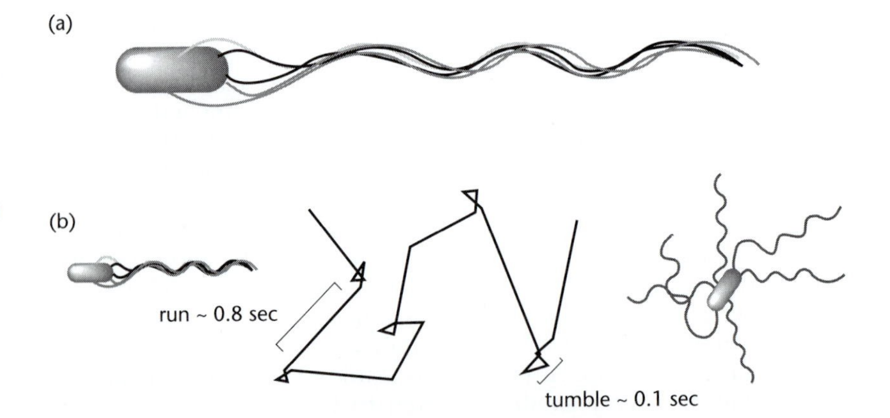


**Figure 1-5** Bacterial swimming. (a) Schematic diagram of *Escherichia coli* swimming. The cell body is a cylinder about 2 μm long and 0.5 μm in diameter and has 6–10 flagella on its surface, each up to 10 μm in length. Coordinated rotation of flagella drives the cell at speeds of about 30 μm/sec through water. (b) Under normal conditions, the bacterium alternates between periods of smooth swimming ('runs') and intermittent chaotic changes in direction ('tumbles').

increased by adding a substance such as methylcellulose. This slows the movements of the cell to the extent that the flagella can be seen to form visible sinusoidal waves that appear to travel backward from the cell.

The higher resolution available in a scanning electron microscope shows the propulsive unit of a bacterium to consist of a bundle of flagella, each a thread about 14 nm in diameter and 10 μm long. In *Escherichia coli*, 6–10 flagella emerge from random points on the side of the body and extend into the surrounding medium. When the cell is swimming smoothly, the flagella collect into a smoothly undulating bundle that drives the cell along (see Figure 1-6).

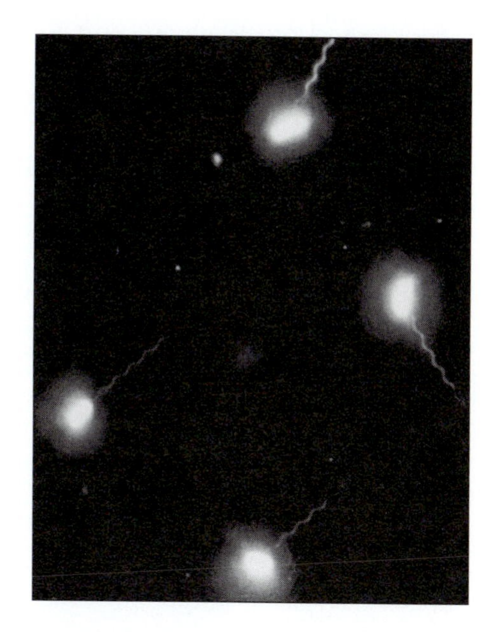

**Figure 1-6** Swimming bacteria seen by dark-field microscopy. The intense illumination reveals the bundle of helical flagella that propel the bacterium through the water. (Courtesy of H. Hotani.)

## *Bacterial flagella are rigid helical structures that rotate*

The cilia and flagella of sperm cells and protozoa are autonomously active structures that propagate bending waves from their base to the tip. Bacterial flagella are smaller and more rigid structures that are rotated by a minute motor in the bacterial cell wall. This was first demonstrated by observing bacteria tethered to a glass slide (or to each other) by treatment with specific antisera. Cells prevented from swimming by this strategy rotate about the point of attachment, continually switching from a counterclockwise to a clockwise direction and then back again (Figure 1-7). Similar behavior was also shown by nonmotile mutant bacteria that lacked flagella, or in which the flagella were straight rather than helical. Experiments such as these demonstrated that the flagella must be turned by a structure at their base—a rotary motor embedded in the cell wall. A detailed description of this motor and its mode of action is given in Chapter 16.

At first sight, a rotating flagellum seems a poor design—a propeller of this shape would indeed be a very inefficient way to drive a submarine. But we must remember the relative importance of viscosity for microorganisms. If each segment of the flagellum is thought of as a short cylindrical rod, then it is clear that this rod will encounter less viscous resistance if it moves end-first rather than sideways through the water (Figure 1-8). Rotating the flagellum will push each rodlike segment in a circular path about the long axis of the flagellum. But each segment will encounter a resistance to motion due to viscosity that will be least if it moves end-first. The net result (summing all of the small rodlike segments together into a helical structure) is motion parallel to the long axis of the flagellum. It is rather as though the bacterium screws itself through the water by rotation of its helical flagella, much as a corkscrew can be forced through a cork simply by being rotated.

## *Flagellated bacteria come in a menagerie of different forms*

*E. coli* and *Salmonella* are not the only bacteria with flagella. Many other bacteria swim by means of flagella, and show great variation in the number, length, and distribution of bacterial flagella they employ. *Proteus mirabilis* has a similar number and distribution of flagella to *E. coli* and

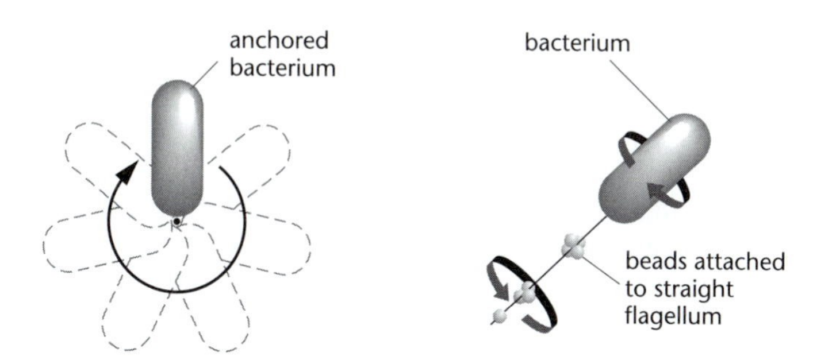


**Figure 1-7** Rotation of bacterial flagella. A bacterium tethered to a surface by its flagellum rotates about the point of attachment. The rotation can also be visualized by attaching small refractile particles to the flagellum (especially effective if the cell is a mutant having a straight flagellum).

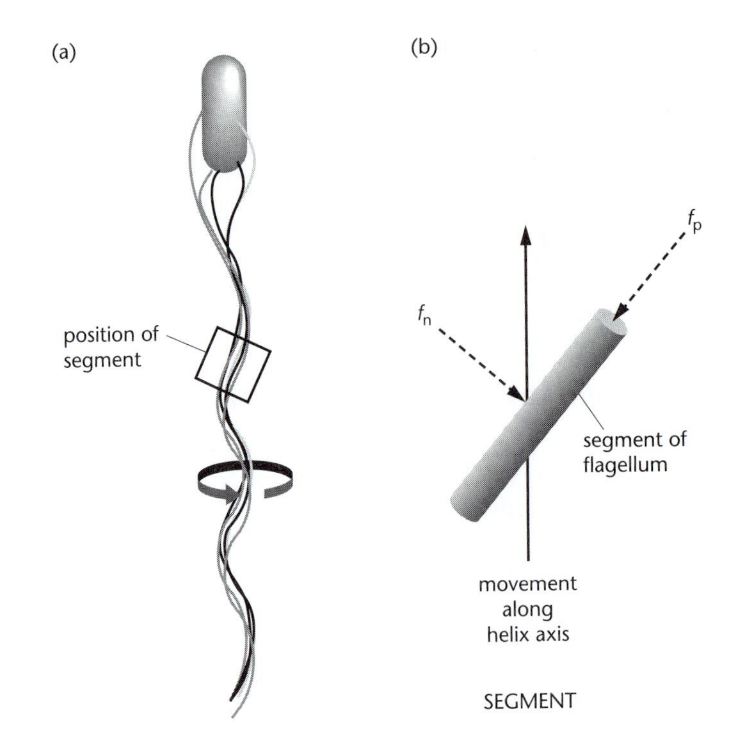


**Figure 1-8** Flagellar dynamics. (a) As a bacterial flagellum rotates, each short segment is carried in a circular path through the water. (b) As it moves, a segment experiences a frictional drag that is lower in a direction parallel to its long axis ($f_p$) than in a direction normal to its long axis ($f_n$). For an infinitely long cylinder $f_n = 2f_p$. If all the short segments in a rotating helical flagellum are summed, the lower resistance $f_p$ results in net movement along the helical axis.

swims in a similar manner when grown in liquid broth. When grown on the surface of a semiliquid agar plate, however, the cells sprout many more flagella and begin to *swarm*—a form of motility that appears to depend on cell–cell contact. Dual motility systems of this kind are common in bacteria that inhabit complex environments, allowing them to move efficiently in surroundings with different physical properties. Another life style is seen in stalked bacteria of the genus *Caulobacter*, which spend most of their life attached to a substratum. When they divide, one daughter becomes a motile cell bearing a single flagellum at one end, while the other cell retains its attachment to the substratum. The salt-loving bacterium *Halobacterium halobium* swims by means of flagella inserted at either pole of the cell. It spontaneously reverses its direction of swimming every 10–15 seconds, the interval being influenced by the intensity of light.[2]

One of the most curious forms of movements is that of spirochetes, which are powered by one or more internal flagella. Spirochetes such as *Treponema pallidium*, the causative agent of syphilis, have a remarkable ability to swim in gel-like media. Their long helical bodies bore through high-viscosity media in a serpentine manner without slippage, as though traveling through a helical tube. These movements are generated by flagella, attached at either end, that rotate within the confines of an outer membrane sheath (Chapter 16).

## *Bacterial flagella and eucaryotic flagella differ in structure and mode of action*

Many eucaryotic cells also swim by means of flagella: this is true of many species of protozoa and algae as well as most types of animal sperm. However, eucaryotic flagella are at least ten times larger than bacterial flagella in both diameter and length, and their structure and mode of action are far more complex. The differences are so great that it seems that they must have arrived at their common whiplike form by convergent evolution: different solutions to the same problem of moving a cell through water at low Reynolds number.

[2] The flagella at either end of a *Halobacterium* cannot be identical in every respect or else the bacterium could not swim. The motors at either pole of the cell probably rotate in different directions (looking out from the cell). Alternatively, if they rotate in the same direction, then the flagella must have opposite handedness, with one a right-handed helix and the other a left-handed helix.

The flagellum of a mammalian sperm, for example, is typically 70 μm long and may be 0.8 μm wide near the cell body, tapering to 0.2 μm at its distal end (Figure 1-9). Unlike a bacterial flagellum, which is an inert structure rotated at the cell body, a eucaryotic flagellum actively propagates bending waves. If a sperm flagellum is severed from the cell body by a focused UV beam or by sonication, it remains able to swim for long periods. Another indication of internal complexity is that the waveform of a eucaryotic flagellum is not always rigidly helical but varies between cell types. The flagellum of mammalian sperm, for example, executes an almost perfectly planar sinusoidal wave.

Although larger than bacterial flagella, eucaryotic flagella still move at low Reynolds number and are therefore dominated by viscous drag. The underlying basis of their forward movement is the greater viscous drag encountered by a thin rod or filament that moves sideways through a fluid rather than moving end-on (see Figure 1-8). From the standpoint of fluid mechanics, the principal difference between the two kinds of flagella is that each small segment of the eucaryotic flagellum is driven by bending motions generated along its length by the mechanochemical interactions of adjoining segments. The segments of a bacterial flagellum move because of a rotation imposed from outside, by the motor in the cell body.

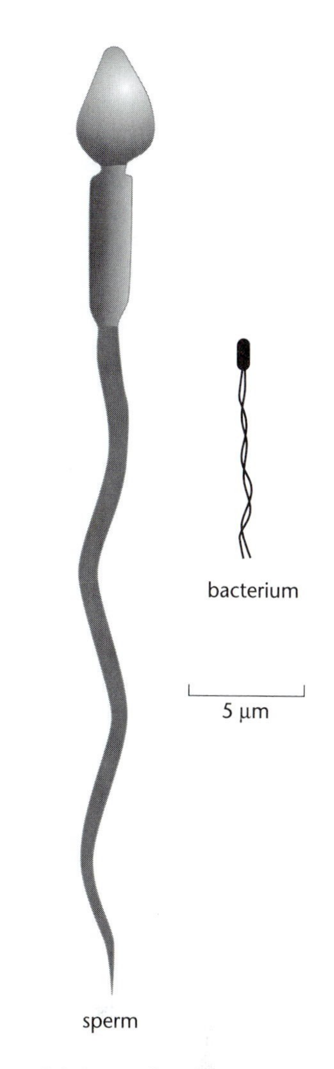

**Figure 1-9** Sperm flagellum and bacterial flagellum compared.

## *Eucaryotic flagella generate a wide variety of waveforms*

The simplest type of waveform seen in eucaryotic flagella is a planar sinusoidal wave that travels steadily toward the rear of the swimming cell. This pattern of beating is seen in many animal sperm that are streamlined for efficient swimming. They typically have a small head region, containing highly condensed chromatin together with enzymes that enable fertilization to occur. The small head has minimal effects on the hydrodynamic performance of the cell, so that sperm flagella show some of the simplest and most regular waveforms.

By contrast, a sinusoidal waveform is only rarely encountered in protozoa, the other major group of eucaryotic cells that swim by means of flagella. The cell body of a protozoan presents much greater resistance to passage though water than the head of a sperm and often plays an important part in the hydrodynamics of swimming.

Take *Euglena gracilis* for example. This is a photosynthetic protozoan (or alga) that swims by means of a single long flagellum. The flagellum is attached to the front end of the long tapered cell; when beating it wraps around the cell to point rearward (Figure 1-10). Helical waves moving from the base of the flagellum to its tip drive the cell forward, and also

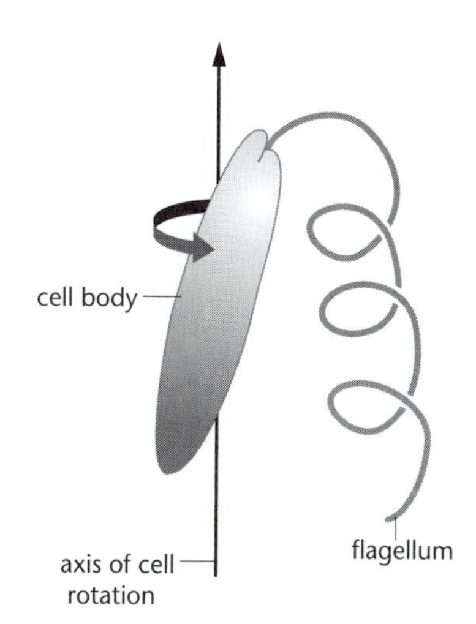


**Figure 1-10** *Euglena* swimming. The single long flagellum generates waves that cause the cell body to spin, moving forward like a single-blade propeller.

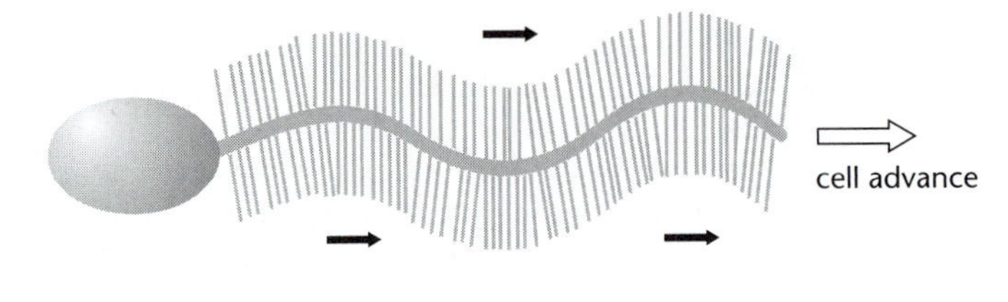


**Figure 1-11** Chrysomonad swimming. This marine organism has a 'tinsellated' flagellum covered in side hairs. The flagellum propagates planar waves from its base to tip, as usual, but the side hairs change its hydrodynamic properties so that it pulls the cell rather than pushes. The cell consequently swims with its hairy flagellum foremost.

exert a torque on the cell. The organism consequently follows a gyrating path through the water in which the anterior of the cell sweeps through a larger radius than the posterior. Spinning with a frequency of about one cycle per second, the cell body acts like a counterbalance against the driving force of the flagellum. A contrasting form of flagellar-driven swimming is seen in the unicellular green alga *Chlamydomonas reinhardtii*, which swims forward using its pair of flagella rather as a human swimmer uses arms in a breast stroke, although in an avoidance response the flagella have an undulating waveform. We will return to *Chlamydomonas* swimming in Chapter 14 to discuss the many mutants of this organism known to be affected in their swimming ability.

The flagella of most mammalian sperm have smooth surfaces, but this is not true of all protozoa. The flagella of some species of *Euglena*, for example, are decorated with a unilateral array of minute filamentous projections that project from points along their entire length and with a tuft of hairs that project at their tips. Other protozoa carry flagella adorned with lateral spines, scales, or bristles. Many of these lateral projections have an important influence on the swimming performance of the cell—even to the extent of reversing the normal direction of swimming! For example, most motile species of *Chrysophytes*—a group of protozoa also termed golden algae—possess a flagellum attached to their front end that propagates a planar wave from base to tip. We would expect such a motion to drive the cell rearward, but the opposite is the case because the flagellum of this organism is covered with stiff hairs that project from its side (Figure 1-11). As this 'tinsel' flagellum bends, the hairs on either side undergo a cycle of movements like side paddles that carry the cell in the forward direction.[3]

Parasitic species of protozoa frequently have unusual arrangements of flagella, possibly because their environment poses special problems of locomotion. The *Trichomonas*, for example, which are parasitic in man and other animals, typically have several flagella, one of which folds back down the cell body and may be associated with an undulating membrane (Figure 1-12). The single flagellum of *Trypanosoma brucei*, the parasitic protozoan that causes African sleeping sickness, is similarly attached to the margin of a delicately undulating membrane. This apparatus is capable of swift reversals, equipping the parasite for movements in the swiftly flowing bloodstream of a mammalian host. In a different stage of its life cycle the trypanosome is confined to the mouthparts of the tsetse fly and the flagellum changes from being a motile organelle to one that provides anchorage to the lining of the insect proboscis. Elsewhere among the protozoa, flagella are found that act as rudders for steerage or act as sensory whiskers, or feelers.

## Many protozoa move by the coordinated beating of cilia on their surface

Cilia are a form of hairlike motile appendage found on a wide variety of eucaryotic cells. They closely resemble eucaryotic flagella in internal structure and mode of action, as discussed in Chapter 14, but they are typically shorter than flagella and are present in larger numbers arrayed

[3] Many *Chrysophytes* possess two flagella, one tinsellated (hairy) and the other smooth; the hairy flagellum advances in front of the cell and is the principal source of motility. A similar flagellar arrangement occurs in the sperm of *Fucus*, a sea weed, where the tinsellated flagellum makes initial contact with the egg.

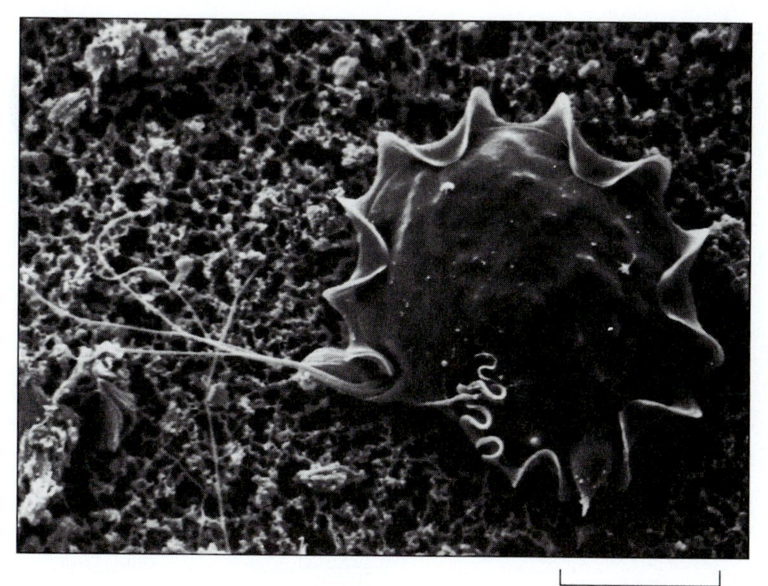


**Figure 1-12** A trichomonad. One of a large family of parasitic flagellates, in this case taken from a termite gut. The cell has several flagella, one of which curves around the cell body and is attached to it by an undulating membrane. (Courtesy of A.V. Grimstone.)

over the cell surface. Their waveform is also more complex consisting of a planar power stroke and a three-dimensional recovery stroke, the shape of which appears to be adapted to the cell's particular hydrodynamic environment (Figure 1-13). A major phylum of protozoa, the *Ciliophora* or ciliates, is characterized by cilia covering large portions of surface.

Ciliates of the genus *Paramecium* are free-living, freshwater protozoa, typically slipper-shaped and 100–200 μm in length. Their surface is covered by thousands of cilia, each about 10 μm long, that beat at a frequency of about 20 cycles per second. They work like small oars to drive water over the cell surface, enabling the cell to swim. Cilia on the surface of paramecia, and indeed most cells, work in near synchrony, showing a slight time lag between the beating of successive rows of cilia, which produces a large-scale ripple pattern on the surface of the cell, the *metachronal wave*. As the cell swims forward, metachronal waves sweep from the posterior left up to the anterior right of the cell (Figure 1-14). The cell consequently rotates as it swims, describing a smoothly rotating path, moving through water with a net rate of up to several millimeters a second.

## Fields of cilia move adjacent layers of water

The coordinated beating of cilia, as we have just seen, carries currents of water over the surface of a cell. Many ciliates take advantage of this action not to swim but to feed. Thus, sessile species of protozoa, such as *Stentor* or *Vorticella*, which spend most of their life anchored to surfaces, use fields of cilia to sweep currents of water into the mouth region of the cell. Particles of food carried in the flow are thereby captured and devoured. Multicellular organisms also take advantage of the ability of cilia to create a continual directed flow of fluid and the apical (outward-facing) surfaces of many epithelial cells are covered by cilia (Figure 1-15). Ciliated cells are

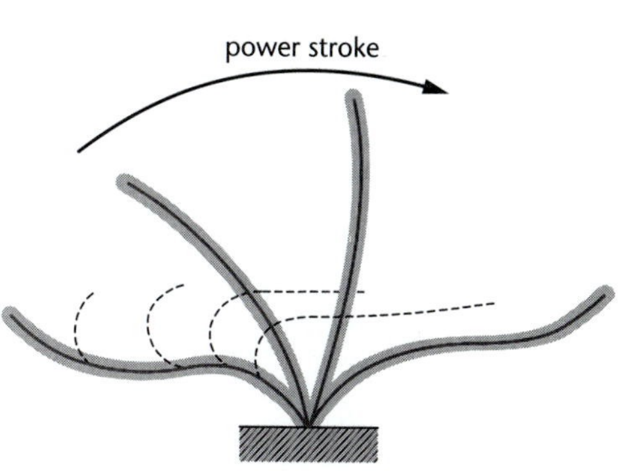


**Figure 1-13** The beating of a cilium. Each cilium performs a repetitive cycle of movements consisting of a power stroke followed by a recovery stroke. In the fast power stroke (solid lines), the cilium is fully extended and fluid driven over the surface of the cell. In the slow recovery stroke (broken lines) the cilium curls back into position with minimal disturbance to the surrounding fluid. Each cycle typically requires 0.1–0.2 seconds and generates a force that is perpendicular to the axis of the cilium.

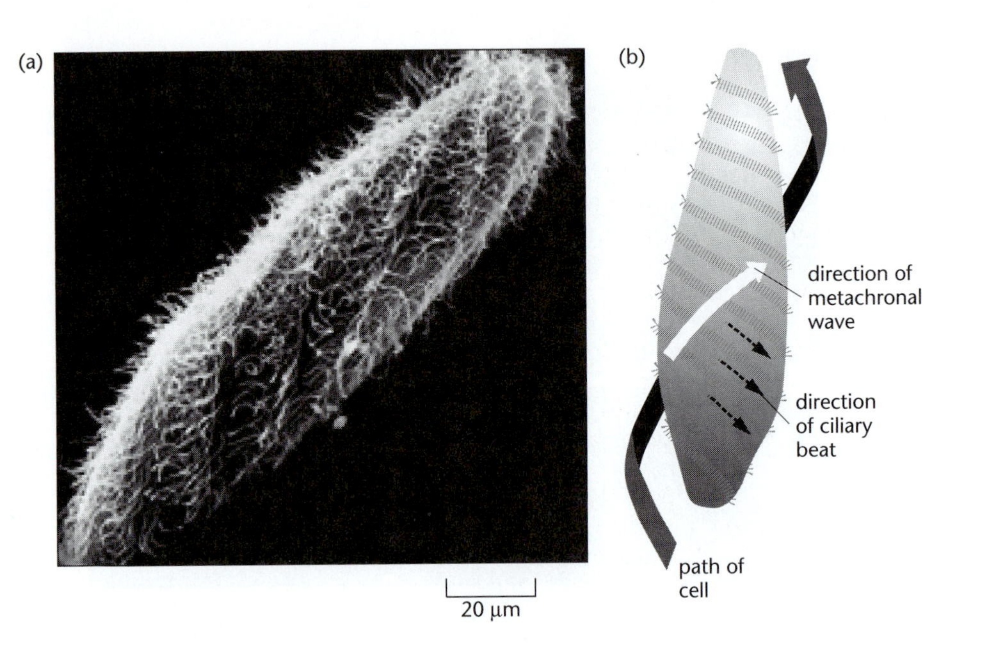


**Figure 1-14** Ciliary beating on the surface of a swimming *Paramecium*. (a) The surface of *Paramecium* is covered by some 5000 cilia, the repetitive beating of which propels the cell through the water. (b) Because of near synchrony in the beating of adjacent cilia, they form waves on the surface. Note that the direction of these metachronal waves is not necessarily the same as that of the individual ciliary beat. (a, courtesy of Sidney Tamm.)

present in enormous numbers on the inner surfaces of the bronchioles of the human lung, and the unceasing coordinated beating of their cilia carries layers of mucus and particles of dust that have been inhaled up to the throat to be swallowed and eliminated. Cilia on the walls of the proximal convoluted tubule of the kidney aid the collection of fluid waste; cilia on ependymal cells lining the ventricles of the brain cause circulatory movements in the cerebrospinal fluid; the mammalian egg is borne by the beating of hundreds of thousands of cilia on the inner wall of the Fallopian tube as the egg migrates from the ovary to the womb.

## *Modified cilia and flagella are used for purposes other than swimming*

Evolution is opportunistic, and once a successful structure has appeared it is often employed in unexpected ways. Thus cilia and eucaryotic flagella form the basis of modified structures that perform functions other than swimming. We have just noted that many ciliated protozoa use fields of cilia to sweep food particles into their mouths for feeding. This typically

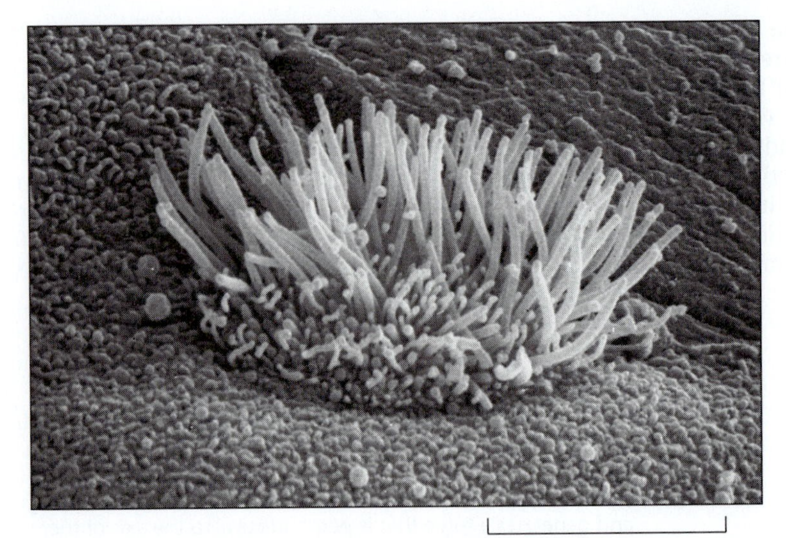


**Figure 1-15** Ciliated cell in an epithelium. Scanning electron micrograph of the esophageal epithelium of a fetal mouse shows an isolated ciliated cell. The apical surface of the cell carries both long motile cilia and shorter, static microvilli. Later in development ciliated cells will proliferate to produce extensive fields of beating cilia. (Courtesy of Raymond Calvert.)

employs highly specific arrangements of cilia and, moreover, often includes cilia fused together to form larger (and presumably more efficient) platelike structures termed *membranelles*. Other free-living protozoa use cilia fused together in structures called *cirri* to perform a form of locomotion similar to walking, as described in Chapter 2.

Parasitic and symbiotic protozoa also employ cilia or flagella to move to, and maintain their position in, a suitable niche within the host organism. Many of these, such as the flagellated cell *Trichonympha*, show an amazing degree of specialization (Figure 1-16). *Trichonympha* inhabits the intestine of wood-eating termites, where it ingests the minute bits of wood in the termite's intestine, transforming them to soluble carbohydrates, a proportion of which can be used by the insect host.

Modified cilia are found also in higher animals, often associated with sensory processes. Olfactory cells, responsible for the sense of smell, are specialized nerve cells that carry (in humans) six or eight extremely long (up to 200 μm) immobile cilia. These project out from the surface of the olfactory epithelium, and their membranes carry receptors for the detection of odors. The photodetector rod cells of vertebrate retina are another highly specialized form of cilium. Between the highly specialized outer segments of the photodetectors, which carry the photoreceptor apparatus, and the inner segment that contains many mitochondria, there is a slender neck region that has an internal structure very similar to that of a cilium. Evidently, during the development of an embryo, specific epithelial cells bearing a single cilium on their surface become progressively modified in form and function to produce the light-detecting cells in the adult eye. Here the genetic program that specifies the structure of a cilium has been taken over and adapted to a different purpose.

**Figure 1-16** The flagellated protozoan *Trichonympha*. This bell-shaped organism ingests particles of wood by means of pseudopodia that protrude from its lower part (not shown). The upper surface of the organism is covered with hairlike flagella that enable the organism to move and position itself in the termite gut, which is its home. (Micrograph courtesy of A.V. Grimstone.)

## *Essential Concepts*

- Cells suspended in water undergo continual random motion due to thermal agitation. Their movements are dominated by viscosity and inertia has little or no influence.
- Many types of free-living cells have evolved whiplike surface appendages known as flagella and cilia. These enable the cell to move in a directed fashion through water—that is, to swim.
- Bacteria swim by means of flagella that are stiff helical rods of protein, and that rotate by means of a motor embedded in the cell wall.
- Eucaryotic flagella are much larger and propagate bending waves along their length. They push or pull the cell along straight or gyrating paths.
- Cilia are closely similar in design to eucaryotic flagella but are shorter and are usually present in large numbers on the surfaces of cells. Many protozoa swim through the coordinated beating of thousands of surface cilia.
- Fields of cilia are also used by cells to circulate water over their surface. Sessile protozoa use cilia in this manner for feeding. Sheets of epithelial cells in higher animals often use cilia to direct the flow of mucous secretions in ducts and passages.
- The basic design of eucaryotic cilia and flagella recurs in different forms specialized for particular tasks, including cell adhesion and sensory detection.

## *Further Reading*


Atkins, P.W. Physical Chemistry. Oxford: Oxford University Press, 1994.

Berg, H.C. Random Walks in Biology. Princeton, NJ: Princeton University Press, 1993.

Feynman, R.P., et al. The Feynman Lectures on Physics. Vol. 1, Chap. 6, Probability. Reading, MA: Addison-Wesley, 1963.

Hille, B. Ionic Channels of Excitable Membranes. Sunderland, MA: Sinauer Associates, 1992.

Lavenda, B.H. Brownian motion. *Sci. Am.* 56: 56–67, 1985.

Nicklas, R.B. A quantitative comparison of cellular motile systems. *Cell Motil. Cytoskeleton* 4: 1–5, 1984.

Perrin, J. Movement Brownien et réalité moléculaire. *Annales de Chimie et de Physique* 18: 5–114, 1909.

Purcell, E.M. Life at low Reynolds number. *Am. J. Phys.* 45: 3–11, 1977.

Silverman, M., Simon, M. Flagellar rotation and the mechanism of bacterial motility. *Nature* 249: 73, 1974.

Singleton, P., Sainsbury, D. Dictionary of Microbiology and Molecular Biology. Chichester, UK: John Wiley & Sons, 1987.

Sleigh, M.A. Protozoa and Other Protists. London: Hodder and Stoughton, 1989.

Van Dyke, M. An Album of Fluid Motion. Stanford, CA: The Parabolic Press, 1988.

Vogel, S. Life in Moving Fluids. Princeton, NJ: Princeton University Press, 1983.

Vogel, S. Life's Devices. Princeton, NJ: Princeton University Press, 1988.

White, C.M. The drag of cylinders in fluids at slow speeds. *Proc. R. Soc. A* 186: 472–479, 1946.

# CHAPTER 2

# Migration of Cells Over Surfaces

Amoebae, diatoms, and some types of cyanobacteria are all single-celled organisms that cannot swim, even though they live in water. Apparently they find it safer and more profitable to clamber over particles of rock or plants than to venture out into the third dimension. Crawling, or gliding, is also the common means of locomotion of cells in a multicellular animal, moving through the minute spaces in tissues and over the surfaces of other cells or the filamentous meshwork of the extracellular matrix. In this way white blood cells migrate through tissues in search of infection; cells in a developing embryo move from one location to another; the tips of nerve axons navigate toward their eventual synaptic targets; and cancer cells spread from an initial focal tumor to multiple sites within the body.

A crawling cell, in contrast to one that is swimming, does not employ a conspicuous motile organelle, like a flagellum, that can be separated from the remainder of the cell and studied in isolation.[1] In general, crawling cells move by means of wormlike cycles of extension and contraction that deform their entire surface, or in some cases they glide like minute slugs without visible means of propulsion. The molecular basis of cell crawling is understood in broad outline (Chapter 8) but we still know next to nothing about the molecular machinery of gliding.

## *Cells close to a surface experience large viscous drag*

In Chapter 1 we saw that viscosity is one of the major forces encountered by a cell swimming through water. Somewhat unexpectedly, viscosity is also important to a crawling cell. This is because viscous resistance increases enormously for a body moving close to a solid surface,

$$F \propto \frac{Av\eta}{h}$$

where $F$ is the viscous force, $A$ is the area of the cell adjacent to the surface, $v$ is the velocity over the surface, $\eta$ is the viscosity of the fluid, and $h$ is the distance between the lower surface of the cell and the substratum. Thus, a crawling or gliding cell moving at submicrometer distances from a solid surface (small $h$) must exert significant force to overcome viscosity. Curiously, this also suggests that a cell does not need to come into molecular contact with the surface in order to pull itself along, since viscous drag

[1] There are exceptions: such as the ciliated protozoan *Stylonichia* that employs bundles of cilia (cirri) to creep and bacteria such as *Proteus* that use flagella to swarm over surfaces.

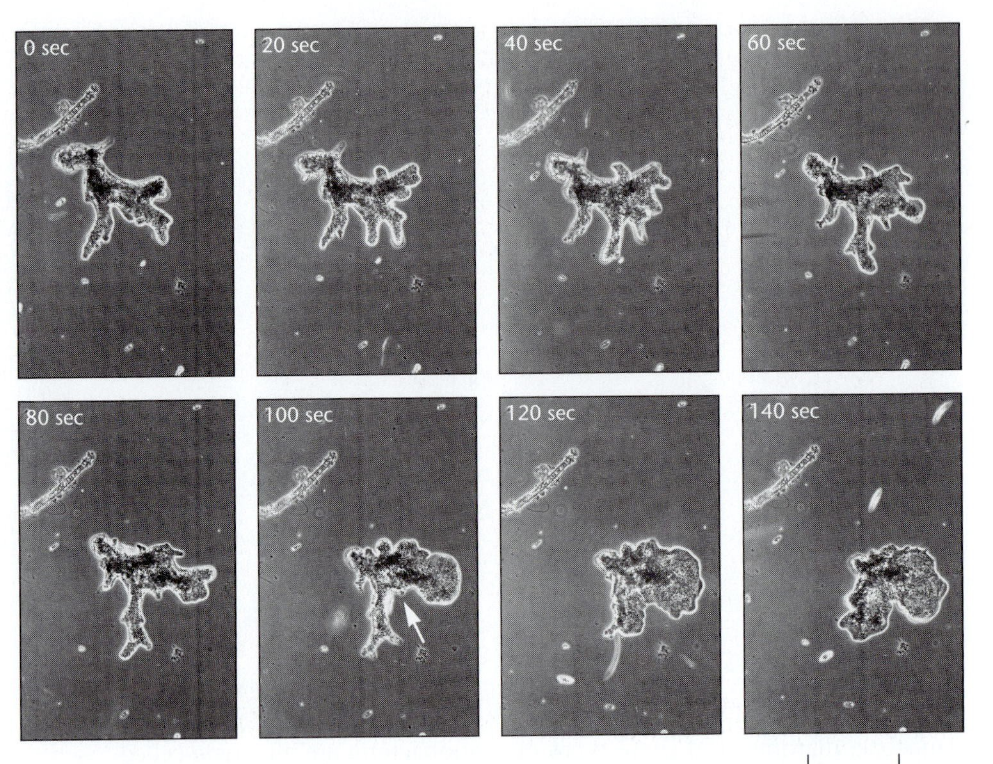


**Figure 2-1** *Amoeba proteus* feeding. The slow random crawling of *A. proteus* is seen in this sequence of photographs taken at 20-second intervals. Although it seems chaotic, *A. proteus* changes its movements when it senses food. A rapidly swimming protozoan that by chance blunders into the amoeba (see arrow at 100 seconds) causes the latter to commence a slow circling movement. In this case the prey escaped, but not all cells are so lucky.

will provide enough traction. Anchorage *is* necessary, however, to prevent the cell's being carried away from the surface by fluid currents or thermal diffusion

Surface drag also affects swimming cells. The viscous resistance experienced by a cylinder dropping in a fluid at Reynolds number of $10^{-4}$ (appropriate for a cilium or a eucaryotic flagellum) is doubled by a wall 500 cylinder diameters away (that is 10 μm or so). A nearby solid surface could therefore have major effects on the viscous drag experienced by a swimming organism. In reciprocal fashion, the movement of a cell through water will also inevitably produce a wide field of disturbance. This could have important consequences for free-living unicellular organisms, for example by acting as a signal to potential predators or prey.

## Amoebae crawl by means of pseudopodia

Large freshwater amoebae of the *Amoeba proteus* type crawl by means of *pseudopodia*. These are blunt protrusions, 100 μm or more in diameter, that require minutes to elongate or to make significant changes in position. As the cell advances, pseudopodia extend and make contact with the substratum while in other parts of the cell other pseudopodia retract and are absorbed into the cell body in a clumsy parody of walking (Figure 2-1).

Although amoeboid motion at first sight seems random and aimless, this impression is removed if these cells are observed as they feed. A chance contact with a ciliate or other small microorganism prompts nearby pseudopodia to start a sinister encircling or grasping motion (see Figure 2-1). Usually the amoeba is too slow to prevent the prey's escaping, but occasionally the smaller cell is trapped: encircling pseudopodia then form a 'cup' around the prey organism and the plasma membranes fuse to produce a food vacuole within which the prey organism is digested.[2]

Various types of pseudopodia are formed by other members of the *Sarcodina*—the subphylum of protozoa that includes amoebae and related organisms. Some pseudopodia are blunt and obtuse like those of *A. proteus*,

[2] There is more to the encounter than appears in the light microscope. Giant amoebae secrete substances that anaesthetize and partially paralyze ciliates that they contact, thereby compensating for the relative slowness of the movements of their pseudopodia.

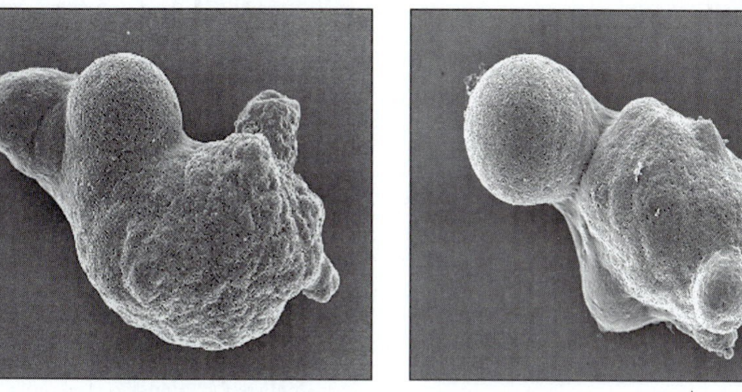


**Figure 2-2** Lobopodia. The parasitic amoeba *Entamoeba histolytica* migrates by sporadically producing bulbous pseudopodia, termed *lobopodia.* (Scanning electron micrographs courtesy of Arturo Gonzalez Robles.)

whereas in the parasitic *Entamoeba* pseudopodia erupt like herniations of the cell surface (Figure 2-2). In amoebae such as *Acanthamoeba castellani* locomotion is produced by flattened fan-shaped pseudopodia (*lamellipodia*) that spread over the surface with a fluid motion. This organism can also burrow through soil or soft agar using its pseudopodia as an excavating tool.

Zoologists also use the terms 'pseudopodia' and 'pseudopods' rather generally to refer to a variety of cell-surface protrusions, even if they are not concerned primarily with locomotion. These include long and slender processes that act as feelers for the cell and others that extend for large distances away from the cell over surfaces and are used principally for feeding (Figure 2-3).

## *A stream of fluid cytoplasm fills a growing pseudopodium*

As each pseudopodium grows on the surface of a giant amoeba, such as *Amoeba proteus*, it is filled by a stream of fluid cytoplasm. Mitochondria and other organelles are carried from the body of the cell, moving within a cylindrical wall of seemingly more rigid cytoplasm (Figure 2-4). The inner compartment—fluid, mobile, and full of organelles—is traditionally termed the *endoplasm* or *plasmasol*, whereas the outer layer, which is transparent and gel-like, is known as the *ectoplasm*, or *plasmagel* (a particularly large and prominent example of the cell cortex, discussed in Chapter 6). These two forms of cytoplasm seem to be interchangeable; in the tip of a growing pseudopodium the stream of fluid plasmasol appears to congeal into gel-like plasmagel. The reverse transformation of gel to sol also takes place as pseudopodia are resorbed. These changes in physical state of the cytoplasm reflect changes in the number, length and degree of cross-linking of actin filaments (Chapter 18).

Cytoplasmic streaming is also a conspicuous feature of the plasmodial slime molds. The life cycle of these curious organisms includes a single-cell amoeboid form and they are usually classified as protozoa, but they also have features in common with fungi. The slime mold *Physarum polycephalum*, for example, forms huge multinucleated 'cells' that can be up to 20 cm in diameter. A huge reticulated network of cytoplasmic strands extends through the cytoplasm of this organism, along which cytoplasm streams in sol-like channels up to a millimeter in diameter. In any particular channel, flow reverses direction repeatedly, but not with any precise periodicity; so that reversals occur at different times at separated regions of the cell in a phenomenon known as *shuttle streaming*. In *Physarum* as in *A. proteus*, gel and sol states of the cytoplasm continually interconvert.

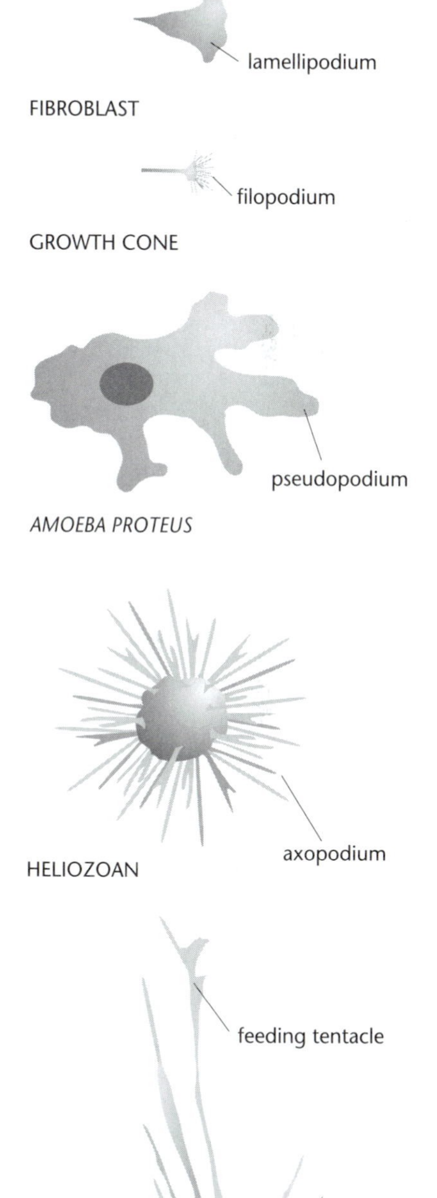


**Figure 2-3** Different types of pseudopodia. Shown schematically, approximately to the same scale.

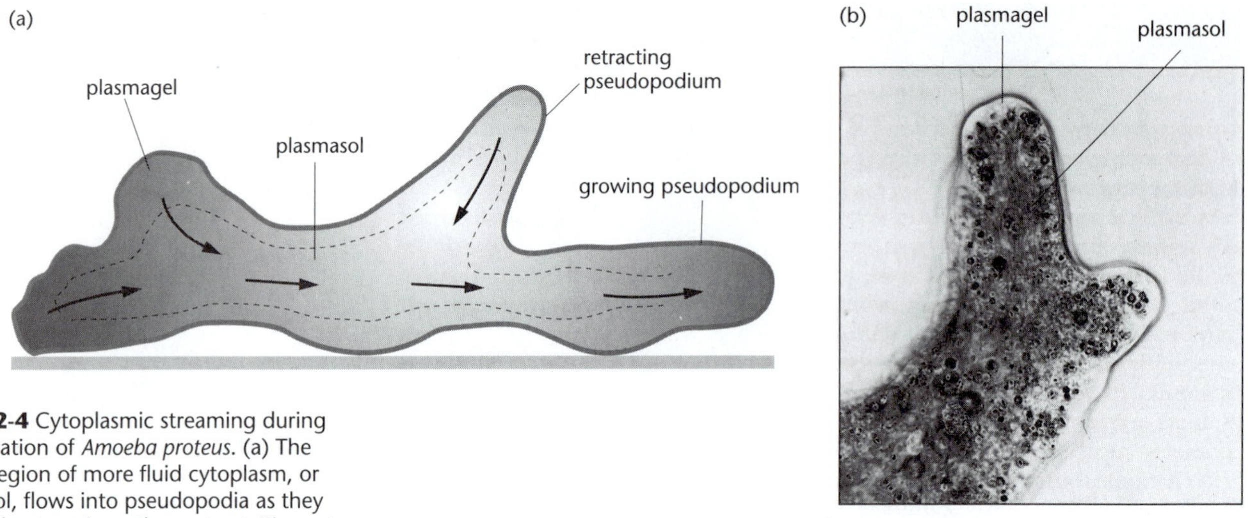

Figure 2-4 Cytoplasmic streaming during the migration of *Amoeba proteus*. (a) The central region of more fluid cytoplasm, or plasmasol, flows into pseudopodia as they form and out again as they retract. The outer shell of more rigid cytoplasm appears to form from the plasmasol at the tip of the growing pseudopodium. (b) Tip of the pseudopodium of *A. proteus*. Note the transparent layer beneath the plasma membrane (plasmagel) that excludes vacuoles, mitochondria, and other organelles.

## *White blood cells migrate freely in the vertebrate body*

Most cells in an adult plant or animal are fixed in place as part of a tissue such as epidermis or muscle and do not move as individuals from one location to another.[3] There are, however, some cell types whose functions require active migration. This is true of sperm cells, for example, and of different kinds of white blood cells.

Blood cells, including lymphocytes, neutrophils, monocytes, and red blood cells, originate in the bone marrow and are carried from that location by passive circulation in the blood. Red blood cells remain confined within blood vessels, but white blood cells continually migrate through the walls of capillaries and small veins to become freely moving cells of the connective tissue. It is there that they carry out their function of defending the organism against infection.

Neutrophils, for example, actively move through the walls of blood vessels and into the surrounding tissues in order to seek and engulf bacteria that cause infection or decay. They are directed to suitable locations by chemical changes in the endothelial cells forming the wall of the blood vessel, which in effect send out 'calls for help.' Changes to the endothelial surface slow and eventually arrest the migration of neutrophils through the vessel (Figure 2-5). The bound neutrophils then change their shape and squeeze through the endothelial layer into the damaged tissue and migrate at speeds up to 30 μm/min toward sites of bacterial infection. Each cell is able to engulf and destroy many bacteria before it itself dies. The large numbers of dead neutrophils that accumulate in cavities excavated from the inflamed tissue are expelled as pus or slowly autolyzed and absorbed by the surrounding tissue.

In the event of a major infection, neutrophils are reinforced by slower battalions of the immune response. Lymphocytes, the most abundant cells of the immune system, circulate continually between the lymph nodes, spleen, and connective tissues, using the blood circulation as a common path. During their migration through connective tissues, they display a form of cell crawling that is closely similar to the motility of neutrophils. Lymphocytes not only seek out antigenic foreign cells, they also interact with macrophages—a type of white blood cell specialized for phagocytosis. Lymphocytes and macrophages, working together, promote the destruction of foreign cells by cytotoxic and phagocytic mechanisms. Macrophages roam freely through connective tissue devouring bacteria, dead cells and the products of tissue damage. Their ability to perform phagocytosis depends on the cortical layer of actin filaments just beneath the plasma membrane, described in Chapter 6.

[3] Cells in epithelial sheets and other tissues can, however, actively change places with their neighbors and this can produce large-scale movements of entire sheets of cells, as described in Chapter 20.

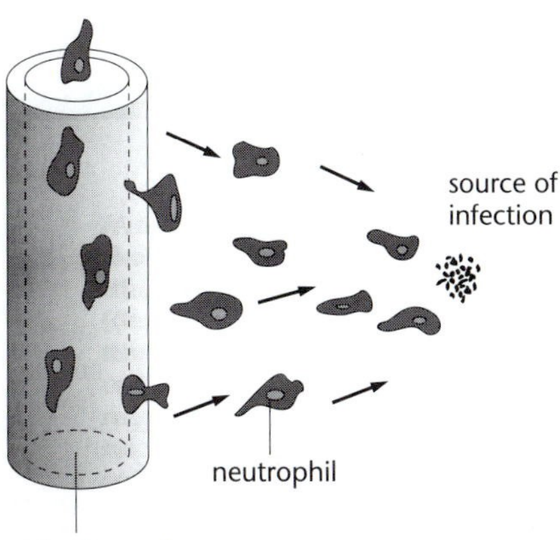


**Figure 2-5** Migration of neutrophils from the bloodstream. Localized damage to tissues causes the cells lining nearby blood vessels to change their surface molecules. Neutrophils carried in the bloodstream bind to these surface molecules, which cause them to slow down and eventually stop. The neutrophils change shape and then squeeze through the endothelial layer into the tissue in search of bacteria, which they engulf and destroy.

## *Fibroblasts move within connective tissue*

Loose connective tissue, which lies beneath the skin, contains a complex network of polysaccharides (such as glycosaminoglycans and cellulose) and proteins (such as collagen) secreted by cells. This is where cell movement habitually occurs in the body of a vertebrate. Connective tissue contains not only occasional wandering macrophages and lymphocytes but also fibroblasts and their relatives—the cells that make collagen fibers and organize them into bundles and networks (Figure 2-6). Fibroblasts are normally sedentary cells, moving only now and then and to a limited extent. In the event of tissue damage, however, they become highly motile, migrating into the wounded area. A deep wound that destroys muscle, bone, blood vessels, and other tissues induces the proliferation and movement of not only fibroblasts and macrophages but also cells of many other kinds.

Since the repair of wounds requires the formation of new tissues, it is not surprising to find that it has some similarity to the embryonic process by which the tissues were initially made. Indeed, cell movements, including various forms of cell crawling, are a necessary and persistent part of normal embryonic development.

## *Cell migration is fundamental to vertebrate embryonic development*

A developing embryo is in continual orchestrated turmoil as cells grow, divide, and change shape. Individual cells move relative to their neighbors until this is prevented by the establishment of cell–cell junctions. Views of the interior of a living fish embryo, for example, made possible by the transparent epithelium of these species, reveal cells migrating deep within the embryo and yolk sac, then aggregating along the embryonic axis (Figure 2-7). At later stages of development there are, in all vertebrate embryos, episodes of large-scale migration that carry specific types of cells to their eventual locations in the adult body.

Germ cells, the precursors of sperm and egg, are determined very early in development and then migrate from their site of origin in the neighborhood of the gut or yolk sac to the epithelium of the genital ridges. They move in a similar fashion to neutrophils, by extending and retracting lamellipodia and filopodia. Precursors of muscle cells in a developing chick limb are also determined very early and migrate into the region of the future limb from segmentally arranged masses of cells known as somites. Having arrived at their destinations, the cells remain inconspicuously in position until the time comes to differentiate and form the muscles of the wing or leg.

The important group of migratory cells known as the neural crest forms from the ectoderm at the site of neural tube closure. The cells break loose from the epithelium and migrate along specific pathways through the embryo to give rise to a variety of tissues, including peripheral neurons, Schwann cells, pigment cells, and various connective tissues in the head. If neural crest cells are marked, either by substituting cells of a related

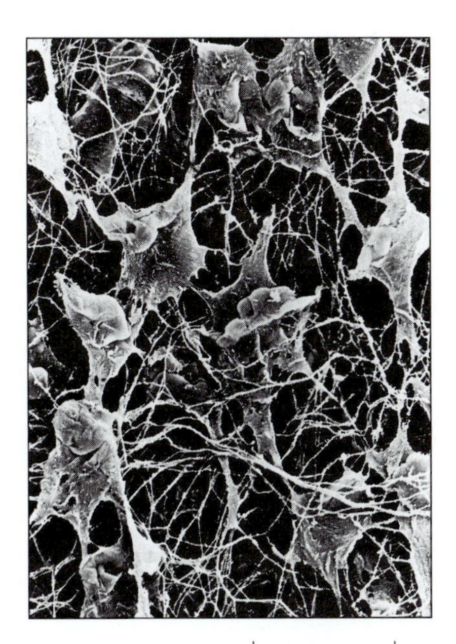


**Figure 2-6** Cells in connective tissue. Fibroblasts are seen embedded in a network of elastin and collagen fibers. (Courtesy of Julian Heath.)

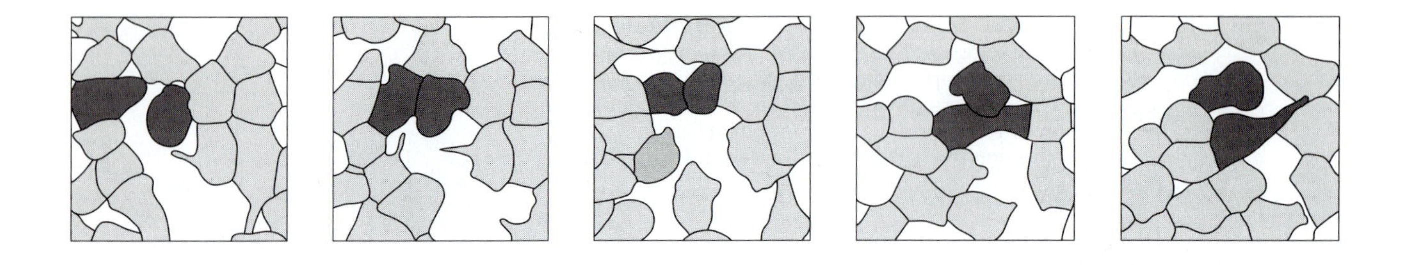

**Figure 2-7** Migrating cells in a fish embryo. Outlines of migrating cells recorded from a video of a *Fundulus* embryo taken at intervals of 3 minutes, with two individual cells marked. Direction of migration is to the right.

species or by radioactively labeling them with tritiated thymidine, then it is possible to trace the paths they follow through the embryo (Figure 2-8). In this way, cells are found to leave the neural crest by two main pathways: one just below the ectoderm and the other leading deep into the body at the somites. The sites that a neural crest cell will colonize depend on its position along the body axis. Cells migrate along pathways defined for them by their host connective tissues and settle wherever these paths may lead; they then differentiate into the type of cell appropriate for that location.

## *Even cells from differentiated tissues have an innate capacity for locomotion*

Most cells in adult tissues, like liver, heart, or brain, are held in place by intercellular junctions. Even these sedentary cells can, however, be stirred to movement by putting them in a tissue culture dish. They can be freed from their connections with neighboring cells by treatment with a protease such as trypsin or with a chelating agent such as EDTA, and then dispersed by gentle pipetting. Single, rounded cells put into tissue culture medium will flatten onto a suitably prepared glass or plastic surface and, in many cases, begin migration. Such cells often retain characteristics of their differentiated cell type: epithelial cells grow as sheets of continuous cells; nerve cells produce long axons; and muscle cells fuse into long multinucleate muscle fibers and begin to make contractile proteins. There are, however, also similarities in their initial response to the substratum and in their early mode of locomotion, suggesting that they share common underlying motile machinery.

The behavior of fibroblasts settling onto a culture dish illustrates the movements common to most vertebrate cells. Following dissociation of the connective tissue—or of a previous population of cells in tissue culture—the cells in suspension appear compact and rounded. Closer examination by scanning electron microscopy shows that their surfaces are not

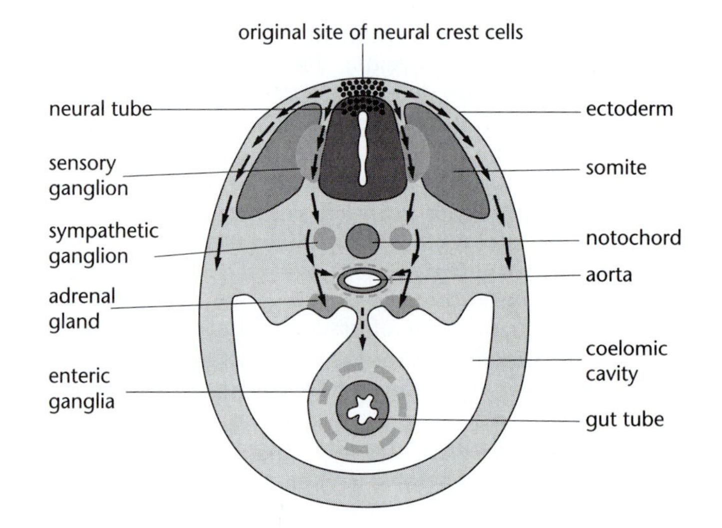


**Figure 2-8** Neural crest migration. A chick embryo is shown in a schematic cross-section. Neural crest cells that take the pathway just beneath the ectoderm will form pigment cells of the skin. Those that take the deep pathways via the somites will form the sensory ganglia, sympathetic ganglia, and parts of the adrenal gland. The enteric ganglia in the wall of the gut are formed from neural crest cells that migrate along the length of the body, originating from either the neck ganglion or the sacral region.

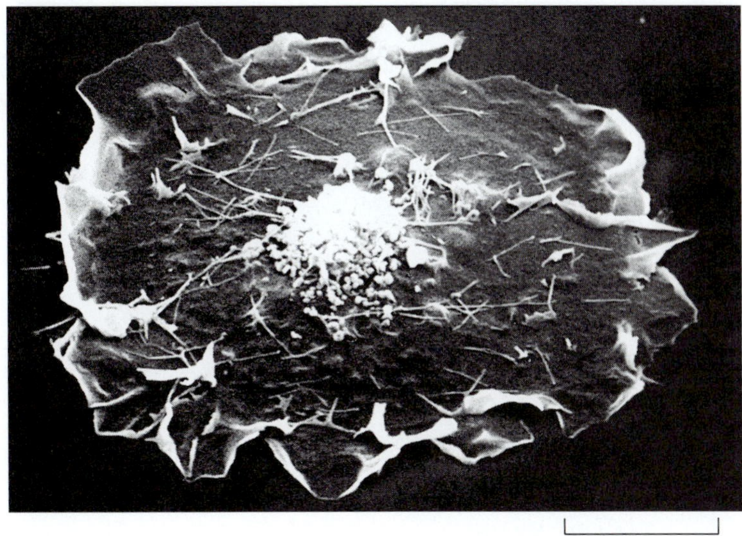


**Figure 2-9** Fibroblast settling onto a tissue culture dish. The dome-shaped central portion covered with small blebs is the remnant of the cell in its rounded form. As the cell settles onto the surface, a skirtlike region of flattened lamellipodia and small filopodia extends outward. (Scanning electron micrograph courtesy of Julian Heath.)

smoothly rounded but are covered with numerous small protrusions (Figure 2-9). The surface protrusions include *filopodia*, which are fine hairlike extensions about 0.1–0.2 nm in diameter and up to 20 μm in length; *lamellipodia*, which are sheets or veils of membrane-enclosed cytoplasm similar in thickness to the filopodia, and hemispherical mounds, or *blebs*, 2–10 μm in diameter, that project from the cell's surface. These dynamic structures form and move around on a time scale of minutes, giving an overall impression of a slow boiling of the cell surface. When the fibroblast comes into contact with the culture substratum, it slowly flattens out to resemble an irregularly shaped fried egg with a rim of flattened lamellipodia. Slowly the cell adopts a distinct polarity and commences directed motion.

## *Fibroblasts crawl by means of lamellipodia*

A fibroblast in culture uses a series of repeated crawling movements analogous in some respects to those shown by giant amoebae, but differing in both in scale and shape. The cell moves smoothly over the culture surface at speeds of about 40 μm/hr (about 1 mm/day) with flattened, spread-out lamellipodia at its front end (Figure 2-10). In a time-lapse video, these lamellipodia surge forward with a smoothly fluid motion, pausing for occasional brief retrograde movements. Lamellipodia that are pulled backward rise up from the dorsal surface of the cell, like waves, to disappear at a region some 10–20 μm back from the leading edge. The continual rearward motion seen in a time-lapse video leaves an impression of a flickering or flamelike movement, commonly termed *ruffling*.

Ruffling is a complex process, with filopodia and lamellipodia appearing and retracting in an unpredictable way. Overall, it results in a rearward movement of these structures toward the tail of the cell, and their collapse close to the central portion of the cell. A continual rearward flow is also occasionally displayed by particles on the exposed cell surface, which are picked up by the ruffling membranes and carried in a straight line backward relative to the direction in which the cell is advancing. The molecular basis of ruffling and its role in cell migration are discussed in Chapter 8.

As a fibroblast moves forward, it is typically drawn into an elongated triangular form. The lamellipodia form one side of this triangle, at the anterior end of the moving cell, and the other two sides project backward as smooth and relatively immobile structures (see Figure 2-10). At its rearward end, the cell is drawn into a long tail, or retraction fiber, which periodically pulls away from the surface and contracts back into the cell

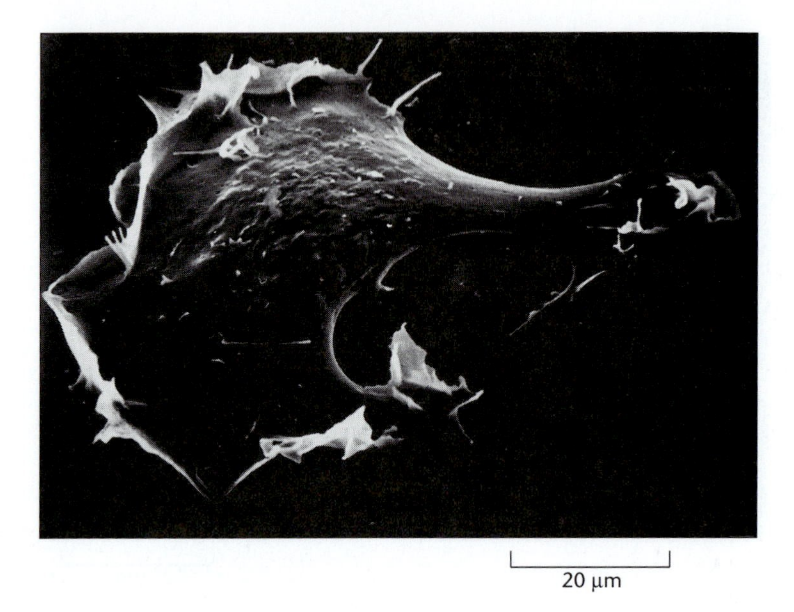


**Figure 2-10** Migrating fibroblast. Scanning electron micrograph of a skin fibroblast migrating to the left. Note the ruffling membrane at its leading margin. (Courtesy of Julian Heath.)

with an accompanying flurry of activity at the leading edge. As lamellipodia extend forward, they form discrete regions of close adhesion to the substratum known as focal adhesions. These are like numerous tiny feet that remain stationary as the cell moves over them but detach from the culture substratum as they reach more proximal positions.

An isolated fibroblast in culture wanders aimlessly over the surface of a tissue culture dish, following a sinuous track (Figure 2-11). Like the path of a swimming cell, its motion can be described as sequence of linked vectors, but the sequence differs from that of a true random walk in that the angles between successive vectors are closer than expected to zero. In other words, the cell shows a persistence in the direction of its movements, this effect being more pronounced the shorter the time intervals that are taken.

## *Filopodia act like feelers for the cell*

The mode of locomotion based on ruffling lamellipodia just described, called *fibroblast locomotion,* is also used by other kinds of vertebrate cells, although the details can vary. For example, when epithelial cells are cultured as single cells, they move by means of lamellipodia like fibroblasts. Soon, however, they attach to each other and begin to move as small groups over the surface of the culture dish. Eventually cells of this kind produce large sheets of flattened cells, similar to the epithelial sheets in the body, and move only at their free boundaries.

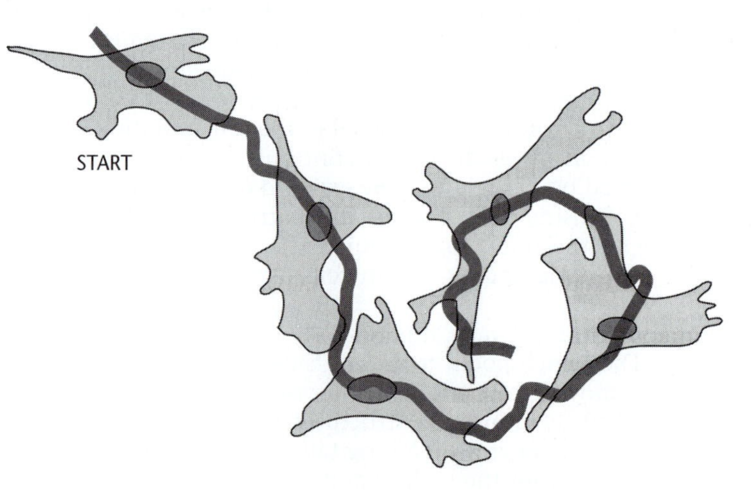


**Figure 2-11** Track of migrating cell. Outlines of a fibroblast migrating over a culture surface drawn at intervals of 120 minutes.

Animal cells produce filopodia in addition to, or instead of, lamellipodia. Sometimes these appear to serve as locomotory organelles, rather like lamellipodia. For example, fibroblasts grown in a three-dimensional matrix of collagen fibers rather than on a planar culture dish become less flattened and produce many filopodia. Cells that wander through fluid-filled spaces in the body, such as neutrophils and the wandering coelomocytes of developing sea urchin embryos, typically employ filopodia rather than lamellipodia as migratory organelles.

We will see in Chapter 6 that lamellipodia and filopodia are related structures. Both are based on actin filaments and they have a similar biochemical composition. However, filopodia often seem to have a sensory rather than a mechanical role. The first contact between two cells in a developing embryo is often by means of their filopodia, which can not only reach out over very long distances but can also easily penetrate the gelatinous extracellular matrix. It has been found, for example, that filopodia extending from neighboring cells in an epithelium initiate formation of cell–cell junctions. Similarly, amoebae of the slime mold *Dictyostelium*, which carry both lamellipodia and filopodia, use the first for migration and the second for functions such as the detection and capture of bacteria.

## Growth cones crawl over surfaces

Filopodia are also a prominent feature of the exploratory tip of a growing axon—the *growth cone*. Even though it remains tethered through its axon to a stationary cell body, the growth cone carries flattened lamellipodia and long whiskery filopodia that extend and retract in a closely similar fashion to the ruffling membranes of a neutrophil or small amoeba (Figure 2-12). Each growth cone can, if severed from its axon, crawl independently over the surface of a culture dish. The capacity of a growth cone to steer a path and respond to multiple directional clues, however, far exceeds that of any independently migratory cell. Growth cones traverse many hundreds of micrometers through the complex terrain of the developing nervous system to find and make synapses with their respective target cells. It is thanks to this path-finding capacity of the growth cones that the basic blueprint or wiring diagram of the nervous system is established.

Growth cones were discovered and described at the end of the nineteenth century by the Spanish scientist Ramon y Cajal, the 'father of neuroanatomy.' Working from static, fixed preparations he deduced many of the dynamic features of the growth of these structures and predicted, among other things, that they could be guided by chemical 'scents' released by target tissues—a form of chemotaxis. Almost exactly a hundred years later this prediction was confirmed by the isolation of proteins secreted by cells that act as diffusible attractants for growth cones. We will return to the fascinating question of how growth cones are guided in the developing brain in Chapter 20.

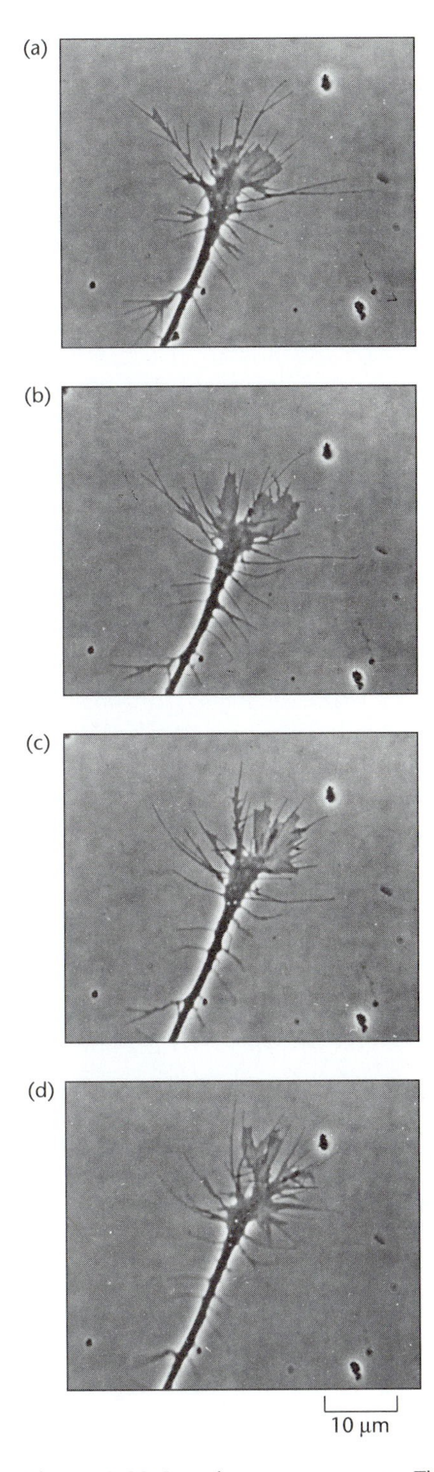


**Figure 2-12** Growth cone movements. The continual, exploratory movements of filopodia at the leading tip of a nerve axon are shown in this sequence of pictures, taken at intervals of 1 minute.

## Myxobacteria, gregarines, and diatoms glide over surfaces

Ameobae, fibroblasts, and other types of vertebrate tissue cells all *crawl* over surfaces, in the sense that they perform a repetitive series of inchwormlike movements. However, cells can also move in a way best described as *gliding*, in which they show smooth continuous motion over a surface, without any evidence of surface distortions. Gliding motility is widespread in bacteria, algae, diatoms, parasitic protozoa, and in the *labyrinthulas* ('net slime molds'). The latter group of eucaryotic organisms secrete extensive networks of tracks of slime, within which cells move by gliding. Filamentous *cyanobacteria*, as we will see in Chapter 16, probably

**Figure 2-13** Fruiting body of myxobacteria. Each fruiting body, packed with spores, is created by the aggregation and differentiation of about a million myxobacteria. (Courtesy of P.L. Grilione and J. Pangborn.)

push themselves over surfaces by streams of secreted mucous, which are directed by pores set at precise positions in their cell wall.

*Myxobacteria* are gliding bacteria that live in the soil and feed on insoluble organic molecules, which they break down by secreting degradative enzymes. They show two distinct forms of gliding movement in which the bacteria move either as single cells or as a large swarm. These two modes depend on different sets of genes, and the organism switches from one set to the other depending on the external conditions (recall from the previous chapter that some flagellated bacteria also have two alternate modes of movement). The swarming mode enables the cells to stay together in loose colonies in which the digestive enzymes secreted by individual cells are pooled, thereby increasing their efficiency of feeding (the 'wolf pack effect'). When food supplies are exhausted, the cells aggregate tightly together and form a multicellular fruiting body within which the bacteria differentiate into spores that can survive even hostile environments (Figure 2-13). When conditions are more favorable, the spores in fruiting bodies germinate to produce a new swarm of bacteria.

Some species of diatoms are also able to glide, despite being enclosed in a tough siliceous cell wall. Diatoms are a large group of algae that occur widely in fresh or sea water. Their silicon-based walls, typically formed from two halves or valves are often elaborately ornamented with pores, slits, ribs, or tubes arranged in symmetric, species-specific patterns. Species capable of gliding are usually bilaterally symmetric and have a slit or *raphe*, which runs the length of their valve. The cells move with the raphe slit adjacent to the substratum, apparently by the polarized secretion of adhesive mucilage.

Gliding motility is common in parasitic protozoa, perhaps because it is an efficient way for them to move within the body cavities of an animal host. The infectious form of *Plasmodium falciparum*, which causes malaria in man, consists of elongated single cells, about 10 μm in length, which are capable of gliding over surfaces at around 1 μm/sec. *Gregarines*, which are relatively huge protozoa (100–300 μm long and up to 40 μm wide) live in insect hosts; they glide over glass surfaces like motorized sausages at 1–5 μm/sec. In these and other instances, it is likely that gliding locomotion is the basis by which these cells invade the tissues of their hosts.

## *Cilia and flagella are also used to move over surfaces*

We mentioned in Chapter 1 that cilia and flagella are sometimes used for purposes other than swimming or driving layers of water over a cell surface. This is true even of the simple flagella produced by bacteria, as demonstrated by the specialized swarmer cells of the bacterium *Proteus mirabilis*. These bacteria are ubiquitous throughout nature, growing on protein wastes that they break down and digest. *Proteus* normally grows in an aqueous environment as small, motile swimmer cells approximately 1 μm wide. On a solid medium, however, they differentiate into elongated swarmer cells, up to 100 μm in length and with hundreds to thousands of flagella over their surface. Swarmer cells are able to move on solid agar surfaces by coordinated movements of their flagella. Groups, or 'rafts' of *Proteus* cells then spread over the solid surface, working cooperatively to secrete digestive enzymes and feeding communally on the products of this digestion. Under some circumstances, the coordinated movements of migrating cells form regular macroscopic patterns, such as rings or even spirals over the surface of a culture dish.

Eucaryotic flagella and cilia are also used to drive cells over surfaces, as illustrated by the curious organisms known as *hypotrichs*. These are ciliates with a flattened cell body whose cilia are confined almost entirely to the lower surface of the cell. The upper surface may bear a few stiff, bristlelike cilia, but on the lower surface cilia are fused together to form tapered clusters, or cirri. The latter structures do not beat rhythmically like ordinary cilia but instead are used like legs as the animal crawls about on vegetation (Figure 2-14).

Flagella that mediate both swimming and gliding are found in the green alga *Chlamydomonas*. The paired flagella of this organism can propagate bending waves, which enable the cell to swim, as already mentioned in Chapter 1 (and discussed in greater detail in Chapter 14). However *Chlamydomonas* can use its flagella in a different way during sexual reproduction; cells catch hold of and draw closer a cell of the opposite mating type. This movement is not due to wave propagation but rather to a distinct type of movement over the surface of the flagella, which can be seen if particles such as small beads are allowed to attach to the flagella. Furthermore, if *Chlamydomonas* settle onto a surface they are able to glide over it by means of their flagella. Under these circumstances the cell splays out its flagella in opposite directions and glides at speeds of several micrometers a second.

*Heliozoa*, are spherical organisms with a central body and a radiating sunburst of needlelike extensions known as axopodia, which they use for both locomotion and feeding. Axopodia are not cilia or flagella but they are related to them structurally by being built around a bundle of cross-linked microtubules. By a coordinated shortening and lengthening of its axopodia, the heliozoan *Actinosphaerium* can roll over a surface like a ball. Axopodia also appear to be sticky and to be able to trap small animals and particles or organic debris. The prey are subsequently drawn into the cell by a steady retraction of the axopodia. The flexible use of such structures is illustrated by a different kind of heliozoan, *Sticholonche*, which uses axopodia for swimming. These organisms possess a bilaterally symmetric arrangement of axopodia, which are embedded at their base in socketlike depressions of the nuclear membranes. Motility results from their moving back and forth like oars.

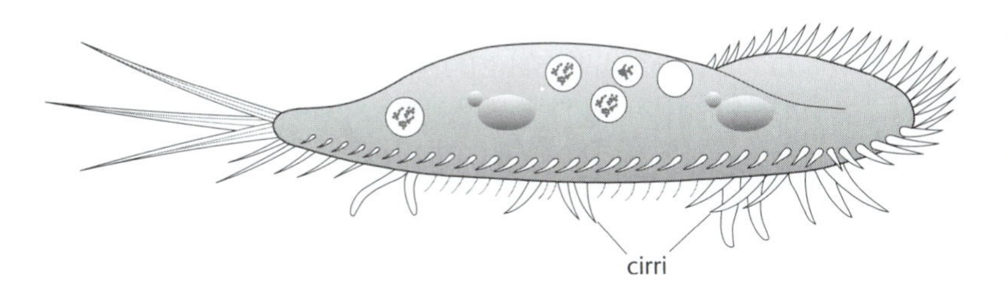


**Figure 2-14** Cilia used as legs. *Stylonychia*, a ciliated protozoan, habitually crawls over surfaces by coordinated movements of clusters of cilia on its lower surface, termed cirri.

## Essential Concepts

- Many cells migrate over solid surfaces by crawling or gliding. This includes free-living organisms such as amoebae and diatoms as well as individual cells in the body of a multicellular animal.
- When a cell crawls or glides over a surface, its movements are dominated by viscous forces, as they are for swimming cells.
- Large free-living amoebae crawl by extending and retracting stubby pseudopodia. As they move, streams of cytoplasm move to fill the growing pseudopodia.
- White blood cells move from the circulation and roam freely within the vertebrate body, searching for infection and decay.
- Fibroblasts in connective tissue become actively motile during episodes of wound healing and tissue repair.
- The developing embryo of most animal species is a site of continual, widespread cell movements. Germ cells and precursors of nerve, muscle, and other tissues migrate to their correct locations.
- Even cells in adult tissues such as liver, heart, or brain, can be stirred into independent movement by growing them in tissue culture.
- Many tissue cells, such as fibroblasts, migrate over the surface of a culture dish by producing motile flattened lamellipodia.
- Specialized cells in many metazoa also extend long thin filopodia that contribute to their migration.
- Cell gliding is shown by myxobacteria, gregarines, diatoms and other organisms. Its mechanism is poorly understood at present.
- Some organisms employ modified cilia or flagella to crawl over surfaces rather than for swimming. Examples include swarming bacteria, hypotrichous protozoa, and heliozoans.

## Further Reading


Abercrombie, M. The crawling of metazoan cells. *Proc. R. Soc. Lond.* 207: 129–147, 1980.

Allison, C., Hughes, C. Bacterial swarming: an example of prokaryotic differentiation and multicellular behaviour. *Sci. Prog. Edinburgh* 75: 403–422, 1991.

Bonner-Fraser, M. Neural crest cell migration in the developing embryo. *Trends Cell Biol.* 3: 392–397, 1993.

Bovee, E.C. Morphological differences among pseudopodia of various small amoebae and their functional significance. In Primitive Motile Systems in Cell Biology. (R.D. Allen, and N. Kamiya, eds.), pp. 189–219. New York: Academic Press, 1964.

Buchsbaum, R. Animals Without Backbones. Chicago: University of Chicago Press, 1976.

Butcher, E.C., Picker, L.J. Lymphocyte homing and homeostasis. *Science* 272: 60–66, 1996.

Edgar, L.A. Diatom locomotion: a consideration of movement in a highly viscous situation. *Br. Phycol. J.* 17: 243–251, 1982.

Grebecki, A. Membrane and cytoskeleton flow in motile cells with emphasis on the contribution of free-living amoebae. *Int. Rev. Cytol.* 148: 37–80, 1994.

Heath, J., Dunn, G.A. Cell to substratum contacts of chick fibroblasts and their relation to the microfilament system. A correlated interference-reflexion and high-voltage electron-microscope study. *J. Cell Sci.* 197: 197–212, 1978.

Mast, S.O. Structure, movement, locomotion and stimulation in *Amoeba*. *J. Morphol. Physiol.* 41: 347, 1926.

Preston, T.M., King, C.A. Strategies for cell-substratum dependent motility among protozoa. *Acta Protozool.* 35: 3–12, 1996.

Shapiro, J.A. Bacteria as multicellular organisms. *Sci. Am.* 256: 82–89, 1988.

Shi, W., Zusman, D.R. The two motility systems of *Myxococcus xanthus* show different selective advantages on various surfaces. *Proc. Natl. Acad. Sci. USA* 90: 3378–3382, 1993.

Shimkets, L.J. Social and developmental biology of the myxobacteria. *Microbiol. Rev.* 54: 473–501, 1990.

Singleton, P., Sainsbury, D. Dictionary of Microbiology and Molecular Biology. Chichester, UK: John Wiley & Sons, 1987.

Trinkaus, J.P., et al. On the convergent cell movements of gastrulation in *Fundulus*. *J. Exp. Zool.* 261: 40–61, 1992.

Zheng, J.Q., et al. Turning of nerve growth cones induced by neurotransmitters. *Nature* 368: 140–144, 1994.

# CHAPTER 3

# Cell Behavior

We have so far described the swimming and crawling movements of animal cells as aimless random walks. But the *purpose* of these movements—the reason they evolved in the first place—is to carry a cell in a specific direction. Bacteria and other microorganisms swim because this takes them closer to sources of food, or farther away from noxious chemicals. Neutrophils and macrophages in our body crawl over surfaces in order to do battle with invading organisms, or to eat and dispose of dead cells. Cells in an embryonic animal perform complex patterns of social movement so they can build the body's shape and position its organs correctly.

This capacity to move in one direction rather than another implies a sort of rudimentary intelligence. The cell expresses preferences; it knows something of its environment's composition and reacts accordingly. As we shall relate in this chapter, the repertoire of behaviors shown by single cells is astonishingly rich and complex. Efforts to understand this repertoire raise many deep problems about the mechanism of integration and control that will occupy us later in the book.

## *Bacteria and amoebae migrate toward distant sources of chemoattractant*

If a micropipette containing a concentrated solution of glucose is placed in a culture of *Escherichia coli*, the bacteria congregate around its open tip (Figure 3-1). They are attracted by the high concentration of glucose diffusing from the micropipette tip and will similarly move toward sources of other sugars, amino acids, and peptides (Table 3-1). Detailed analysis of the path of a bacterium in this situation shows that the pattern of its swimming is changed by the contents of the micropipette. As already described in Chapter 1, a swimming bacterium normally switches

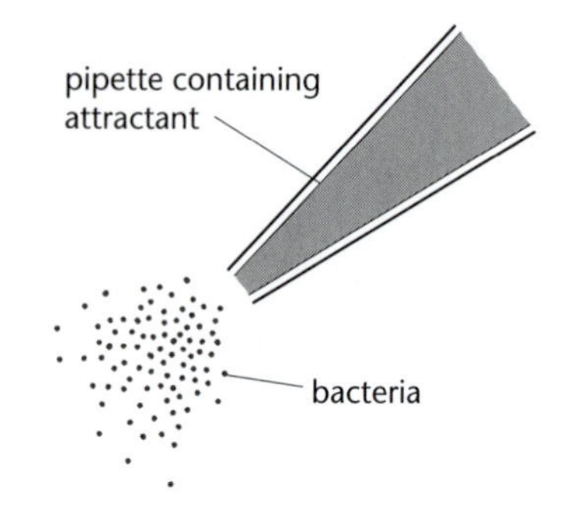


**Figure 3-1** Bacteria congregating about a pipette tip. Attractant molecules diffusing from the pipette create a concentration gradient, which the motile bacteria detect and move toward.

**Table 3-1 Cell attractants and repellents**

| Cell type | Substance |
|---|---|
| bacteria (*E. coli*) | Attractants: aspartate; serine; maltose; ribose.<br>Repellents: nickel ions; cobalt ions; ethanol; benzoate; butyrate. |
| fungi (yeasts and others) | Attractants: sex pheromones (steroids); mating factors (proteins). |
| slime mold (*Dictyostelium*) | Attractants: cyclic AMP. |
| neutrophils, macrophages | Attractants: *N*-formylated peptides; growth factors; soluble collagen. |
| nerve cells | Attractants: secreted proteins such as Netrin; Slit; Semaphorin. |

A list of some of the substances that act as attractants or repellents for living cells. Threshold concentrations for detection of chemoattractants are typically in the range $10^{-6}$ M to $10^{-8}$ M; repellents often have a higher threshold. Receptors on the cell surface may be specific for a single substance or they may recognize a group of related structures; in some cases the same substance attracts some cells and repels others (e.g., Netrin).

rapidly between periods of smooth swimming, in which it persists for a second or so in a given direction, and brief periods of tumbling, during which its direction changes in a highly erratic fashion. But if the swimming cell by chance swims into a region containing dissolved attractants, then its tumbles are suppressed and the duration of its smooth swimming runs is extended. Conversely, if the concentration of attractant becomes less, then the cell tumbles more frequently and smooth runs become shorter (Figure 3-2). In this way a bias is introduced into the otherwise random motion of the cell, carrying it in a direction that should be advantageous for its growth and proliferation.

Two terms commonly used to describe the movement of an organism in response to external stimuli are a *taxis*, in which the organism orients in the direction of the stimulus and moves toward it, and a *kinesis*, in which the speed of movement of the organism changes with the stimulus but the organism does not actually orient. According to this definition, the response of bacteria to chemical gradients is strictly speaking a kinesis, although we normally call it 'chemotaxis.' In fact, calculations show that, because of their small size and the rapidity of diffusion, bacterial cells would be unlikely, to be able to sense a gradient over their surface. The same is not true, however, of the many eucaryotic cells that sense chemical gradients, most of which perform a true chemotaxis in which they sense differences in concentration over their surface and orient in response.

Indeed, an ability to 'smell' nutrient molecules and move toward them seems to be a common feature of all free-living, single-celled organisms. Carnivorous cells, such as *Amoeba proteus*, use chemotaxis to crawl after prey. Chemical gradients of cyclic AMP are used by amoeboid cells of the slime mold *Dictyostelium discoideum* as they congregate to produce a multicellular body. Chemical signals are also used in sexual reproduction, for example in the migration of male and female gametes of many species of brown algae that are stimulated and guided by a diffusible sex attractant.

(a) uniform concentration

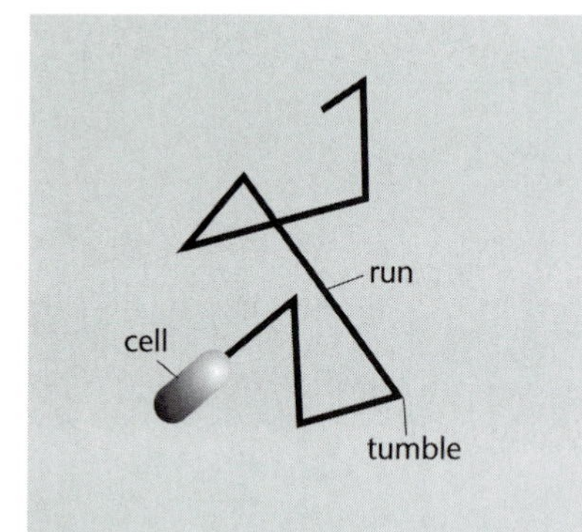


(b) gradient

**Figure 3-2** Swimming behavior of a bacterium with and without a gradient of attractant. (a) In a uniform concentration of chemoattractant the cell wanders randomly by alternating 'runs' and 'tumbles'. (b) A gradient of chemoattractant biases the random walk by making the runs persist for longer periods when the cell is traveling toward a higher concentration.

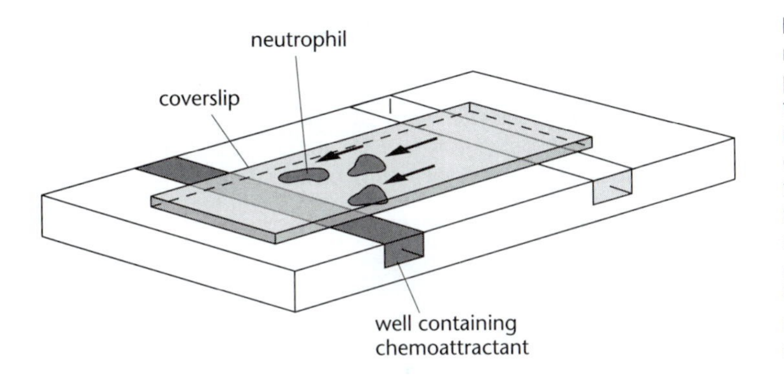


**Figure 3-3** Chamber for observing neutrophil chemotaxis. A glass coverslip is placed onto a modified glass slide containing two reservoirs, each filled with buffer. One reservoir contains a high concentration of chemoattractant. Neutrophils are placed in a thin film of buffer sandwiched between the two glass surfaces; the thin film of liquid is in contact with the two reservoirs so that a gradient of chemoattractant is established. Cells are observed in a microscope as they migrate toward the reservoir containing chemoattractant.

## *Cells migrate over long distances in the body*

Cells in multicellular organisms also move toward distant sources of diffusing chemicals. One of the best-studied examples is that of neutrophils—white blood cells that move through vertebrate tissues in search of infection and decay. If a population of blood cells is placed on a microscope slide, neutrophils will move toward and engulf dead or dying cells, a sequence that can be provoked at will by destroying a selected red blood cell by a laser microbeam.

Many substances act as chemoattractants for neutrophils (see Table 3-1). Bacterial infection is signaled by small N-terminal blocked peptides, such as *N*-formyl-methionyl-leucyl-phenylalanine (fMet-Leu-Phe).[1] The response can be analyzed quantitatively by enclosing the cells in a flat chamber in which a gradient of peptide has been established (Figure 3-3). Responsive cells turn in the direction of the highest concentration of the attractant and pursue a wandering path in its direction, their behavior in gradients of different steepness shows that they are able to detect a concentration difference of less than 10% from one side of the cell to the other.

Neutrophils exposed to a sudden, uniform increase in the concentration of chemoattractant produce more lamellipodia and other types of protrusion from their entire surface. By contrast, a micropipette containing chemoattractant induces the cells to extend pseudopodia only from the region of the cell closest to the micropipette; subsequently, they migrate toward the source of the attractant. A similar response is shown by tumor cells exposed to epidermal growth factor, and may be involved in the invasion of breast cancer cells into neighboring tissues (Figure 3-4). Polarized migration in both cases requires a selective induction of actin-rich pseudopodia in one part of the cell, a topic we will return to in Chapter 8.

## *Adaptation is an essential part of the chemotactic response*

The process of diffusion carries molecules over long distances, generating concentration gradients that can be long-range but very shallow. Cells have to be sensitive to very low concentrations of attractants because that determines the distance over which they can sense the distant source. But they must not lose their way as they close in on the source, where the concentrations are much higher. Cells of *E. coli*, for example, can respond to aspartate at concentrations from less than 0.1 μM to more than 1 mM—a factor of over 10,000-fold! This remarkable capacity to respond to gradients over a wide range of absolute concentrations is a consequence of

[1] Procaryotic proteins have a formyl-methionine residue at their N-terminus and eucaryotic proteins do not. Peptides such as fMet-Leu-Phe derived from bacterial proteins therefore 'smell' like bacteria.

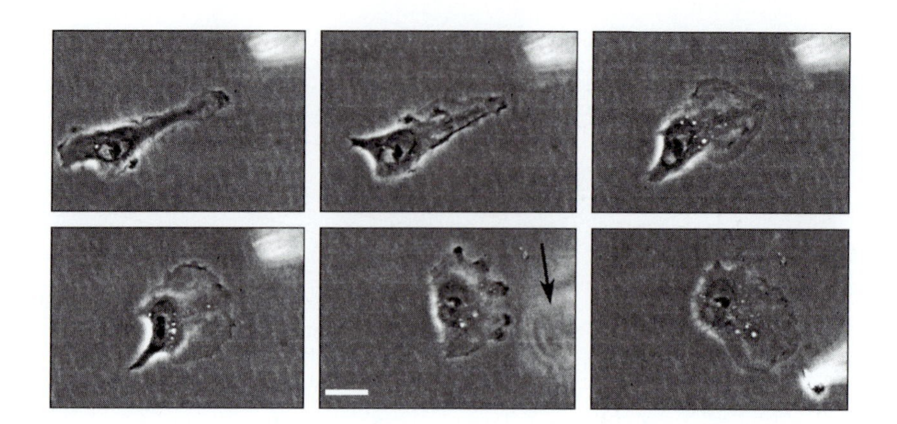

**Figure 3-4** Oriented response of a tumor cell to growth factor. A pipette containing epidermal growth factor was introduced into a culture containing mammary adenocarcinoma cells. One cell extends pseudopodia in the direction of the pipette and begins to migrate in its direction, following the pipette as it is moved (arrow). A similar process is thought to promote breast cancer invasion. Scale bar = 20 μm. (From M. Bailly, et al., 1998. © Academic Press.)

*adaptation* in the sensory system. As the cell moves toward a chemoattractant source, its sensory system continually adjusts the baseline against which it compares further increases in concentration.

Adaptation during bacterial chemotaxis can be measured experimentally by exposing bacteria to sudden increments in the concentration of an attractant. The normal pattern of intermittent tumbling is suppressed and the cell swims in a smooth fashion (Figure 3-5). After a short period, however, the tumbling frequency returns to the pre-stimulus level, the cells having in effect become desensitized to the attractant. Adaptation begins as soon as the attractant is added and the time it takes depends on the strength of the stimulus—if the change in attractant concentration is large, then it may take many minutes before the bacteria return to their original pattern of swimming. Removal of attractant produces the opposite sequence of changes: initially a very high frequency of tumbles is seen, which slowly returns to the original level. Repellents, as one might expect, produce the opposite response, so that removal of a repellent causes a period of smooth swimming.

Adaptation is an almost universal feature of behavioral responses, including those of higher animals. Its molecular basis is particularly well understood in the case of bacterial chemotaxis, as we will see in Chapter 16.

## Cells are sensitive to weak electric fields

Chemotaxis is one kind of directed movement shown by living cells, but there are others. Bacteria are especially versatile in this respect because of the enormous range of environments they inhabit. Different species of bacteria move in response to light, temperature, magnetic fields, osmotic pressure, oxygen tension, and gravity. Protozoa also have a wide repertoire of tactic behavior. We will see below, for example, that the green alga *Chlamydomonas*, is very sensitive to light, showing a violent avoidance

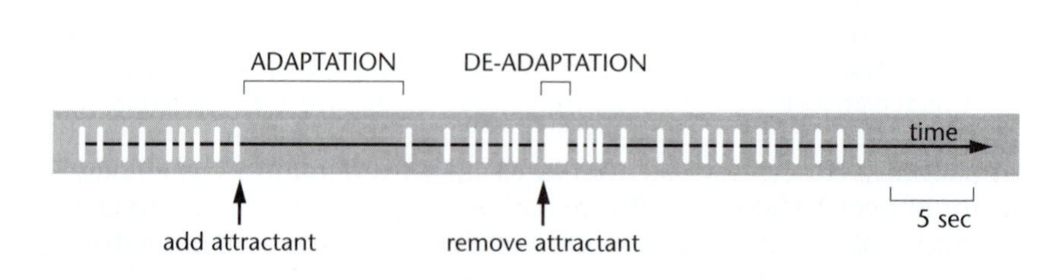


**Figure 3-5** Adaptation in a swimming bacterium. The time course of a swimming bacterium is shown, with each vertical line representing a tumble, during which the bacterium changes its direction. Following addition of attractant, tumbling is suppressed for a while but soon returns to its original frequency. If attractant is then removed, a transient burst of tumbling takes place as the cell changes its direction in a chaotic fashion. (Repellents produce a similar, but inverted, response—repellent addition causing a transient burst of tumbles while repellent removal suppresses tumbling.)

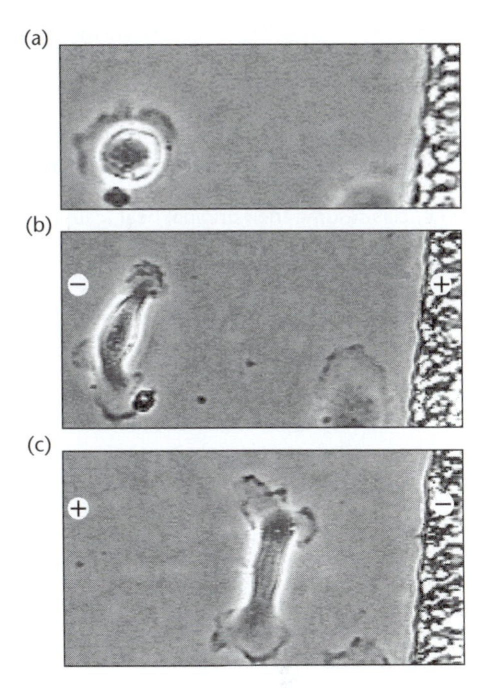


**Figure 3-6** Epithelial cell responding to an electric field. (a) Before exposure to the field, the cell is flattened but rounded. (b) After 1 hr of exposure to an electric field of 150 mV/mm with the indicated polarity the cell becomes elongated at right angles to the field and starts to move toward the negative pole (the cathode). (c) After a further 6 hr with reversed polarity the cell has reversed its direction and moved in the direction of the new cathode. (Courtesy of Colin McCaig.)

(photophobic) to bright light, and a more gentle attraction (phototactic) to dim light.

The range of tropic responses shown by vertebrate cells is more limited, but it includes the curious phenomenon of *galvanotaxis*, or migration in an electric field. Nerve cells growing in tissue culture show a detectable response to electric fields as weak as 0.1 V/cm, their growth cones turning in the direction of the negative pole and away from the positive pole. Similarly, fibroblasts and cells from the neural crest move toward the negative pole in a steady electric field, whereas epithelial cells have been found to orient their plane of cell division at right angles to weak electric fields (Figure 3-6).

A field as weak as 0.1 V/cm is unlikely to cause major perturbations to transmembrane voltages or ion fluxes. A more potent response may occur in the plane of the membrane, where membrane proteins, many of them negatively charged, can move in the lipid bilayer in response to a voltage gradient—a form of electrophoresis. Experiments have shown that certain types of membrane glycoproteins accumulate at one side of a cell exposed to an external electric field and the asymmetry generated in this fashion could influence the cell's behavior.

At first sight these responses to externally applied electric fields seem artificial and of little relevance to a living animal. However, surprisingly large electric fields do exist naturally in living tissues. They can be measured between one side and the other of an epithelial layer, for example, or at the cut ends of regenerating amphibian limbs. Sizable currents exist close to nerve axons and nerve synapses (and very large currents exist in the extreme case of the electric organ of the marine ray or freshwater electric eel). Migrating cells produce small electrical currents that are polarized in the direction of their migration.

## *Paramecia avoid obstacles by reversing their cilia*

Many cells have specific mechanisms for avoiding obstructions that lie in their paths. Ciliates, for example, avoid an obstacle by a transient reversal of ciliary beating. A swimming *Paramecium* that encounters a solid barrier executes a stereotyped sequence of movements (Figure 3-7). First it swims backward for a short distance, then it rotates by a small amount (a movement termed 'skittling'), then it resumes forward swimming in a new direction, the entire sequence being repeated until the obstacle is avoided. Analysis of this behavior shows that it is caused by transient ciliary reversal—contact with the barrier triggers the cilia to execute a power stroke in the direction opposite to that previously used, so the cell swims backward. Recovery of the normal direction of beat takes place asynchronously over the surface of the cell, leading to a transient rotation of the cell.

Ciliary reversal can be provoked in a variety of ways, such as by the application of certain chemicals to the cell or by electrical or mechanical stimulation. There is experimental evidence that all these responses are mediated by a change in intracellular $Ca^{2+}$ levels, which in turn changes the direction of the ciliary power stroke (Chapter 14). Spontaneous reversals of ciliary beat occur even during normal swimming, leading to

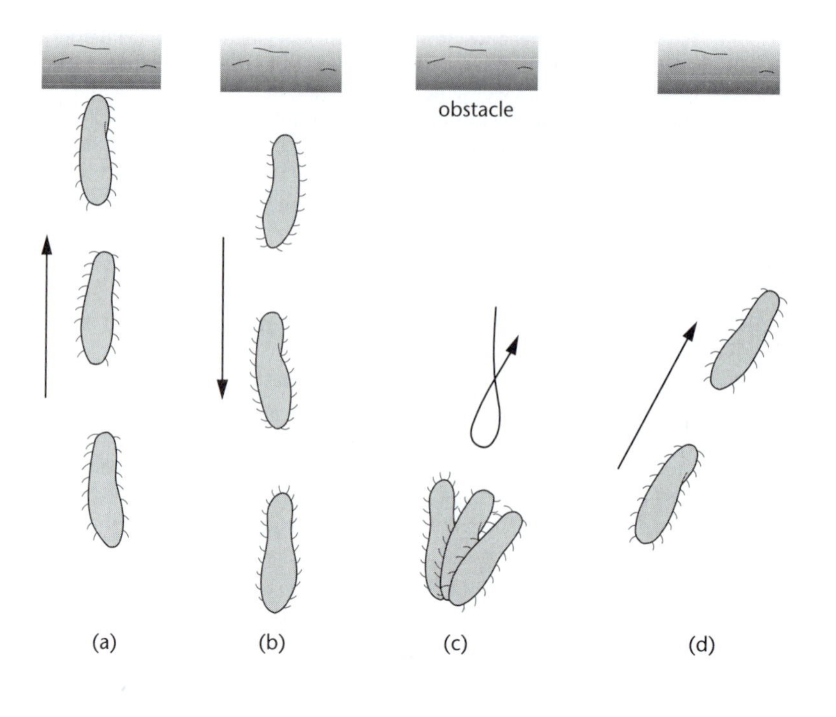


**Figure 3-7** Avoidance of an obstacle by a swimming *Paramecium*. (a) Cell swims into an obstacle, causing cilia on the cell surface to reverse the direction of their beating. (b) Because of the reversal of ciliary beating the cell swims in the reverse direction, away from the obstacle. (c) After a short period of time, cilia resume their original beat, causing the cell to rotate around its posterior end. (d) The cell resumes forward swimming in a new direction.

intermittent changes in direction. As for swimming bacteria discussed above, this underlying random motion can be modified to allow the cell to swim in the direction of chemical or other tropic stimuli.

## *Cells are guided by their substrata*

Crawling cells are guided by the terrain over which they move. Although a surface is necessary to provide lamellipodia with a firm anchorage, not all surfaces are equivalent for this purpose. Indeed, animal cells have specific receptors on their surface, similar to the receptors that bind to chemoattractant molecules, by which they recognize molecules bound to the substratum. Some surface-attached guidance molecules promote migration by triggering the growth of additional pseudopodia or other extensions of the cell surface. Other components of the extracellular matrix may inhibit migration by causing the retraction of portions of the cell. The molecular mechanisms by which contact with the substratum influences the machinery of cell movement, and hence provide guidance for a migrating cell, are discussed later in the book (see Chapters 8 and 20).

Cell migration is influenced by physical, as well as chemical, features of the terrain. In tissue culture, cells will move onto tracks of artificial adhesive material, such as polylysine or silicon oxide. They prefer these surfaces even though, if given no alternative, they would have been able to attach to and grow on the less favorable surface. Fibroblasts on a surface coated with a uniformly increasing gradient of a charged substance turn and move in the direction of the increasing adhesiveness. Spatial contours are also important and cells will align themselves with long cylindrical surfaces (such as glass fibers) or with grooves cut into the substratum and move along them (Figure 3-8).

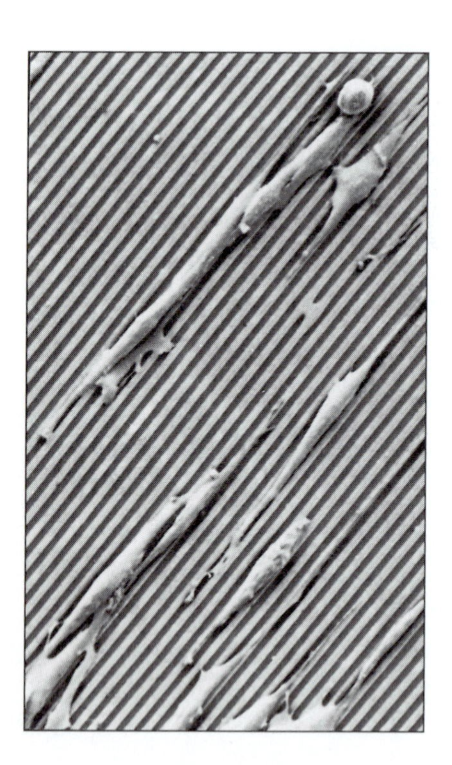

**Figure 3-8** Alignment of cultured cells with their substratum. Fibroblasts growing on a grooved surface align in the direction of the grooves. The grooves in this case are 2 μm deep, 3 μm wide, and spaced 3 μm apart; both the depth and spacing of the grooves influence the degree of alignment of the cells. (Courtesy of Peter Clark and Adam Curtis.)

Examples of both chemical and physical guidance occur during the migration of cells in a developing embryo. The migration of neural crest cells already mentioned, for example, follows pathways of extracellular matrix rich in collagen fibers and an associated protein known as *fibronectin*. In tissue culture, neural crest cells will follow tracks of fibronectin laid down on the substratum, preferring such tracks to surfaces coated with collagen alone.

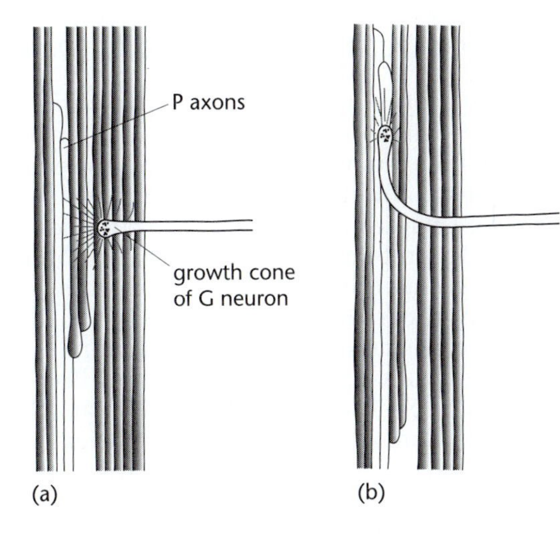


**Figure 3-9** Selection of pathway by an identified growth cone. (a) The tip of an axon from the G neuron in a grasshopper embryo crosses the nervous system to the region at which it selects a longitudinal axonal bundle. The filopodium of this growth cone makes contact with about 100 axons in 25 fascicles, and makes selective contact with only the P axons. (b) Ten hours later in development, the G axon has joined the A/P fascicle and is closely associated with the P axons. An identical sequence of events is repeated on either side of the segment and in each of 17 segmental ganglia of the developing insect.

## *Nerve cells have a highly developed capacity for pathfinding*

An even greater degree of pathway selection occurs during the growth of nerve axons. Axons grow from individual neurons by the extension of a cell process and then advance into the surrounding tissue, using a specialized form of cell locomotion. The motile structure at the tip of each axon crawls forward by means of filopodia and lamellipodia rather as in the ruffling membrane of a fibroblast (see Figure 2-10). Since growth cones must establish the initial synaptic connections of the central and peripheral nervous system, they must be able to follow precisely defined paths, which are often long and tortuous.

The behavior of a particular set of neurons in a developing grasshopper embryo illustrates the high degree of pathway guidance that can occur. These neurons are all progeny of a single neuroblast and differentiate at a stage when the surrounding neuroepithelium already consists of a forest of axons that emerged from other neurons (Figure 3-9). The previously formed axons are grouped into transverse and longitudinal bundles, or fascicles, forming a roughly orthogonal ladderlike scaffold. Growth cones of the newly differentiated cells first grow to the opposite side of the embryo by following a single runglike bundle of axons that runs transverse to the anterior/posterior axis. When they reach the other side of the embryonic nervous system, however, each growth cone chooses a different longitudinal fascicle to follow. Thus, a particular growth cone, from a cell designated G, always selects one fascicle of axons out of a total of about 25 fascicles that it makes contact with en route. Furthermore, within the fascicle of axons, only the two posterior directed axons are recognized: the other two axons that run in an anterograde direction can be ablated by UV damage without affecting the development of the G cell process.

At the same time that the G cell axon is growing out, many hundreds of other neurons in the same region of the grasshopper nervous system are also searching for their paths. Each growth cone follows molecular clues in its surroundings, many of them associated with its neighboring cells, and will eventually lay down a unique axonal connection to a synaptic target. The ability to follow the development of individual nerve cells in an insect (and other invertebrates) throughout their development offers a wonderful opportunity to investigate the mechanisms of pathway guidance. As we will see later, this approach is especially powerful when imaging techniques are combined with sophisticated genetics and molecular biology techniques, as in studies of neuronal development in the fruit fly *Drosophila*.

## *Contacts between cells modify their migratory behavior*

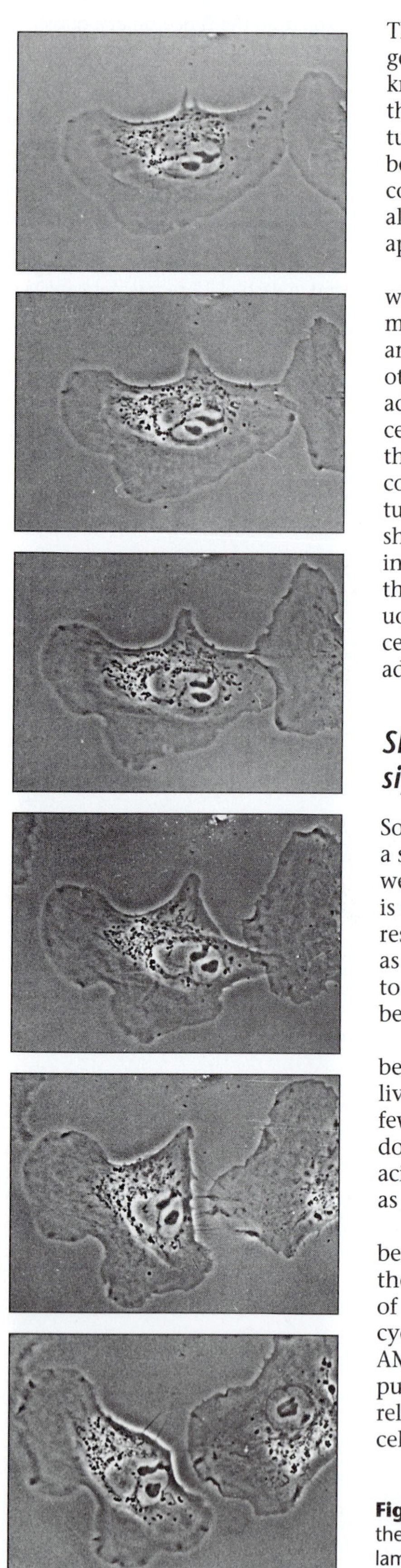


Figure 3-10 Close encounter between two fibroblasts. The cell in the center makes contact with, and is repelled by, the lamellipodium of a second cell moving into the field from the right. The figure shows pictures of chick embryo fibroblasts taken at 5-minute intervals.

There is another way in which growth cones can be guided to their targets—by being inhibited if they wander in the wrong direction. It is known, for example, that axons from the embryonic spinal cord pass only through the anterior half of the somites (the precursors of body musculature, see Chapter 10), avoiding the posterior half. This has been shown to be due to proteins on the surfaces of somite cells that inhibit growth cones, causing them to collapse and retract. Inhibitory mechanisms may also play a part in guiding axons from the retina of the eye to their appropriate locations on the optic tectum.[2]

In cultures of fibroblasts, contact inhibition of migration is seen when one cell in culture collides with another. Where their two ruffling membranes make contact, a quiescent region forms where ruffling ceases and the cells seem to stick to each other. Meanwhile ruffling continues at other regions and after a period of about 10 minutes the cells break their adhesion and move away along new directions (Figure 3-10). Epithelial cells show an even more profound contact inhibition of migration, since the cells do not move away from each other following contact. Sequential collisions with other cells result in small colonies being formed and eventually a sheet of contiguous cells is established similar to the epithelial sheets found in the body. This phenomenon is presumed to be important in wound healing, during which sheets of epithelial cells move out from the margins of a wounded area by extending lamellipodia. Once a continuous sheet of epithelial cells has been established over the wounded area, cell migration ceases and cell junctions are formed between the newly adjacent cells.

## *Slime mold amoebae congregate in response to chemical signals*

Some of the most complicated patterns of behavior are shown by cells in a social setting as they respond to signals from their neighbors. Indeed, as we will see in Chapter 20, the shaping of multicellular plants and animals is the direct result of the behavior of many different individual cells, each responding to its individual setting. Even single-celled organisms, such as the *Myxobacteria* mentioned in the previous chapter, have the capacity to congregate into large multicellular structures and show complex behavioral changes as they do so.

The life cycle of the slime mold *Dictyostelium discoideum* is one of the best-studied examples of this kind. The amoeboid cells of this organism live in soil or leaf mold, feeding on bacteria and fungi and dividing every few hours. In this condition they act like carnivorous individuals, hunting down bacteria by means of the smell of bacterial products, such as folic acid. They are repelled by the smell of their own sort and disperse as widely as possible.

If their food supply is exhausted, however, *Dictyostelium* amoebae become more gregarious. They stop dividing and, finding the smell of their neighbors suddenly attractive, start to huddle together. The nature of the attracting signal is well understood and it consists of the nucleotide cyclic AMP, released in pulses from individual cells. Each pulse of cyclic AMP induces surrounding cells both to move toward the source of the pulse and to secrete their own pulse of cyclic AMP. This new pulse, released after a slight delay, orients and induces a pulse of cyclic AMP from cells just beyond and so on. Regular waves of cyclic AMP concentration

**Figure 3-11** Formation of a *Dictyostelium* fruiting body. The aggregation of single amoebae gives rise to a multicellular slug, which in its natural environment migrates upward through the soil. Once on the surface of the soil, the slug develops a long stalk topped with a fruiting body containing spores. Eventually the fruiting body ruptures, and each spore released produces a new amoebae, ready to divide and start the cycle again. (Montage of a series of scanning electron micrographs courtesy of M.J. Grimson and R.L. Blanton.)

travel out from each aggregation center, causing more distant amoebae to move inward in surging concentric or spiraling waves.

At the center of each aggregating population, cells stick to each other, eventually forming a relatively huge (1–2 mm) wormlike structure that crawls about as a glistening slug. The slug can migrate for up to 20 days and in this condition is attracted both to sources of light and to heat (usually the soil surface). When it eventually stops moving it commences a sequence of complex morphological changes (which entail yet another set of directed cell movements, in this case confined to the body of the slug) that result in the formation of a stalk and a spore capsule. Spores released from this capsule then germinate to form amoebae and thereby complete the life cycle (Figure 3-11).

## Single cells respond to light

Sunlight is a ubiquitous environmental stimulus and many free-living cells have evolved the capacity to move either toward or away from it. The fundamental nature of this response is illustrated by the fact that the same primary detector of light—the protein rhodopsin carrying a molecule of the small pigment molecule retinal—is used by bacteria, protozoa, and humans to detect light. In humans, rhodopsin is the primary light-sensitive pigment in the retina.

The salt-loving bacterium *Halobacterium salinarium* contains arrays of rhodopsin molecules in its membrane that act as specialized receptors for light. Excitation of the photoreceptors by blue light generates repellent signals and orange light generates an attractant signal. As a result the cells swim, by means of their flagella, away from harmful ultraviolet light at the surface of the brine pools in which they live, and congregate at depths where the orange light provides optimal stimulation of energy production.

The unicellular alga *Chlamydomonas* shows a variety of responses to light, including both positive and negative phototaxis as well as a light-induced 'shock' response in which it reverses swimming direction. These responses are mediated by an orange-red eyespot 1–3 μm long located near the equator of the cell. The eyespot is a specialized chloroplast containing an accumulation of rhodopsin positioned close to the plasma membrane. When light falls on the eyespot it causes a conformational change in rhodopsin (as it does in our eyes) and generates internal signals (including transient changes in $Ca^{2+}$ concentration) that affect the beating of its two flagella.

During normal forward swimming, *Chlamydomonas* rolls around its longitudinal axis so that its eyespot, which is situated on one side of the body, continuously scans the environment through 360°. If the cell

[2] Growth cone migration is also inhibited by contact with oligodendrocytes (the cells that make myelin in the central nervous system), which may be one reason nerve axons fail to regenerate following damage to the brain or spinal cord.

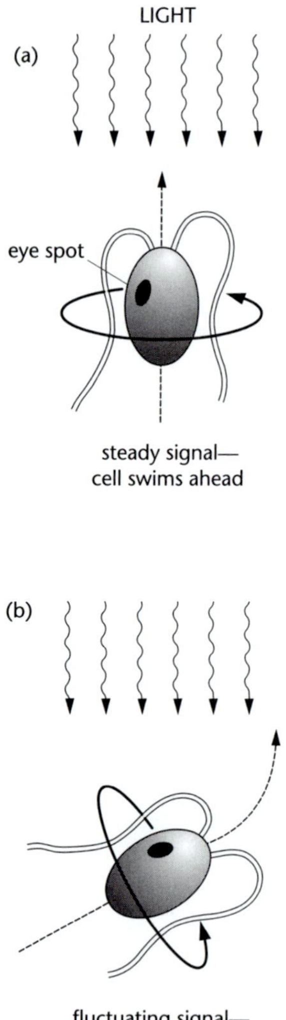


**Figure 3-12** Photoresponse of *Chlamydomonas.* (a) If the cell is swimming directly toward the light source, then its eye spot will receive a constant illumination, even though the cell spins as it swims. (b) If the cell is obliquely oriented to the light, however, the eye spot receives a fluctuating level of illumination as it rotates into and out of the shadow cast by the cell body. The fluctuating signal thus generated is used by the cell to selectively enhance the amplitude of one of the flagella—like a rowboat steered by oars, the cell now turns to the light.

happens to be swimming perpendicular to the direction of light stimulus, then the eyespot will be alternately illuminated and shaded as the eyespot faces toward or away from the light. The fluctuating signal thereby produced causes the cell to turn toward the light. Once the cell is swimming parallel to the stimulus beam, the amount of light falling on the photoreceptor remains relatively constant as the cell rolls and no further correction is necessary (Figure 3-12).

The photoshock response is produced by a stronger stimulus than that sufficient to elicit phototaxis. In this case, light causes $Ca^{2+}$ channels in the membrane of the flagella themselves to open. $Ca^{2+}$ ions flood in and cause the flagella to switch from a 'breast stroke' to a flagellar type of beat that causes the cell to move backward (Chapter 14).

## Protozoa carry out complex sequences of actions

Before leaving this chapter it is useful to consider some relatively complex sequences of movements performed by single cells to show how they can be synchronized and coordinated into a complex and seemingly 'intelligent' mode of action. Not surprisingly, the best examples come from protozoa, which in both the complexity of their internal structure and in the patterns of their behavior rival many multicellular organisms.

Even the relatively undifferentiated cell of *Amoeba proteus* shows complicated responses. To simple stimuli, such as contact with a solid object, it reacts either negatively or positively. A negative response is usually a change in direction just sufficient to avoid the objectionable agent. This is repeated until success is achieved by trial and error. A positive response to a particle of food, such as the cyst of another protozoan, is analogous to that of a higher animal. The cyst is rolled until it is either successfully ingested or the baffling cyst is lost. The chase may last several minutes and, to an observer at the other end of a microscope, seems almost sentient in its motivation and flexibility.

Ciliates are even more versatile in their behavior. The normally sessile ciliate, *Stentor*, for example, shows a series of reactions of increasing intensity to a noxious stimulus, such as a light touch with a glass needle, or the application of a cloud of carmine particles to the feeding vortex (Figure 3-13). Like the turning response of *Paramecium* described above, these responses are produced by a reversal of ciliary beating in selected regions of the cell surface. The initial response of *Stentor* is to turn away from the stimulus: the cell remains anchored to the substratum but twists about its lower axis. The second response is a brief reversal of the direction of ciliar beating, which has the effect of driving water away from the animal. If this does not eliminate the stimulus, the response can be repeated many times. Thirdly, if the irritation persists, *Stentor* will rapidly contract and slowly extend, repeatedly if necessary. Finally, if all else fails, the cell releases its holdfast and swims to seek a new environment.

There is obviously a risk of reading too much into responses of this kind. Protozoa should not be confused with higher animals, whose actions are controlled by a nervous system. But neither should we dismiss such behavior as trivial because we have some analytical knowledge of the molecular basis of ciliary beating or pseudopodial extension. Coordinated, integrated responses of the kind shown by *Stentor* can occur only if

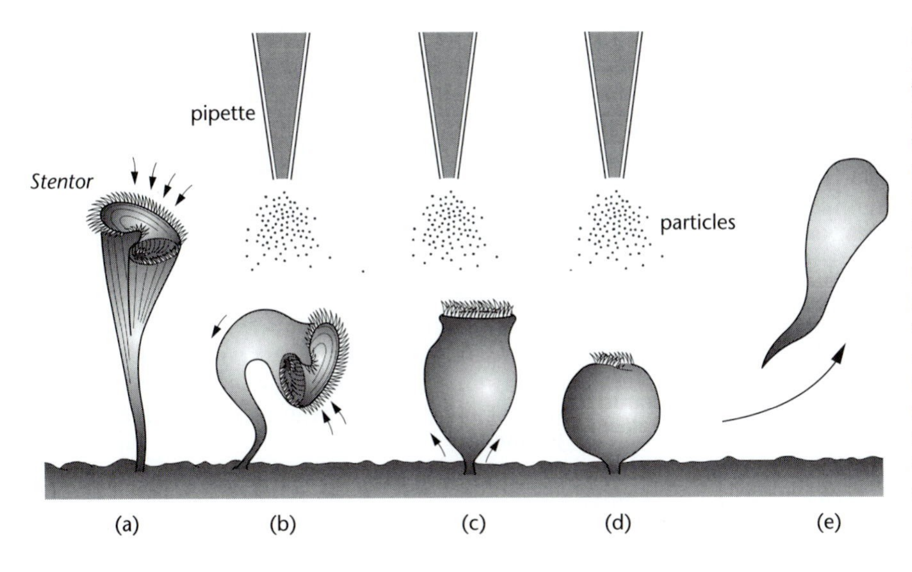


**Figure 3-13** Graded response of *Stentor* to an objectionable stimulus. In its normal feeding position (a) *Stentor* is anchored to the substratum by its stalk and generates feeding currents by the coordinated beating of cilia on its surface. The currents sweep particles into its mouth region. Pipetting a cloud of fine particles into its feeding vortex (b) causes the animal to twist on its stalk and bend away from the obnoxious stimulus. If the stimulus continues (c) then the feeding vortex stops and the body cilia reverse their direction of beat, thereby driving water away from the animal. If this maneuver is still unsuccessful (d) then the cell contracts sharply and eventually (e) releases the attachment of its stalk to the substratum and swims to another location.

the cell contains an information processing device of some sophistication. We will see later in the book that this processing device is a network of intercellular signaling reactions, catalyzed by proteins that have specific locations in the cell. This network of proteins acts in a manner that is directly comparable to the nervous system of higher animals, initiating movements and specifying the sequence of actions in time and space.

## *Essential Concepts*

- Physical and chemical factors in a cell's environment affect its otherwise random swimming or crawling, causing it to move in a specific direction.
- Bacteria detect and swim toward distant sources of sugars, amino acids and other nutrients.
- Neutrophils are attracted by a variety of proteins, lipids and low-molecular-weight peptides that signal a bacterial infection or tissue destruction.
- Many different cells show galvanotaxis—a directed response to an externally applied electrical field.
- The physical terrain over which cells travel also influences their path. Crawling cells are responsive to subtle interactions with surface molecules, while swimming cells can avoid obstacles by having specific avoidance responses.
- Collisions between migrating cells often generate specific responses, such as local paralysis of migration or the formation of cell aggregates. These may form the basis of complex social behavior, as illustrated in the life cycle of the cellular slime molds.
- Many single cells detect and respond to light, often avoiding very intense light of short wavelengths but being attracted by softer, warmer lights. Interestingly, bacteria and algae use the same pigment molecule (rhodopsin) to detect light as we do in our eyes.
- Single-celled amoebae and ciliates sometimes perform long, detailed sequences of actions in which they appear to make decisions and show motivation. Their behavior is controlled by protein-based biochemical circuits which serve a similar function to a nervous system in multicellular animals.

## Further Reading


Adler, J. Chemotaxis in bacteria. *Annu. Rev. Biochem.* 44: 341–376, 1975.

Bailly, M., et al. Regulation of protrusion shape and adhesion to the substratum during chemotactic responses of mammary carcinoma cells. *Exp. Cell Res.* 241: 285–299, 1998.

Berg, H. How bacteria swim. *Sci. Am.* 233: 36–44, 1975.

Broadie, K., et al. From growth cone to synapse: the life history of the RP3 motor neuron. *Dev. Suppl.*: 227–238, 1993.

Bryant, P.J. Filopodia: fickle fingers of cell fate? *Curr. Biol.* 9: R655–R657, 1999.

Cooper, M.S., Schliwa, M. Motility of cultured fish epidermal cells in the presence and absence of direct current electric fields. *J. Cell Biol.* 102: 1384–1399, 1986.

Gibbs, D. The daily life of *Amoeba proteus*. *Am. J. Psychol.* 19: 230–241, 1908.

Hallett, M.B. Controlling the molecular motor of neutrophil chemotaxis. *BioEssays* 19: 615–621, 1997.

Hinrichsen, R.D., Schultz, J.E. *Paramecium*: a model system for the study of excitable cells. *Trends Neurosci.* 11: 27–32, 1988.

Lackie, J.M., et al. Cell Behaviour: Control and Mechanism of Motility. Eynsham, UK: Portland Press, 1999.

McCaig, C.D., Zhao, M. Physiological electrical fields modify cell behaviour. *BioEssays* 19: 819–826, 1997.

Raper, J.A., et al. Pathfinding by neuronal growth cones in grasshopper embyos: I Divergent choices made by the growth cones of sibling neurons. *J. Neurosci.* 3: 20–30, 1983.

Tartar, V. The Biology of *Stentor*. Oxford: Pergamon Press, 1961.

Trinkaus, J.P. Cells into Organs: The Forces That Shape the Embryo. Reading, MA: Prentice-Hall, 1984.

Witman, G.B. *Chlamydomonas* phototaxis. *Trends Cell Biol.* 3: 403–408, 1993.

# CHAPTER 4

# The Cytoskeleton

So far we have described moving cells as if seen in a light microscope, detailing form and physical constraints but not underlying mechanisms. The time has now come to probe more deeply—to open the skin of the cell, so to speak, and expose its muscles, bones, and nerves.

We will see in the chapters that follow that large-scale cell movements are driven by the *cytoskeleton*: a cohesive meshwork of protein filaments extending throughout the cytoplasm of plant and animal cells. The cytoskeleton is the primary determinant of cell shape and movement and to this end it operates according to its own functional 'logic' or set of rules.[1] From a mechanical standpoint, the cytoskeleton acts like a set of struts and girders that support the form of the cell and constrain its movements according to engineering parameters such as elasticity and bending modulus. From a biochemical standpoint, the cytoskeleton performs a repertoire of characteristic reactions repeating these over and over again as living cells move. These reactions include the polarized growth and shrinkage of protein polymers, their association through multiple linking of proteins into larger structures, and the directed motion of motor proteins walking along protein polymers.

In later chapters we will focus in greater detail on particular parts of the cytoskeleton and attempt to explain in molecular detail how they produce movements. Here, we will give a broader perspective of the cytoskeleton as a whole and introduce some of the strategies by which it is used by cells.

[1] There are also a myriad molecular-scale movements that have no direct connection with the cytoskeleton, such as the spinning of an ATP synthase molecule in the plasma membrane, or the progression of a replication complex along a DNA molecule.

## *Shapes and movements are generated by the cytoskeleton*

The simplest way to find out which components of the cell are responsible for movement is to perform a 'dissection.' This can be done by literally cutting pieces from a cell to see whether they retain the capacity for movement, or it can be performed in a more figurative sense by extracting selected components of the cell, such as the plasma membrane.

Microsurgery has been performed on very large cells, such as *Amoeba proteus* or *Stentor*, and on the far smaller vertebrate tissue cells growing in tissue culture. Cells have been cut with fine glass needles held in a micromanipulator or by laser beams focused through the objective of the microscope. Cells in tissue culture may also become fragmented following certain kinds of drug treatment (Figure 4-1). The plasma membrane generally reseals in such operations, generating self-contained fragments of the cell containing as little as 5% of the original volume. The fragments often retain the capacity for surface ruffling, cytoplasmic streaming, and migration over the substratum. Experiments of this kind show that the nucleus, Golgi apparatus, and endoplasmic reticulum are not necessary to sustain cell movements, at least for a period of several hours. The machinery for movement must therefore be located in another part of the cell.

Cytoplasm free of the plasma membrane has been collected from very large cells. For example, if *Amoeba proteus* is drawn into a into a narrow glass tube that is subsequently broken, or if the contents of a giant squid axon are squeezed out by means of a miniature roller, they yield tiny droplets of naked cytoplasm. Under the microscope, these droplets show persistent movements such as cytoplasmic streaming and the saltatory movements of small particles.

Another way to 'dissect' a cell is to treat it with a nonionic detergent. These soapy substances disperse lipid bilayers and allow metabolites and proteins to leak from the cell, but without disrupting normal protein–protein interactions and leaving many enzyme activities unimpaired. A cell treated with a dilute solution of a detergent such as Triton X-100 becomes more contrasty in appearance under a phase contrast microscope as soluble proteins are lost. Eventually it becomes a ghost of a cell, which has lost its plasma membrane and half of its protein but retains its original shape.

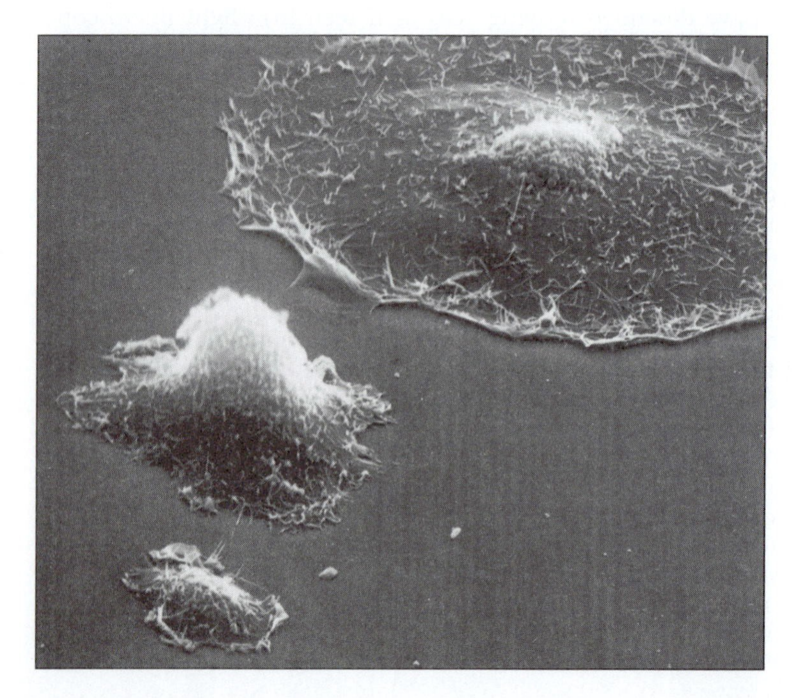

**Figure 4-1** Motile fragments of a cell. A fibroblast in tissue culture is seen in this scanning electron micrograph beside two cell fragments generated by chemical treatment (cytochalasin followed by trypsin). The smaller fragment is about 10% of the volume of the intact cell. Fragments of this kind have been shown to remain intact, generate surface lamellipodia and filopodia, and migrate over the surface for several hours. (Courtesy of Guenter Albrecht-Buehler.)

Significantly, detergent-treated cells often retain the ability to move. This is most dramatically evident in the case of muscle contraction and ciliary beating, both of which are generated by highly organized, compact structures. Thus myofibrils, which are large bundles of actin and myosin filaments, are able to contract and relax repeatedly even after removal from a muscle cell. Sperm flagella, with a filamentous core of microtubules, can continue to swim, even after the sperm head has been removed and the plasma membrane stripped away. Other less highly-organized forms of movement may also continue after removal of the plasma membrane, including particle transport and some stages in mitosis.

Putting these observations together we conclude that the machinery of cell movement is located principally outside the nucleus and major membrane-bound organelles; it is left behind after soluble components are extracted. This cell residue corresponds to the cytoskeleton.

## *The cytoskeleton contains actin filaments, microtubules, and intermediate filaments*

If we treat a cell with nonionic detergent, dissolving its membranes and allowing small molecules to diffuse away, what is left? In a light microscope the residue appears as a transparent ghost with a distinct nucleus but little internal structure. In an electron microscope, however, an irregular haystack of interlinked protein filaments is revealed. Filaments radiate from a position close to the nucleus out into the cytoplasm, extending to the limits of its periphery, where they delineate the form of the cell. At higher magnification, regional differences can be seen, perhaps the most important being the distinction between the periphery of the cell and its more central region close to the nucleus.

In a typical animal cell, the peripheral region consists largely of a dense three-dimensional meshwork of thin filaments (Figure 4-2). This mesh is composed of actin filaments (microfilaments), which are thin, flexible filaments with a diameter of about 8 nm made of the protein actin. Microfilaments are often cross-linked into a dense, three-dimensional weave and they can also be more regularly arranged into parallel bundles, such as those that form the core of filopodia on the cell surface. The cortical meshwork contains small amounts of other material, thought to represent soluble proteins of the cytoplasm, but relatively few membrane-bounded organelles.

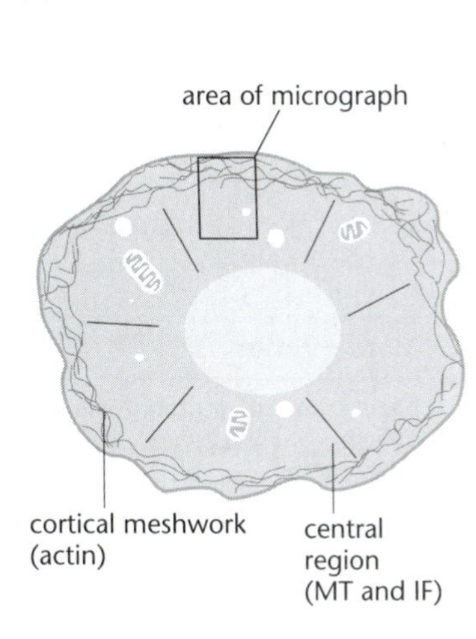


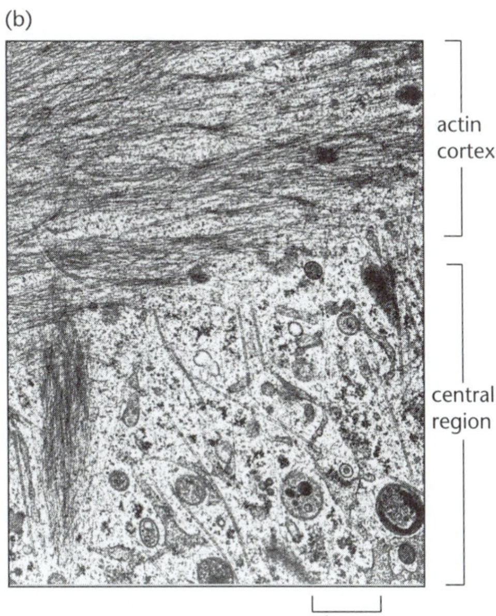


**Figure 4-2** The cytoskeleton. (a) Schematic section of an animal cell showing the location its peripheral actin-rich cortical layer and the more open central region containing microtubules (MT) and intermediate filaments (IF). (b) Electron micrograph of a region of a fibroblast cytoplasm showing the two distinct cytoskeletal regions. (b, courtesy of Julian Heath.)

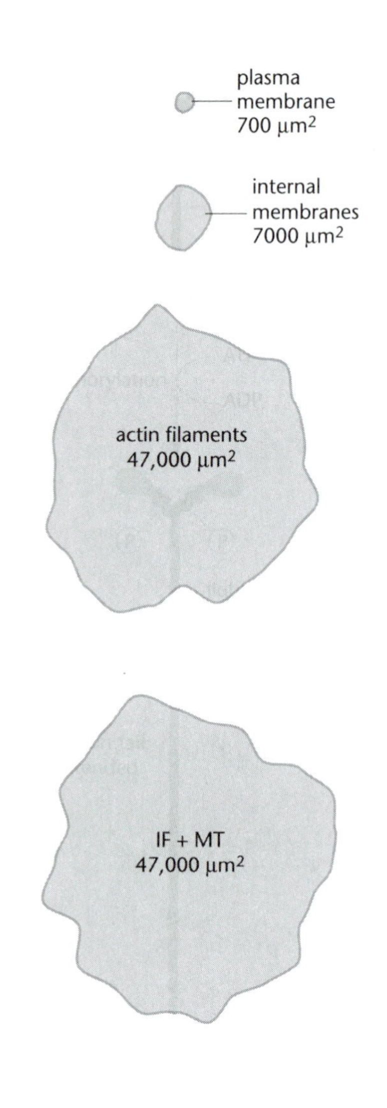


**Figure 4-3** Surface area of the cytoskeleton. Areas exposed to the cytosol (given in $\mu m^2$) are calculated for a typical animal tissue cell about 20 μm in diameter. IF, intermediate filaments; MT, microtubules. (Data from P.A. Janmey, 1998.)

By contrast, the central region of the cell, close to the nucleus, has a more open construction with fewer actin filaments and a smaller number of long *microtubules*, interspersed by abundant granular material and membrane organelles, including mitochondria. Microtubules are long polymers of the protein tubulin that stand out because they are relatively thick and inflexible. In thin sections they are even more conspicuous because of their hollow cylindrical form: a microtubule has an outer diameter of 25 nm and a central canal, or lumen, of about 15 nm diameter.

A third category of protein filament, widely found in animal cells and often located in the central region together with microtubules, is *intermediate filaments*. Their distribution is often similar to that of the cell's microtubules, but they can be distinguished from the latter by their smaller diameter (which is close to 10 nm). Intermediate filaments terminate on the matrix of protein that encloses the nucleus and radiate from there into the surrounding cytoplasm, often impinging on membrane-associated junctions with neighboring cells (Chapter 17). They are irregular, flexible ropes, composed of a family of proteins that vary a great deal from one cell type to another.

Note that the cytoskeleton undoubtedly has other functions than cell shape and movement. Filaments of the cytoskeleton form a continuous, dynamic connection between nearly all cellular structures, presenting an enormous surface area to which proteins and other cytoplasmic components can attach. To put this aspect in perspective, the surface area created by the filaments of the cytoskeleton in a typical animal cells is more than *ten times* that of the total surface of their membranes (Figure 4-3). This large surface, along with the strong tendency of macromolecules of all kinds to associate in the cytoplasm due to competition for water (discussed in Chapter 18), gives the cytoskeleton a major role in the organization of cell contents, even those normally thought of as being soluble.

## *Most eucaryotic movements depend on actin filaments and microtubules*

Actin filaments and microtubules form the basis of almost all large-scale movements in eucaryotic cells. This includes ciliary beating, muscle contraction, mitosis, phagocytosis, and cell migration together with a variety of internal movements (Table 4-1). In broad terms, actin filaments

**Table 4-1 Types of filament responsible for cell movements**

| Cell movement | Filament | Filament subunit | Motor/process |
|---|---|---|---|
| bacterial swimming | flagellum (procaryotic) | flagellin | flagellar motor (proton driven) |
| protozoan swimming, sperm swimming | cilia or flagella (eucaryotic) | tubulin | dynein |
| bacterial gliding, Gregarine gliding | none | not known | ? secretion |
| nematode sperm crawling | MSP filament | major sperm protein (MSP) | ? polymerization |
| amoeboid locomotion, fibroblast crawling | actin filament | actin | myosin, polymerization |
| muscle contraction | actin filament | actin | myosin |
| *Vorticella* contraction | spasmin filament | centrin | calcium-induced conformational change |
| organelle transport | actin filament, microtubule | actin, tubulin | kinesin, dynein, myosin |
| cytoplasmic streaming (*Nitella*) | actin filament | actin | myosin |
| chromosome movement | microtubule | tubulin | kinesin, dynein, polymerization |

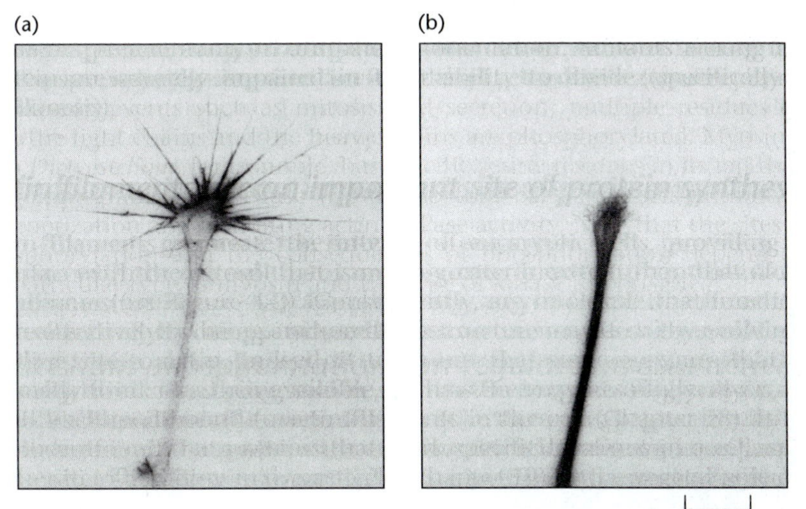


**Figure 4-4** Distribution of actin filaments and microtubules in a nerve cell. (a) Nerve cell in culture stained with a reagent that binds selectively to actin filaments. The stained regions are abundant in the motile growth cone at the tip of the axon and especially in the peripheral regions of filopodia and lamellipodia. (b) The same cell stained for microtubules shows these to be largely absent from the growing tip but to extend along the length of the axon, where they provide the basis for rapid internal movements (axonal transport). (Micrographs courtesy of Peter Hollenbeck.)

associated with the plasma membrane are responsible for most movements of the cell surface, whereas microtubules dominate its intracellular movements (Figure 4-4).

It is no accident that one of the most fundamental characteristics of the large and complex cells of plants and animals is that they show continual movements in their cytoplasm. A large and complex cell, like a major city, requires an effective transportation system. The more rapidly and precisely molecules can be carried from one organelle to another in the cell, the more highly specialized and efficient those organelles can become. We will see later in the book that internal movements are crucial for many vital processes in eucaryotic cells, including the replication of chromosomes (Chapter 13), the maintenance of internal membrane-bound compartments by secretion and endocytosis (Chapter 6), and the development of asymmetric cell shapes (Chapter 19).

Although actin and tubulin dominate the shape and movements of animal and plant cells, it is important to note that movements can be generated in other ways. Bacteria swim by means of a proton-driven rotary motor attached to a helical filament made of a procaryotic-specific protein called flagellin; *Cyanobacteria* glide over surfaces but, so far as is known, not by means of actin filaments or microtubules. Nematode sperm lack flagella and crawl like amoebae over surfaces by means of a single pseudopodium containing, in place of the usual cytoskeletal components, a meshwork of filaments, 2–3 nm in diameter. (The mechanism of this cell movement is discussed in Chapter 8.) Further afield, plant cells movements are based on the flow of water in and out of the cell, such as the opening and closing of stomata and the rapidly wilting leaf of a *Mimosa*.

It is very difficult, in biology, to establish that a particular process is *not* present. So, although they have not so far been found, we should not rule out the possibility that movements based on proton pumps, on other types of filaments, or on water movement may also be used by the cells of higher vertebrates.[2]

## Accessory proteins hold the cytoskeleton together and link it to membranes

By themselves, the three main types of filaments would make a house of straw with neither form nor strength. The shape of a cell and its ability to move in a coordinated fashion depend on a large retinue of accessory proteins that link the cytoskeletal filaments to each other and to other parts of the cell. These accessory proteins form the myriad side-arms and

[2] Certain cells in the mammalian cochlea undergo very rapid, voltage-driven changes in shape, which contribute to the amplification of sound vibrations entering the ear. These movements are driven by proteins in the plasma membrane and probably do not involve the cytoskeleton directly.

**Figure 4-5** Filament cross-links. (a) Electron micrograph of the cytoplasm of a nerve axon showing the longitudinally arrayed neurofilaments (a type of intermediate filament) linked along their length by thin transverse cross-bridges of protein. (b) The engineering principle by which protein filaments can be strengthened by cross-linking. Protein links convert bending forces, which protein filaments are able to withstand only poorly, into compressive and tensile forces, which are more easily resisted by the protein filaments. (Micrograph courtesy of Nobutaka Hirokawa.)

(a)

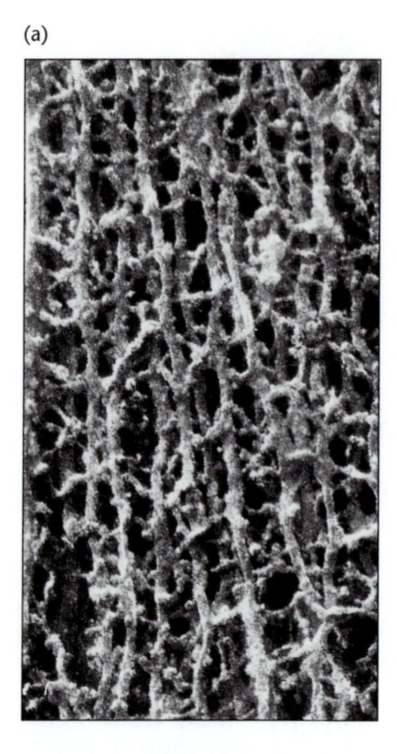

(b)

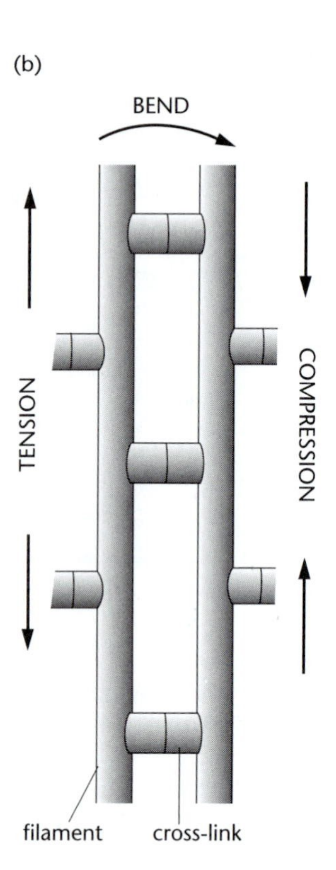


cross-bridges seen in electron micrographs projecting from the main filaments (Figure 4-5a). They give rise to a complex pattern of proteins characteristic of the type of cell under study and even the region of cytoplasm from which they were taken.

Accessory proteins add enormously to the strength of a cytoskeleton. They strengthen individual filaments by binding along their length, making them thicker, more stable and more rigid; they can act as cross-links and flexible ties that join filaments together into larger frameworks (Figure 4-5b). The latter strategy allows rigid structures to be built from elements such as actin filaments that are themselves unable to sustain large bending forces. Accessory proteins are also essential for the controlled assembly of bundles of filaments at particular locations of the cell (enabling the cell to change its shape) and in the generation of internal movements, as we discuss below.

Although some cytoskeletal proteins interact directly with a phospholipid bilayer, most protein filaments associate with membranes indirectly, by binding to membrane proteins. They are often linked laterally at points along their lengths, an arrangement that allows interactions that are individually weak to produce a substantial attachment in aggregate. Although filaments sometimes bind directly to an integral membrane protein (one that is firmly embedded in the phospholipid bilayer) more commonly they bind the membrane proteins indirectly through one or more intermediary proteins forming an attachment complex on the inside surface of the plasma membrane.

## *Why are protein filaments used in cell shape determination and movements?*

First of all, protein filaments are easy to make. If a protein by chance evolves a binding site for its own surface, then unless it forms a closed aggregate (such as a head-to-head dimer) it will usually produce a helical filament (Figure 4-6). The ease with which such structures arise is illustrated by the fact that they occur gratuitously. Both the enzyme glutamate dehydrogenase and the hormone glucagon spontaneously, and for no obvious purpose, produce long polymeric filaments in cell-free extracts. Sickle cell hemoglobin, the result of a single mutation in the human β-globulin gene, exhibits an unwanted polymerization inside the cell. In its deoxygenated form, sickle cell hemoglobin tends to polymerize into long helical polymers, which distort the plasma membranes of red blood cells, giving them a distinctively spiky, crescent shape (Figure 4-7).

Secondly, protein filaments offer a ready way to construct a space-enclosing framework for the eucaryotic cell. Intermediate filaments made like twine from overlapping fibrous elements are ideally suited to sustain tensile force (in fact, hair and wool are formed from intermediate filament proteins). Microtubules, because of their globular subunits and cylindrical form are better able to sustain compression and bending forces. Actin filaments, which by themselves are relatively weak, can be strapped together with accessory proteins to produce rigid elements and can produce contraction when combined with myosin. We will see in Chapter 18 that these mechanical features allow the three types of protein filaments to build open, strong, mechanically integrated and self-stabilizing structures.

## *Cytoskeletal filaments are produced by the polymerization of subunits*

As well as providing cells with a mechanically strong scaffold, protein filaments also help solve another fundamental problem of cellular life—that of organization. The molecules that make up the cell are incredibly diverse. A typical cell in a mammalian tissue might make over 10,000 different kinds of protein, for example, which are further diversified by

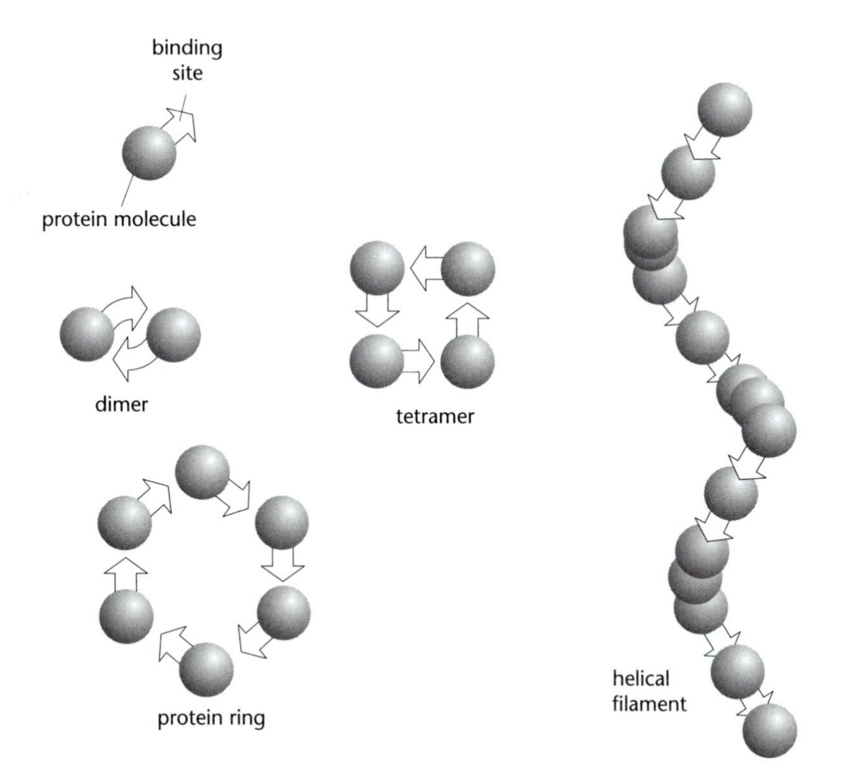


**Figure 4-6** Filament evolution. A protein molecule that, in the course of evolution acquires a binding site for another region of its own surface will produce a head-to-head dimer, a closed ring, or a helical filament. Helical filaments are therefore easy to make, and indeed are found widely in the cell.

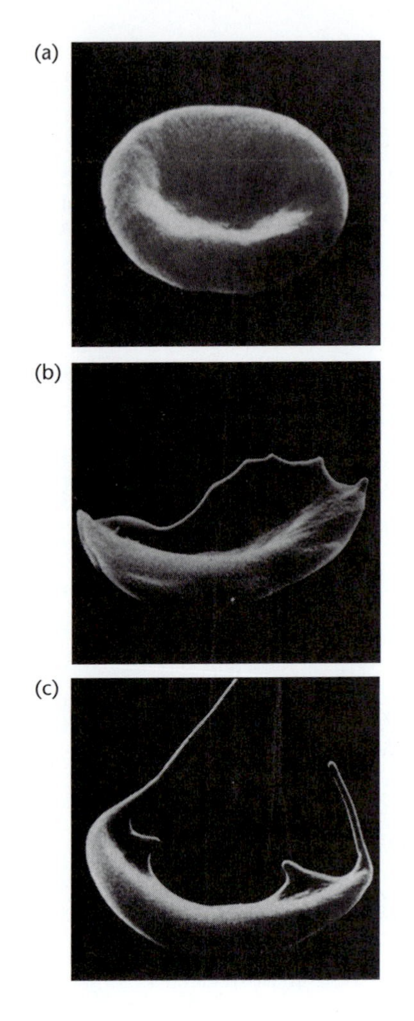


**Figure 4-7** Filaments distorting a cell. A mutation in the β-hemoglobin gene causes it to assemble into polymers when the concentration of oxygen is low. (a) Human red blood cell with a normal biconcave discoid shape. (b) and (c) Red cells from a patient with sickle cell anemia. Rods of hemoglobin in the cytoplasm distort the cell shape of the red blood cells, making them fragile and prone to block fine capillaries throughout the body. The prevalence of the sickle cell gene, which is as high as 40% of the population in some regions of Africa, can be attributed to the fact that possession of just one copy of the gene confers increased resistance to malaria. (Courtesy of M. Bessis, *Nouvelle Revue Francaise d'Hematologie* 12: 721–746, 1972.)

the action of post translational modifications. In order to function as a coherent and integrated structure, the cell has to organize its protein and other molecules spatially, grouping them according to their function. But how can a eucaryotic cell, with a diameter of 10 μm or more, be spatially organized by protein molecules that are typically less than 5 nm—that is, two thousand times smaller in linear dimensions?

The answer lies in *polymerization*. For each of the three major types of cytoskeletal protein, thousands of identical protein molecules assemble into linear filaments that can be long enough, if necessary, to stretch from one side of the cell to the other. Such filaments connect protein complexes and organelles in different regions of the cell, in many cases serving as tracks for transport and communication between them. They establish the organizational framework of a eucaryotic cell.

During the polymerization of cytoskeletal filaments, actin, tubulin, or intermediate filament protein molecules, freely soluble in the cytosol, interact with the ends of their cognate protein filaments at rates that are limited only by diffusion. Proteins at the filament ends can also dissociate and become soluble again (Figure 4-8). Polymerization proceeds, as in the case of glucagon or sickle cell hemoglobin, until the concentration of soluble protein has dropped to a point that the rate of subunit addition to the fiber ends exactly balances the rate of subunit loss, resulting in a polymerization equilibrium.

## *Polymerization of actin and tubulin is controlled by nucleotide hydrolysis*

Although actin and tubulin polymerize in a way that is superficially similar to glucagon or sickle cell hemoglobin, their assembly actually shows subtle and unique features. It is now known that the formation of actin filaments and microtubules is regulated by the hydrolysis of the energy-containing metabolites ATP or GTP, and that this enables the cell to exert a powerful control over when and where these filaments are made.

The precise mechanism by which actin filaments and microtubules polymerize is a subject for later chapters (see especially Chapter 11). But in broad terms, the protein subunits in these polymers (actin or tubulin) exist in two different conformations depending upon whether they are bound to a nucleotide triphosphate (NTP) or a nucleotide diphosphate

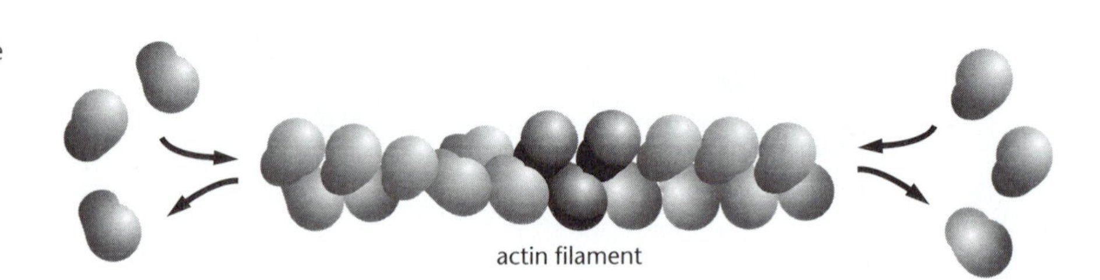


**Figure 4-8** Protein polymerization. Soluble protein molecules (polymer subunits) add reversibly to the free ends of the polymer (in this case an actin filament).

**Figure 4-9** Protein polymerization as a thermal ratchet. Hypothetical scheme by which the growth of a protein polymer (here an actin filament) might exert a force. Thermally driven movements of the plasma membrane at the tip of the cell extension occasionally allow a subunit molecule to diffuse onto the growing end of the filament. The leading margin of the cell will therefore advance in increments by a ratchetlike mechanism.

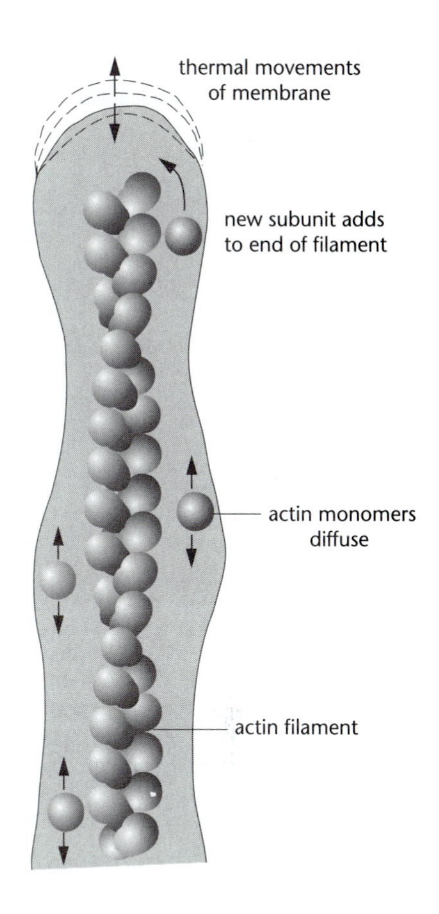


(NDP). These two conformations tend to assemble together in different ways, so that the NTP conformation readily assembles into a polymer but the NDP form does not.

Subunits with bound NTP first assemble into a polymer and then, after some delay, the nucleotide is hydrolyzed to NDP. As each subunit changes its conformation it fits less well into the polymer, creating stored elastic energy that can be released if the polymer depolymerizes. The bulk of the polymer therefore becomes unstable, and maintains its assembled form only as a result of kinetic or other factors that block disassembly. The kinetic factors result in a polymer that fluctuates rapidly in length, alternating between a growing and a shrinking state—a phenomenon known as *dynamic instability* (discussed in Chapter 11). The 'other factors' include numerous proteins that bind to actin or tubulin and thereby change their ability to form a stable polymer.

Note that nucleotide hydrolysis is not the only way to escape equilibrium constraints and to promote polymerization dynamics. Intermediate filaments also turn over in living cells and in this case their turnover is controlled by reversible phosphorylation—a kinase-phosphatase system serves to couple the energy released by ATP hydrolysis into the filament system.

## *Polymerization is a form of movement*

The elongation of a protein filament as subunits are added one by one is itself a form of movement that in principle can be harnessed by the cell. Thus, if a population of red blood cells from a patient with sickle cell anemia is exposed to a low concentration of oxygen, then the mutant hemoglobin in the cytoplasm rapidly forms long fibers. As they grow longer in the cytoplasm, these fibers press against the plasma membrane and produce spiky protrusions on the cell surface. The movement in this instance is irreversible and unwanted, but a similar process underpins many deliberate movements of living cells.

We will see in later chapters that the production of surface extensions such as filopodia, the movement of chromosomes during cell division, and the migration of infectious bacteria through the cytoplasm are all driven, at least in part, by the ongoing addition of protein subunits onto the ends of protein filaments. It has been speculated that such polymerization-driven movements may have been among the first to appear during evolution, since they can be produced by a single protein, and do not require complicated motor molecules.

Precisely how protein molecules adding to the end of a protein filament can push against an object is still not fully established, but we can at least point to some obvious possibilities. The simplest mechanism is a *thermal ratchet* in which small, random motions of the object to be moved, such as a local region of membrane, create transient gaps between it and the end of the polymer (Figure 4-9). Subunit molecules may then diffuse into the small gap and add to the end of the polymer; as each subunit is added, the polymer increases in length and the thermal fluctuations are displaced by a small increment (hence the term 'ratchet'). In effect, random thermal movements are biased in one direction by polymerization.[3]

[3] In Chapter 13 we will see that a similar mechanism, operating in reverse can couple the *depolymerization* of a filament to movement. Some form of attachment must exist in this case between the object and the polymer.

The rate of movement produced by a thermal ratchet is dictated by the size and frequency of thermal movements, which are beyond the control of the cell. It is, however, possible to imagine mechanisms by which the assembly process might be regulated by proteins situated at the end of the polymer. There are even ways, as we will discuss in Chapter 18, in which the random ratchets of many filaments can work in concert to drive very large structures forward.

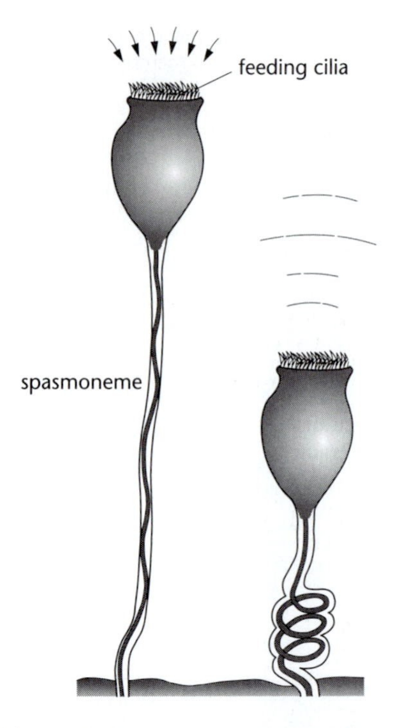


**Figure 4-10** Spasmoneme contraction. Sketch of the protozoan *Vorticella* before and after the contraction of its long stalk, or spasmoneme.

## *Changes in protein conformation can produce large-scale movements*

All protein molecules in living cells have a well-defined shape because their polypeptide chains fold in a stable three-dimensional arrangement, held in place by weak, noncovalent forces. Moreover many proteins—and this is true of many proteins involved in cell movements—possess two or more different conformations and can switch from one to the other depending on external factors. One conformation might be favored by binding to specific ions or small molecules or by covalent modifications, such as the addition of a phosphate group to one of its amino acids. Conversion from one shape to another then has the potential of producing movements, which can become hugely amplified if many molecules are linked together.

A dramatic example is seen in certain ciliates such as *Vorticella* and *Zoothamnium*. These single-celled organisms have a bell-shaped body that is firmly attached to the substratum by a slender stalk containing a spiral contractile fiber. When undisturbed, the bell is poised on the end of a long straight stalk with its feeding cilia in rapid action. However, the least disturbance causes the stalk to change its shape, coiling into a helix whose contour length is essentially the same as that of the extended stalk (Figure 4-10).

Some species have stalks as long as 1 mm and may reduce their end-to-end length by 50% in less than 10 msec (making this shortening of end-to-end length faster than muscle contraction). The mechanically active part of the stalk, called the *spasmoneme*, contains a mass of parallel protein filaments, 2–4 nm in diameter, interpenetrated by a system of membrane tubules. The proteins that make up the spasmoneme filaments bind calcium and change in conformation, thereby producing cell movement.[4] Thus, if cells of *Vorticella* are extracted with glycerol (a treatment that removes the plasma membrane) and exposed to a rise in concentration of free calcium ions (say, from $10^{-8}$ to $10^{-6}$ M), then the spasmoneme contracts. The contraction can be reversed and the cycle repeated many times, simply by manipulating the levels of calcium in the solution bathing the lysed cells. In the living organism the concentrations of free calcium are triggered by stimuli impinging on the plasma membrane and spread to the interior of the stalk by an excitable mechanism similar to that seen in muscle cells (Chapter 9).

There are other examples of springs based on protein conformational changes, such as the acrosome of *Limulus* sperm, a coil of actin filaments that extends explosively during fertilization (Chapter 6). The storage of mechanical strain energy in the subunits of a microtubule, mentioned above, illustrates the same principle in a less dramatic form.

## *Protein motors move along actin filaments and microtubules*

By far the most common way for eucaryotic cells to generate rapid and powerful movements is by means of *motor proteins*. Myosin, kinesin, dynein, and related proteins all have the ability to advance along actin filaments or microtubules, using the hydrolysis of ATP to drive their movements. Stated in broad terms, the motor protein alternates between

[4] The spasmoneme is unique to *Vorticella* and related protozoa, but the protein it is made of belongs to the widely distributed *centrin* family of calcium-binding proteins found in centrosomes, ciliary rootlets, and other locations in most eucaryotic organisms (Chapter 15).

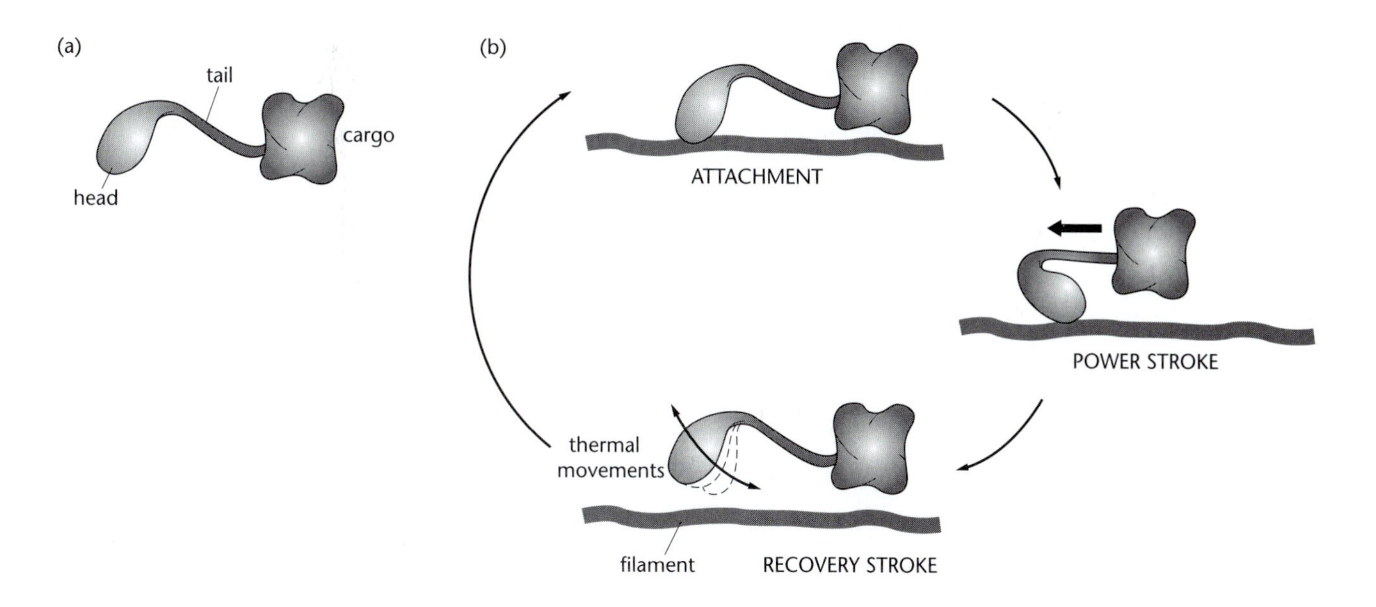


**Figure 4-11** Motor proteins move along protein filaments. (a) Cartoon of a typical motor protein showing its active head region, which hydrolyzes ATP and generates movements, and its tail region, which attaches the motor to the cargo to be moved (such as a vesicle or other organelle). (b) Three main steps in the progression of a motor protein. Only the power stroke consumes ATP energy.

two conformations, each stabilized by the binding of a different nucleotide molecule—intact ATP in one conformation and the hydrolysis products ADP and inorganic phosphate in the other. The changes in shape, occurring cyclically as ATP molecules are hydrolyzed, carry the motor protein along the filament in a series of steps. If the motor protein is attached by its 'tail' to another protein filament or to a membrane vesicle, it can pull this cargo along as it goes, thereby performing useful work in the cell.

Molecular motors enlarge our view of the function of cytoskeletal filaments, which are evidently not just 'structural supports' for the cell, but also 'pathways' or 'tracks' through the cytoplasm along which organelles can travel. A corollary of this view is that filaments need to be not only strong and straight but also *directional*—they must have a polarity or else the traveling protein would wander equally in opposite directions along the fiber. Of the three kinds of protein filament commonly found in animal cells, only actin filaments and microtubules have a structural polarity and only these two are used by motor proteins. Cells are now known to be full of motor proteins, and their interaction with actin filaments or microtubules is the basis for many of the best-understood movements of eucaryotic cells (see Table 4-1).

## *Motor proteins are driven by ATP hydrolysis*

The detailed molecular workings of motor proteins are of great interest from the standpoint of cell movements. They are also important at a more fundamental level since they should teach us how protein machines convert chemical energy to mechanical energy with high efficiency. Because of this convergence of interest, many experiments have been done and many analyses performed on myosins (the proteins responsible for muscle contraction) and more recently on dynein and kinesin (responsible for ciliary beating, mitosis and movements along microtubules). It is now clear that, while each motor protein has unique features that suit it to its particular task in the cell, there are also underlying features that they all share in their mechanism of action.

All motor proteins so far examined have at least two structurally distinct parts: a globular head region that engages with the filament and actively moves along it, and a tail region, usually filamentous, through which the motor protein is attached to its cargo (Figure 4-11a). Progression along a filament depends on cyclic movements of the head region that are sometime described as 'walking,' but this is an imperfect analogy. When a dog or a human walks, it actively drives both the forward and the backward movements of its legs, with different sets of muscles being employed for the two stages. For myosin, dynein, and kinesin, however, at least part of the cycle (probably the forward motion) is driven not by the protein itself but by thermal motion.

During the backward thrust of the motor head, termed the *power stroke*, the protein changes its conformation while it is firmly attached to the filament. The effect of this change is to develop tension in the tail of the motor protein and—if the force is great enough—to move the tail and whatever is attached to it forward. By contrast, the forward movement, or *recovery stroke*, takes place when the protein is loosely attached, if at all, to the filament (Figure 4-11b). The head region slides along the latter through thermal motion, in effect by diffusion. These two movements, occurring in alternation as the motor protein proceeds along the filament, constitute its mechanical cycle. They are driven by a cycle of chemical changes in which an ATP molecule binds to the head region and is hydrolyzed to ADP and inorganic phosphate, which are then released. Precisely how these two cycles are coupled, and at what stage ATP is hydrolyzed, are issues that are still debated and may be different for different motor proteins. The myosin cycle in skeletal muscle is described in greater detail in Chapter 9.

## *Shape and movement are regulated by ions and small diffusible molecules*

The cytoskeleton provides the hardware for cell shape and movement, but it does not function in isolation from the rest of the cell. It is connected physically to other parts of the cell, such as the plasma membrane and the nucleus, through the accessory proteins already mentioned. It is also 'chemically connected' to the reactions of metabolism and to the complex functional networks of small molecules and enzymes that carry signals in the cytoplasm, such as calcium ions ($Ca^{2+}$), cyclic AMP (cAMP), phosphatidylinositol bisphosphate ($PIP_2$), G proteins, and protein kinases (Figure 4-12). These signaling molecules also regulate intermediary metabolism, protein synthesis, gene expression, and many other processes, thereby integrating the movements and shape changes produced by the cytoskeleton with other events in the cell.

Of the many different signaling molecules in cells, calcium ions appear to have a special relationship to large-scale movements. The normal intracellular concentration of free $Ca^{2+}$ is low (about 0.1 μM) but it can rise by a factor of 100 or more following an inward leak from outside the cell (where calcium is usually abundant) or following release from internal stores. Such concentration changes are normally transient, thanks to membrane-associated pumps that use the energy available from ATP hydrolysis to pump $Ca^{2+}$ out of the cell or back into special internal membrane compartments. The sudden pulse generated in this way can trigger a rapid movement or other change involving the cytoskeleton. When a muscle contracts or a field of cilia changes its direction of beat, or when a migrating cell responds to an external stimulus, the event is typically initiated by a transient increase in $Ca^{2+}$. The increases are often highly localized, corresponding to a brief 'puff' or 'spark' a few hundred nanometers in diameter and lasting perhaps a few milliseconds, but they are sufficient to initiate an entire cascade of cytoskeletal changes.

How does a change in calcium ion concentration affect the cytoskeleton? One way is if $Ca^{2+}$ binds directly to a cytoskeletal protein and changes its conformation, and in fact many cytoskeletal proteins do possess calcium-binding domains. Alternatively, $Ca^{2+}$ could work indirectly—by

**Figure 4-12** (*opposite*) A few of the many signaling molecules that control the cytoskeleton.

## Protein kinases

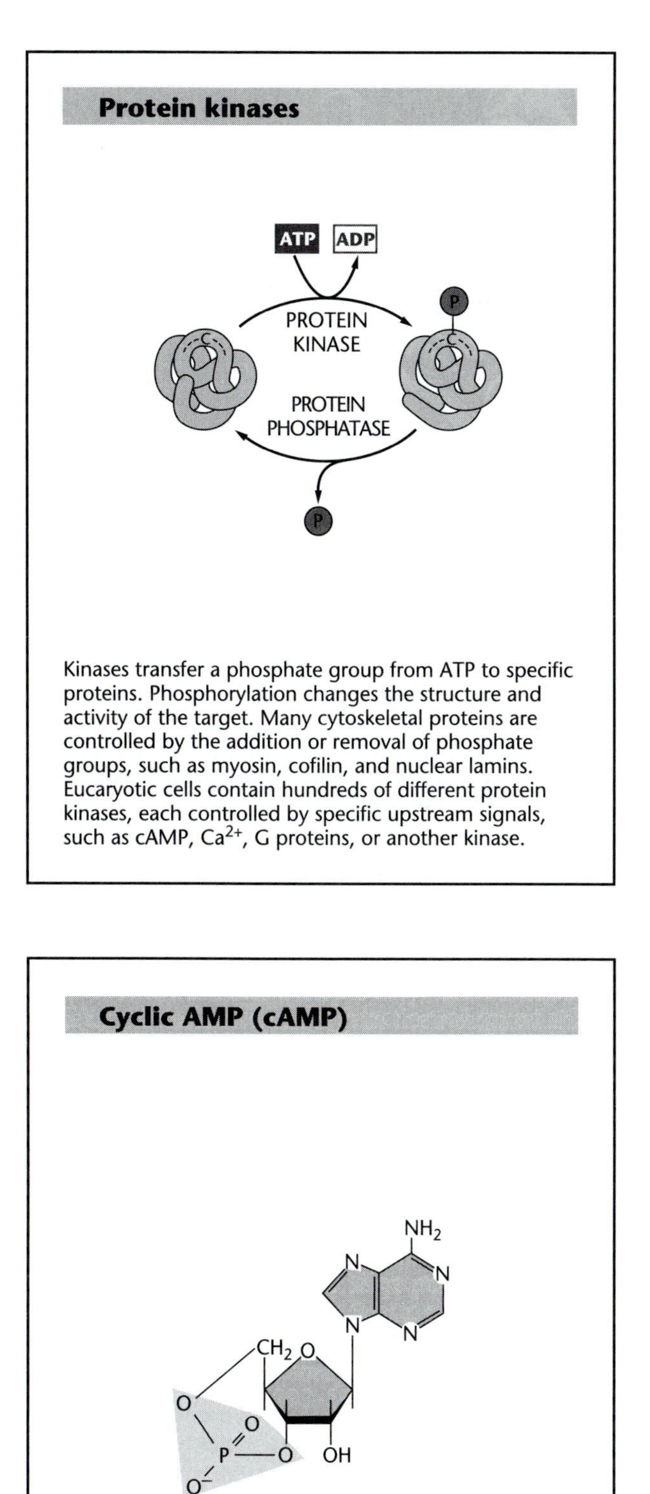


Kinases transfer a phosphate group from ATP to specific proteins. Phosphorylation changes the structure and activity of the target. Many cytoskeletal proteins are controlled by the addition or removal of phosphate groups, such as myosin, cofilin, and nuclear lamins. Eucaryotic cells contain hundreds of different protein kinases, each controlled by specific upstream signals, such as cAMP, $Ca^{2+}$, G proteins, or another kinase.

## Cyclic AMP (cAMP)

Nucleotide that is generated from ATP in response to hormonal stimulation of cell surface receptors. cAMP acts as a signaling molecule by activating a specific kinase (PKA) which in turn phosphorylates many proteins. Also secreted by the slime mold *Dictyostelium* as an aggregation-inducing chemoattractant.

## Phosphatidylinositol bisphosphate ($PIP_2$)

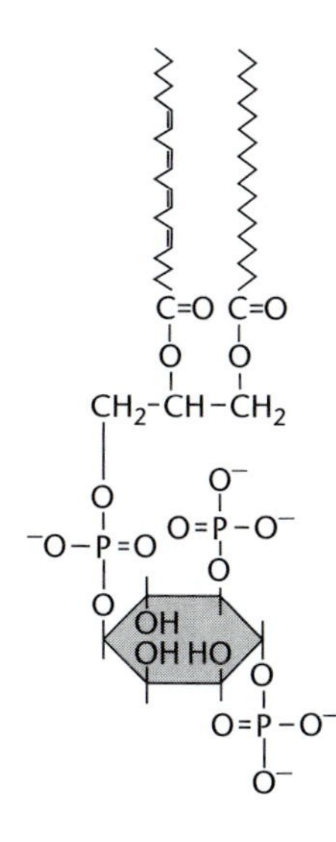


A minor phospholipid of the inner leaflet of the plasma membrane that changes in concentration (due to hydrolysis) in response to the binding of ligands to specific cell surface receptors. $PIP_2$ regulates the activity of a number of proteins in the actin cortex, such as profilin and gelsolin.

## Calcium ions ($Ca^{2+}$)

Universally employed as a signal in eucaryotic cells. The resting concentration of $Ca^{2+}$ is normally less than $10^{-7}$ M but can rise briefly to $10^{-6}$ M or more during a signal. $Ca^{2+}$ influences many cytoskeletal processes in the cell, such as muscle contraction and chemotaxis. $Ca^{2+}$ works by binding to proteins such as calmodulin or troponin and changing their activity.

## G proteins

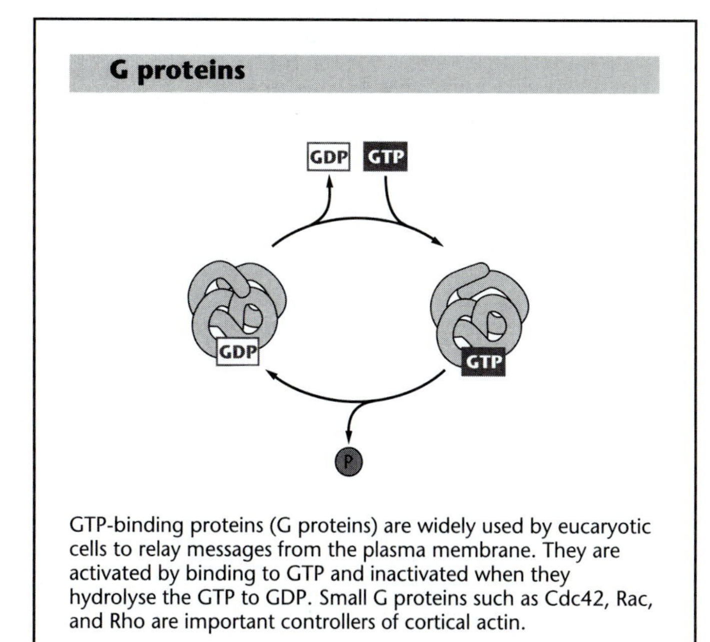


GTP-binding proteins (G proteins) are widely used by eucaryotic cells to relay messages from the plasma membrane. They are activated by binding to GTP and inactivated when they hydrolyse the GTP to GDP. Small G proteins such as Cdc42, Rac, and Rho are important controllers of cortical actin.

influencing the activity of an enzyme that in turn produces a signal in the cell and influences movement. For example, changes in calcium ions modulate production of the diffusible nucleotide cyclic AMP and of phosphoinositides, both of which are important signaling molecules.

## Cytoskeletal responses are coordinated by protein kinases

Yet another way in which $Ca^{2+}$ can affect the cytoskeleton is by activating a protein kinase. A family of *$Ca^{2+}$-activated protein kinases* exists in eucaryotic cells, able to add phosphate groups to specific sites on cytoskeletal proteins. Sometimes the new phosphate group changes the conformation and activity of the protein directly. In other cases the phosphate group works like a 'tag' or 'label' to which another protein binds and produces a change. Cells also contain phosphatases that remove phosphate groups from proteins, thereby making the action of many protein kinases cyclic. The more rapidly a phosphate group turns over, the faster the concentration of a specific phosphorylated protein can change in response to cell signals. Such a phosphate group can therefore act as a signal in much the same way as $Ca^{2+}$ ions or cyclic nucleotides.

There are hundreds of different protein kinases that together phosphorylate perhaps a third of the 10,000 to 20,000 proteins of a typical eucaryotic cell. Some kinases are themselves phosphorylated, which further increases the biochemical complexity. Cytoskeletal proteins are phosphorylated, as a rule, and the number and position of phosphate groups influences their assembly or disassembly, their interaction with other components, their enzymatic activity, and their turnover.

The spatial and temporal integration that can be achieved by protein kinases and phosphatases is vividly demonstrated by the cytoskeletal changes that accompany mitosis. As we will see in Chapter 13, every landmark event in this complex sequence of movements is controlled by protein kinases.

## A motile cell is an 'intelligent' cell

It is no surprise that animals have a nervous system whereas plants do not. Organisms that move through their environment continually encounter new situations—sometimes life-threatening at other times potentially rewarding—to which they must respond. Competition with other organisms gives an advantage to those that can move more quickly or show greater wisdom in their selection of direction.

As with organisms, so with cells. There are, as we have already discussed, many single-celled organisms that move independently through their environment. Crawling amoebae, swimming ciliates, gliding bacteria, all must fulfill certain basic requirements in terms of regulatory interactions. They have to continually monitor features of their environment such as temperature, turbulence, physical obstacles, sources of nutrition, predators and prey, and be ready to respond to them quickly. This in turn implies a minimal level of information-processing circuitry composed of biochemical reactions that, like the nervous system of a higher animal, control and integrate the cell's responses.[5]

One of the most striking features of both cytoplasmic signaling pathways and the motile machinery is their extensive interconnection. Cross-talk exists at every level. A G protein that triggers a range of different cell processes, for example, might itself be fired by many different membrane receptors. A protein kinase might act on many different cytoskeletal proteins, phosphorylating them sometimes singly, at other times in combinations. An actin-binding protein may participate in several different cytoskeletal assemblies, each of which also contains other proteins of similar but not identical function. Thus, if we try to imagine this circuitry as a whole, we see an extensively interconnected network in

[5] Even cells that cannot move, such as plant cells, contain highly interconnected networks of signaling reactions and show highly sophisticated *chemical* responses to stimuli. Perhaps we overestimate the 'intelligence' of a moving cell because of our primitive association of movements with volition? Perhaps tulip cells are also intelligent, in their own fashion?

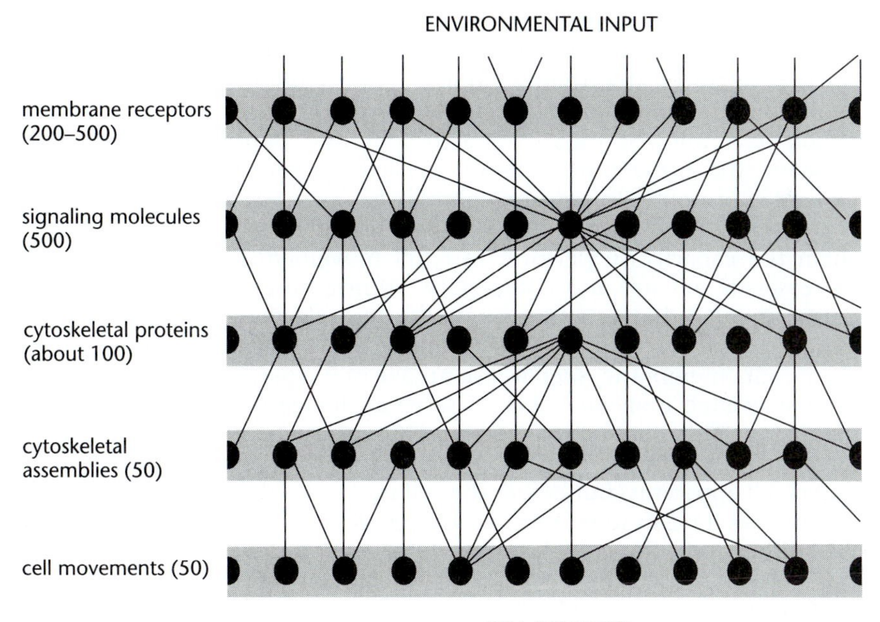


**Figure 4-13** The motile physiology of an animal cell seen as a parallel network. Signals received on the outside of the cell, usually by specific membrane receptors, lead to changes in the concentration of intracellular signaling molecules. These include, most importantly, $Ca^{2+}$ ions, phosphatidylinositol bisphosphate ($PIP_2$), cyclic AMP, protein kinases, and GTP-binding proteins. Changes in signaling molecules then affect cytoskeletal proteins, changing their activity through allosteric interactions or by posttranslational modifications and hence modulate the structure and activity of subcellular assemblies such as ciliary axonemes, stress fibers, or contractile bundles. Finally, the subcellular assemblies act together to generate a particular sequence of movements or shape changes. Order-of-magnitude estimates of the numbers of elements at each level are indicated in parentheses.

which signals travel from the cell surface to the motile machinery along many parallel pathways (Figure 4-13). The extensively parallel networks, analogous to those used in computer-based pattern recognition, may have collective properties that would be extremely useful to a living cell. Perhaps, in the future, it will be by a combination of molecular genetics and computer simulation that we will finally understand how cells move.

## *The cytoskeleton was probably crucial for the evolution of eucaryotic cells*

A cytoskeleton is one of the features, like mitochondria or the nucleus, that distinguishes eucaryotic cells from procaryotic cells. Essentially every cell in a plant, animal, or protozoan contains actin filaments and microtubules, whereas these proteins are absent from bacteria. As we will relate in later chapters, the proteins that make up actin filaments and microtubules are among the most highly conserved of all eucaryotic proteins in sequence. The mechanisms of motility and cytoplasmic organization they mediate are similarly conserved, and it seems likely that these proteins arose just once in the course of evolution, roughly at the same time that eucaryotic cells first appeared. Was this coincidence accidental, or could the 'invention' of dynamic polymers have been crucial for the appearance of eucaryotic cells?

Life arose on Earth more than 3.5 billion years ago. For the first 2 billion years, there were only bacteria (procaryotes) until, about 1.5 billion years ago, eucaryotic cells appeared. These had many distinguishing features, including the possession of a nucleus, the possession of a cytoskeleton, and the ability to form complex multicellular organisms, Which of these came first we cannot say, but they were evidently mutually supportive.

Development of a cytoskeleton would have freed primitive eucaryotic organisms from their dependence on a cell wall for protection against physical stress and allowed them to change their shape. Surface plasticity would have brought with it the potential for phagocytosis, an actin-dependent movement that enables efficient feeding, so that an organism could capitalize on the biosynthetic work of others. This process would, in turn, have set the stage for acquisition of endosymbiotic organelles, leading eventually to mitochondria and chloroplasts. Even the nucleus, it has been suggested, could conceivably have arisen by phagocytosis of a separate organism that then became an endosymbiont. At a later stage in evolution, the appearance of a mitotic spindle based on microtubules would have allowed the genome to split into multiple chromosomes, allowing the genetic information carried by eucaryotic organisms to expand enormously in size and complexity.

Actin filaments and microtubules would have also brought with them the capacity to perform internal movements, with concomitant increases in cell size and complexity. From a purely geometrical standpoint, any cell that gets bigger must suffer a decrease in the ratio of its surface area to its volume. Consequently, any function that depends on the plasma membrane, such as the uptake of nutrients or the synthesis of lipid-associated molecules, has to serve an increasingly large volume of cytoplasm. This may be one reason why eucaryotic cells contain such a large quantity of internal membranes, such as the endoplasmic reticulum and Golgi apparatus. But before the cell can make efficient use of internal membranes it needs to be able to carry vesicles to and from the cell surface. Free diffusion in the cytoplasm would be far too slow and inaccurate for this purpose, especially for particles the size of membrane vesicles (see Table 1-1).

## Bacteria have an organized cytoplasm

Although the three filament-forming proteins of the cytoskeleton are unique to eucaryotic cells, distantly related proteins have been found in bacteria. Comparison of the amino acid sequence of actin with that of other proteins reveals common elements with a number of existing bacterial proteins, including the enzyme hexokinase, which catalyzes the phosphorylation of sugars, and with heat-shock proteins, which aid in the folding of other proteins through association with their hydrophobic domains. Both hexokinase and heat-shock proteins bind to and hydrolyze ATP as part of their normal function, although neither forms filaments.

We will discuss in Chapter 13 a group of proteins involved in septation (cell division) in bacteria. These include one protein with a possible similarity to actin and another, the protein FtsZ, that is closely similar in three-dimensional structure to tubulin. FtsZ can polymerize into filaments, hydrolyzes GTP, and has every indication of being a cytoskeletal protein. Control over the location of this protein is exerted by another set of proteins (the product of the *min* locus) that appear able to position the plane of division in the midplane of the cell. Another bacterial protein with a potential to form filaments is elongation factor EF-Tu, which interacts with ribosomes in an essential step in protein synthesis and it is one of the most abundant proteins of the bacterial cytoplasm (about 5% of the total). EF-Tu can form long branching filaments in the test tube, although their significance for the living cell is presently unclear.

These are all hints that the bacterial cytoplasm may have a degree of internal order, conferred perhaps by filamentous proteins, that is similar to the eucaryotic cytoskeleton. It is also worth recalling that protein polymerization has the potential at least of generating directed movements. The structure of this framework, how it is formed, and whether it supports directed movements are questions for future research.

## The cytoskeleton carries epigenetic information

The cytoskeleton can function for long periods in the complete absence of the nucleus. A cell that has lost its nucleus will continue to ruffle, migrate, and ingest particles for many hours (see Figure 4-1). Small fragments of cells can move over surfaces and can even 'grow' in the sense of assembling further cytoskeletal structures. A developing nerve axon cut

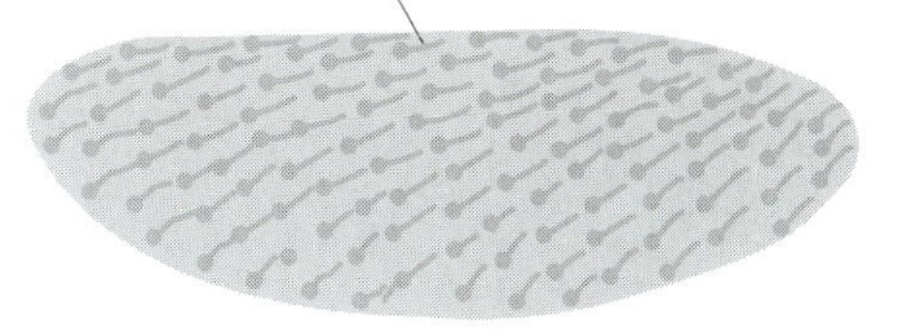

**Figure 4-14** Inheritance through the cytoskeleton. Schematic diagram of the rows of cilia on the surface of a normal *Paramecium* and on a *Paramecium* in which rows of cilia have been inverted so that they beat in the opposite direction. Such altered patterns are propagated indefinitely as the cell divides, although the information for this heritable characteristic is carried in the cytoskeleton, not in DNA.

from its cell body continues to elongate for several hours due to the assembly of cytoskeletal and membrane precursors.

In principle, therefore, any cytoskeletal structure passed from one cell to its daughter without being totally deconstructed to its component molecules could act as a seed to nucleate the growth of similar structures. In such a case the cytoskeleton would carry *epigenetic information*—hereditary information that is not present, or not present in a readily accessible form, in the genome. Since all present-day eucaryotic cells are related by a continuous line of descent to the first primordial eucaryotic cell, it is conceivable that structures such as cytoskeletal filaments have never had to polymerize in the complete absence of nucleating structures.

A classic example of this general idea is seen in *Paramecium*. As we saw in Chapter 1, this single-celled organism has rows of cilia on its surface, which normally beat in the same direction enabling effective cell motility. By treating the cells with a mechanical insult or a drug that interferes with their normal replication, it is possible to disturb this pattern and produce some inverted rows of cilia that beat in the direction opposite to that of their neighbors (Figure 4-14). Once established, the altered patterns are passed on to the cells of successive generations, apparently forever. This form of heredity has nothing to do with DNA: the modified cells inherit a particular pattern of ciliary rows via their cytoskeleton.

Inheritance of a particular cytoskeletal arrangement can have far-reaching consequences for morphogenesis, such as the shaping of the adult body plan of an organism. A clear example is seen in the development of the mud snail, whose coiled shell is genetically determined to be either a right-handed or a left-handed spiral. The handedness of the adult is observable as early as the eight-cell stage of development as a result of the asymmetric patterns of cell divisions known as spiral cleavage (Figure 4-15). This handedness of the embryo is determined by a

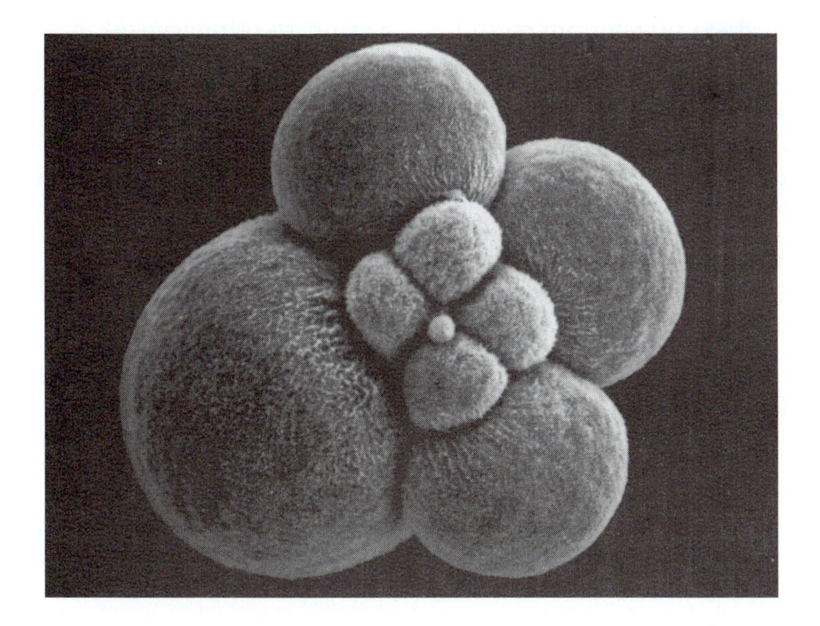

**Figure 4-15** Spiral cleavage. Scanning electron micrograph of an eight-cell embryo of the mud snail, *Ilyanassa obsoleta*, viewed from the animal pole. The small spherical object in the center is one of the residual polar bodies. The largest cell is the immediate recipient of morphogenetic factors that ultimately influence larval organogenesis. (Courtesy of Michael M. Craig, *Cell* 68 (2): cover photo, 1992.)

slightly skewed position of the mitotic spindle during the early cleavages. But what determines the position of the mitotic spindle? It appears that a factor produced by the mother snail during the course of oogenesis influences the orientation of the embryonic spindle. Moreover, the handedness of the egg spindle, whether right- or left-handed, reflects the genotype of the mother, not the embryo itself. So it is possible to have a 'right-handed' spindle leading to a 'right-handed' shell in a snail whose genetic make-up is left-handed.

## *Cytoskeletal proteins show overlapping redundancy in their functions*

One of the most powerful ways to probe the complex mechanism of cell movements is by molecular genetics. Methods of clonal analysis using recombinant DNA make it possible not only to isolate and sequence cytoskeletal proteins but also to manipulate their level of expression. Thus, a gene for a specific protein can be physically introduced into a cultured cell and there expressed, allowing the structural and motile consequences to be analyzed. Similarly, it is possible to substitute a normal gene for one that is defective—containing a mutation that either makes the protein nonfunctional or modifies its performance in a specific way. These methods greatly extend, in scope and facility, classical genetic methods of analysis.

A paradoxical consequence of such manipulations when applied to cytoskeletal proteins is that gene removal often has little effect. A good example of this is seen when proteins binding to actin are deleted from the slime mold *Dictyostelium*. The amoeboid cells of this organism have a similar shape and move in a similar manner to the cells of higher organisms, but being haploid they are readily manipulated by molecular genetic or classical genetic methods. Thus it was possible, by the techniques just mentioned, to mutate or delete genes carried by this organism that code for actin-binding proteins and then to examine the behavioral consequences of their removal. But although these proteins are an integral part of the actin cytoskeleton and presumably have a function, this function often proves to be difficult to detect. Mutant *Dictyostelium* lacking such proteins as severin and α-actinin crawl and follow chemical clues in a dish as well as their wild-type cousins. The cells crawl at the same rate, show the same morphological characteristics as the wild-type, and are as good as wild-type cells in the detection and response to gradients of chemoattractant.

If these proteins are part of the machinery of movement, how can they be removed from the cell without arresting its migration? The simplest explanation is that the functions of these proteins are, to some degree, duplicated in the cell. This is especially easy to believe in the case of cell crawling, because many actin-associated proteins have similar functions, such as capping, fragmenting, or bundling. However, under rigorous conditions of their normal habitat, which for a *Dictyostelium* amoeba is soil, with all the risks of starvation, desiccation, infection, and predation, such mutants will presumably not perform as well as their healthy cousins. Otherwise, why would the proteins have evolved in the first place?

## Essential Concepts

- The structures that define cell shape and generate movement reside chiefly in an insoluble network of protein filaments called the cytoskeleton.
- In higher animals, the cytoskeleton contains three principal types of filaments: actin filaments, microtubules, and intermediate filaments, each with a distinctive protein composition and ultrastructure.
- These filaments are linked to each other and to membranes by accessory proteins forming a three-dimensional scaffolding that gives the cell its shape and capacity for movements.
- The filaments forming the cytoskeleton are produced by polymerization reactions in which a single type of protein subunit adds repetitively. For actin filaments and microtubules, the process is coupled to nucleotide hydrolysis, thereby enabling the cell to control when, where, and the extent to which it occurs.
- The directed growth of a protein filament is a form of movement that can be harnessed by the cell. This was probably the basis of the earliest eucaryotic cell movements to appear during evolution.
- However, the majority of known movements in present-day eucaryotic cells are driven by motor proteins that progress along either actin filaments or microtubules by undergoing cyclic ATP-driven changes in shape.
- Different parts of the cytoskeleton are functionally connected to each other and to other components of the cell by protein links and by chemical signals carried by the changing concentrations of intracellular ions and small molecules.
- The cytoskeleton is unique to eucaryotic cells and it plays a vital role in its most distinctive processes. The appearance of a dynamic cytoskeleton was probably one of the crucial steps in the evolution of eucaryotic cells.
- Cytoskeletal filaments almost always grow from preexisting cellular structures. The cytoskeleton therefore carries information independently of the nucleus—this information being expressed in the position and structure of its components.

## Further Reading


Albrecht-Buehler, G. Autonomous movements of cytoplasmic fragments. *Proc. Natl. Acad. Sci. USA* 77: 6639–6643, 1980.

Amos, W.B. Structure and coiling of the stalk in the peritrich ciliates *Vorticella* and *Carchesium*. *J. Cell Sci.* 10: 95, 1972.

Amos, L.A., Amos, W.B. Molecules of the Cytoskeleton. London: Macmillan Education, 1991.

Beisson, J., Sonnenborn, T.M. Cytoplasmic inheritance of the organization of the cell cortex in *Paramecium aurelia*. *Proc. Natl. Acad. Sci. USA* 53: 275–282, 1965.

Bershadsky, A.D., Vasiliev, J.M. Cytoskeleton. New York: Plenum Press, 1988.

Bray, D. Intracellular signalling as a parallel distributed process. *J. Theor. Biol.* 143: 215–231, 1990.

Buckley, I.K., Porter, K.R. Cytoplasmic fibrils in living cultured cells. *Protoplasma* 64: 349, 1967.

Clark, P., et al. Topographical control of cell behaviour. II. Multiple grooved substrata. *Development* 108: 635–644, 1990.

Doolittle, R.F. The origins and evolution of eukaryotic proteins. *Phil. Trans. R. Soc. Lond. B.* 349: 235–240, 1995.

Erickson, H.P. FtsZ, a prokaryotic homolog of tubulin? *Cell* 80: 367–370, 1995.

Ingber, D.E. The architecture of life. *Sci. Am.* 278: 30–39, 1998.

Janmey, P.A. The cytoskeleton and cell signaling: component localization and mechanical coupling. *Physiol. Rev.* 78: 763–781, 1998.

Kreis, T., Vale, R. Guidebook to the Cytoskeletal and Motor Proteins. Oxford, UK: Oxford University Press, 1999.

Mitchison, T.J. Evolution of a dynamic cytoskeleton. *Phil. Trans. R. Soc. Lond. B.* 349: 299–304, 1995.

Moriyama, Y., et al. High-speed video cinematographic demonstration of stalk and zooid contraction of *Vorticella convallaria. Biophys. J.* 74: 487–491, 1998.

Norris, V., et al. The *Escherichia coli* enzoskeleton. *Mol. Microbiol.* 19: 197–204, 1996.

Peskin, C.S., et al. Cellular motions and thermal fluctuations: the brownian ratchet. *Biophys. J.* 65: 316–324, 1993.

Poulsen, N.C., et al. Diatom gliding is the result of an actin-myosin motility system. *Cell Motil. Cytoskeleton* 44: 23–33, 1999.

Searcy, D.G., Hixon, W.G. Cytoskeletal origins in sulfur-metabolizing archaebacteria. *BioSystems* 25: 1–11, 1991.

Sonnenborn, T.M. Gene action in development. *Proc. R. Soc. Lond. B.* 176: 347–366, 1970.

Thompson, G.M., Wolpert, L. Isolation of motile cytoplasm from *Amoeba proteus. Exp. Cell Res.* 32: 156–160, 1963.

Vale, R.D., Oosawa, F. Protein motors and Maxwell's demons: does the mechanochemical transduction involve a thermal ratchet? *Adv. Biophys.* 26: 97–134, 1990.

Zheng, J., et al. Prestin is the motor protein of cochlear outer hair cells. *Nature* 405: 149–152, 2000.

# PART TWO

## Molecular Basis of Cell Movements

# CHAPTER 5

# Actin Filaments

Actin is the archetype of cytoskeletal proteins. Abundant and amazingly versatile, it participates in movements of every known eucaryotic cell. Phagocytosis, cytokinesis, cell crawling, cytoplasmic streaming, muscle contraction, the folding of embryonic epithelia—all depend on structures built from actin. Actin molecules associate into static bundles, contractile bundles, two-dimensional networks, and three-dimensional gels. Each assembly has a characteristic structure, specified by a menagerie of different actin-binding proteins, and a distinctive location and function in the cell. But at the heart of each structure is the actin molecule itself, globular, highly conserved, and able to self-associate into linear filaments.

The study of cell movements received a major impetus in 1990 when a group led by Ken Holmes reported the structure of the actin molecule at atomic resolution. This achievement immediately provided a framework on which to place previous decades of biochemical and physiological studies on actin. Furthermore, as the atomic structure of proteins that associate with actin became available, researchers could fit (in a virtual sense) these molecules onto actin using computer graphics, and learn about the details of their interaction. Even more important, they could design rational mutations and informative derivatives for studies of function and mechanism in living cells.

In this chapter we consider the structure of the actin molecule and the subtle process by which it self-assembles into filaments. The ways in which the cell controls filament formation are described, including the remarkable example of *Listeria* bacteria, which highjack the polymerization machinery for their own ends. Lastly we introduce the family of actin-binding proteins and mention some of their functions that underpin cell movements.

## Actin is abundant in most eucaryotic cells

Actin was originally isolated from muscle by Straub and colleagues working at the University of Szeged in Hungary during World War II. They developed a procedure in which minced muscle was dehydrated with acetone, thereby denaturing unwanted protein, and then extracted with a low-ionic-strength buffer. The clear watery solution obtained showed a striking increase in viscosity on the addition of salt: the first indication of the ability of actin to assemble into long filaments.

For 20 or so years after this initial discovery, actin was regarded as a distinctive, if not unique component of muscle. Then, in the late 1960s, reports began to emerge of the presence of an actinlike protein in many nonmuscle cells, such as blood platelets and brain cells. Today, actin is recognized as one of the most versatile and abundant proteins of the eucaryotic cell. It has been found in every animal and plant cell so far examined, including fungi and yeast, and typically represents 5–10% of the total protein (20% of the protein in muscle). However, skeletal muscle remains the most convenient source of this protein for biochemical and structural studies and actin is still commonly extracted by a similar procedure to that described by Straub in 1942.

## Actin is a highly conserved protein

Actin is a globular protein composed of a single polypeptide chain of 375 amino acids (374 in some variants) and with a molecular weight close to 42 kDa. It has a typical amino acid composition for a globular protein, with a slight excess of acidic residues resulting in an isoelectric point of about pH 5.5. In most cells that have been examined, actin molecules are modified after translation by the acylation of their N-terminal residues and the methylation of one of their histidines.

The actin monomer (also known as globular actin, or G-actin) has two peanut-shaped domains lying side by side and connected at one end, resulting in a hinged molecule with a deep cleft. Within the cleft are binding sites for actin's essential cofactors: ATP and $Ca^{2+}$ (Figure 5-1). The cofactors make extensive contacts with both domains, which may explain the stabilizing effect of bound nucleotide and $Ca^{2+}$ on actin structure. The two halves of the molecule are relatively rigid and capable of some movement relative to each other—their ability to flex apart may be important for the exchange of bound nucleotide. The folds of the molecule as a whole show a remarkable similarity to those of the glycolytic enzyme hexokinase and part of the heat shock protein HSP-70. The possibility exists that these three proteins could have a common ancestor, although their amino acid sequences are quite different.[1]

Actins from different species even as far apart as plants, yeasts, and mammals are closely similar in amino acid sequence. For example, actin from *Acanthamoeba* differs from mammalian actin in only 15 out of 375 amino acids, while a comparison of skeletal muscle actin from various species of warm-blooded vertebrates fails to reveal any amino acid change whatever. This striking evolutionary conservatism, which is shown by few other proteins, is sometimes explained by the very large number of protein–protein interactions actin has to fulfill (see below). It may be difficult, it is argued, for such a protein to change its amino acid sequence without adversely affecting one or more of its multiple tasks in the cell. Certainly mutagenesis experiments with actin often show multiple biochemical effects resulting from each mutation.

## Vertebrates have a family of actin genes

Some lower eucaryotes such as yeasts have only one actin gene encoding a single protein. However, all higher eucaryotes have multiple isoforms of actin encoded by a family of actin genes and at least six types of actin are

[1] Most molecular biologists agree that the three-dimensional structures of proteins can reflect their evolutionary origin independently of their amino acid sequences. Another example discussed later in the book is the similarity in structure of the GTP-binding domains of tubulin and the signaling protein Ras (Chapter 11).

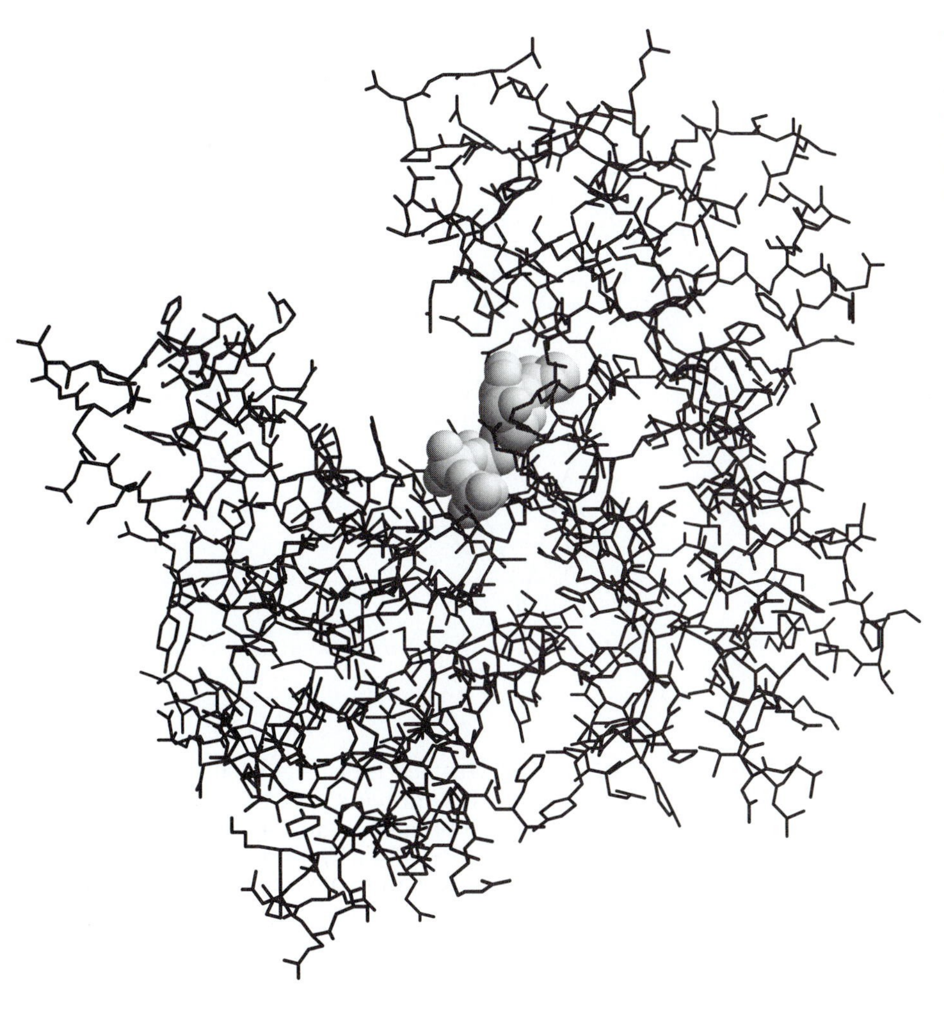

**Figure 5-1** Size and shape of an actin monomer. The protein itself is shown as a wireframe model with the positions of non-hydrogen atoms connected by lines. The ATP molecule and calcium ion associated with actin are represented in space-filling format.

present in mammalian tissues. The latter fall into three classes depending on their isoelectric point: α (the most acidic), β and γ. Each actin isoform has a distinctive tissue distribution (Table 5-1), although individual cells can contain more than one variant. Differences in amino acid sequence between the various actin isoforms are small, muscle and cytoplasmic actins typically differing by some 23 out of 375 amino acids. This difference is still greater than that between mammalian cytoplasmic γ-actin and the actin of *Acanthamoeba* or *Dictyostelium* (15 differences). Indeed, comparative sequence analysis suggests that the entire muscle gene family is derived from a single gene that arose at some time before the appearance of the chordates (Figure 5-2).

**Table 5-1 Major isoforms of mammalian actin**

| Tissue source | Designation | N-terminal sequence |
|---|---|---|
| skeletal muscle | $\alpha_{sk}$ | Asp–Glu–Asp–Glu–Thr... |
| cardiac muscle | $\alpha_{c}$ | Asp–Asp–Glu–Glu–Thr... |
| smooth muscle (aorta) | $\alpha_{sm}$ | Glu–Glu–Glu–Asp–Ser... |
| nonmuscle tissue | β | Asp–Asp–Asp–Ile... |
| smooth muscle (stomach) | $\gamma_{sm}$ | Glu–Glu–Glu–Thr... |
| nonmuscle tissue | γ | Glu–Glu–Glu–Ile... |

α-, β- and γ-actins were originally detected by isoelectric focusing. Their acidity ranking depends on the number of acidic amino acids in their N-terminal region and whether they are aspartate or glutamate (Asp is slightly more acidic than Glu).

Figure 5-2 Evolutionary origins of actin genes deduced from sequence studies. In present-day mammals, the nonmuscle actins are closest in sequence to the ancestral actin gene, whereas the muscle actins are offsprings of a single side branch. This diagram does not show the position of invertebrates, like insects, which branched off from the rest of animals very early in evolution and in which the muscle genes evolved from nonmuscle actin in a separate event.

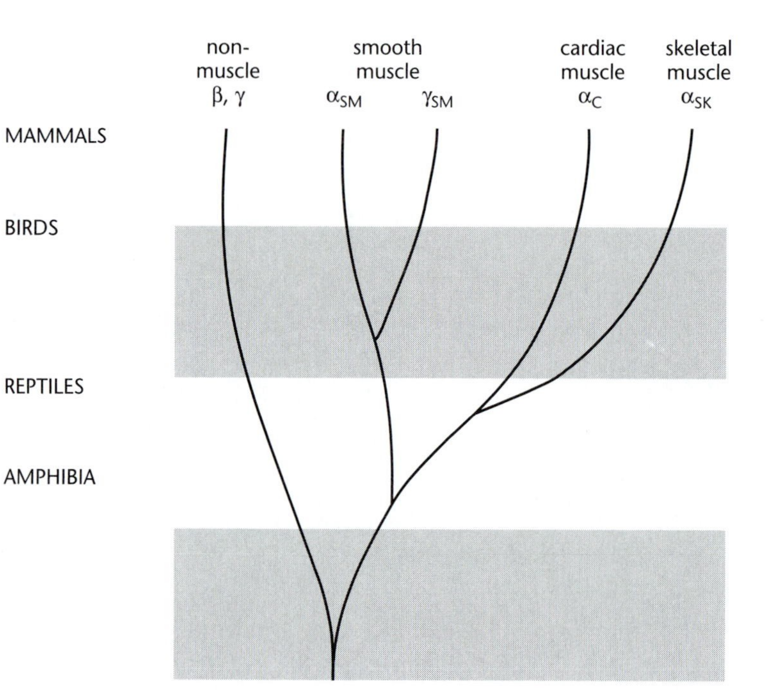


Different actin isoforms are virtually indistinguishable in their ability to polymerize or interact with other proteins, so why have they diverged during evolution? One possibility is that they are a developmental convenience: the differentiation of smooth muscle in the wall of a blood vessel, for example, could require a separate set of proteins including actin to be coordinately expressed. Subtle functional differences also probably exist between the various actins that have so far escaped detection in cell-free assays—for example, it has been found that the skeletal α-actin cannot substitute for cardiac α-actin in the mouse heart. Similar questions arise in regard to the different isoforms of tubulin (Chapter 11).

## *Actin assembles spontaneously into filaments*

Each globular actin (G-actin) monomer has binding sites on its surface through which it associates with other monomers, eventually generating a polymeric filament that may be several micrometers in length. The actin filament (also known as filamentous actin, or F-actin) can be described as a two-stranded helix in which each actin monomer contacts four other monomers, with the strongest contacts occurring along each of the two strands (Figure 5-3). Atomic models of the actin filament based on the known structure of the monomer reveal extensive hydrophobic contacts and potential hydrogen bonds between subunits, especially along the long-pitch helix.

Actin filaments have a unique polarity due to the orientation of each of their constituent actin molecules. This polarity can be detected in negatively stained images of pure actin filaments but is most readily

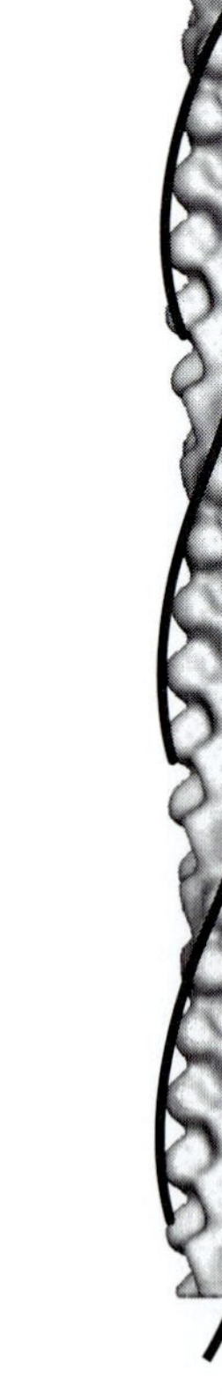


**Figure 5-3** Actin filament structure. The filament is a linear chain of actin monomers, each making contact with four neighbors. The arrangement can viewed as a two-stranded right-handed double helix that twists around itself every 37 nm. Note, however, that isolated actin filaments are not rigid bodies and that the pitch can change, for example when the filament is decorated with a second protein. (Courtesy of J.R. Bamburg, A. McGough and S. Ono, *Trends Cell Biol.* 9: 364–370, 1999.)

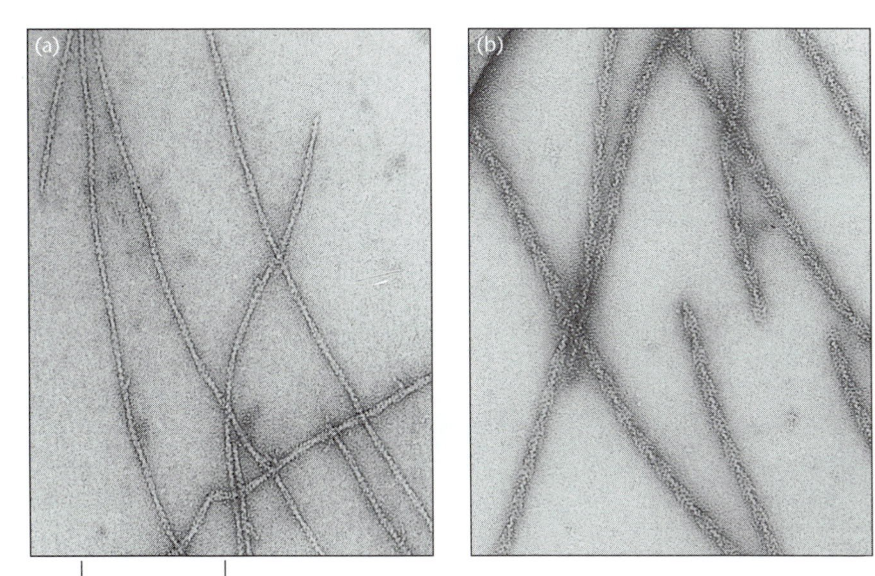


**Figure 5-4** Actin filaments decorated with myosin heads. (a) A solution of actin filaments was applied to a grid covered with a thin plastic film, stained with uranyl acetate, and viewed in an electron microscope. Note the periodic helical crossovers every 37 nm. (b) A grid carrying actin filaments, as above, was exposed to a solution of myosin before staining. Molecules of myosin bind at an angle to each actin monomer forming a series of arrowheads that point from the barbed end of the filament. (Courtesy of Roger Craig.)

demonstrated by allowing the filaments to react with fragments of myosin. Fragments of myosin containing its active 'head' region (Chapter 7) bind to the actin molecules in a filament in a sterically precise fashion, thereby forming a series of lateral projections that are tilted in one direction. Viewed in the electron microscope, these projections appear as arrowheads enabling two ends of the actin filament to be distinguished (Figure 5-4).

It may be seen in Figure 5-3 that the two strands of the actin 'double helix' are arranged with parallel polarity. This has the geometrical consequence that the two ends of an actin filament are structurally different—that is, they expose different portions of the actin monomer surface to the solvent. Thus we find that there are proteins that recognize one end or the other of an actin filament, and that the two ends have different rates of polymerization. But the same geometry that makes the two ends different also dictates that the grooves on either side of the actin filament are sterically identical. Side-binding proteins, such as tropomyosin, therefore must of necessity bind equally to both sides of the helix. By comparison, in a DNA double helix which is antiparallel, the two ends are identical but the grooves on either side (called major and minor grooves) are different and bind to different sets of proteins.

The strength of binding between actin monomers is influenced by ambient conditions such as ionic strength, so that actin preparations can be shifted between the unpolymerized and polymerized state at will. This is a useful feature in purifying actin, since filaments can be separated by physical means, such as centrifugation, from actin-binding proteins or protein contaminants. It also permits the kinetics of the polymerization process to be analyzed, using optical probes or viscosity measurements.

## *Actin filaments grow at their free ends by a reversible association of actin molecules*

A typical actin polymerization curve is shown in Figure 5-5. A solution of G-actin containing ATP is exposed to a sudden increase in ionic strength, which causes it to polymerize. There is a short delay, or *lag phase*, lasting seconds to minutes depending on the actin concentration. After this, the concentration of polymer rises steadily to a plateau value at which a low concentration of actin monomers, known as the *critical concentration*, exists in equilibrium with actin filaments. The critical concentration,

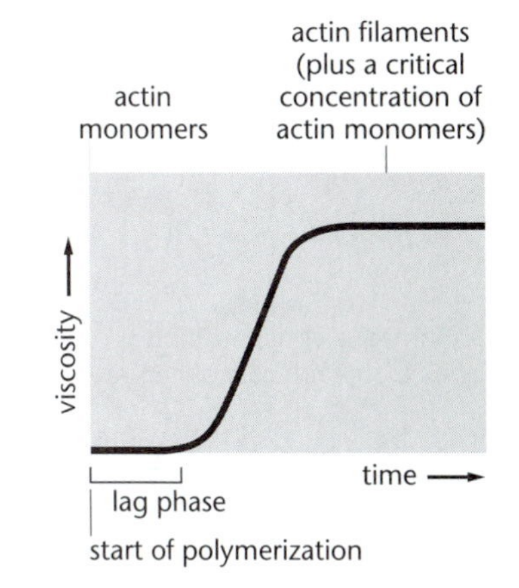


**Figure 5-5** Time course of actin polymerization as measured by viscosity. Polymerization can be induced by adding monovalent (KCl) or divalent ($MgCl_2$, $CaCl_2$) salts. As indicated here, polymerization usually proceeds after an initial lag phase of a few minutes.

**Figure 5-6** Two stages of actin polymerization. (a) The first, and usually the slower of the two stages, is nucleation—formation of an oligomer that acts as a nucleus for the further addition of monomer. (b) Once the nucleus has formed, actin monomers can add sequentially at either end to generate a long filament.

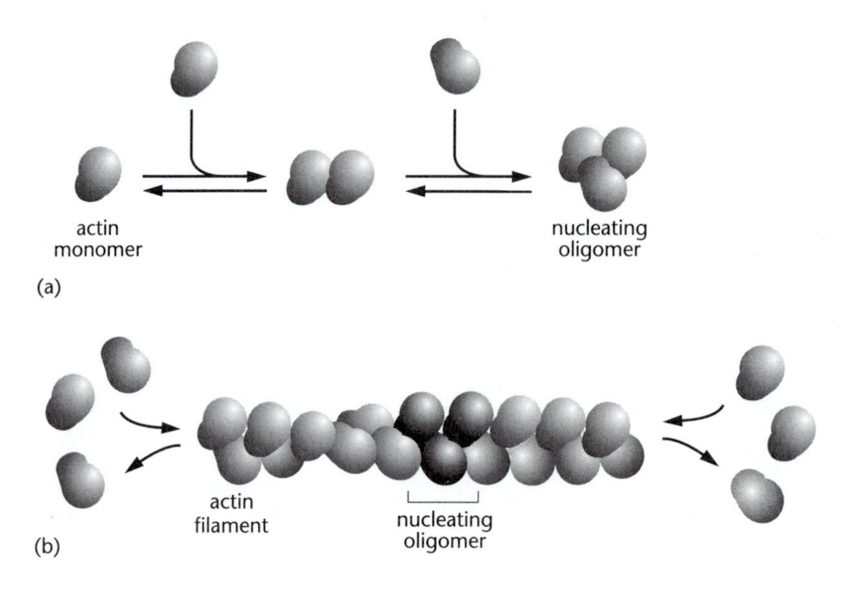


typically around 8 μg/ml or 0.2 μM, is also the threshold above which monomeric actin can spontaneously polymerize into filaments.

The mechanism of actin polymerization can conveniently be divided into two stages: (1) the assembly of monomers into nucleating seeds—short oligomers of actin that must form before the filaments can grow rapidly; (2) successive addition of monomers to the ends of the growing filament (Figure 5-6). Under most conditions, the first stage (formation of a nucleating oligomer) is rate limiting and consequently accounts for the lag phase. As one might expect from this explanation, the lag phase can be eliminated by adding preformed nuclei, such as fragments of actin filaments produced by sonication.[2]

Once actin polymerization is underway (stage 2), then the process can be described fairly simply as the reversible association of monomers at either end of the polymer. The greater the concentration of actin monomers at the start of polymerization, the faster they begin to form filaments (rate of growth = $k_{ON}$ × concentration − $k_{OFF}$). However, as growth proceeds, monomers are used up, being used to make filament, and their concentration consequently falls. Eventually monomeric actin reaches the critical concentration ($C_c$), at which monomers and filaments are at equilibrium and there is no further net polymerization.

Going in the other direction, if we start with a mixture of pure actin filaments with no monomers, the filaments will initially dissolve—that is, the initial rate of polymerization will be negative. As subunits leave the actin filaments, the concentration of actin monomers will rise until it again reaches $C_c$. Figure 5-7 summarizes the behavior of actin filaments growing at different concentrations of actin.

**Figure 5-7** Kinetic equations for polymer growth. (a) Schematic diagram showing addition of an actin monomer to the end of an actin filament.
Rate of monomer addition to filament = $k_{ON} \times C$
Rate of loss of monomer from filament = $k_{OFF}$
Net growth of the filament = $k_{ON} \times C - k_{OFF}$
where $C$ is the concentration of actin monomers. (b) Rate of growth of actin filaments at different concentrations of free actin molecules. Below the critical concentration, no filaments are formed and any that are present undergo negative growth (i.e., they depolymerize). Above the critical concentration, filaments are formed and grow by the addition of actin molecules to their ends. As the free actin molecules are used up, their concentration falls until it reaches the critical concentration. At this point the net growth is zero and free actin molecules and filaments are at equilibrium. Since at equilibrium, ON and OFF rates balance:

$$k_{ON} \times C - k_{OFF} = 0$$

or

$$C = k_{OFF}/k_{ON}$$

At equilibrium, the concentration $C$ corresponds to the critical concentration $C_c$, that is,

$$C_c = k_{OFF}/k_{ON} = 1/K$$

Where $K$ is the affinity constant of subunits for the end of the polymer.

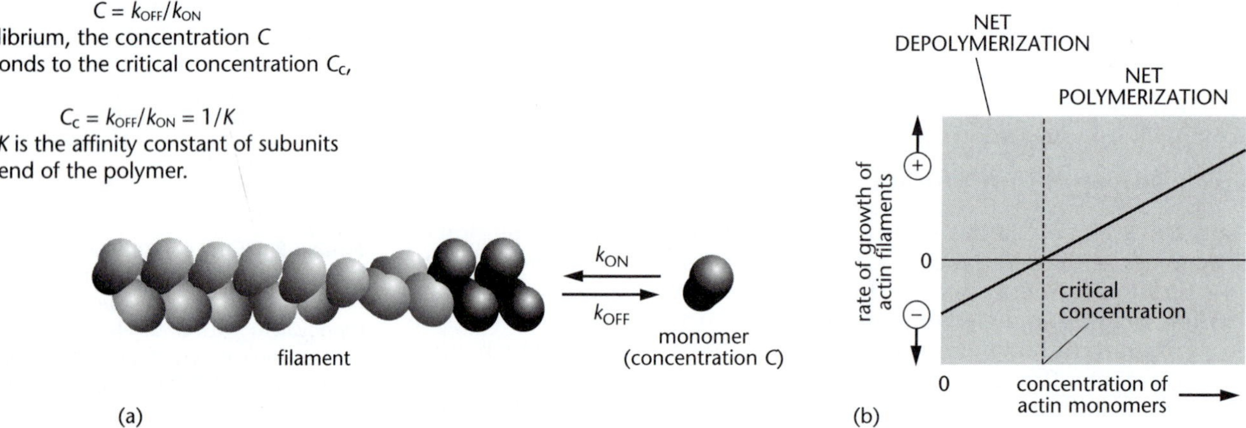

## *The two ends of an actin filament grow at different rates*

Growth at the two ends of an actin filament can be distinguished by electron microscopy. For example, if short lengths of filament decorated with myosin heads are put into a polymerizing mixture, then growth is typically 10 times faster at the barbed end than at the pointed (Figure 5-8). The same difference exists in filaments that have not been exposed to myosin heads, and the terms *barbed end* and *pointed end* are still used to refer to their fast-growing and slow-growing ends, respectively.[3]

Why do the two ends grow at different rates? The answer to this question takes us into the fascinating topic of how cells control the formation of their microfilaments and microtubules. To begin with, measurements of the rates of association and dissociation (the $k_{ON}$ and $k_{OFF}$ mentioned above) show that *both* rates are faster at the barbed end. Typical values are $k_{ON} = 5\ \mu M^{-1}s^{-1}$ and $k_{OFF} = 1\ s^{-1}$ for the barbed end, and $k_{ON} = 0.1\ \mu M^{-1}s^{-1}$ and $k_{OFF} = 0.2\ s^{-1}$ for the pointed end. Moreover, as these numbers reveal, critical concentrations ($C_c$) for the two ends also differ. Using the above equations we see that for the barbed end $C_c = 0.2\ \mu M$ while for the pointed end $C_c = 2\ \mu M$.

What is the consequence of a tenfold difference in critical concentration at the two ends? If we imagine a population of actin filaments growing from a solution of monomers, we see that they will continue to add subunits at both ends until the concentration of actin monomers falls below 2 µM, which is the critical concentration of the pointed end. At this stage the pointed end will stop growing, and indeed will begin to shrink, whereas the barbed end will continue to grow. Evidently the system will come to an equilibrium at a concentration between the two critical concentrations (actually 0.2–0.3 µM) (Figure 5-9).

If the barbed end is growing while the pointed end is shrinking, then actin molecules within the filament should move steadily from one end to the other! Indeed, experiments in which radioactively labeled actin molecules are added to an equilibrium mixture of actin monomers and filaments confirm that this movement does occur. A radioactive actin

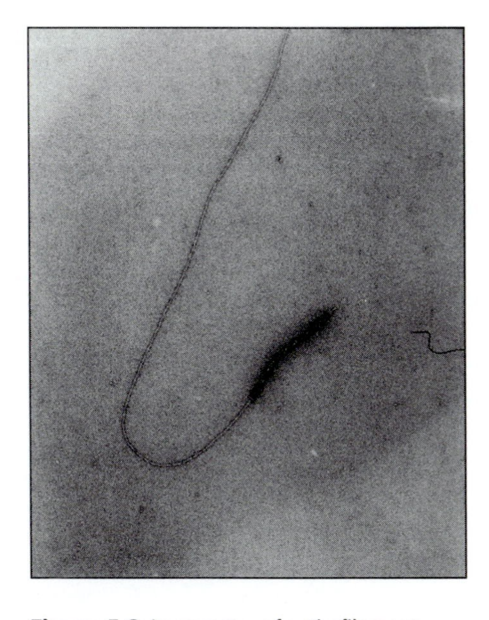

**Figure 5-8** Asymmetry of actin filament growth. If a short actin filament decorated with myosin heads is used to nucleate actin polymerization, assembly occurs much more rapidly at the barbed end of the seed filament than its pointed end. (Courtesy of M.S. Runge and T.D. Pollard.)

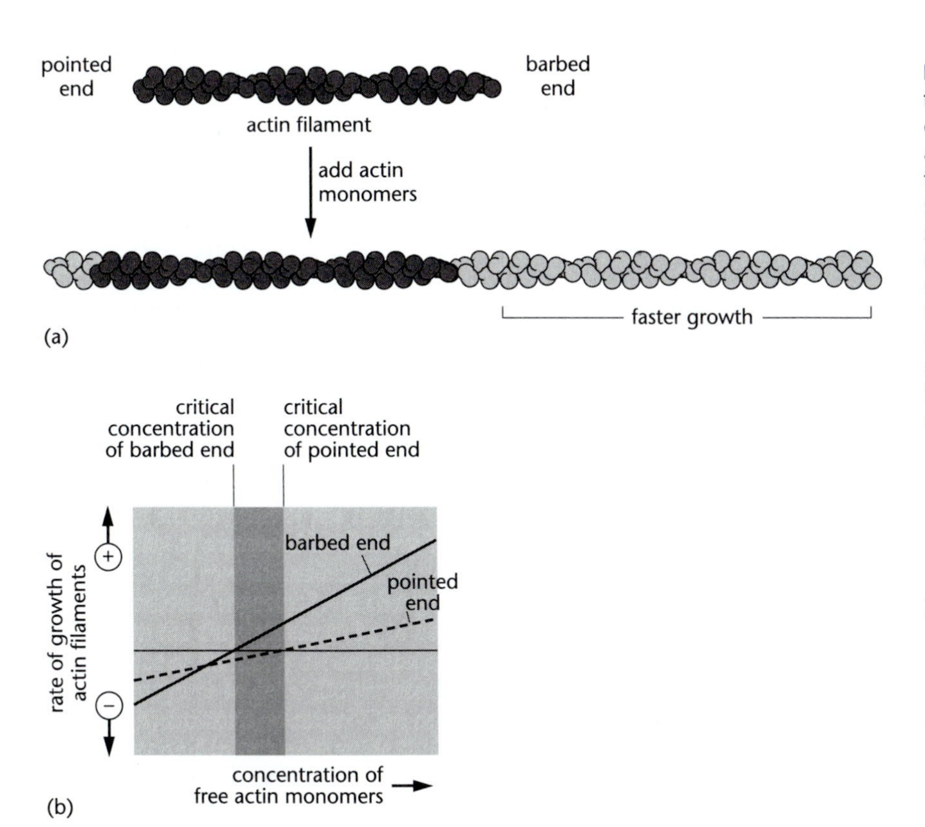


**Figure 5-9** Polymerization of actin into filaments. (a) Schematic diagram showing different rates of growth at two ends of an actin filament. (b) Rate of growth of actin filaments at different concentrations of free monomers. This graph is similar to that shown in Figure 5-7 except that the two ends of the filament have different growth rates. At a concentration of free actin above the critical concentration, the rate of growth at one end (the barbed end) is greater than at the other end (the pointed end). Note that the affinity constants (*K*) of the two ends of the filament are now different.

[2] End-to-end association of actin filaments, known as *annealing* can also take place if the concentration of filament ends is sufficiently high and may be of great significance in the cell cortex.

[3] The terms 'plus end' and 'minus end' are also used for actin filaments and microtubules. In the case of actin, *plus end* is the fast-growing, or barbed end, whereas *minus end* is the slow-growing, or pointed end, of the filament.

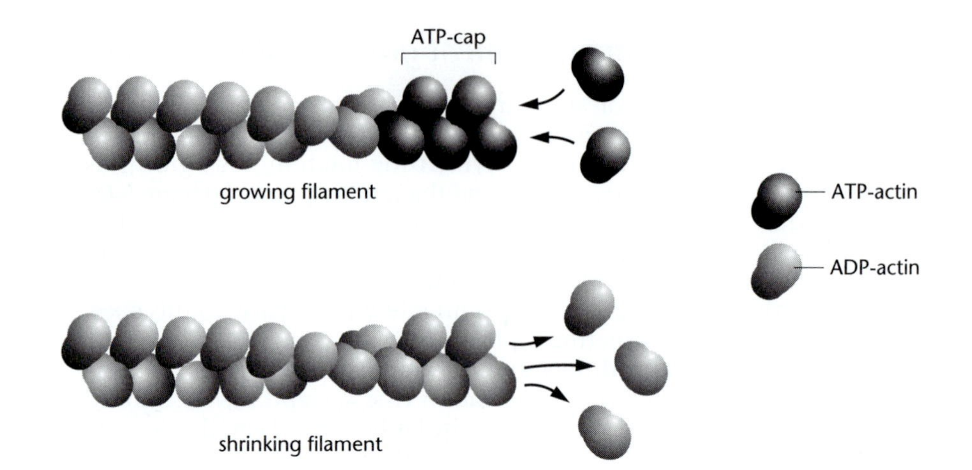


**Figure 5-10** Forming an ATP-cap at the end of an actin filament. In the rapidly growing filament, actin monomers carrying ATP add to the end of the filament faster than the ATP they carry can be hydrolyzed. A terminal segment of ATP-containing actin molecules, or ATP-cap, is therefore formed, which is itself favorable for further subunit addition. In the shrinking actin filament, the ATP-cap has been lost due to ATP hydrolysis, exposing a terminal region composed of ADP-actin. The latter has a higher critical concentration for growth and therefore tends to depolymerize.

molecule adding to the barbed end will eventually fall off its pointed end. Between these two events the labeled molecule will progress from one end of the filament to the other, a movement termed *treadmilling.*

How treadmilling occurs, and where the energy to drive it comes from, are our next concerns.

## *ATP hydrolysis changes the rate of growth of actin filaments*

ATP hydrolysis plays a curious role in actin polymerization that was not understood for many years. It is not essential for filament formation as such, since actin carrying the 'low-energy' form ADP (or even actin without any bound nucleotide) can assemble into filaments. But the rate of filament formation is much slower with ADP than ATP, and actin filaments do not treadmill if only ADP is present. In the absence of ATP, filaments behave as equilibrium polymers with identical $C_C$'s for both ends. It now seems likely that the bound nucleotide has a subtle influence on the conformation of the actin molecule. Of the two conformations, ATP-actin packs most stably into the lattice of the filament and polymerizes most readily, especially at the barbed end.

Consider just one end of an actin filament. At all concentrations above the critical concentration, actin-ATP molecules will add to a free end faster than the ATP they carry can be hydrolyzed. The newly-added subunits will therefore create a region at the end of the growing filament made of actin molecules bound to ATP—an *ATP-cap* (Figure 5-10). But as the concentration of free actin monomers decreases (being used to make filaments) the rate of addition of new monomers will decrease. As the hydrolysis reaction catches up with the addition of new monomers, the ATP-cap becomes smaller in size. If it is actually lost, then ADP-actin subunits will be exposed at the filament end and are likely to dissociate.

The details are complex and not fully understood. But in broad terms it seems that the barbed end, because of its more rapid rate of subunit addition, has an ATP-cap even at steady state. The rate of dissociation from this end will be relatively slow, reflecting the superior packing of ATP-actin into the filament lattice. At the pointed end, by contrast, hydrolysis will be faster than subunit addition, so no ATP cap will form.[4] This end will lose subunits because ADP-actin does not pack so well into the filament.

We see that the conformational change induced by nucleotide hydrolysis causes the two ends of the actin filament to undergo essentially different binding reactions, with different critical concentrations for actin subunits. This difference in binding reactions will be maintained by the hydrolysis of actin-bound ATP, which therefore provides the energy necessary to drive the treadmilling of subunits through the polymer.

[4] In reality, each ATP-cap grows and shrinks in a stochastic fashion as monomers come and go and bound nucleotides are hydrolyzed. These fluctuations are small and usually masked by the activity of actin-binding proteins. However, similar fluctuations in growing microtubules are much greater in extent and have a profound influence on their assembly in the living cell (Chapter 11).

**Table 5-2 Toxins that bind to actin**

| Compound | Source | Action |
|---|---|---|
| cytochalasins A, B, C, D, E | A family of alkaloids secreted by various fungi. | Bind to the ends of actin filaments and prevent further polymerization. |
| phalloidin | One of a family of toxic components from the mushroom *Amanita phalloides.* | Phalloidin binds to actin filaments and stabilizes them against depolymerization. A fluorescent derivative, rhodamine phalloidin, is used to stain actin filaments in permeabilized cells. |
| latrunculin | One of a family of cyclic compounds from the Red Sea sponge *Latruncula magnifica.* | The latrunculins disrupt actin filaments by binding to actin monomers, forming a tight 1:1 complex. |

## *Specific toxins bind to actin and affect its polymerization*

Clear evidence for the important role played by actin polymerization in cell movements comes from the action of a number of actin-binding toxins. In the test tube, these substances bind to actin and disrupt the normal polymerization kinetics: when applied to a cell they paralyze normal movements and cause rapid changes in cell morphology. These toxins include the cytochalasins, the phalloidins, and the latrunculins (Table 5-2 and Figure 5-11).

The *cytochalasins* are a group of closely related fungal metabolites secreted by different species of molds. They bind to the barbed ends of actin filaments and prevent elongation, an effect that is specific to actin filaments—cytochalasins do not bind to microtubules, for example. When added to cells at low concentrations (typically 1 μg/ml or less) cytochalasins rapidly paralyze certain kinds of vertebrate cell movements, including cell locomotion and the division of the cell in cytokinesis. They do not, however, inhibit mitosis, which depends mainly on microtubules, nor do they prevent muscles from contracting, since this process involves actin filaments in a stable form and therefore does not entail assembly and disassembly.

*Phalloidin*—one member of a group of cyclic peptides from poisonous mushrooms—promotes actin polymerization rather than preventing it. This compound binds to actin molecules in a filament more strongly than to actin monomers; model building indicates that each phalloidin contacts three actin subunits simultaneously, thereby providing stability to the filament. Phalloidin shifts the equilibrium between filaments and monomers toward filaments, lowering the critical concentration for polymerization by 10-fold to 30-fold. Unlike the cytochalasins, phalloidin is not readily taken up by living cells and has to be injected before it can

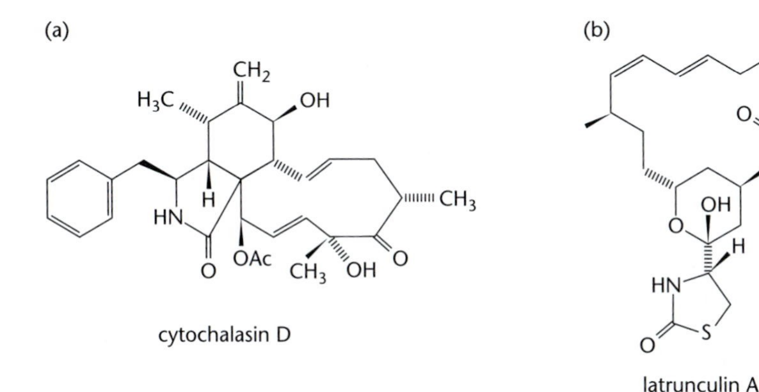


**Figure 5-11** Chemical structure of cytochalasin and latrunculin.

take effect. When this is done, however, it is found that the same range of movements that are affected by cytochalasin are also inhibited by phalloidin, showing that depolymerization is just as important as polymerization for the generation of these movements. A fluorescent derivative of phalloidin, called rhodamine phalloidin, is often used to stain actin filaments.

A number of compounds that act on actin have been isolated from a Red Sea sponge. The first of these to be characterized was *latrunculin*, which is taken up by intact vertebrate and yeast cells and acts as rapidly as cytochalasins on processes that depend on actin filament polymerization. However, latrunculin does so by depolymerizing actin filaments directly, forming tight 1:1 complexes with actin monomers, rather than capping their free ends like cytochalasin. Related compounds from the same sponge also affect actin polymerization, which has evidently been an important target of action for this organism (as for pathogenic bacteria, see below).

## Thymosin sequesters unpolymerized actin

The concentration of unpolymerized actin in the cytosol of most cells is very high—typically around 5 mg/ml, or 100 μM. By contrast, the critical concentration for polymerization in a simple ionic solution is almost a *thousandfold* smaller—around 8 μg/ml or 0.2 μM. Moreover, turnover of subunits with actin filaments is 100–200-fold faster in cells than for pure actin solutions. These huge differences are due to proteins in the cytoplasm that regulate actin polymerization, including several proteins that bind to actin monomers.

In higher animals, the most abundant monomer-binding factor is *thymosin*, a small, highly charged protein (or, according to some definitions, a peptide) composed of 43 amino acids. Thymosin was originally isolated from calf thymus, but has subsequently been found in most types of animal cells, often at astonishingly high concentrations (over 0.5 mM in blood platelets). Thymosin sequesters a large pool of G-actin within cells in such a way that it can be readily released when needed for the polymerization of actin filaments. It is able to do this because of its high concentration and because of its relatively weak binding to monomeric actin ($K_d$ about 0.7 μM). Thus thymosin is able to buffer the actin monomer concentration rapidly even during episodes of rapid actin polymerization.

The type of thymosin found in platelets, known as thymosin-$\beta_4$, has been extensively studied and its structure determined by NMR. In solution, thymosin-$\beta_4$ has a poorly defined structure, probably the result of multiple interconverting conformations, but when it forms a complex with actin it adopts a defined structure. This disorder–order transition has subtle but important implications for the energetics of complex formation and provides an intriguing insight into the basis of its buffering capacity. Because the dissociation reaction is accompanied by an increase in the number of accessible states (an increase in entropy) it occurs more readily than would otherwise would be the case.[5]

## Profilin and cofilin regulate actin assembly into filaments

One of the first actin-monomer-binding proteins was isolated by Tilney and colleagues in the 1970s from the sperm of the sea cucumber *Thyone*. A membrane sac situated above the nucleus contains high concentrations of monomeric actin in a complex with a small second protein. When the head of the sac bursts and the sperm contacts the jelly surrounding the egg, the sperm rapidly grows a long spiky process filled with actin filaments that penetrates the egg jelly. This spike, which is 50–90 μm in length and 0.05 μm in diameter, and contains 60 or so longitudinally aligned actin filaments, grows out in under 10 seconds.

[5] A similar situation exists in certain types of antibody molecules, such as the T-cell receptor. As with thymosin, the flexibility of the nonliganded antibody molecule enables its binding to be both specific and weak, a rare combination of properties.

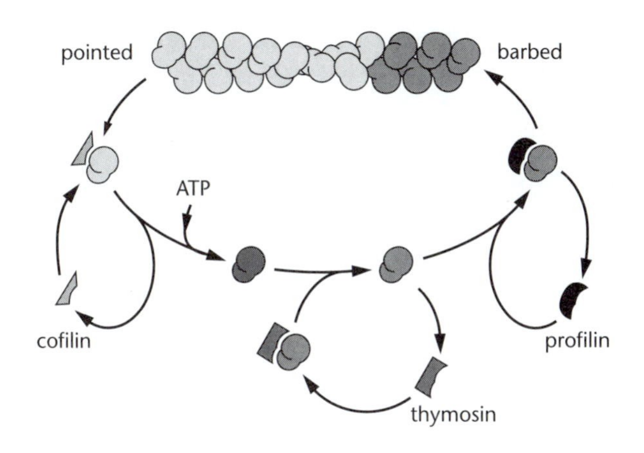


**Figure 5-12** Actin polymerization in the cell. Cofilin (ADF) works mainly at the pointed end, stimulating dissociation, whereas profilin speeds up monomer addition to the barbed end. Thymosin maintains a reserve, or buffer, of actin monomers.

The complex of proteins in the *Thyone* sperm was given the punning name 'profilactin' because of its association with fertilization and apparent role in regulating actin filament formation. The protein associated with actin in this complex was therefore called *'profilin.'* Many other profilins—all small proteins, with molecular weights around 14–15 kDa—have since been found widely distributed in both vertebrate and invertebrate cells. One of the most bizarre locations is in the pollen grains of certain plants, in which profilin is so abundant that it is the principal cause of pollen allergies.

Although profilin forms a dimeric complex with actin monomers, there is only enough of it in most vertebrate cells to sequester a small fraction of the actin pool. Furthermore, the association has interesting properties that imply a more subtle role than that of simple actin buffering. Profilin catalyzes exchange of actin-bound nucleotides (ADP and ATP) and promotes transfer of actin subunits from thymosin to the barbed (plus) end of actin filaments. Furthermore, profilin has a binding site for $PIP_2$, a phospholipid extensively used in membrane signaling (see Figure 4-12), and actin binding is inhibited in the presence of this lipid. Many cell biologists believe that the main function of profilin is to control the growth of actin filaments close to membranes, as discussed in the next chapter.

The third major category of actin-monomer binding protein in vertebrate cells[6] is *cofilin* (also known as actin-depolymerizing factor, or *ADF*). Like profilin and thymosin, cofilin forms a dimeric complex with G-actin, but it differs from these two in the strength of its binding. In fact, cofilin associates so strongly to actin that it changes the helical pitch of actin filaments, producing a helix with crossovers every 27 nm instead of 37 nm, and can even sever the filaments. The activity of cofilin is regulated by phosphoinositides, pH, and phosphorylation, the addition of a phosphate group causing an inhibition of actin binding. Inhibition (phosphorylation) is catalyzed by a specific kinase, which itself is stimulated by a small GTP-containing protein called Rac (discussed in the next chapter) and activation of cofilin is due to the action of a phosphatase. Cofilin seems to have its primary action in stimulating *disassembly* of actin monomers from the pointed end of actin filaments. In this view, cofilin and profilin have complementary roles in controlling assembly of actin monomers onto filaments, whereas thymosin maintains a reserve of monomers (Figure 5-12).

## *Pathogenic bacteria provide a simplified model of actin nucleation*

Although addition of actin molecules to the end of a filament is an important factor controlling actin polymerization, it is by no means the only one. In particular, it may be recalled that for pure actin the slowest

[6] Other examples exist. For example, the enzyme DNAase I, made in large amounts in the pancreas, binds tightly and specifically to G-actin. The function of this interaction is not fully understood, but probably has more to do with controlling the enzyme (which is inhibited in the complex) than controlling actin.

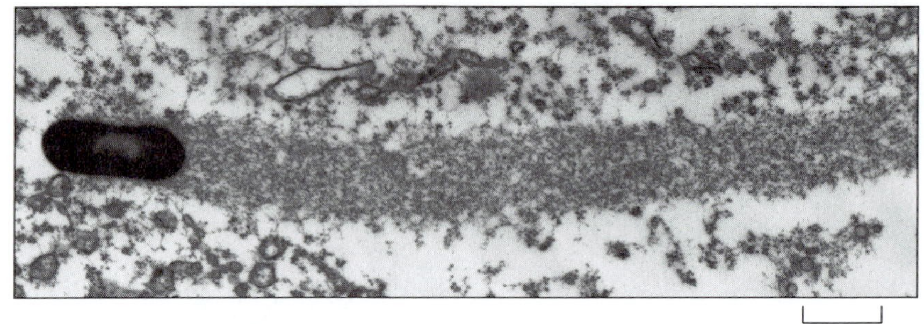


**Figure 5-13** Actin-based movement of a bacterium. The bacterium *Listeria monocytogenes* spreads from cell to cell by inducing the assembly of actin filaments in the host cell cytosol. In this electron micrograph the rod-shaped bacterium is at the apex of a comet tail of interconnected actin filaments. (Courtesy of Pascale Cossart.)

step is usually nucleation, so that once a small cluster of actin monomers has formed then polymerization proceeds relatively rapidly (see Figure 5-6). In a living cell, as we have just seen, unwanted nucleation in the cytoplasm is suppressed by thymosin and profilin. But there are specific locations (typically close to the plasma membrane) where the kinetic barrier to polymerization is overcome by clusters of proteins. By catalyzing formation of a nucleating oligomer of actin, clusters of this kind can determine precisely where and when filaments form. Intriguing clues to the composition of these nucleating complexes have come, unexpectedly, from an analysis of the mechanism of infection used by pathogenic bacteria.

*Listeria monocytogenes*, a bacterium that infects cattle and can cause a severe form of food poisoning in humans, enters cells by being phagocytosed. It escapes into the host cell cytosol by secreting enzymes that break down the membrane of the phagosome, and then hijacks the actin-based motility system of the host cell to spread to adjoining cells. By nucleating actin filaments at one region of its surface, an individual bacterium is able to move through the cytosol at rates of 10 μm/min or more, leaving behind a tail of actin filaments (Figure 5-13). When the moving bacterium collides with the plasma membrane of the host cell, it keeps moving outward, inducing the formation of a long, thin filopodium with the bacterium at its tip. Engulfment of this projection by a neighboring cell allows the bacterium to enter the new cytoplasm without exposure to the extracellular environment, thereby avoiding recognition by antibodies produced by the host.

## *VASP and Arp2/3 nucleate actin filaments*

Some of the most important developments in biology have come from unexpected sources. Thus, the study of the infectious cycle of *Listeria* has provided essential insight into the general question of the polymerization of actin filaments and the generation of cell extensions. The bacterial system is uniquely accessible to analysis because it can be studied in cell-free extracts and because of the ease with which bacterial genotypes can be manipulated. It is also constitutive, so that there is no need to consider complicated signaling reactions that regulate other kinds of actin-based movements. The only protein the bacterium itself has to synthesize in order to induce actin filaments is ActA, which then recruits proteins from the cytoplasm to form nucleation centers on the bacterial surface.

One host-encoded protein shown to be essential for *Listeria* motility is VASP (vasodilator-stimulated phosphoprotein)—a protein originally identified in blood platelets, where it is involved in control of actin polymerization. This protein contains several polyproline sequences to which profilin binds tightly (in fact, profilin can be purified by passing extracts through a polyproline column). VASP itself binds to ActA on the surface of the bacterium and thus stimulates actin filament polymerization. A second family of proteins also found in the nucleation complex is called Arps (actin-related proteins). These proteins are related in structure and

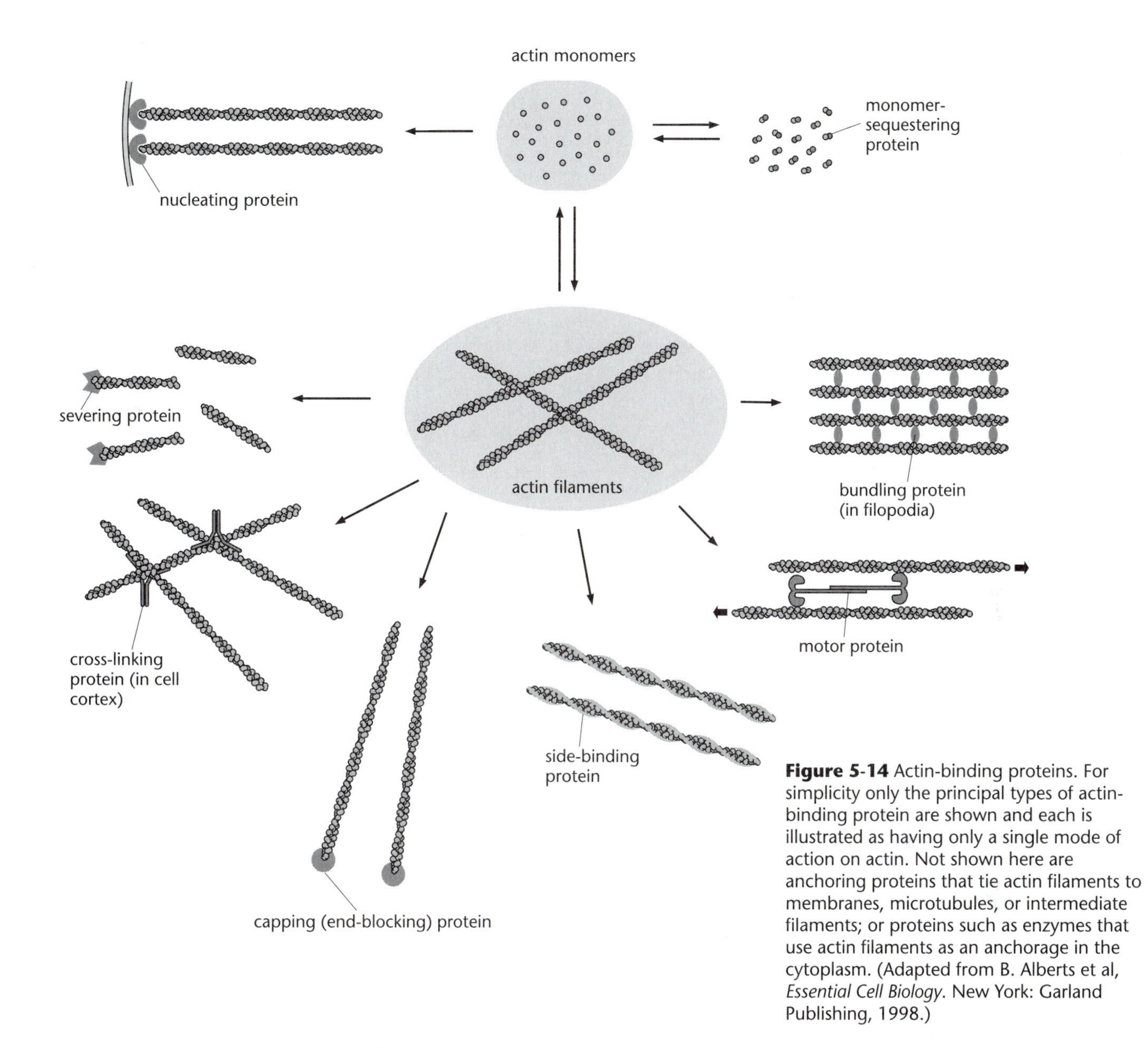


**Figure 5-14** Actin-binding proteins. For simplicity only the principal types of actin-binding protein are shown and each is illustrated as having only a single mode of action on actin. Not shown here are anchoring proteins that tie actin filaments to membranes, microtubules, or intermediate filaments; or proteins such as enzymes that use actin filaments as an anchorage in the cytoplasm. (Adapted from B. Alberts et al, *Essential Cell Biology*. New York: Garland Publishing, 1998.)

sequence to actin itself, as their name suggests, and a complex containing two Arps and five other subunits forms a 'nucleation machine' on the surface of the bacterium.

Since all of the proteins except for ActA come from the host cell, it seems likely that they will be involved in actin polymerization elsewhere. Indeed, VASP and the proteins of the Arp2/3 complex can all be detected in noninfected cells, often located in a region near the plasma membrane. In migrating cells, these proteins are believed to provide the essential step in nucleating actin filaments at the leading edge of the cell, the basis by which the cell advances (Chapter 8).

## *Actin-binding proteins have a modular construction*

Monomer binding and filament nucleation do not exhaust the repertoire of actin-binding proteins. As we will see in the next few chapters, cells also contain proteins that cap the ends of actin filaments, proteins that sever filaments into short fragments, and proteins that stabilize filaments by lateral associations or link neighboring filaments together (Figure 5-14). More than 60 actin-binding proteins have been identified in animal cells (Table 5-3), and these can work in concert in their association with actin,

**Table 5-3 Actin-binding proteins in vertebrates**

| Protein | Effect on actin | MW (kDa) |
|---|---|---|
| MONOMER BINDING | | |
| thymosin | buffers G-actin | 5 |
| profilin | regulates barbed end | 16 |
| cofilin/ADF | regulates pointed end, severs filament | 15–19 |
| END-BINDING | | |
| capping protein | caps barbed end of filament | 30 + 34 |
| gelsolin | caps barbed end, severs filament | 90 |
| tropomodulin | caps pointed end | 40 |
| villin | severs and cross-links in microvilli | 95 |
| Arp2/3 | caps pointed end, nucleates | 44/50 |
| SIDE-BINDING | | |
| tropomyosin | stiffens actin filaments | 35(× 2) |
| troponin | calcium regulation in muscle | |
| α-actinin | cross-links filaments | 100(× 2) |
| fimbrin | binds actin filaments | 68 |
| MOTORS | | |
| myosin I | cell motility/microvilli function | 120 + 17(× 3) |
| myosin II | muscle contraction/cytokinesis/cell polarity | 200(× 2) + 17(× 2) |
| myosin V | vesicle transport | 200 |
| MEMBRANE-BINDING | | |
| spectrin | links to membranes/ankyrin | 250 + 280 |
| ezrin/moesin/radixin | links to membranes/band 4.1 | 68 |
| band 4.1 | links to membranes/spectrin | 80 |
| ponticulin | links actin to membranes | 17 |

Partial list of some of the most abundant actin-binding proteins present in vertebrate cells; many others exist. Those listed are often present in multiple varieties or isoforms. Molecular weights are given as the size of individual polypeptide chains multiplied by the number of copies and are only approximate.

with one either inhibiting or encouraging binding of another. The binding of cofilin and that of tropomyosin, for example, are mutually exclusive, whereas tropomyosin *promotes* interaction with the motor protein myosin (Chapter 7).

One way to make sense of the plethora of actin-binding proteins is to examine their domain structure. All actin-binding proteins, by definition, possess at least one actin-binding domain. Many also possess a regulatory domain that binds $Ca^{2+}$ or the specific phospholipid $PIP_2$. Combinations of these and other functional domains by genetic exchange are believed to have led to the evolution of the large family of actin-binding proteins.

Proteins that contain at least two actin-binding sites for F-actin are able to form a link between two subunits, either within a single filament or between one filament and its neighbor. In the latter case, the filaments will become bundled or cross-linked. The difference between bundling and cross-linking is not great but depends on the length of the protein between the two actin-binding domains and their strength of binding. Thus the small protein *fimbrin* forces actin filaments to lie in parallel arrays, whereas the long and flexible protein *filamin*, which is composed of a series of long flexible subunits that incorporate a dimerization domain at one end and an actin-binding domain at the other, can form loose three-dimensional gels of actin filaments. *Spectrin* and *α-actinin* are like filamin in that they tend to produce loose networks in which adjacent actin filaments have unrelated (or in some cases antiparallel) orientations.

Interestingly, an actin-binding protein called *cortexillin* from the slime mold *Dictyostelium* combines binding domains similar to spectrin with a coiled-coil domain that forms a parallel dimer. Cortexillin is enriched in the cortex of *Dictyostelium* amoebae and has the capacity to bundle actin filaments preferentially in antiparallel orientation. Mutants lacking this protein are severely impaired in their ability to divide (specifically in cytokinesis).

## *Actin filaments are an important site of protein synthesis*

Actin filaments permeate the interior of eucaryotic cells, providing an interface with the cytosol that is much greater in extent than that of cell membranes (see Figure 4-3). Consequently, any molecule that functions more effectively by being anchored to a structure may do so by evolving a binding site to actin. Indeed, it turns out that some enzymes that we normally think of as being soluble, such as the enzymes of glycolysis, are in all likelihood bound to actin filaments in the cell (Chapter 18). It has also been found that proteins that mark specific locations in a cell, such as the site of budding in a yeast cell (Chapter 19) or the vegetal pole of a frog egg (Chapter 20), are typically associated with actin filaments in the cell cortex.

A similar argument holds for protein synthesis. In the early 1980s, experimenters discovered that if vertebrate cells are extracted with nonionic detergent then much of the cell's messenger RNA (mRNA) is left behind in the residue. In other words, many mRNA molecules are linked to the cytoskeleton—an association now realized to be of general significance. In a typical situation, mRNA molecules are transported in a compact and nonfunctional form along microtubules or actin filaments to a specific location. Once delivered to their target, the mRNA associates with ribosomes, tRNAs, and translation factors, thereby directing the synthesis of the protein it encodes.[7]

The molecular mechanism of RNA targeting is still being worked out. Many localized mRNAs contain specific targeting sequences ('zip codes') in their 3′ untranslated region, which are both necessary and sufficient to direct mRNAs to their target. Proteins that recognize these sequences, transport the RNA, anchor it, and control its expression and stability are now being identified.

Which cell proteins are made by polysomes attached to actin filaments? The list includes cytosolic proteins, membrane-associated proteins, and cytoskeletal proteins. Most newly-synthesized actin molecules in a fibroblast, for example, appear first on the meshwork of actin filaments in the cell cortex. In a developing muscle cell that makes both muscle and nonmuscle isoforms of actin, these are synthesized at different locations, β-actin being made in the leading lamellipodia and α-actin close to the nucleus. Another example is seen in oligodendrocytes, the cells in the central nervous system that make myelin. The mRNA encoding abundant myelin basic protein is localized to the peripheral processes of these cells, where myelination actually occurs. In these and other cases, mRNAs are targeted to their correct destinations by proteins that recognize specific sequences in the messenger RNA molecule. What these proteins are and how they work are areas of intense investigation.

## *Outstanding Questions*

*How do different actin isoforms differ in function (if at all)? What are the detailed molecular events occurring at the ends of an actin filament in a solution containing actin monomers and ATP? By what mechanism does an actin molecule hydrolyze its bound ATP? Why do specific fungi and sea sponges make substances that inhibit actin polymerization? Why does pancreatic DNase bind to actin? How are different populations of actin filaments in the same cytoplasm distinguished? What competitive and cooperative interactions occur between actin-binding proteins? How do specific mRNA molecules associate with actin filaments and direct protein synthesis?*

[7] It is intriguing that elongation factor EF-1, one of the most abundant proteins of the eucaryotic cytosol (and a relative of the bacterial EF-Tu) binds strongly to actin filaments and can produce actin bundles under defined conditions.

## *Essential Concepts*

- Actin is abundant in all eucaryotic cells. Vertebrate species have a family of closely similar actin genes, expressed selectively in different tissues.
- Actin exists as a monomer in low-ionic-strength solution but spontaneously assembles into filaments on addition of salt.
- Polymerization is regulated by ATP, which binds to (and stabilizes) actin monomers. Bound ATP is hydrolyzed to ADP as polymerization proceeds.
- The kinetics of polymerization reveal a simple mechanism of addition of monomers to filament ends, with the two ends of an actin filament growing at different rates.
- Major shifts in actin polymerization occur during certain cell shape changes. There is evidence, from the use of drugs such as cytochalasin, that growth and shrinkage of actin filaments occur continuously in the course of many movements.
- Three types of protein control the pool of actin monomers in vertebrate cells. Thymosin sequesters large quantities of actin in a reserve form; profilin suppresses unwanted nucleation and promotes filament growth at the barbed end; cofilin accelerates depolymerization at the pointed end.
- In the cell, nucleating complexes of proteins determine where and when actin filaments form. Important clues to the identity of these proteins come from studies of pathogenic bacteria that commandeer the cell's actin machinery for their own infective cycle.
- The length and number of actin filaments and their association with other components of the cell are controlled by a remarkably large collection of associated proteins, each of which binds to actin and influences its behavior in specific ways.
- Many proteins are made by mRNA molecules targeted to actin filaments. These include actin itself and proteins responsible for cell polarity.

## *Further Reading*


Bamburg, J.R. Proteins of the ADF/cofilin family: essential regulators of actin dynamics. *Annu. Rev. Cell Dev. Biol.* 15: 185–230, 1999.

Bassell, G., Singer, R.H. mRNA and cytoskeletal filaments. *Curr. Opin. Cell Biol.* 9: 109–115, 1997.

Bubb, M.R., et al. Swinholide-A is a microfilament disrupting marine toxin that stabilizes actin dimers and severs actin-filaments. *J. Biol. Chem.* 270: 3463–3466, 1995.

Carlier, M.F., Pantaloni, D. Actin assembly in response to extracellular signals: role of capping proteins, thymosin $\beta_4$ and profilin. *Semin. Cell Biol.* 5: 183–191, 1994.

Chen, H., et al. Regulating actin filament dynamics *in vivo*. *Trends Biochem. Sci.* 25: 19–23, 1999.

Cooper, J.A. Effects of cytochalasin and phalloidin on actin. *J. Cell Biol.* 105: 1473–1478, 1987.

Elzinga, M., et al. Complete amino-acid sequence of actin of rabbit skeletal muscle. *Proc. Natl. Acad. Sci. USA* 70: 2687–2691, 1973.

Geese, M., et al. Accumulation of profilin II at the surface of *Listeria* is concomitant with the onset of motility and correlates with bacterial speed. *J. Cell Sci.* 113: 1415–1426, 2000.

Hatano, S., Oosawa, F. Isolation and characterization of plasmodium actin. *Biochim. Biophys. Acta.* 127: 488–498, 1966.

Ishikawa, H., et al. Formation of arrowhead complexes with heavy meromysin in a variety of cell types. *J. Cell Biol.* 43: 312–328, 1969.

Kabsch, W., et al. Atomic structure of the actin:DNase I complex. *Nature* 347: 37–44, 1990.

Kislauskis, E.H., et al. Isoform-specific 3′-untranslated sequences sort α-cardiac and β-cytoplasmic actin messenger RNAs to different cytoplasmic compartments. *J. Cell Biol.* 123: 165–172, 1993.

Kocks, C., et al. *L. monocytogenes*-induced actin assembly requires the ActA gene product, a surface protein.

*Cell* 68: 521–531, 1992.
Kreis, T., Vale, R. Guidebook to the Cytoskeletal and Motor Proteins. Oxford, UK: Oxford University Press, 1999.
Lappalainen, P., Drubin, D.G. Cofilin promotes rapid actin filament turnover *in vivo*. *Nature* 388: 78–82, 1997.
Lenk, R., et al. A cytoskeletal structure with associated polyribosomes obtained from HeLa cells. *Cell* 10: 67–78, 1977.
Loisel, T.P., et al. Reconstitution of actin-based motility of *Listeria* and *Shigella* using pure proteins. *Nature* 401: 613–616, 2000.
McGough, A., et al. Cofilin changes the twist of F-actin: implications for actin filament dynamics and cellular function. *J. Cell Biol.* 138: 771–781, 1997.
Pollard, T.D. Rate constants for the reacions of ATP- and ADP-actin with the ends of actin filaments. *J. Cell Biol.* 103: 2747–2754, 1986.
Rosenblatt, J., et al. *Xenopus* actin depolymerizing factor/cofilin (XAC) is responsible for the turnover of actin filaments in *Listeria monocytogenes* tails. *J. Cell Biol.* 136: 1323–1332, 1997.
Safer, D., Nachmias, V.T. Beta thymosins as actin binding peptides. *Bioessays* 16: 473–479, 1994.
Sheterline, P., et al. Actin. Oxford, UK: Oxford University Press, 1998.
Spector, I., et al. Latrunculins—novel marine macrolides that disrupt microfilament organization and affect cell growth. *Cell Motil. Cytoskeleton* 13: 127–144, 1989.
Spooner, B.S., et al. Microfilaments and cell locomotion. *J. Cell Biol.* 49: 595–613, 1971.
Straub, F.B. Studies from the Institute of Medical Chemistry University of Szeged. Vol. 2. New York: Karger, 1942.
Sun, H.Q., et al. Actin monomer binding proteins. *Curr. Opin. Cell Biol.* 7: 102–110, 1995.
Theriot, J.A. The cell biology of infection by transcellular bacterial pathogens. *Annu. Rev. Cell Dev. Biol.* 11: 213–239, 1995
Tilney, L.G., Inoué, S. The acrosomal reaction of *Thyone* sperm. II. The kinetics and possible mechanism of acrosomal process elongation. *J. Cell Biol.* 93: 820–827, 1982.
Vandekerckhove, J., Weber, K. Chordate muscle actins differ distinctly from invertebrate muscle actins. *J. Mol. Biol.* 179: 391–413, 1984.
Wegner, A. Head-to-tail polymerization of actin. *J. Mol. Biol.* 108: 139–150, 1976.
Welch, M.D., et al. Interaction of the human Arp2/3 complex and the *Listeria monocytogenes* ActA protein in actin filament nucleation. *Science* 281: 105–108, 1998.

# CHAPTER 6

# Actin and Membranes

Although actin is present throughout the cell, it is most highly concentrated in the region close to the plasma membrane. As a broad generalization (to which there are many exceptions), actin can be thought of as forming a peripheral layer around the cell, like an elastic stocking or a muscular skin. This layer, known as the *cell cortex*, gives the outer surface of the cell mechanical strength and enables it to move. Thanks to its actin cortex, an animal cell can produce extensions of its surface, crawl, engulf particles, and deform its external shape, as it does when it undergoes division. The invention of the cortex during cellular evolution, like the appearance of an opposable thumb in the evolution of primates, provided the necessary basis for the emergence of an entire range of properties. Indeed, all of the most distinctive features of higher animals, such as their complex internal anatomy, their immunity to infection and their complex nervous systems, depend on the actin cortex.

In this chapter we describe the principal molecular components of the cell cortex, and some of the specialized structures it forms. We will find that this part of the cytoskeleton is highly complicated in structural and biochemical terms and that there is still much to learn.

## *Filamin and α-actinin cross-link actin filaments*

The largest and most clearly defined membrane cortices are seen in free-living amoebae and in those vertebrate cells that are 'free living' in the sense that they migrate independently through tissues. The cortex of a human macrophage, for example, is a layer some 0.2 to 0.5 μm thick beneath the plasma membrane that excludes most cytoplasmic organelles (Figure 6-1). It is made of a dense three-dimensional network of filaments tightly adherent to the plasma membrane that can be co-isolated with it

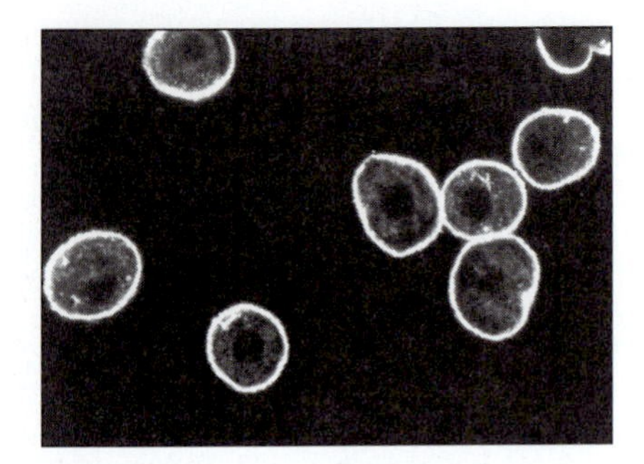

**Figure 6-1** Actin cortex of rat liver cells. The cells were isolated from the liver and attached to a glass coverslip before being stained with rhodamine-phalloidin, a fluorescent dye that binds to actin filaments. The sharp plane of focus of the confocal microscope reveals the thick actin cortex beneath the plasma membrane as a peripheral ring of fluorescence. (Courtesy of I. Meijerman et al, *Biochem. Biophys. Res. Commun.* 240: 697–700, 1997. © Academic Press.)

from disrupted cells (Figure 6-2). Within the cortex, actin filaments are linked together in a three-dimensional network with small pores (about 0.1 μm wide). Only small vesicles can pass through the network, thereby restricting the entry of larger organelles and granules, and this probably accounts for the hyaline (that is, 'glassy') appearance of the cell periphery in the light microscope.

The gel-like properties of the actin cortex are due to proteins that cross-link actin filaments, the most abundant of which, in vertebrate cells, are α-actinin and filamin. *α-Actinin* is composed of two polypeptide chains, each about 100 kDa in molecular weight. Each chain has three domains: an actin-binding domain at the N-terminus, a flexible central domain with similarities to spectrin (discussed below), and a calcium-binding domain. The two chains of α-actinin associate in an antiparallel fashion, bringing the actin-binding domain of one chain close to the calcium-binding site on the other chain. *Filamin*, also known as actin-binding protein (ABP-280), is similar to α-actinin, except that it is a larger protein and it forms a more open dimer, the subunits being attached mainly at their C-terminal ends (Figure 6-3). Perhaps because of its more open construction, filamin provides a flexible link between adjacent actin filaments.

Even small amounts of α-actinin or filamin have a dramatic effect on the physical properties of a solution of actin filaments, changing it from a viscous solution to a semisolid gel. This is especially true of the long, flexible molecule filamin. One filamin dimer per 200 molecules of actin monomers (about the ratio found in macrophages) is enough to create a large, continuous network whose pore size is further reduced by X-shaped entanglements between filaments. Synthetic actin gels made in this way have interesting mechanical properties, preserving their form when subjected to a sudden force but readily deforming under a low, steady pressure. This may be because the links have time to dissociate when the gel is subjected to a low, steady pressure.

α-Actinin and filamin are members of a large family of related actin cross-linking proteins. These all share a sequence of about 226 amino

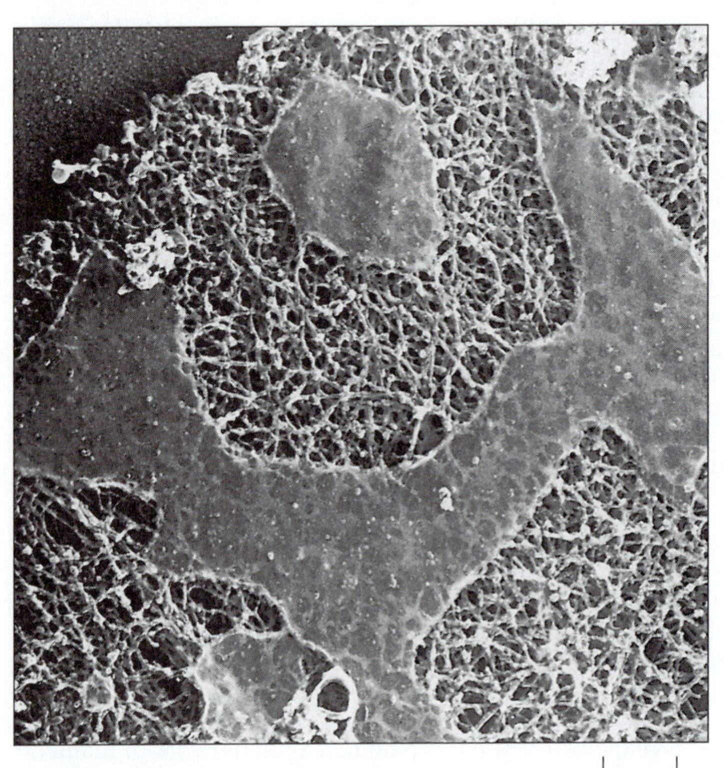


**Figure 6-2** Network of actin filaments in the cell cortex. Actin cortex of a cultured fibroblast exposed by partial extraction with a nonionic detergent and viewed by fast-freeze, deep-etch electron microscopy. Portions of the plasma membrane have been dissolved away revealing the underlying network of filaments. Filaments in the network are linked by proteins such as filamin and α-actinin into a three-dimensional network with the properties of a gel. (V. Small, *Electron Microscopy Rev.* 1: 165, © 1998 with permission from Elsevier Science.)

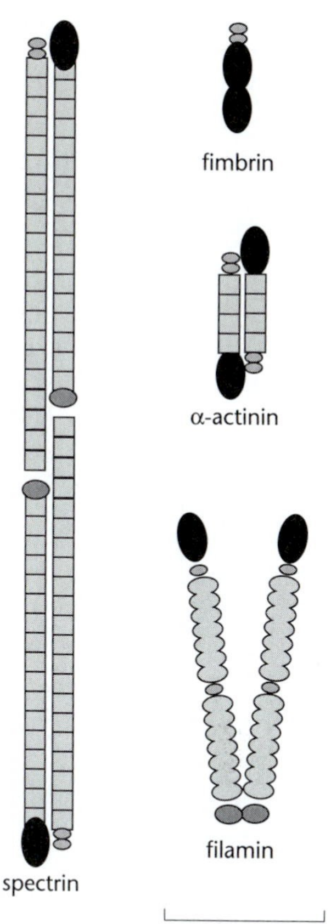


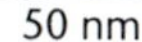

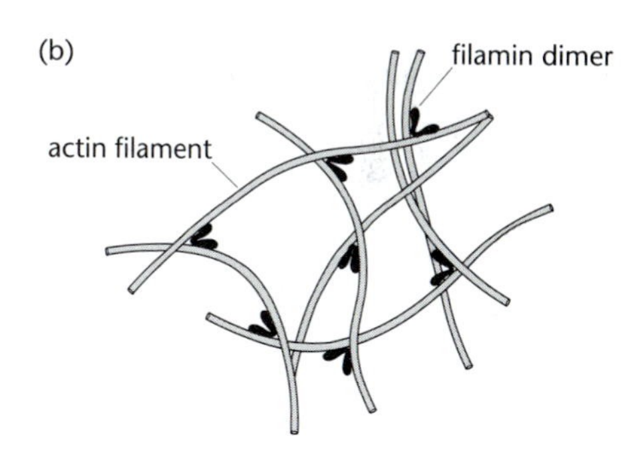


**Figure 6-3** Actin cross-linking proteins. (a) Summary of the modular structures of four actin cross-linking proteins. Each of the proteins shown has two actin-binding sites (black) that are related in sequence. Fimbrin has two directly adjacent actin-binding sites, so that it holds its two actin filaments very close together (14 nm apart), aligned with the same polarity. The two actin-binding sites in α-actinin are more widely separated and are linked by a somewhat flexible spacer 30 nm long, so that it forms actin filament bundles with a greater separation between the filaments (40 nm apart) than does fimbrin. Filamin has two actin-binding sites that are very widely spaced, with a V-shaped linkage between them. Spectrin is a tetramer of two α and two β subunits, and the tetramer has two actin-binding sites spaced about 200 nm apart. The spacer regions of these various proteins are built in a modular fashion from repeating units that include α-helical motifs, β-sheet motifs, and $Ca^{2+}$-binding domains. (b) How a few filamin molecules can link actin filaments together, changing a solution of actin filaments into a gel.

acids that forms their actin-binding domain, but differ in the way these domains are linked together (see Figure 6-3). Other members of the family include spectrin and dystrophin, highly flexible actin cross-linking proteins found closely associated with the plasma membrane, and fimbrin, which is less flexible and is found mainly in parallel bundles of actin filaments. In cortexillin, a novel member of this family recently found in *Dictyostelium discoideum*, the two actin-binding domains are linked by a two-stranded coiled-coil domain.

## *Actin-fragmenting proteins rapidly reorganize the cortex*

If a dense suspension of white blood cells is homogenized in cold buffer lacking calcium ions, centrifuged to remove nuclei and large pieces of debris, and the fluid extract then warmed, it can be seen to set to a gel. Gels prepared in this way, which are composed largely of actin, have interesting physical properties: they can be induced to dissolve, contract, or produce streaming movements by changing the temperature and the concentration of ions and small molecules (especially calcium ions and ATP). These transformations mimic changes that take place in the cortex as cells move or change their shape. Such changes are produced by the action of multiple actin-binding proteins (in addition to α-actinin and filamin already mentioned). They include myosin—a motor protein described in detail in the following chapter that generates movement in combination with actin filaments—and gelsolin, one of a family of actin-fragmenting proteins.

*Gelsolin* is a compact protein of about 80,000 Da (780 amino acids) found in most vertebrate cells. Purified gelsolin has three distinct effects on actin: (1) it binds to actin monomers to promote actin nucleation; (2) it caps the fast-growing barbed end of actin filaments, preventing addition of further monomers to that end; (3) it severs actin filaments by breaking bonds between adjacent actin monomers. The effects of the protein are regulated by $Ca^{2+}$ ions and by phosphoinositides, especially $PIP_2$. A transient rise in calcium ions promotes binding of gelsolin to actin and severing of filaments, whereas $PIP_2$ detaches gelsolin from the ends of actin filaments. Repeated cycles of this kind fuel the generation of short actin filaments, for example in the leading lamellipodium of a migrating cell (Chapter 8).

Like many proteins associated with the cytoskeleton, gelsolin is built on a modular plan, composed of a series of independently functioning domains. It has six similar domains, all of which have some affinity for actin. Most of the severing activity resides within the first two domains. Truncated versions lacking these two domains exhibit

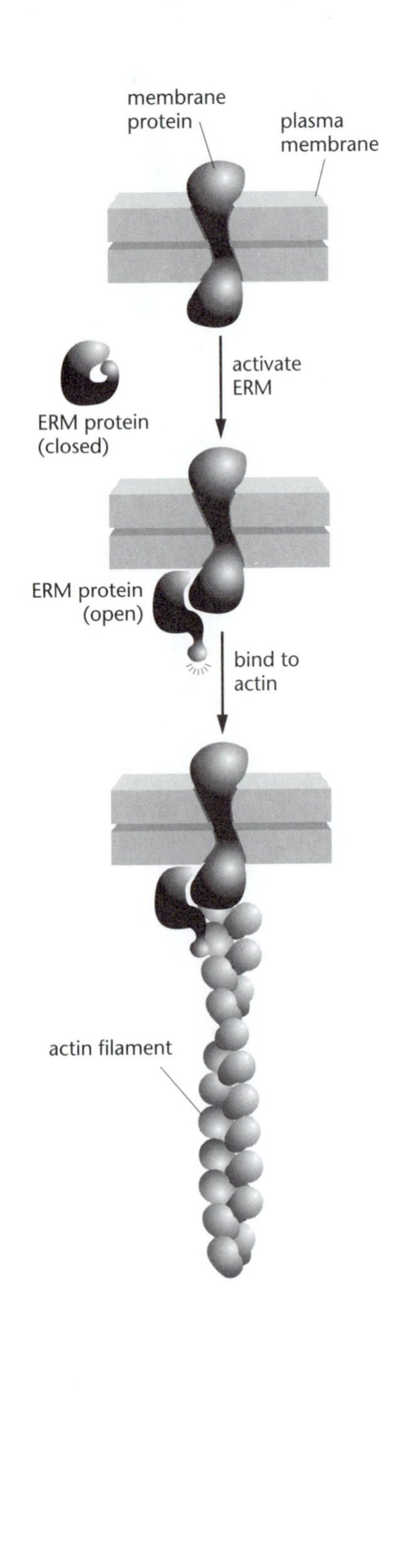


**Figure 6-4** ERM proteins link actin filaments to membranes. ERM proteins have a globular head domain that interacts with integral membrane proteins and a tail domain that binds to actin filaments. In the inactive state, the molecule is folded into compact conformation in which the head and the tail domains inhibit each other. In the active, open state both membrane-binding and actin-binding domains are exposed. The closed-to-open conformational change is triggered by phosphorylation of the ERM protein.

calcium-independent severing, and injection of this molecule into living cells causes a rapid, calcium-insensitive destruction of actin filaments.[1] A number of other related proteins are known that also have actin severing activity, including severin and fragmin from protozoa, and villin, a major protein of intestinal microvilli described below.

## Actin filaments attach, usually indirectly, to the plasma membrane

Actin can be linked to the plasma membrane in a variety of ways. The simplest mode of attachment is by direct binding to an integral membrane protein (that is, a protein so firmly embedded in the plasma membrane that it can be removed only by dissolving the lipid bilayer). In the slime mold *Dictyostelium*, for example, a 17 kDa integral membrane protein called *ponticulin* binds directly to actin filaments. A second protein, *cortexillin*, is able both to bundle actin filaments and to link them to the plasma membrane. Vertebrate cells also contain proteins, such as *calpactin* and *lipocortin*, that associate both with actin and phospholipids and might therefore mediate the direct association of actin to membranes.

However, the more usual interaction is indirect, and many proteins are known to act as linkers, or adapters, between actin filaments and membrane proteins. Both α-actinin and filamin can act in this way, as well as numerous other specific proteins in red blood cells, in cell junctions, and in actin-rich extensions of the cell surface. Each membrane linker is adapted to function in its particular location but all must have at least two distinct domains, one that binds to actin and another that binds to an integral membrane protein.

A good example is seen in the group of proteins known collectively as *ERM proteins* (for *e*zrin, *r*adixin, and *m*oesin, the principal members of the group). These proteins are widespread in vertebrate cells, found in cleavage furrows, microvilli, ruffling membranes, and other locations. ERM proteins are short molecules with a globular N-terminal domain and a rodlike, α-helical tail (Figure 6-4). The globular head interacts with integral membrane proteins whereas the C-terminal tail binds to actin. Interestingly, head and tail domains of ERM proteins are mutually inhibitory so that the protein can switch itself off by folding into a hairpin conformation—a switch-like feature shared by other cytoskeletal proteins such as vinculin and myosin. The ERM switch is probably activated by phosphorylation of the protein by a kinase that is under the control of the small G protein Rho described below.

## The red blood cell membrane is supported by a network of spectrin and actin

One of the best-characterized associations of actin with the plasma membrane is seen in the thin membrane cortex of red blood cells. Mammalian red blood cells (erythrocytes) are highly specialized cells that lack internal organelles, microtubules, and intermediate filaments.[2] Their membrane is

[1] Gelsolin is a major substrate for caspase, one of the proteases responsible for programmed cell death (apoptosis). Proteolysis with caspase releases a gelsolin fragment that severs actin filaments in an uncontrolled manner perhaps causing dying cells to round up and detach from the substratum.

[2] The red blood cells of birds and amphibians have a more complicated structure. They retain their nucleus and their shape is influenced by a bundle of microtubules that runs around their perimeter (described in Chapter 19).

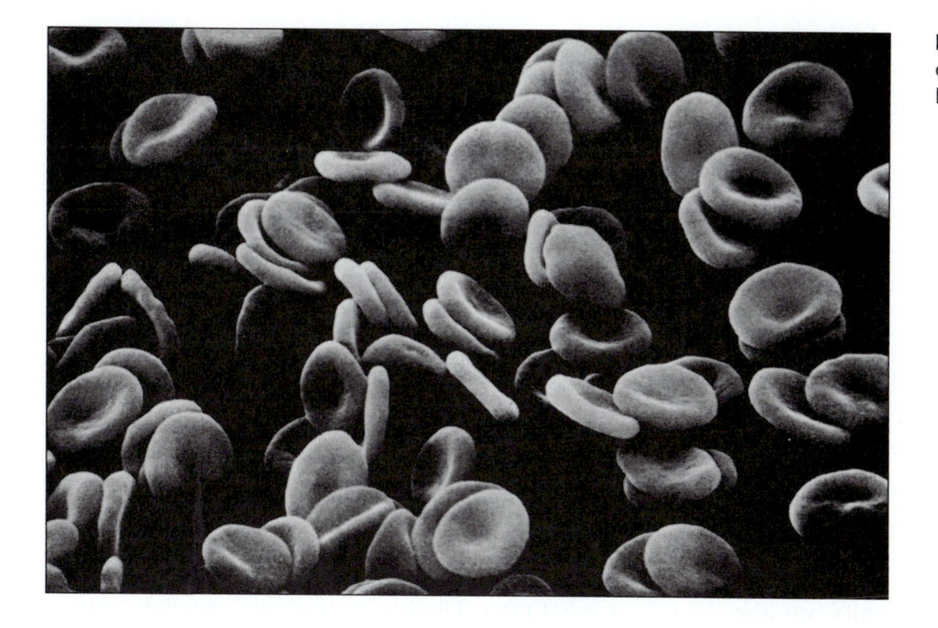

**Figure 6-5** Scanning electron microscopy of human erythrocytes. (Courtesy of Bernadette Chailley.)

therefore the only means these bags of hemoglobin have to maintain their distinctive biconcave shape (Figure 6-5). The cell 'skin,' about 20 nm thick, comprises the plasma membrane together with a layer of associated protein. Although it is very flexible, this skin must also be tough: during its lifetime of 120 days, a human erythrocyte travels approximately 300 miles through narrow and crowded blood vessels.

The erythrocyte cortex can be obtained by lysing the cell by osmotic shock and collecting the empty sacs by centrifugation. Known as *erythrocyte ghosts* because of their transparency, these flexible sacs reseal under appropriate conditions and can regain their original shape if ATP is present. The membrane skeleton evidently plays an important part in this shape recovery since, if the membrane skeleton is detached from the inner face of the plasma membrane by mild proteolysis, the ghost cell does not return to its original biconcave disk form.

If the lipid component of the membrane is extracted by detergent, it leaves behind a shell of protein composed of a two-dimensional network of interconnected threads (Figure 6-6; Table 6-1). A major component of this network is a large dimeric protein, called *spectrin* for its ghostly origin. Spectrin is made of two large polypeptides, α-spectrin (280 kDa) and β-spectrin (246 kDa), each a flexible beads-on-a-string chain of repeating units containing about 106 amino acids. The repeating units consist of three α-helical strands wrapped together in a short coiled-coil (similar

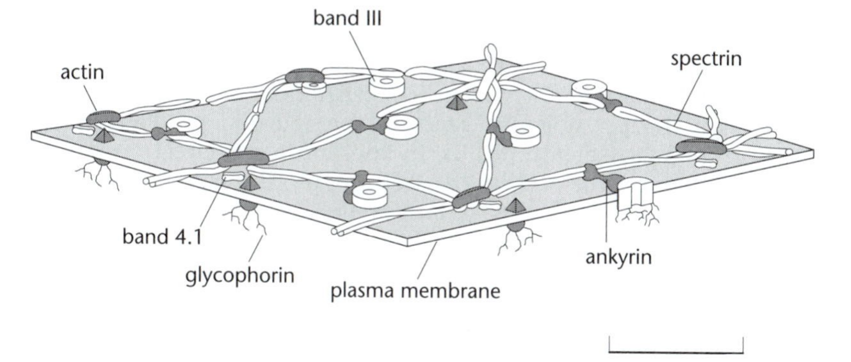


**Figure 6-6** Schematic diagram of the spectrin-actin network on the inner face of the red blood cell membrane. Two regions of attachment to the plasma membrane are indicated, one mediated by ankyrin and band III, and the other by band 4.1 and glycophorin. The actual network is far more compact and complex than shown here.

**Table 6-1 Proteins of the red blood cell cortex**

| Protein | Size (kDa) | Copies per cell | Relative number | Number per $\mu m^2$ |
|---|---|---|---|---|
| INTEGRAL MEMBRANE PROTEINS | | | | |
| band III | 105 | 1,200,000 | 12 | 8000 |
| glycophorin dimer | 50 | 1,000,000 | 10 | 6600 |
| CORTICAL PROTEINS | | | | |
| spectrin, dimer | 280/246 | 200,000 | 2 | 1300 |
| ankyrin | 206 | 100,000 | 1 | 700 |
| actin, monomer | 43 | 500,000 | 5 | 3300 |
| tropomyosin | 29/27 | 80,000 | 0.8 | 530 |
| band 4.1 | 78 | 200,000 | 2 | 1300 |
| adducin $\alpha/\beta$ | 103/97 | 30,000 | 0.3 | 300 |

Calculations based on a surface area for human erythrocytes of 150 $\mu m^2$. The human erythrocyte membrane skeleton also contains calmodulin, CapZ, tropomodulin, caldesmon, dematin, and other proteins.

units are also found in α-actinin). Dimers of spectrin associate into a tetramer with a length of 200 nm, which is the form found in the erythrocyte cortex. The ends of spectrin tetramers in the red blood cell cortex are linked to short lengths of actin filament (each containing about 12 actin monomers, which is enough to form one half-turn of the filament helix).

The importance of spectrin is revealed in a number of genetic disorders that affect the shape and stability of red blood cells. For example, hereditary anemia in both mice and humans is caused by a deficiency in the synthesis of spectrin. Humans with this condition may have as little as 5% of the normal amount of spectrin; their red blood cells are spherical in shape and extremely fragile.

Spectrin was originally thought to be unique to erythrocytes but is now known to belong to a family of closely related spectrin genes expressed in cells as diverse as neurons and amoebae. It is usually associated with membranes but is often enriched in specific locations, such as regions of cell–cell contact and in regions at which exocytosis occurs. Different forms of spectrin are present in the axon compared to the dendrites and cell bodies of mature nerve cells, in skeletal muscle, and (in birds) in the intestinal brush border.[3]

## *Ankyrin and band 4.1 link spectrin to the plasma membrane*

Spectrin is attached to the erythocyte plasma membrane through ankyrin and band 4.1 (see Figure 6-6). *Ankyrin* links spectrin to the anion transporter, also known as band III protein, which is held firmly in the lipid bilayer by its hydrophobic interactions. *Band 4.1*, related to the ERM family of proteins mentioned above, binds to the C-terminus of spectrin chains, to actin, and also to the cytoplasmic tail of *glycophorin*—a heavily glycosylated protein with a single transmembrane domain.

Significantly, band 4.1 attaches to glycophorin only when the signaling phospholipid $PIP_2$ is present. The concentration of this lipid changes rapidly with the physiological state of the cell and produces multiple effects on the cytoskeleton. Thus, in addition to an effect on band 4.1, $PIP_2$ also influences the activity of α-actinin, profilin, and gelsolin.

Just as spectrin has close relatives that are found in diverse cell types, so proteins similar to erythrocyte ankyrin and band 4.1 are widely distributed in vertebrate tissues. For example, synapsin-1, a prominent component of the cytoskeleton of neurons, is related to band 4.1. The binding of this protein to membrane-associated actin may play a part in the release of transmitter at synapses.

[3] Much of the diversity of mammalian spectrins arises from alternative mRNA splicing, in which the same gene produces multiple proteins. Variants of band 4.1 and many proteins of skeletal muscles are also produced in this manner (Chapter 10).

## *How does the cell regulate assembly of its spectrin network?*

It is interesting to ask how the cortical layer of spectrin, ankyrin, and other proteins is built by the cell. Indeed, this is a general problem, since the cytoskeleton contains many complex assemblies of proteins each of which is built to precise specifications, with constituent proteins held in the correct numbers and locations. In the case of the spectrin network and others that have been analyzed in detail, it seems that posttranslational modifications and the selective degradation of unwanted molecules are both employed extensively.

As red blood cells form in the bone marrow, their membrane skeleton slowly matures from an undifferentiated and relatively labile network to a highly specific stable network of proteins. In the course of this refinement, crucial proteins are synthesized at different times, each contributing to the final fully stable network. The short segments of actin filaments associated with spectrin, for example, probably arise from the progressive depolymerization of longer filaments in a more open and irregular network. In a mature erythrocyte, these short actin rods are capped at either end by specific proteins (adducin and tropomodulin) which are also involved in the binding interactions that hold the network together.

Spectrin, the major component of the membrane skeleton, changes during erythropoiesis from the αγ isoform (also known as fodrin) common to many cell types, to the αβ isoform unique to red blood cells. The changeover is a gradual process characterized by a progressive accumulation of erythrocyte spectrin and the downregulation and eventual disappearance of fodrin. Throughout the transitional period, both spectrin isoforms continue to be made, to assemble onto the plasma membrane, and to be degraded by endogenous proteases (perhaps calpain)—the α chain being at all times made in large excess.

The reason for this wasteful mode of assembly apparently lies in the high α-helical content of the spectrin molecule. All spectrin molecules have multiple repeating stretches of triple α-helices, 106 amino acids long, which interact with a heterologous spectrin to form large, antiparallel coiled-coil molecules. Since dimerization is substantially more rapid than synthesis, large quantities of unwanted oligomers such as αα- and ββ-spectrin are initially formed, possibly assembling on the polysomes. With the formation of such homodimers, spectrin molecules are effectively trapped because of the high energy of activation required to unwrap the coiled-coil interactions. The only way to remove them may be by proteolysis.

A major factor in the stabilization of spectrin dimers is their anchorage to the plasma membrane. At an early stage of differentiation, the primary source of anchorage is ankyrin, which is able to interact directly with the phospholipid bilayer through the posttranslational addition of a fatty acid residue. Later, band III provides the principal anchorage for ankyrin and its attached spectrin. Other components such as band 4.1 and the membrane protein glycophorin all contribute to the final stable network in a complicated sequence of interactions.

## *Adherens junctions provide a stable anchorage for actin filaments*

One of the most conspicuous sites of interaction between actin and the plasma membrane is an *adherens junction*. This is a location in an animal tissue cell where actin filaments come into close contact with the membrane and are linked, through integral membrane proteins, either to another cell or to the extracellular matrix.

In cell–cell adherens junctions the integral membrane proteins are *cadherins*, a family of $Ca^{2+}$-dependent cell adhesion proteins able to form homophilic (that is, self–self) associations at their extracellular domain. Opposing faces of the junction are held together by the extracellular domains of cadherins, which bind each other. On the inside of the membrane, the cytoplasmic domains of cadherin molecules insert into a cluster of other proteins that itself is attached to actin filaments. In this way the adherens junctions forms a mechanically strong linkage between one cell and the next, holding them together in a tissue.

Adherens junctions are found between heart cells, at synapses in the central nervous system, and between folds of a myelin sheath. In epithelial cells they form a continuous adhesion belt (or zonula adherens) around each of the interacting epithelial cells between the apical and basal domains of the cell (Figure 6-7). A belt of actin filaments on the cytoplasmic side of the adhesion belt is potentially contractile. We will see in the final two chapters that it confers on the epithelial sheet the capacity to develop and change in remarkable ways, binding or rolling up into a tube or developing into a cup-shaped cavity, and so on.

Adherens junctions have a complicated structure that is still under investigation. There are, to begin with, multiple different subtypes of cadherins, each characteristic of the particular cell that makes them. Different subtypes of cadherins combine with different affinities, and this confers much of the specificity in cell–cell contacts. Proteins that attach to the cytoplasmic domains of cadherins are also highly specialized, the list including ERM proteins and α-actinin, as well as specialized proteins such as catenins and plakoglobins (Table 6-2).

Why should a 'simple' mechanical attachment contain so many different proteins and have such a complicated structure? One reason is that mechanical properties are actually not simple. The strength, elasticity, shear strength, and viscoelasticity of different parts of a cell are crucially important for their functioning and must be closely controlled in a dynamic way. The second reason is that, as we will discuss shortly in the context of focal adhesions, cell adhesions have not only a mechanical role but also a signaling one. Because of their privileged position immediately subjacent to the plasma membrane, junctional proteins relay

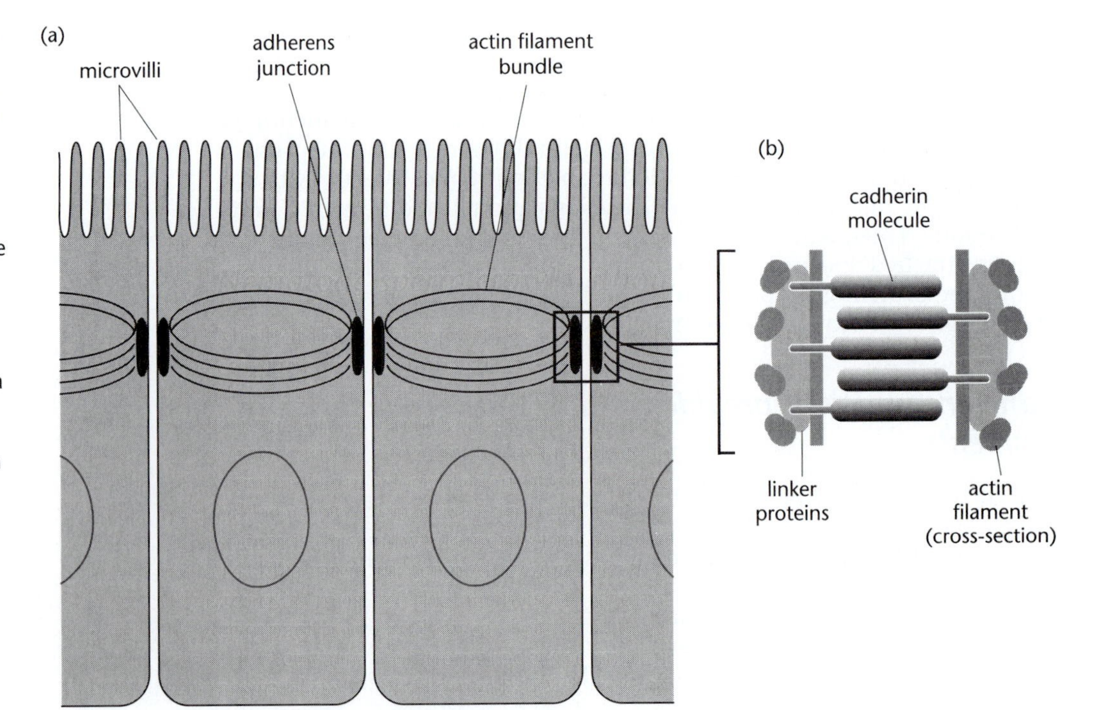


**Figure 6-7** Adhesion belts between epithelial cells in the small intestine. (a) This beltlike anchoring junction encircles each of the interacting cells. Its most obvious feature is a contractile bundle of actin filaments running along the cytoplasmic surface of the junctional plasma membrane. (b) Actin filaments at an adherens belt are joined to those in a neighboring cell by transmembrane linker proteins (cadherins), which bind to identical cadherin molecules on the adjacent cell.

**Table 6-2 Major proteins of cell–cell adherens junctions**

| Protein | Size (kDa) | Function |
|---|---|---|
| actin, monomer | 43 | major structural component |
| cadherin | ~ 80 | homophilic binding/integral membrane proteins |
| catenin | 88/113 | link cadherins to the cytoskeleton, β-catenin activates gene transcription |
| plakoglobin | 82 | binds cadherins and α-catenin |
| tensin | 186 | caps and cross-links actin filaments |
| vinculin | 117 | binds tensin and α-actinin |
| α-actinin | 103 | cross-links actin filaments, binds vinculin |
| zyxin | 82 | binds α-actinin |
| radixin | 82 | caps actin filaments |
| Src ($pp60^{c\text{-}src}$) | 43 | protein tyrosine kinase |
| ena/VASP | 46 | nucleates actin filaments |

information on the mechanical and chemical properties on the outside of the cell to its interior. We will see in Chapter 20 that the formation of cell-to-cell junctions is a vital stage in the formation of an embryonic tissue, which triggers many specific downstream events.

## *Cultured cells attach to their substratum at focal adhesions*

A different type of adherens junction enables cells to attach to the extracellular matrix. Cultured fibroblasts migrating on a surface coated with extracellular matrix molecules, for example, form specialized *focal adhesions* where the plasma membrane is held approximately 15 nm from the substratum. Each focal adhesion is the site of attachment of a bundle of actin filaments, termed a *stress fiber* that terminates in a cluster of proteins attached to the plasma membrane. An important difference between these cell–substratum attachments and the cell–cell adherens junctions mentioned above is that the anchorage is mediated by integrins rather than cadherins. We will see below that integrins have binding sites for components of the extracellular matrix.

Focal adhesions may be visualized by reflection interference microscopy in which visible light is directed onto the 'underbelly' of cells growing on a thin glass coverslip (Figure 6-8). The junctions then appear as dark oval patches, typically 1–2 μm long but often longer, which form at the leading margin of the cell, remain fixed in position as the cell passes overhead, and detach farther back on the cell. Fibroblasts torn from the culture surface with a microelectrode or with a jet of fluid leave their focal adhesions behind, demonstrating that they are not only points of close contact but also sites of mechanical anchorage.

Formation of focal adhesions depends on the cell substratum and is typically promoted by extracellular matrix proteins such as fibronectin, laminin, and collagen. These multidomain proteins adhere to the glass or plastic surface of the tissue culture dish and present specific sequences for recognition by cell-surface receptors. The short sequence Arg-Gly-Asp (or 'RGD' using single-letter notation) found in the cell-binding domain of both fibronectin and laminin is especially important.

As just mentioned, focal adhesions are built around *integrins*: a large and diverse family of transmembrane proteins. Each integrin molecule is

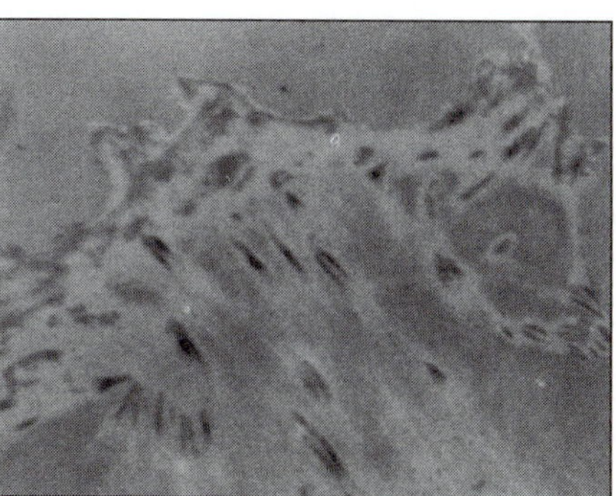


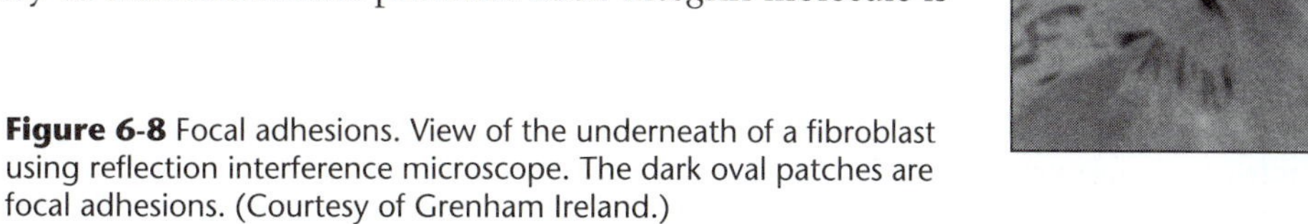
**Figure 6-8** Focal adhesions. View of the underneath of a fibroblast using reflection interference microscope. The dark oval patches are focal adhesions. (Courtesy of Grenham Ireland.)

a noncovalently associated heterodimer of two distinct, high-molecular-weight polypeptides called α- and β-integrin, which act as transmembrane linkers in a variety of cells. Many subtypes of both α- and β-integrin exist, and these can combine to produce a huge variety of heterodimers, each with a selective affinity for a particular extracellular matrix molecule. Integrins are also found on blood platelets, where they are involved in blood clotting, and in lymphocytes and macrophages, where they play a role in the crucially important matrix interactions of these cells.

One domain of the integrin molecule is exposed on the cell surface and binds to molecules such as fibronectin or vitronectin in the extracellular space. At the other end of the integrin molecule, its cytoplasmic domain associates with a complex cluster of proteins. Some of these are similar to those found in cell–cell adherens junctions (α-actinin, vinculin, tensin), while others, such as talin, are found only in focal adhesions. Actin filaments from the cell cortex insert into this protein plaque, thereby becoming mechanically linked to the extracellular matrix on the other side of the plasma membrane (Figure 6-9).

## *Focal adhesions send and receive signals*

Cell junctions are not simply sites of mechanical anchorage of the cell. They also function in cell communication. A cell that has formed focal adhesions knows that it has attached firmly to a suitable substratum, so it can now grow or differentiate. In reciprocal fashion, focal adhesions are less likely to be produced if a cell is actively dividing, or in the process of changing its shape. There is also evidence that the strength of attachment of focal adhesions to the extracellular matrix can be modulated by the cell, providing a clutchlike mechanism by which it can regulate its migratory speed (Chapter 8).

An interesting example of the cell's control over focal adhesions occurs during transformation by tumor viruses. Transformation typically changes the flattened form of a cell such as a fibroblast into a rounded

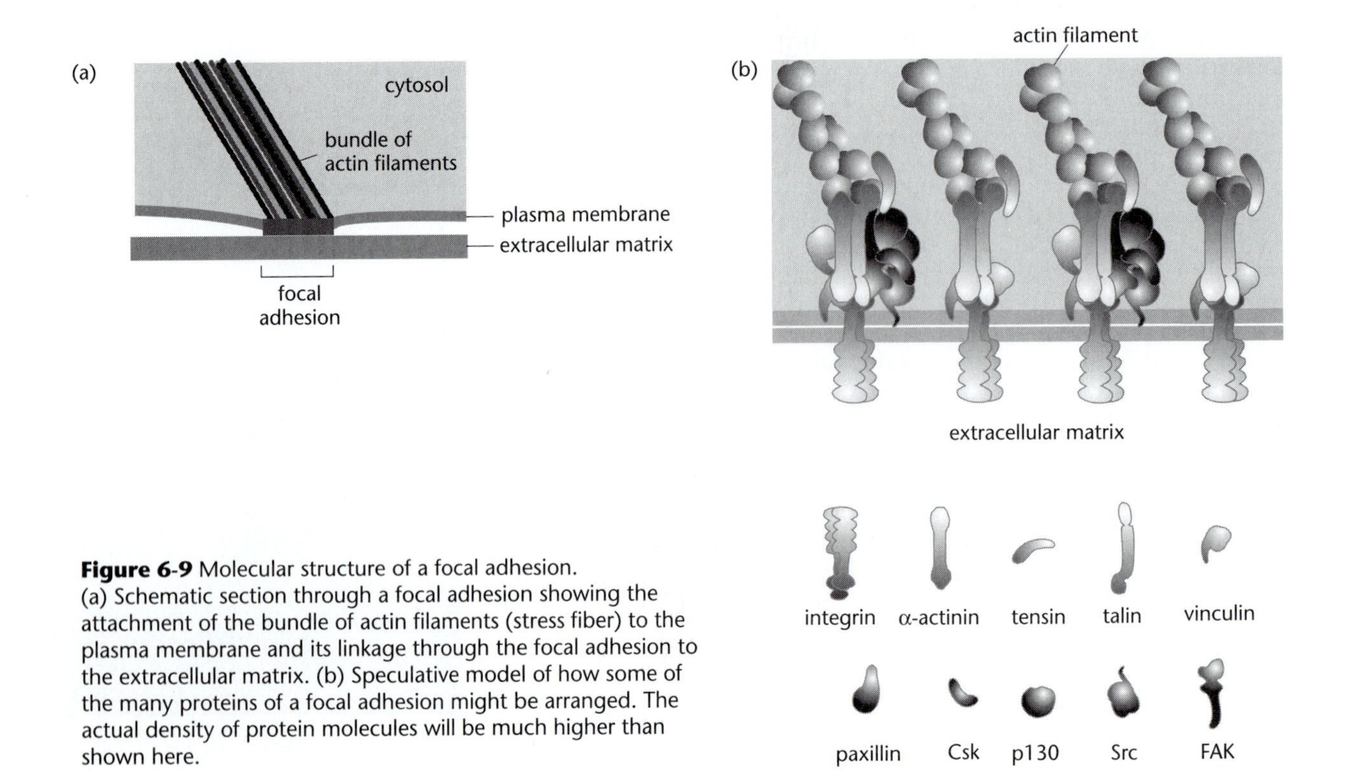


**Figure 6-9** Molecular structure of a focal adhesion. (a) Schematic section through a focal adhesion showing the attachment of the bundle of actin filaments (stress fiber) to the plasma membrane and its linkage through the focal adhesion to the extracellular matrix. (b) Speculative model of how some of the many proteins of a focal adhesion might be arranged. The actual density of protein molecules will be much higher than shown here.

morphology, at the same time altering growth properties so that the cells are no longer responsive to contact inhibition by adjoining cells (and are therefore more likely to grow into a tumor). In the well-studied case of transformation of avian cells by the Rous sarcoma virus, changes are initiated by a specific tyrosine kinase (v-Src) produced by the virus. This kinase localizes to the focal adhesions and triggers their breakup and eventual degradation. In uninfected cells, a less active tyrosine kinase, known as c-Src helps to control the normal turnover of focal adhesions during cell motility.

Some proteins in a focal adhesion have a signaling rather than a mechanical or structural role (Table 6-3). The focal adhesion kinase, or FAK, for example, is an enzyme with a remarkably wide range of functions. It binds to structural components of the focal adhesion such as integrins as well as to the tyrosine kinase Src and the adapter protein Grb2. FAK phosphorylates tyrosines in various proteins in the focal adhesion, including itself, thereby causing other proteins with binding domains for phosphorylated tyrosines to cluster in this region. The complex both sends and receives signals passing between the focal adhesion and targets in the cytoplasm and nucleus.

We see that even though a focal adhesion is a large cytoskeletal-based structure, it also acts like a signaling complex. It receives multiple inputs from components of the extracellular matrix, mechanical tension, and phosphorylation signals, integrates them, and then produces multiple outputs that influence the growth and division of the cell as well as its state of differentiation.

## *Focal adhesions are triggered by both external and internal influences*

How cells make focal adhesions is a topic of great interest. One of the earliest steps is thought to be the side-by-side association of integrin molecules to form a cluster in the membrane. In a fibroblast that has made contact with a suitable surface (one coated with vitronectin, for example), integrin molecules diffusing in the membrane become trapped when they bind to the extracellular matrix. A set of tethered integrin molecules is produced that under normal conditions then matures into a focal adhesion.

However, a suitable surface for the attachment of integrin molecules is not sufficient by itself. It is possible to prevent focal adhesions from forming even after integrin molecules have collected together. Thus, if cells are exposed to inhibitors of protein phosphorylation, or if certain signaling components are inactivated, then focal adhesions will not form. A widely used experimental protocol employs fibroblasts attached to

**Table 6-3 Major proteins of focal adhesions**

| Protein | Size (kDa) | Function |
|---|---|---|
| actin | 43 | major structural component |
| integrin (α and β) | ~ 100 | integral membrane proteins, bind extracellular matrix molecules |
| talin | 270 | binds vinculin, actin, and integrins |
| FAK | 125 | protein tyrosine kinase, binds integrins |
| tensin | 186 | caps and cross-links actin filaments |
| vinculin | 117 | binds α-actinin and talin |
| α-actinin | 103 | cross-links actin filaments, binds vinculin |
| paxillin | 68 | binds vinculin and FAK |
| Src ($pp60^{c\text{-}src}$) | 43 | protein tyrosine kinase |
| p130 | 130 | substrate for Src, binds tensin |

Focal adhesions also contain fimbrin (actin-bundling), VASP (actin-nucleating), $PI_3$-kinase (phospholipid signaling), calpain (proteolysis), and many other proteins.

surfaces in media lacking serum. Such cells lack focal adhesions (which can be detected by staining for vinculin) but make them rapidly following the injection of specific signaling components. The most important of the latter is the small GTP-containing protein Rho, which, as we will see shortly, has a central role in the control of many actin–membrane associations.

The final ingredient needed to make a large, mature focal adhesion is mechanical tension. If a fibroblastic cell in culture is stretched mechanically by the stroking action of an electrode on its upper surface, then focal adhesions on its lower surface become larger and oriented in the direction of the stretching. Evidently mechanical forces lead directly to the recruitment of more focal adhesion proteins and the insertion of more actin filaments. Functionally this is exactly what one would expect, since the purpose of a focal adhesion is to resist tensile forces experienced by the cell. But the synergism it reveals between biochemical and mechanical forces is impressive, and one that we will encounter repeatedly in the operation of the cytoskeleton. Mechanical aspects of the cytoskeleton are the subject of Chapter 18.

## *Rho GTPases control actin's association with the membrane*

The protein *Rho* belongs to a large family of over 40 GTP-binding proteins (*G proteins*) known as the Ras superfamily, which relay intracellular signals and regulate many processes in cells. Many human cancers, for example, can be produced by oncogenic mutations in Ras genes. In general, these proteins function as self-inactivating molecular switches that exist in either an active state, when GTP is bound, or an inactive state, when GDP is bound ('active' in this context meaning that the molecule triggers downstream signaling events in the cell). GTP addition and loss of GTP are catalyzed by other proteins, which in this way, control the rate of switching of the G protein. Downstream of the G protein pathway, Rho and its relatives interact with kinases and other proteins, and thereby control a variety of processes, many of them associated with the actin cortex.

Exactly how Rho acts is not fully known at present but it is certain to be complex. A family of serine/threonine kinases has been found that is activated by Rho and these have multiple targets in the cytoskeleton. With regard to the linkage of actin to the plasma membrane, for example, one of the direct actions of Rho may be through the ERM family of proteins. Rho controls a protein kinase that phosphorylates, and thereby activates, ERM proteins. Activation of ERM proteins by phosphorylation (see Figure 6-4) may contribute to the assembly of a focal adhesion. Rho also affects the level of inositol phospholipid $PIP_2$ in the membrane, which in turn is known to act on several cytoskeletal proteins.

Possibly the most important link between Rho and focal adhesions, however, is the development of mechanical tension. Rho activation causes an increase in the phosphorylation of myosin light chains, which, as we will see in Chapter 7, causes bundles of myosin and actin to form and contract. We have already mentioned that tension development is essential for focal adhesion formation. Thus when actin filaments are attached to integrins, Rho-induced contractions can draw the integrins together into a tight patch. A cascade of phosphorylation and other reactions could be initiated, resulting in a mature focal adhesion.[4]

Two closely related G proteins are also intimately involved with the actin cortex. We have just seen that injection of quiescent, serum-starved, fibroblasts in culture with Rho leads to the rapid formation of both stress fibers and focal adhesions. Injecting a closely related G protein, called Rac, into serum-starved fibroblasts produces not stress fibers and focal

[4] Tension development at focal adhesions is also subject to negative regulation. Caldesmon, a protein abundant in smooth muscle and in focal adhesions, has the capacity to inhibit myosin, and microtubules in the vicinity of a focal adhesion may also oppose the development of tension.

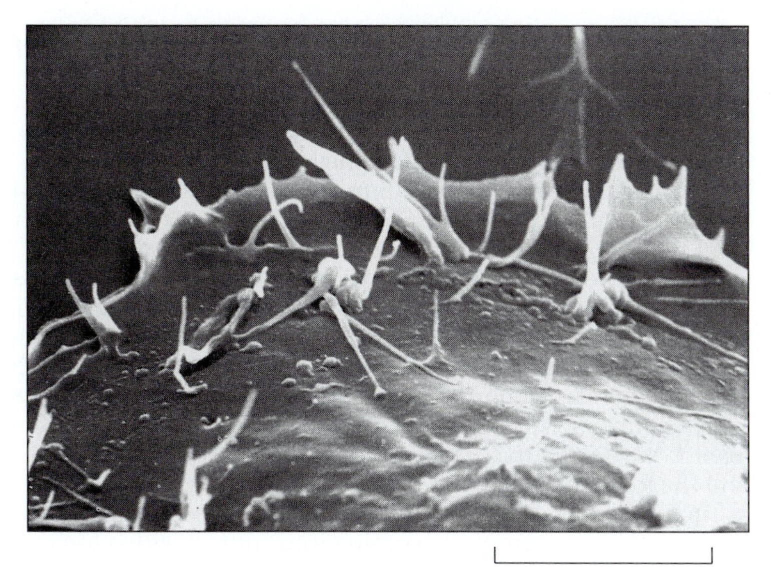


**Figure 6-10** Filopodia on the cell surface. Scanning electron micrograph of a fibroblast in tissue culture. (Courtesy of Julian Heath.)

adhesions but multiple actin-rich lamellipodia or ruffles over the cell surface. A third G protein, Cdc42, under the same conditions, causes long thin filopodia rather than lamellipodia to be produced (Figure 6-10). Although each G protein is distinct in its action, some cross-talk occurs, so that Cdc42 activates Rac, and Rac (after a significant delay) activates Rho. Taken together, these observations suggest that members of the Rho GTPase family are part of a regulatory network controlling the association of surface receptors with the actin cytoskeleton (Figure 6-11).

The effects of Rac, Rho, and Cdc42 are not restricted to fibroblasts. These proteins also operate in nerve cells, epithelial cells, lymphocytes, and macrophages, and the detailed consequences of their activation are dependent on the particular cellular context.

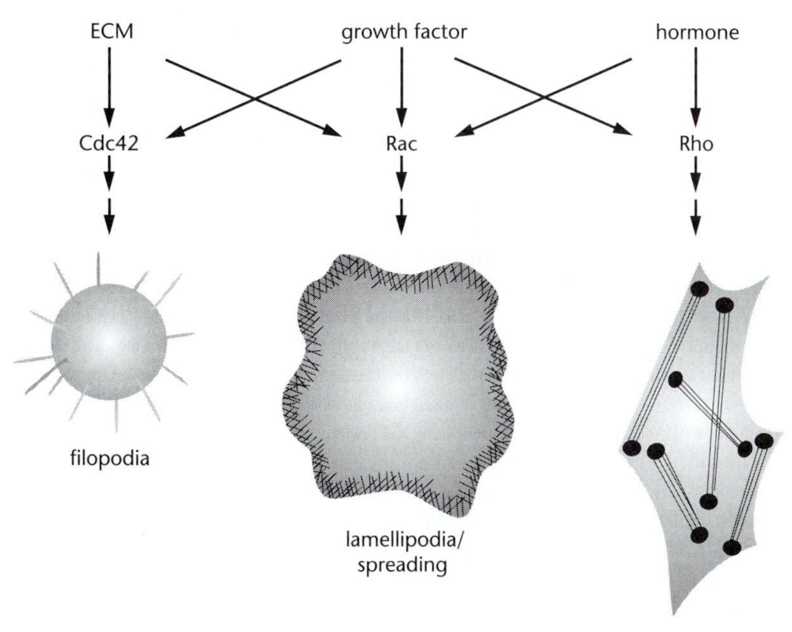


**Figure 6-11** Rho proteins and the actin cortex. Injection of activated Cdc42, Rac, or Rho into quiescent, serum-starved fibroblasts causes changes in surface morphology and the production of the actin-based structures shown. Each G protein can also be selectively activated by specific hormones or growth factors acting on cell-surface receptors (ECM, extracellular matrix).

## *Lamellipodia, filopodia, and microvilli form in response to external stimuli*

The surface structures generated in response to Rac and Cdc42 contain actin filaments and are part of the cortical cytoskeleton. In filopodia, actin filaments are collected into a loose bundle and cross-linked by bundling proteins such as fimbrin. In lamellipodia they form nets rather than bundles, terminating at the leading margin in a dendritic filigree of filaments. These structures occur widely in cells both in culture and in the animal and are often found on cells that are moving, or dividing or undergoing a sudden increase in growth. Thus, neuronal growth cones produce filopodia and lamellipodia as they crawl toward distinct synaptic targets, and neutrophils or amoeboid cells of *Dictyostelium* generate lamellipodia and filopodia on their surface when exposed to chemotactic agents. We discuss in Chapter 8 the relationship between these structures and the forward crawling of the cell as a whole.

Another role for surface ruffling is displayed during the *phagocytosis* performed by a macrophage coming into contact with a foreign organism or a dead cell. The macrophage develops a local accumulation of actin-rich cortical material at the site of contact that develops into pseudopodia (the nonspecific, generic term) rich in actin. These surround and eventually engulf the particle, which is then attacked by lysosomal enzymes and eventual digested. Perversely, some pathogenic bacteria such as *Salmonella* deliberately induce a similar surface response in intestinal epithelial cells, and use it to gain entrance to the cells they infect (Chapter 16).

A different kind of cortical response is shown by some oocytes following fertilization. In the large eggs of sea urchins, for example, contact with a sperm causes microvilli on the egg surface to change in length. Before fertilization, the surface of a sea urchin egg is covered with some 130,000 short microvilli. Upon fertilization these increase in average length from about 0.3 μm to 1.0 μm. In the region of the sperm head, a cluster of microvilli becomes even longer, extending up and around the head and thereby aiding its attachment and fusion. These rapid changes follow, and may be triggered by, a wave of release of calcium ions from intracellular stores. The $Ca^{2+}$ wave, which can be displayed by injecting specific fluorescent dyes into the egg, starts at the point of fertilization and spreads around the large cell. As it progresses, the wave is accompanied by increasing tension of the surface and the changes in length of microvilli just mentioned.

## *Filopodia grow by controlled actin polymerization*

The structural core of actin-rich extensions of the surface such as filopodia is a loose bundle of actin filaments. These are arranged with their fast-growing, barbed ends close to the plasma membrane, and when the protrusion grows on the cell surface it is because actin monomers are adding to the barbed ends of the filaments. Recall that in the acrosomal reaction of *Thyone* sperm, a pool of G-actin polymerizes explosively on contact with the egg and produces a long thin membrane-enclosed spike (Chapter 5). Recall also that as *Listeria* and related bacteria spread through and between the cells they infect, they are driven by the controlled assembly of actin filaments at their barbed ends (Chapter 5).

The force required to push the plasma membrane outward could in principle come from the free energy released during polymerization. It can be shown experimentally that if G-actin is enclosed in a large lipid vesicle and then induced to polymerize, it can distend the surface. However, actin does not operate in isolation and in the cell many different molecules and molecular processes collaborate to regulate polymerization. Monomeric actin itself is held in reserve, sequestered in complexes with proteins such as thymosin. There are also hints that, in at least some cells, concentrated accumulations of unpolymerized actin may exist close to the membrane, like reserve depots.

The point of actual assembly of actin monomers—the barbed end of the nascent filament—is the location of clusters of actin-binding proteins. These might be created by gelsolin-mediated fragmentation of existing actin filaments or by the formation of nucleating complexes such as the Arp2/3 complex close to the plasma membrane in response to external signals. We will describe in Chapter 8 how the continual activation of nucleating complexes just beneath the leading margin of a migrating cell drives its lamellipodia forward, and a similar mechanism is likely to exist for filopodia. It is also likely that the characteristic exploratory function of filopodia has a molecular correlate and that specific proteins will be found at their tips.

## *Bundling proteins increase the rigidity of cell extensions*

Once a filopodium or microvillus has grown out of the cell by actin polymerization, its strength and stiffness rely on proteins that hold actin filaments together. The strategy of cross-linking to increase stiffness, mentioned in Chapter 4, is widely used by cells, which employ for this purpose a variety of actin bundling and cross-linking proteins. Some of the most bizarre examples are found in invertebrate sperm, which produce long actin-containing spikes (acrosomal processes) when they make contact with the egg. In the rapidly made *Thyone* acrosomal process mentioned above, the filaments are only loosely cross-linked, probably by a form of spectrin. But the acrosomal process of *Limulus*, the horseshoe crab, is premade and much more tightly bundled. A bundle of actin filaments some 60 μm long, and cross-linked by a distinctive bundling protein called scruin, lies coiled up around the base of the head of the unreacted sperm. Contact with the egg causes the actin filaments in the bundle to change their relative packing, probably triggered by a calcium-induced change in conformation of scruin. The bundle uncoils and extends, causing the process literally to screw itself into the egg jelly.

Oocytes also show rearrangements of their actin cortex. Unfertilized sea urchin eggs, for example, contain numerous short microvilli that within an hour of fertilization rearrange to form filopodia 5–10 μm long over their entire surface. The actin filaments within these projections are linked together in regular hexagonal bundles by a protein called fascin, producing distinctive 12 nm traverse striations. Fascins are widespread in other tissues of both invertebrates and vertebrates and can be detected in both stress fibers and filopodia.

Actin filaments in tightly-packed structures such as microvilli, acrosomal processes, and stereocilia are almost always arranged in longitudinal alignment. In other words, the crossovers of adjacent helices lie in a series of planes at right angles to the axis of the bundle (Figure 6-12). This feature, which often results in cross-striations in electron micrographs, probably arises from the fact that bundling proteins are usually short dimeric

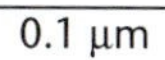


**Figure 6-12** Packing of helical filaments into bundles. (a) Short cross-links between helical filaments tend to force them into longitudinal alignment, often generating regular striations across the bundle. The precise geometry of a bundle of actin filaments is more complicated because the filaments are not usually in a planar array and because each twist of an actin filament does not usually contain an integral number of subunits. (b) Cross-section of the bundle of actin filaments in a stereocilium showing regular striations about 12 nm apart. (Micrograph courtesy of Lewis Tilney.)

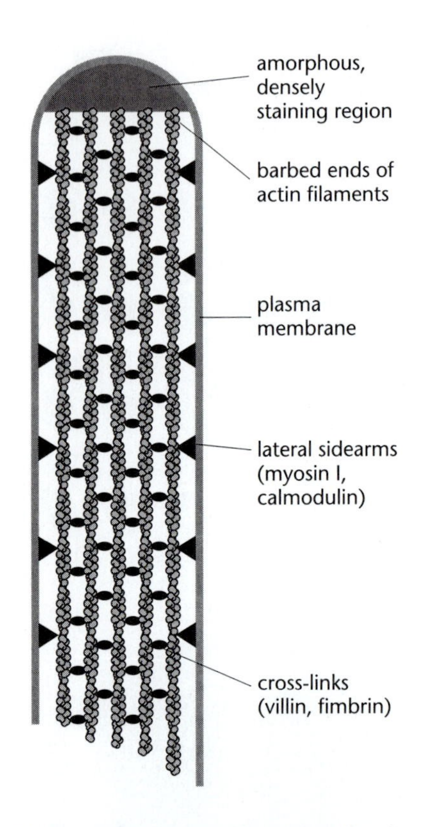


**Figure 6-13** A microvillus. A bundle of parallel actin filaments held together by the actin-bundling proteins villin and fimbrin forms the core of a microvillus. Lateral arms (complexes of myosin I and the $Ca^{2+}$-binding protein calmodulin) connect the sides of the actin filament bundle to the overlying plasma membrane. The barbed ends of the actin filaments are embedded in an amorphous densely stained substance of unknown composition.

molecules, and tend to tie corresponding regions of actin filaments together. How the filaments pack in the other dimension (at right angles to the axis of the bundle) is a more difficult question to answer. The helically arranged subunits in a filament do not automatically fit into a regular geometrical lattice, so some flexibility is needed, either in the cross-links or in the twist of the helices themselves.

## *Intestinal microvilli are held together by fimbrin and villin*

One of the best-characterized forms of stable actin filament bundle is seen in the intestinal brush border. Intestinal *microvilli* are cylindrical projections, 1–2 μm long and about 0.1 μm in diameter (see Figure 6-7). They are profuse on the apical surfaces of epithelial cells lining the intestine, forming a close-packed layer called a *brush border*. This has an absorptive function, increasing the area of the cell surface exposed to the intestine 10–20-fold. The microvillar membrane carries many specialized enzymes involved in the breakdown and membrane transport of food, including those responsible for hydrolysis of sucrose and other disaccharides, as well as proteins required for the transport of fatty acids from the intestinal cavity, and the $Na^+$ cotransport of sugars and amino acids.

Each microvillus contains a bundle of 30 or so actin filaments held together by cross-linking proteins, notably fimbrin and villin (Figure 6-13). *Fimbrin* is a widely occurring protein also present in filopodia and stereocilia that has the ability to bundle actin filaments into a tight, almost crystalline array. *Villin*, is an actin-fragmenting protein related to gelsolin. At low calcium concentrations (less than 0.1 μM $Ca^{2+}$) villin causes actin filaments to form bundles. However, if calcium rises above 5 μM, villin causes fragmentation of actin filaments, by a similar mechanism to that used by gelsolin. Villin-induced breakdown of the actin-rich core may be an essential step allowing large vesicles of membrane to be sloughed off and released into the intestine, where the digestive enzymes then continue their action.

As for any large cytoskeletal structure, microvilli are enormously complex in detailed molecular composition and structure (Table 6-4). Apart from actin filaments, fimbrin, and villin, there are also specific proteins at the tip of the structure (around the barbed ends of the actin filaments). A single-headed myosin (one of a large family of motor proteins discussed in the next chapter) with calmodulin light chains forms the side-arms that extend at regular intervals of 33 nm from the actin filaments to make contact with the plasma membrane. The presence of myosin is an indication that microvilli are not static structures. In fact, both the cells and

**Table 6-4 Major proteins of intestinal microvilli**

| Protein | Size (kDa) | Function |
|---|---|---|
| actin, monomer | 43 | forms microvillus core |
| fimbrin | 68 | bundles actin filaments |
| villin | 95 | cross-links actin filaments, severs them in response to $Ca^{2+}$ |
| myosin I | 110 | anchors membrane to actin core |
| calmodulin | 17 | associated with myosin I |
| spectrin (α and β) | 240/235 | anchors actin filaments to membrane at microvillus base |
| myosin II | 200 | cross-links actin filaments at base, generates cortical tension |
| paxillin | 68 | binds vinculin and FAK |
| Src (pp60 $^{c\text{-}src}$) | 43 | protein tyrosine kinase |
| p130 | 130 | substrate of Src, binds tensin |

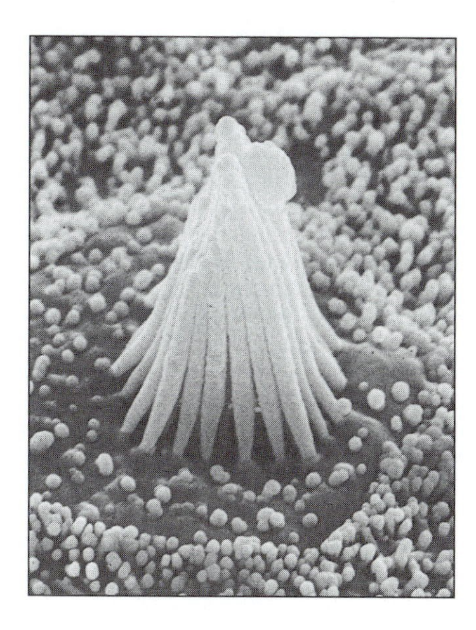

**Figure 6-14** Stereocilia. Regular array of stereocilia on a hair cell in the inner ear of a bullfrog. The hair bundle comprises about 50 stereocilia and one true cilium (the kinocilium), which in this species has a bulbous swelling at its end. The longest stereocilia are 8.4 μm in length. Note the stubby microvilli on the surfaces of the surrounding cells. (Scanning electron micrograph courtesy of R.A. Jacobs and A.J. Hudspeth.)

their microvilli are in a state of continual change. Most of the proteins in a micovillus *turnover* (that is, they are degraded and replaced by newly synthesized molecules) within an hour or so, and membrane components move continually toward the microvillus tip as part of the continual sloughing of surface membrane.

## Stereocilia are specialized microvilli that perform a sensory function

Microvilli, like chocolates, come in many shapes and flavors. All are made in basically the same way, but they differ in the length and number of actin filaments they contain, how these are bundled, and how the filaments are attached to the membrane. Possibly the most elaborate structures of this kind, at least in mammals, are the *stereocilia* present in the hair cells of the ear (Figure 6-14).

Hair cells are found in the cochlea of the vertebrate ear where they are primary detectors of sound vibrations, and in the vestibular apparatus where they respond to the position and movements of the head. In fish, hair cells are present in the lateral line organs along the flank, which monitor the progress of the animal through the water. Stereocilia have even been found on the surface of some species of protozoa, acting as delicate triggers for the discharge of poisoned darts known as nematocysts.

Stereocilia can be up to 30 μm long and 1 μm wide. They are cylindrical and taper at their base where they meet the surface of the hair cell, like pencils resting on their points. In a typical stereocilium there may be several thousand actin filaments in parallel array tied to each other through proteins such as fimbrin (see Figure 6-12).[5] Actin filaments in the center of the core bundle are longer than the others, and extend into the base of the stereocilium. Here a cuticular plate of dense filaments extends laterally to make contact with adjacent supporting cells.

Hair cells are amazingly sensitive to mechanical displacement. The mammalian ear, for example, reliably measures mechanical stimuli whose average amplitude at threshold is about the diameter of a hydrogen atom. This performance is believed to be due to thin threads that extend from the membrane at the tip of one stereocilium to that of an adjacent stereocilium, where they are physically attached to ion channels. The rigid structure of the stereocilium might thus enable minute movements to be amplified into a change in electric potential.

## Cells control the number, length, and position of actin filaments in their cortex

The length and position of individual stereocilia are precisely determined, often to an amazing degree. In the cochlea of the chick, for example, each hair cell has a bank of stereocilia arranged in hexagonal pattern in which sequential rows increase stepwise in height. The array as a whole has a definite orientation on the hair cell and detailed measurements of the number, size, and arrangement of stereocilia show that these vary systematically from one region of the cochlea to another. Variations are so

[5] In cross-section the filaments in a stereocilium appear disordered, like a liquid crystal. In other types of actin bundle, the arrangement can be more regular, however, showing hexagonal or square lattices.

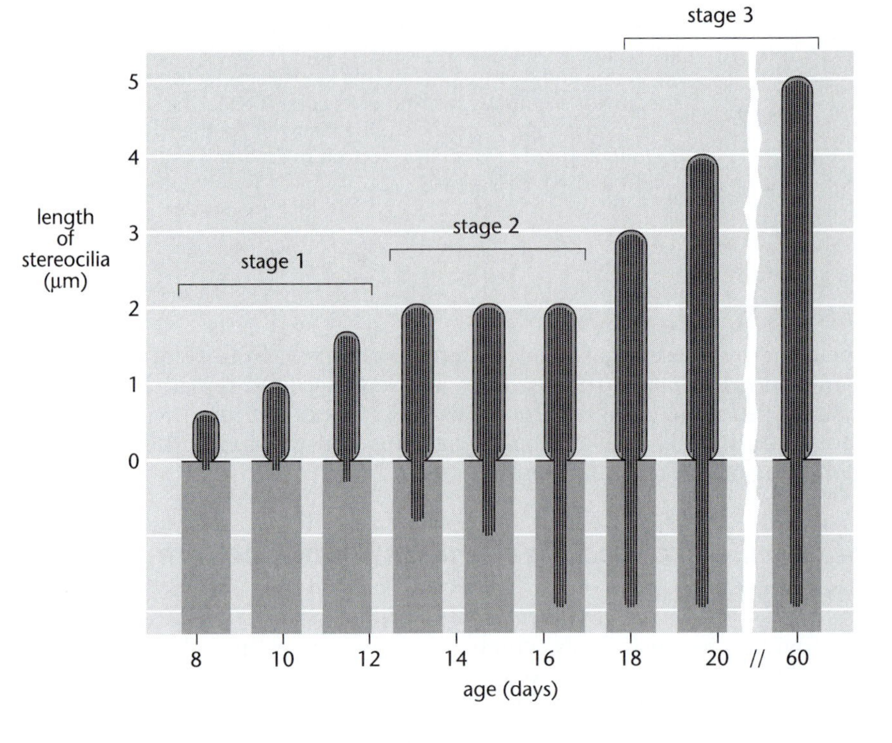


**Figure 6-15** Phases of actin growth in stereocilia. In *stage 1*, actin filaments in stereocilia grow at their barbed ends and become more ordered by cross-linking. In *stage 2*, new actin filaments are added to make the stereocilium wider, and subunits are added to the pointed end of core actin filaments to lengthen the rootlets. In *stage 3*, actin filaments in the stereocilium again elongate at their barbed ends. All stereocilia go through the same three stages, but the time of initiation and duration of growth vary both among the stereocilia on a single hair cell and from one hair cell to another. In this way a remarkably precise gradation of filament lengths is produced.

precise and reproducible from animal to animal that one could tell the position of a cell by the number, length, and orientation of the stereocilia it carries.

Formation of the cochlea presents a remarkable demonstration of the ability of cells to control the length, number, and position of actin filaments in their cortex (Figure 6-15). Three stages of growth of stereocilia have been distinguished. (1) Stereocilia begin to grow on each hair cell, emerging in a staggered sequence that generates a truncated version of the final organ-pipe array. Growth during this phase is believed to be at the barbed end of actin filaments. (2) New actin filaments accumulate on the outside of the core bundle to make the stereocilia thicker while the central filaments elongate, apparently at their pointed ends, to form rootlets. (3) Actin filaments elongate at their barbed ends and grow to an extent that depends on their position in the cochlea.

## Outstanding Questions

*What is the composition of the actin cortex? How does it vary from one location of the cell to another? How are these differences created and what is their function? Which molecules are unique to filopodia and how do they contribute to the exploratory action of these structures? What determines the biconcave shape of a red blood cell? Is it the spectrin network—if so, how? What is the composition and spatial arrangement of proteins in adherens junctions? How is their growth controlled by mechanical tension? What tells a cell that it has made an effective adhesion to another cell or to the substratum? Do conformational changes propagate across the membrane, through cadherin or integrin molecules? How do Rho GTPases interact with adherens junctions and focal adhesions? What is the network of signal reactions controlling the actin cortex? How does the* Limulus *sperm acrosomal process elongate? Is it by propagated conformational changes through the structure? Do intestinal microvilli move? If so how, and why? How does a cell control the position and lengths of stereocilia on its surface? How important are the mechanical properties of stereocilia? Are they purely passive structures?*

## *Essential Concepts*

- All animal cells possess a cortical layer of actin filaments on the inner face of the plasma membrane. This layer supports the flimsy lipid bilayer mechanically and is responsible for many cell surface movements.
- The cortex is made of actin filaments attached at their barbed (plus) ends to the membrane and cross-linked into a three-dimensional web, or matrix, by proteins such as α-actinin and filamin.
- Dynamic restructuring of the cortex occurs in response to calcium changes, mediated by severing proteins such as gelsolin.
- Actin filaments are usually anchored to the plasma membrane by proteins with binding sites for both actin (or an actin-binding protein) and an integral membrane protein.
- In the membrane skeleton of vertebrate red blood cells, short lengths of actin filament cross-linked by flexible spectrin molecules are attached to the plasma membrane by specific proteins such as ankyrin and band 4.1.
- Adherens junctions between cells comprise clusters of cadherin molecules, which provide anchorage between one cell and the next, and bundles of actin filaments attached to their cytoplasmic face.
- Focal adhesions on the lower surface of cells in culture form in response to extracellular matrix molecules such as fibronectin. Integrin molecules bind to matrix proteins on the outside of the cell and to talin, vinculin, α-actinin, and other proteins on the inside.
- In addition to a simple anchoring function, adherens junctions and focal adhesions also have a signaling role and contain a number of protein kinases and their substrates.
- Cortical actin also assembles in specific regions in response to external cues. Examples include the region of contact between a phagocytic cell and a suitable particle, where the accumulation of actin gives rise to pseudopodia that extend and engulf the particle.
- Small G proteins, Rho, Rac, and Cdc42, have an important controlling influence over the association of actin with the plasma membrane, as in the formation of focal adhesions and filopodia.
- Rapid formation and elongation of surface extensions occurs in many cells in response to surface signals, for example following fertilization of an egg or chemotactic stimulation of neutrophils or in the activation of blood platelets.
- Actin filaments that form during these changes are assembled locally from actin monomers sequestered in the cytoplasm. Protein complexes near the membrane, formed in response to external cues, nucleate actin polymerization at specific locations.
- Once formed, surface extensions can be stabilized by extensive cross-linking of their actin core and the linking of the latter to the plasma membrane, a well-characterized example being intestinal microvilli.
- The packing of actin filaments in such stable extensions can be extremely regular. Hair cell stereocilia, present in amazingly precise arrays in the cochlea, illustrate the fine control a cell exerts over its cortical actin structures.

## *Further Reading*


Agre, P., et al. Deficient red-cell spectrin in severe, recessively inherited spherocytosis. *N. Eng J. Med.* 306: 1155–1161, 1982.

Bennett, V. Ankyrins: adaptors between diverse plasma membrane proteins and the cytoplasm. *J. Biol. Chem.* 267: 8703–8706, 1992.

Bennett, V., Gilligan, D.M. The spectrin based membrane skeleton and micro-scale organization of the plasma membrane. *Annu. Rev. Cell Biol.* 9: 27–66, 1991.

Blanchard, A., et al. The structure and function of alpha-actinin. *J. Muscle Res. Cell Motil.* 10: 280–289, 1989.

Bretscher, A. Regulation of cortical structure by the ezrin-radixin-moesin protein family. *Curr. Opin. Cell Biol.* 11: 109–116, 1999.

Burridge, K., et al. Focal adhesion assembly. *Trends Cell Biol.* 7: 342–347, 1998.

Burtnick, L.D., et al. The crystal structure of plasma gelsolin: implications for actin severing, capping and nucleation. *Cell* 90: 661–670, 1997.

Cowin, P., Burke, B. Cytoskeletal-membrane interactions. *Curr. Opin. Cell Biol.* 8: 56–65, 1996.

DeRosier, D.J., et al. A change of twist of actin provides the force for the extension of the acrosomal process in *Limulus* sperm: the false discharge reaction. *J. Cell Biol.* 93: 324–327, 1982.

Faix, J., et al. Cortexillins, major determinants of cell shape and size, are actin-bundling proteins with a parallel coiled-coil tail. *Cell* 86: 631–642, 1996.

Fowler, V.M. Regulation of actin filament length in erythrocytes and striated muscle. *Curr. Opin. Cell Biol.* 8: 86–96, 1996.

Furukawa, R., Fechheimer, M. The structure, function, and assembly of actin filament bundles. *Int. Rev. Cytol.* 175: 29–90, 1997.

Gerisch, G., et al. Chemoattractant-controlled accumulation of coronin at the leading edge of *Dictyostelium* cells monitored using a green fluorescent protein–coronin fusion protein. *Curr. Biol.* 5: 1280–1283, 1995.

Grum, V.L., et al. Structures of two repeats of spectrin suggest models of flexibility. *Cell* 98: 523–535, 1999.

Hall, A. Rho GTPases and the actin cytoskeleton. *Science* 279: 509–514, 1998.

Hartwig, J.H., et al. Thrombin receptor ligation and activated Rac uncap actin filament barbed ends through phosphoinositide synthesis in permeabilized human platelets. *Cell* 82: 643–653, 1995.

Jockusch, B.M., et al. The molecular architecture of focal adhesions. *Annu. Rev. Cell Dev. Biol.* 11: 376–416, 1995.

Kaibuchi, K., et al. Regulation of the cytoskeleton and cell adhesion by the Rho family GTPases in mammalian cells. *Annu. Rev. Biochem.* 68: 459-486, 1999.

Kothakota, S., et al. Caspase-3-generated fragment of gelsolin: effector of morphological change in apoptosis. *Science* 278: 294–298, 1997.

Lazarides, E., Woods, C. Biogenesis of the red blood cell membrane-skeleton and the control of erythroid morphogenesis. *Annu. Rev. Cell Biol.* 5: 427–452, 1989.

Luna, A.J., et al. The *Dictyostelium discoideum* plasma membrane: a model system for the study of actin–membrane interactions. *Adv. Cell Biol.* 3: 1–33, 1990.

Machesky, L.M. Complex dynamics at the leading edge. *Curr. Biol.* 7: 164–167, 1997.

Mangeat, P., et al. ERM proteins in cell adhesion and membrane dynamics. *Trends Cell Biol.* 9: 187–192, 1999.

Mills, R.G., et al. Slow axonal transport of soluble actin with actin depolymerizing factor, cofilin, and profilin suggests actin moves in an unassembled form. *J. Neurochem.* 67: 1225–1234, 1996.

Miyata, H., et al. Protrusive growth from giant liposomes driven by actin polymerization. *Proc. Natl. Acad. Sci. USA* 96: 2048-2053, 1999.

Mogilner, A., Oster, G. Cell motility driven by actin polymerization. *Biophys. J.* 71: 3030–3045, 1996.

Nobes, C.D., Hall, A. Rho GTPases control polarity, protrusion, and adhesion during cell movement. *J. Cell Biol.* 144: 1235–1244, 1999.

Pearson, M.A., et al. Structure of the ERM protein moesin reveals the FERM domain fold masked by an extended actin-binding tail domain. *Cell* 101: 259–270, 2000.

Rosenblatt, J., et al. Xenopus actin depolymerizing factor/cofilin (XAC) is responsible for the turnover of actin filaments in *Listeria monocytogenes* tails. *J. Cell Biol.* 136: 1323–1332, 1997.

Schafer, D.A., Cooper, J.A. Control of actin assembly at filament ends. *Annu. Rev. Cell Dev. Biol.* 11: 497–518, 1995.

Schoenwaelder, S.M., Burridge, K. Bidirectional signaling between the cytoskeleton and integrins. *Curr. Opin. Cell Biol.* 11: 274–286, 1999.

Schutt, C.E., et al. The structure of crystalline profilin–β-actin. *Nature* 365: 810–816, 1993.

Shaw, R.J., et al. RhoA-dependent phosphphorylation and relocalization of ERM proteins into apical membrane/actin protrusions in fibroblasts. *Mol. Biol. Cell* 9: 403–419, 1998.

Svitkina, T.M., Borisy, G.G. Arp2/3 complex and actin depolymerizing factor/cofilin in dendritic organization and treadmilling of actin filament array in lamellipodia. *J. Cell Biol.* 145: 1009–1025, 1999.

Terry, D.R., et al. Misakinolide A is a marine macrolide that caps but does not sever filamentous actin. *J. Biol. Chem.* 272: 7841–7845, 1997.

Tilney, L.G., et al. Actin filaments, stereocilia, and hair cells of the bird cochlea. V. How the staircase pattern of stereociliary lengths is generated. *J. Cell Biol.* 106: 355–365, 1988.

Tsukita, S., et al. Molecular linkage between cadherins and actin filaments in cell–cell adherens junctions. *Curr. Biol.* 4: 834–839, 1992.

Tsukita, S., et al. ERM proteins: head-to-tail regulation of actin-plasma membrane interactions. *Trends Biochem. Sci.* 22: 53–58, 1997.
Vacquier, V.D. Dynamic changes of the egg cortex. *Dev. Biol.* 84: 1–26, 1981.
Viel, A., Branton, D. Spectrin: on the path from structure to function. *Curr. Opin. Cell Biol.* 8: 49–55, 1996.
Way, M., Matsudaira, P. The secrets of severing? *Curr. Biol.* 3: 887–890, 1993.
Winter, D., et al. The complex containing actin-related proteins Arp2 and Arp3 is required for the motility and integrity of yeast actin patches. *Curr. Biol.* 7: 519–529, 1997.
Yap, A.S., et al. Molecular and functional analysis of cadherin-based adherens junctions. *Annu. Rev. Cell Dev. Biol.* 13: 119–146, 1997.
Zamir, E. et al. Dynamics and segregation of cell-matrix adhesions in cultured fibroblasts. *Nat. Cell Biol.* 2: 191–196, 2000.
Zigmond, S.H. Signal transduction and actin filament organization. *Curr. Biol.* 8: 66–73, 1996.

It might seem at first sight that plant cells have no place in a book on cell movements. Trees and flowers are rooted in one spot and their cells are so encased in thick polysaccharide cell walls that surface deformations and dynamic changes in shape are not possible. However, the interior of a plant cell is a different story. Here the light microscope reveals continual widespread and often rapid movements of organelles, which move unceasingly, like traffic in a city, from one location in the cell to another. In the giant cells of the freshwater algae *Nitella* or *Chara*, streams of membrane organelles follow the spiral rows of chloroplasts in a continuous belt around the cell. These movements, termed cyclosis, astounded the scientific community when they were first observed, by Corti in 1774.

We now know that cytoplasmic movements in general are driven by motor proteins, and that those in *Nitella*, in particular, are driven by myosin. As mentioned briefly in Chapter 4, myosins are proteins able to move along actin filaments, obtaining the energy they need from the hydrolysis of ATP. This basic mechanism of force generation is harnessed by the cell to drive many different movements, including organelle transport, deformations of the cell surface, and contractions. Myosins and other motor proteins are at the heart of cell movements. They epitomize the intimate connection between the structure of biomolecules and the movement of living organisms.

In this chapter we will first describe the structure and motile properties of myosin II (conventional myosin), which drives muscle contraction, as well as numerous contractile processes in nonmuscle cells. We will then move on to the families of 'unconventional' myosins, which come in a rich variety of shapes and sizes and have often unexpected functions. Defects in human vision and hearing, for example, can result from mutations affecting particular myosins. In the course of the chapter we will have the opportunity to highlight the importance of recombinant DNA studies and genetics in the analysis of the cytoskeleton.

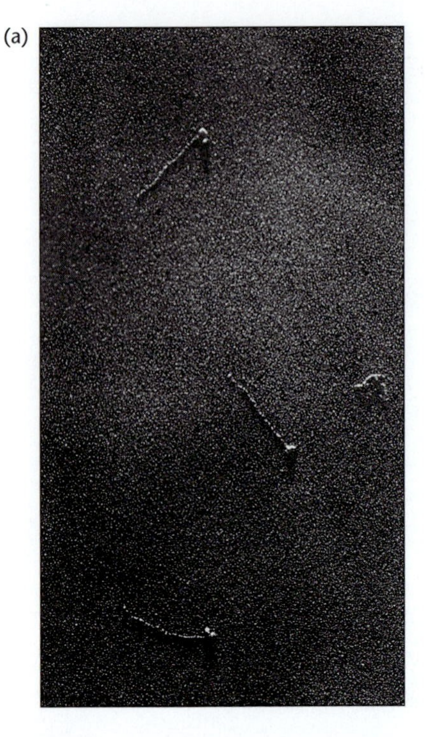

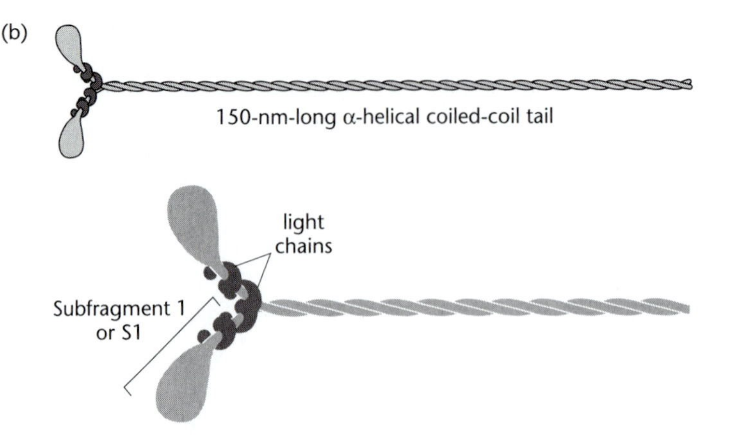


**Figure 7-1** Myosin II. (a) Electron micrograph of myosin II molecules from rabbit skeletal muscle. (b) Anatomy of the molecule, which comprises two heavy chains, each about 2000 amino acids long, and four smaller light chains. Each heavy chain folds into a globular head region at its N-terminal end and intertwines with the other heavy chain at its C-terminal end, forming a rigid coiled-coil tail. The light chains wrap around the α-helical region linking each head to the tail, one pair of light chains per neck. (a, courtesy of Gerald Offer.)

## Muscle myosin is a two-headed, long-tailed molecule

Myosin has been known for longer than any other cytoskeletal protein. In 1864, Kühne described how concentrated salt solutions could extract a substance from muscle that he called 'myosin.' It was not until 1939 that Englehardt and Ljubimowa made the startling discovery that Kühne's myosin, previously considered a structural protein, possessed the ATPase activity believed to be the energy source for the contraction of muscle. Shortly thereafter, Albert Szent-Gyorgyi and his colleagues in Hungary demonstrated that artificial fibrils made from myosin and a second protein, actin, contracted when ATP was added.

Muscle myosins and their homologues in nonmuscle cells are large proteins (500 kDa) consisting of two polypeptides, each with a globular head joined to an extended α-helical tail that winds around the tail of its partner in a rigid coiled-coil (Figure 7-1). This form of myosin is referred to as *myosin II*, or *conventional myosin* to distinguish it from the more recently discovered 'unconventional' myosins described later in the chapter. Each head of myosin II has an actin-binding site and actin-activated ATPase activity. The tails of multiple myosin molecules can associate to form a multimolecular filament.

Myosin II molecules are built up from six polypeptide chains (three different gene products): two identical heavy chains of about 200 kDa and two pairs of light chains of about 20 kDa each (see Figure 7-1). The C-terminal halves of the two heavy chains assemble into a coiled-coil rod, while their N-terminal sequences fold to form two globular heads. One of each of the two types of light chain is associated with each globular head.

Limited exposure of myosin to proteolytic enzymes such as trypsin or papain cleaves the molecule at specific sites, thought to be the location of flexible regions of the polypeptide chain. Two domains commonly produced by such treatment are heavy meromyosin (HMM), which contains the two globular heads (plus their light chains) linked by a short region of tail, and subfragment 1 (S1) which consists of single heads.

## Myosin heads bind to actin filaments and hydrolyze ATP

The head region of the myosin molecule is the site where the transformation of chemical energy to mechanical energy takes place. In chemical terms, it is an actin-activated, $Mg^{2+}$-dependent ATPase (Figure 7-2). Myosin heads (S1 fragments) work only slowly, each head taking about 30 seconds to split successive ATP molecules. However, this rate is powerfully

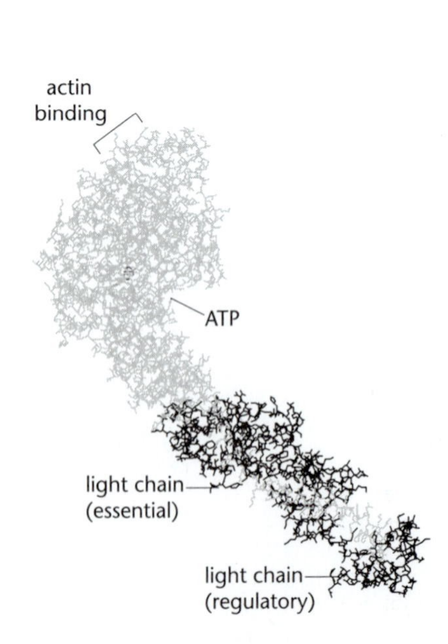


**Figure 7-2** Myosin head. Structure of the S1 fragment of scallop myosin II shown as a wire-frame model. The two light chains clamped around the myosin neck are shown in darker gray. (Data from protein database file 1B7T.)

stimulated by actin filaments, each myosin head now hydrolyzing 5 to 10 molecules of ATP every second. We will see in Chapter 9 how this hydrolysis drives the movements of the myosin head and produces contractile force.

Light chains in the myosin head are not directly required to hydrolyze ATP or to bind to actin, but may regulate these properties. All myosin light chains belong to the same family as the ubiquitous calcium-binding protein calmodulin (Chapter 9). The light chains bind to short amino acid sequences, rich in isoleucine (I) and glutamine (Q) and known as *IQ motifs*, in the region of the molecule between the head and the tail. Two major subtypes of light chain are found in muscle myosin, regulatory and essential, and each S1 fragment contains one molecule of each kind. Some regulatory light chains bind calcium ions and in molluscan muscle this interaction seems to affect the activity of myosin directly. In vertebrate skeletal muscle the principal effect of $Ca^{2+}$ is due to its action on the thin filaments, as detailed in Chapter 9. In the form of myosin II found in the contractile ring and elsewhere in nonmuscle cells, the regulatory light chain is also the substrate for kinases, which regulate its activity. The 'essential' light chain (misnamed because it can in fact be removed without loss of ATPase activity) stabilizes the extended heavy chain α-helix in the light chain domain.

In the absence of ATP, S1 fragments bind to actin molecules in a filament, forming the regular structure known as a decorated filament, already described in Chapter 5 (see Figure 5-4). Adding ATP to this complex causes the S1 heads to detach and the decoration to be lost, but, in a muscle, heads are held in such a position that they soon reattach, and hence progress along the actin filament toward its barbed end (Chapter 9).

## *Motility assays probe the interaction between myosin and actin*

The usual way to assess the quantity or activity of an enzyme is to measure the rate of the reaction it catalyzes. But how do you measure the activity of myosin or another motor protein whose principal function is to produce movements on a nanometer scale? At a crude, macroscopic, level, one can observe the contraction of artificial fibrils made from actin and myosin. But over the past decade, biologists have developing vastly superior techniques by which the movements of individual motor molecules can be visualized directly, and measured with great accuracy.

Assays of myosin function grew out of the study of streaming in *Nitella* cells mentioned in the introduction. Streaming movement in these enormous algal cells takes place on parallel actin cables, each about 0.2 µm in diameter, consisting of hundreds of actin filaments with identical polarity. An extensive network of endoplasmic reticulum (ER) in the cytoplasm slides against the actin cables, being propelled by a specific form of myosin and carrying with it other components of the viscous cytoplasm. In the first type of assay, cells of *Nitella* were slit open longitudinally and latex beads coated with myosin or fragments of myosin were applied through a micropipette. When provided with ATP, the myosin-coated beads were seen moving at steady rates due to repeated cycles of interaction with the aligned actin filaments. The beads move steadily toward the barbed end of the actin filaments—the same direction shown by the cytoplasmic streaming *in situ* before the cell was dissected.

Many new kinds of cell-free motility assay have been developed since the initial work with *Nitella*. In one variant, myosin molecules (or the head portions of myosin molecules) are fixed to a glass surface and the ATP-dependent movement of actin filaments is observed instead. Fluorescently labeled actin filaments now crawl like snakes over the bed of myosin (Figure 7-3). The velocity of movement in such assays depends on the rate of cycling of the myosin heads on actin filaments. For example, it

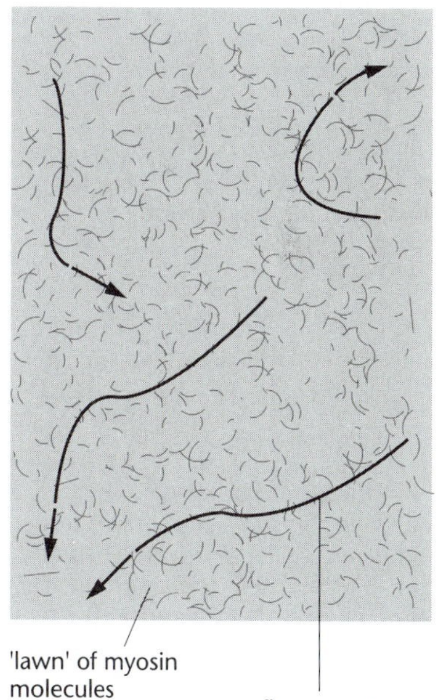


**Figure 7-3** Motility assay. Fluorescent actin filaments are allowed to attach to a lawn of myosin molecules (or HMM fragments) on a microscope cover glass. The reiterated action of myosin heads causes the filaments to slide over the surface, a movement that can be readily measured. Note that the myosin heads are randomly distributed over the surface—persistent movement of a filament in one direction is possible because only those myosin heads with the correct orientation can attach and move.

is found that myosin molecules (or S1 fragments) from skeletal muscle cause more rapid movement than myosins from smooth muscle, and myosin from platelets is even slower.

## *Single myosin molecules can be measured at work*

The most exciting aspect of motility assays is that they allow single motor molecules to be measured. Molecular motors such as myosin, and the microtubule-based motors described in Chapter 12, have sizes measured in tens of nanometers and take steps of a few nanometers. The forces exerted by these molecular machines are typically a few piconewtons (a newton being the force required to accelerate a mass of one kilogram by 1 m per second per second). Thus one of the challenges has been to design microscopes with detection systems capable of measuring these tiny forces and displacements with a temporal resolution in the millisecond range (an ATPase cycle is about 30 msec). In most cases the motor itself is not visualized but an object attached to it—a cytoskeletal polymer or a latex bead or a glass needle. Although the conventional resolution limit of visible light is approximately 200 nm and therefore too large to be useful, the relative position of an object can be followed with nanometer accuracy by means of a photodiode detector or by computational means. Dual-beam interferometry can be used to extend motion detection limits to a resolution of less than a nanometer.

Traditionally, force measurements were made by holding a muscle fiber at either end in a tensiometer. The force can be feedback-controlled and quick-release steps can be used to measure tension recovery with millisecond timing. The force exerted by myosin molecules can in similar fashion be measured by attaching actin filaments to a flexible glass needle. By measuring the maximal bending of the needle caused by the motor, and knowing the needle's stiffness, the isometric force generated by the myosin can be measured.

Another strategy for applying a counteractive force has been the introduction of laser-based optical traps and allied techniques. A focused spot of laser light tends to draw into its center any small (micrometer-sized) object, such as a polystyrene bead. The bead is trapped by light energy in the center of the spot and if the spot is moved it carries the bead with it, just as if the bead were held in a minute pair of tweezers. This technique, sometimes referred to as *laser tweezers*, has been used in assays of increasing sophistication and accuracy to maneuver molecules into position and hold them so that their interaction can be detected. Servomechanisms controlled by electronic feedback then allow one to measure with enormous accuracy the force and displacements caused by the actin–myosin interactions (Figure 7-4). It has even been possible to combine

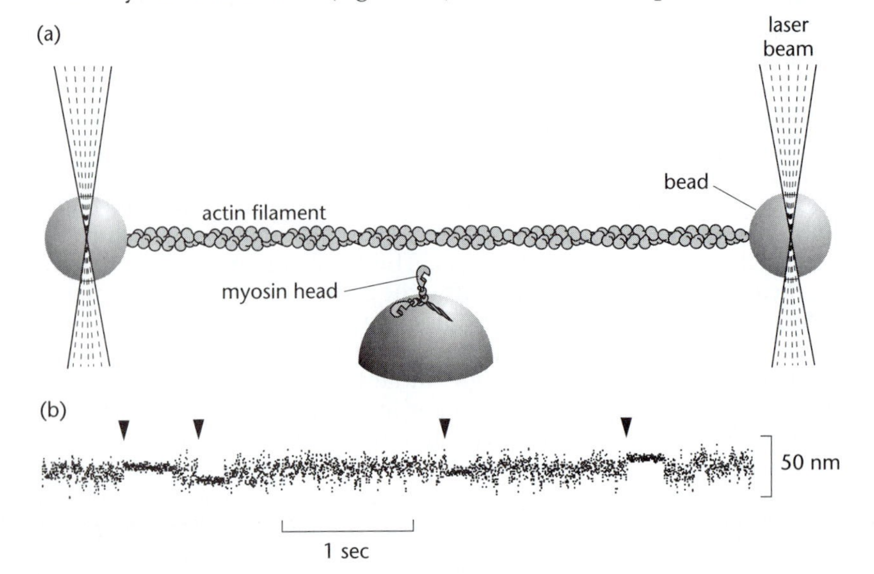


**Figure 7-4** Measuring a single myosin molecule. (a) In this setup (not drawn to scale) a single actin filament is attached at either end to a bead held in a laser trap. The filament is pulled taut and lowered onto a silica bead firmly fixed to a microscope coverslip, The bead is sparsely covered with myosin II molecules (actually the truncated head regions of these molecules). (b) Typical record of the movements of a filament in such an experiment. Large displacements due to thermal noise are reduced when the filament is trapped by a myosin head (arrowheads). Detailed analysis of these periods of attachment yields information on the step size of the myosin head (about 10 nm under conditions of low load) and the force it generates (3–4 pN). The work done by the motor, calculated as the force × distance, is around 10kT—about 40% of the maximum energy available from the hydrolysis of a single ATP molecule. (Recording courtesy of Justin Molloy.)

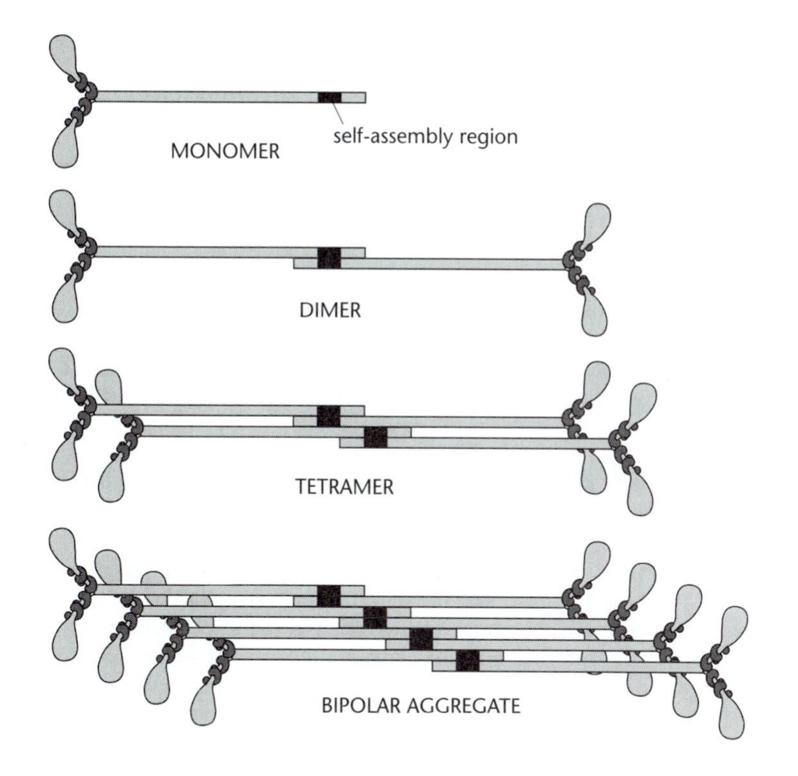


**Figure 7-5** Bipolar aggregate of myosin II molecules. Specific sequences in the long tail of *Acanthamoeba* myosin cause the molecules to associate into bipolar aggregates, as shown.

these mechanical measurements with high-resolution fluorescence techniques that record the binding and dissociation of a fluorescent analogue of ATP. In principle, we can now measure the movement, force generation, and ATP hydrolysis of a single myosin molecule!

## *The tail of myosin II self-assembles into a bipolar filament*

If the myosin head provides the driving force for movement, then its tail must be attached to whatever is to be pulled. As we will see later in the chapter, different myosin families are normally classified according to their head sequences. But within a family, tails vary greatly, each adapted to the particular function of the molecule. Myosin II is characterized by having a long, stiff α-helical tail, which enables many molecules to associate into a large bipolar aggregate (or filament), with heads pointing away from the midline (Figure 7-5). Indeed, this configuration is vital for the function of the molecule. Each myosin head binds to actin filaments in a stereospecific way and moves along the filament in a specific direction (toward the barbed end). Thus, two heads, linked together but pointing in opposite directions will bind to actin filaments of opposite polarity and attempt to move in opposite directions along them. The ensuing tug-of-war produces a force tending to pull the two actin filaments together—that is, a contraction (Figure 7-6).

The tail region of the myosin II molecule is built from a common structural motif known as a *coiled-coil*. This rodlike structure comprises two α-helices with a characteristic spacing of hydrophobic residues that form a 'stripe' of hydrophobicity that winds around the α-helix and helps

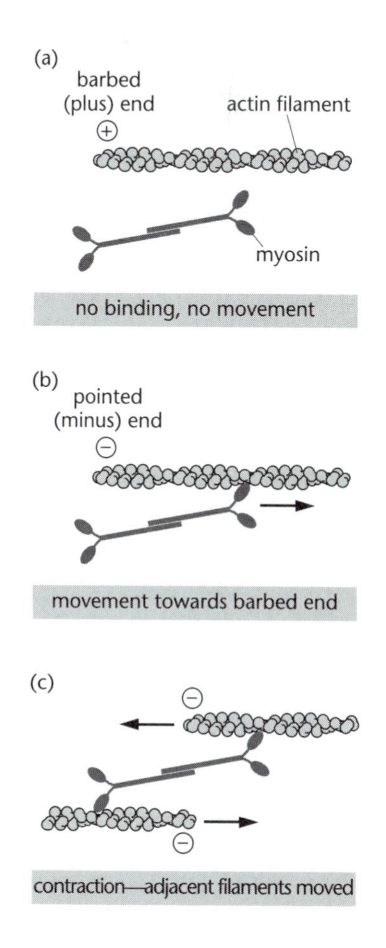


**Figure 7-6** How a bipolar myosin filament causes contraction. (a) Binding of myosin heads to actin filaments is stereospecific and does not occur if the actin filament is in the incorrect orientation. (b) If the actin filament is correctly oriented, the myosin moves along it toward the barbed end. (c) If binding takes place at either end of a bipolar aggregate, a tug-of-war ensues in which two actin filaments are moved against each other—the molecular basis of contraction.

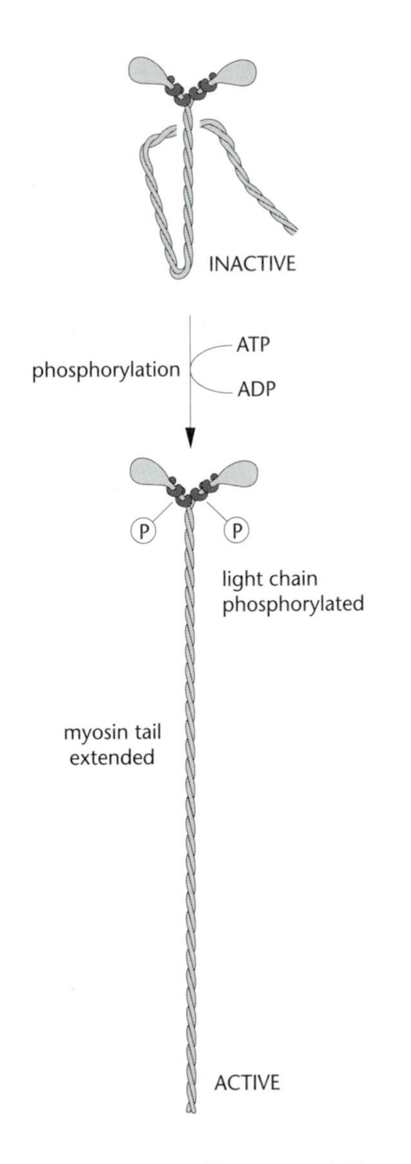


**Figure 7-7** Phosphorylation control. The controlled phosphorylation of one of the light chains has at least two effects: it causes a change in the conformation of the myosin head, exposing its actin-binding site, and it releases the myosin tail from a sticky patch on the myosin head, thereby allowing the myosin molecules to assemble into bipolar filaments.

them coil around each other. In myosin and many other cytoskeletal proteins, the two helices run in parallel (that is, in the same direction from N-terminus to C-terminus), giving rise to a filament with a diameter of about 2 nm.

While the structure of individual myosin II molecules depends on ionic and hydrophobic interactions between the two α-helical heavy chains, association of myosin molecules with each other depends on both hydrophobic and ionic interactions between their tails. All species of myosin II have a strong seven-residue repeat characteristic of a coiled-coil α-helix, and in addition a 28-residue repeating pattern of charged amino acids. There are also distinct regions in particular myosins where the repeating pattern is less strong, and these may be the origin of specific sites of bending and sensitivity to proteolysis seen in these molecules.

## *Myosin II molecules cause bundles of actin filaments to contract*

We see that myosin II molecules assemble into bipolar filaments that can cause actin filaments to contract if supplied with ATP. The strength of contraction and its direction and duration will then depend on the spatial arrangement of actin and myosin molecules. In an artificial mixture of actin filaments and myosin molecules, contraction takes the form of local clumping. Clots of tangled actin filaments, drawn together by the action of the myosin heads, expel fluid as they contract—a process known as *syneresis*. Something analogous to syneresis may occur in the actin cortex, for example in the tail of a migrating cell.

Efficient, spatially oriented contractions, however, require highly organized filament systems, especially bundles of actin filaments arranged with alternating polarity. Actin bundles of this kind can be found in nonmuscle cells, for example in the stress fibers in fibroblasts that end in focal adhesions (Chapter 6). Although structures of this kind are not built with the precision of a skeleton myofibril, they do have a rudimentary musclelike organization and are capable of contraction, albeit relatively slow and weak.

How are organized bundles of actin and myosin produced by the cell? How does the cell control precisely when contraction takes place? Our next concern is to examine these questions.

## *Assembly of myosin into filaments is influenced by phosphorylation*

Evidently, the spontaneous self-assembly of myosin II filaments into large bipolar aggregates must be controlled in the living cell. This is especially the case in nonmuscle cells, where contractile assemblies of actin and myosin are produced only in certain locations and at specific times. For these forms of myosin II, the principal form of control is exerted by protein kinases that phosphorylate the protein at specific locations.

If nonmuscle myosin II from human platelets is treated with a phosphatase to remove covalently bound phosphate groups from the light chains, it becomes freely soluble. Sedimentation analysis reveals single myosin molecules with a compact folded configuration, and the electron microscope shows such molecules to have their tail regions folded back, as though adhering to a sticky patch near the myosin head (Figure 7-7). In this configuration myosin molecules are unable to assemble efficiently into filaments. The effect of adding back phosphate groups to a specific site on the myosin head is to make it less sticky: the tail can extend and associate lengthwise with the tails of other myosin molecules (thereby generating bipolar myosin filaments). It may be recalled from the previous chapter that a similar 'foldback inhibition' strategy is also used by ERM proteins to control their interaction with the plasma membrane. Another example will be discussed shortly in regard to the myosin light chain kinase.

Phosphorylation sites are also found on the heavy chain of myosins, and during events such as mitosis and secretion, multiple residues on both the light chains and the heavy chains are phosphorylated. Myosin II from *Dictyostelium*, for example, has specific serine residues in its tail that, when phosphorylated, cause the tail to fold back on itself, inhibiting polymerization and inhibiting actin-ATPase activity. Note that the sites of phosphorylation and the consequences of phosphorylation depend on the particular type of myosin and the type of kinase responsible.[1]

## Light chain phosphorylation can regulate myosin activity

Phosphorylation influences more than just the aggregation of myosin molecules, it can also regulate their ATPase activity. For example, a specific serine residue in the regulatory light chain of smooth-muscle myosin II has to be phosphorylated before it can cause contraction. Phosphorylation is catalyzed by the enzyme *myosin light chain kinase* (*MLCK*) and is a relatively slow process with maximum contraction requiring nearly a second. Note that the very much faster activation of myosin II from skeletal muscle, which becomes active within a few milliseconds, is controlled not by phosphorylation but by an actin-based mechanism, discussed in Chapter 9.

Myosin light chain kinase has a catalytic domain, containing the active site required for phosphorylation, and a regulatory domain. The latter works as a 'pseudo-substrate' which, in the nonstimulated enzyme occupies the active site, thereby preventing the kinase from acting on its normal substrate (another example of foldback inhibition). When $Ca^{2+}$-calmodulin binds to the regulatory domain, the latter loses its affinity for the active site, thereby releasing the activity of the enzyme.

For example, if a smooth-muscle cell receives neuronal or hormonal stimulation, its intracellular calcium ion concentration rises from approximately 140 nM in the resting cell to 500–700 nM. As a consequence of this elevated concentration, four $Ca^{2+}$ ions bind to each calmodulin molecule, inducing a conformational change that exposes hydrophobic sites for interaction with a number of proteins, including MLCK. The complex of $Ca^{2+}$-calmodulin and MLCK then catalyzes the transfer of a phosphate group to the myosin light chain, triggering the cycling of myosin along actin filaments with the development of force and contraction of the muscle.

A second enzyme in smooth-muscle cells, a phosphatase, removes the phosphate group from myosin light chains. Its activity is independent of calcium ions but is subject to indirect control by the GTP-binding protein Rho, mentioned in the previous chapter. In its active state, Rho inhibits the phosphatase responsible for removal of phosphate groups from the myosin light chain and also activates another kinase (Rho-kinase) able to phosphorylate the myosin.

## Tropomyosin helps to stabilize actin filaments for contraction

For a cell to make a large contractile structure such as an adhesion belt or stress fiber, it needs more than just myosin. Obviously, actin filaments have to be there, not only to give the myosins something to pull against but also to help build the contractile structure. Bipolar filaments of myosin form faster if actin filaments are present, suggesting a cascade of mutually reinforcing steps in the formation of contractile bundles in the cell.

Moreover, as we saw in Chapter 6, other proteins, such as α-actinin, filamin, and fimbrin, also influence the spatial arrangement of actin

[1] Phosphorylation of the myosin head can affect the function of the tail, and vice versa: the different parts of the molecule are more interdependent than suggested by the simple diagram in Figure 7-1.

filaments. Specific combinations of actin-binding proteins both cooperate and compete to produce ordered patterns of interactions with specific, and perhaps profound, consequences for the cell. Interactions among the components of the network are further modulated by changes in the concentrations of ions in the cytosol, by protein phosphorylation, and by physical forces that stretch or compress the network and hence move its constituent molecules relative to each other. We are only just beginning to understand this crucial aspect of cytoskeletal organization.

In many large cytoskeletal assemblies, there is usually a key protein that is synthesized in limiting amounts and that regulates the formation of the entire complex. In the contractile assemblies containing myosin, a protein called *tropomyosin* often has this role. Tropomyosin is a rigid rodlike molecule composed of two fully α-helical chains in a coiled-coil dimer. We saw above that a similar structural motif is present in myosin molecules, so the two proteins give similar x-ray diffraction patterns (hence the similarity in name) but they have very different functions. Tropomyosin binds along the length of actin filaments, stabilizing them and modifying the types of other proteins that can bind. In particular, tropomyosin enhances the interaction of actin filaments with myosin II but inhibits, or at least limits, actin's association with actin-bundling and actin-fragmenting proteins.

Most cells contain multiple isoforms of tropomyosin that differ in sequence and molecular weight. Humans have four tropomyosin genes, for example, the gene products of which are further diversified by alternative promoter activities and alternative splicing. High-molecular-weight forms, typified by the 36 kDa form present in striated muscle, bind most strongly to actin. They align with actin filaments, each tropomyosin interacting with seven actin monomers along its length and overlapping in a head-to-tail fashion with the next tropomyosin molecule (Figure 7-8). At the other end of the range, a small isoform of tropomyosin with a molecular weight of 30 kDa, abundant in nonmuscle cells, binds less strongly to actin filaments. It interacts with only six actin monomers, and shows no overlap.

Because higher-molecular-weight isoforms of tropomyosin associate more strongly with actin, they are usually found in structures that are more stable, such as stress fibers. The low-molecular-weight isoforms are abundant in dynamic regions of the cell such as lamellipodia. Changes in the number of stress fibers in a cell (following viral transformation, for example) occur in parallel with, and perhaps are caused by, changes in the types of tropomyosin present. In fission yeast a specific tropomyosin is associated with the transient bundle of actin filaments that forms in these cells as they enter division, which is our next topic.

## *Large animal cells divide by a contractile ring of actin filaments*

One of the best known contractile bundles is that employed by certain large animal cells to divide their cytoplasm into two (Figure 7-9). The first division of a sea urchin egg, for example, employs a ring of actin filaments associated with myosin II molecules just beneath the plasma membrane at the position of cleavage. The force exerted by this *contractile ring* has been measured by means of flexible needles inserted into the egg. The

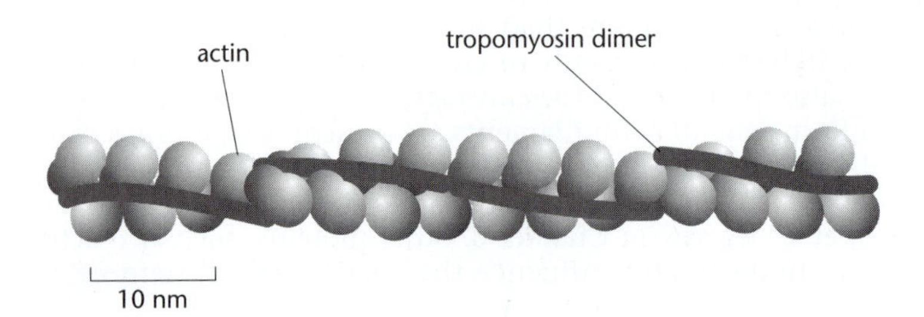


**Figure 7-8** Tropomyosin and actin. The tropomyosin molecule is a long, rigid coiled-coil dimer that fits into the groove of an actin filament. This association makes the filament stiffer and also modifies its association with other proteins. Thus, tropomyosin–actin filaments are not cross-linked by filamin or α-actinin, but their interaction with myosin II is enhanced.

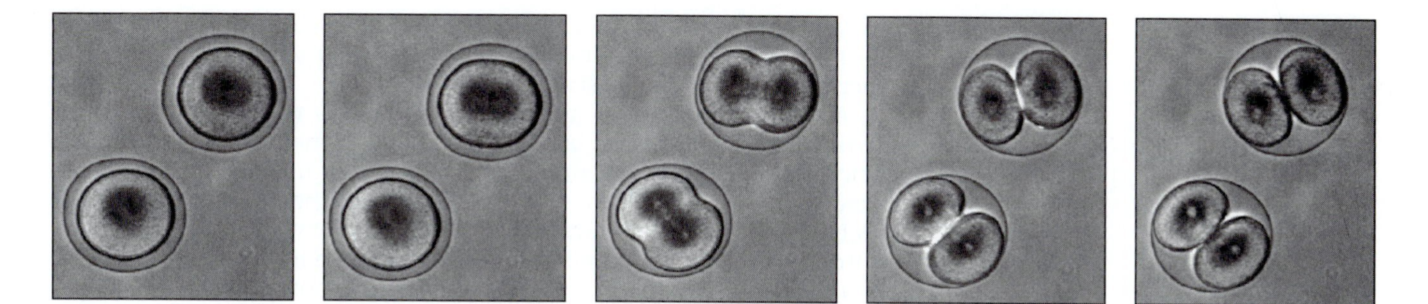

**Figure 7-9** Cytokinesis. Two sea urchin cells (*Lytechinus variegatus*) divide in this sequence of frames about 1 minute apart, taken from a time-lapse video sequence. (Courtesy of Julie Canman.)

contractile ring assembles at a late stage of mitosis (nuclear division) and if the actin-specific drug cytochalasin is added at this stage, cortical filaments are disrupted and the furrow retracts, leaving the cell uncleaved. If cytochalasin is rapidly removed by washing, the furrow reforms and splits the cell.

As it constricts the cytoplasm, the contractile ring maintains the same thickness in cross-section, so that its total volume and the number of filaments it contains must decrease steadily (Figure 7-10). Eventually the contractile ring disappears entirely, just before the two daughter cells separate. It is therefore not a permanent structure but one that is made and dismantled as required.

The contractile ring is not in fact a universal feature of cell division. Plant cells do not have one, and recent studies with vertebrate tissue cells, *Dictyostelium* amoebae, and yeast cells show that these can also divide without an accumulation of either actin or myosin II at their midline. Moreover, as we discuss shortly, mutants of *Dictyostelium* lacking the gene for myosin II can perform cytokinesis, although their efficiency is impaired. It now seems more likely that in its most generic form, animal cell division is driven by a flow of actin from the opposing poles toward the center of the cell, driven by regional differences in actin assembly and disruption. According to this view, myosin II-containing contractile rings are needed only where large forces have to be developed, as when a large volume of cytoplasm has to be partitioned. We will return to the cortical flow of actin, and the forces that drive it, in the following chapter.

One of the most remarkable aspects of cell division is its precision. Some machinery in the cell selects the appropriate location and then directs actin filaments and myosin II and other molecules to that site. There is evidence that the microtubule-based mitotic spindle has such a role: the division plane always bisects the mitotic spindle, and if the spindle is experimentally moved to another region of the cell, the contractile ring moves with it. Furthermore, furrowing always begins in the region of the cortex closest to the mitotic apparatus. We will discuss other aspects of the machinery of cell division in Chapter 13 and return to the fascinating question of cross-talk between actin filaments and microtubules in Chapter 19.

## *Genetic analyses probe the function of myosins*

Genetics today makes an increasingly important contribution to our understanding of cell movements, providing a means both to test the function of known components and to identify new ones. Two distinct approaches are widely employed: (1) forward or 'classical' genetics, which begins with a mutant cell or organism and then traces the gene or genes that cause the observed defect, and (2) reverse genetics in which investigators begin with a known protein and then inactivate it by mutation.

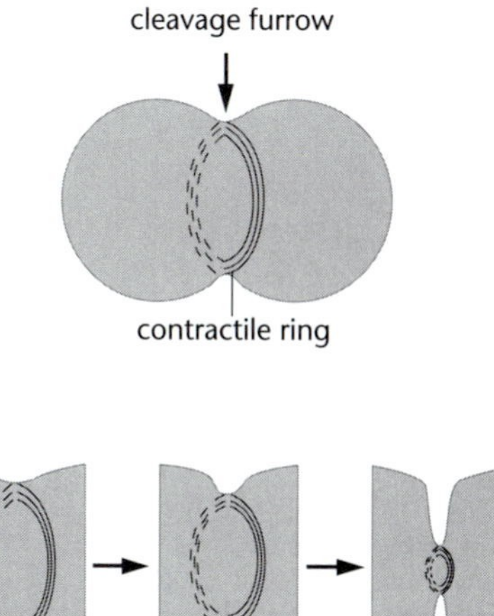


**Figure 7-10** Contractile ring. (a) The toroidal bundle of actin filaments and myosin II molecules forms just beneath the plasma membrane. (b) Contraction of the ring pulls the membrane inward to produce a furrow, and squeeze the contents of the cytoplasm into two.

Typical organisms used in such studies are the slime mold *Dictyostelium*, the fruit fly *Drosophila*, and the yeast *Saccharomyces*, all of which are amenable to genetic analysis.

An early triumph of the genetic approach was the isolation of mutants of *Dictyostelium* lacking myosin II. Amoeboid cells of such mutants are defective in a number of respects but unexpectedly still able to crawl over surfaces. The cells can also divide, although not by the usual mechanism. Instead of dividing in suspension by means of a typical contractile ring, the large multinucleated cells of the mutant achieve fission by pulling themselves apart while on a surface. In contrast, genetic studies show that myosin II is necessary for normal division of cells in the embryo of a nematode worm and for cell division in a yeast.[2] In *Drosophila*, myosin II appears to be needed not only for conventional cell division but also for the cortical furrowing of the multinucleate blastomere of this organism (discussed in Chapter 20).

Since the amino acid sequence and three-dimensional structure of the myosin II molecule are known, it is possible to target mutations to particular locations of the molecule to see what part they play in cytokinesis. Another trick, again based on genetics, is to fuse the myosin heavy chain to the small protein green fluorescent protein (GFP), making it possible to track the localization of the myosin in cells undergoing division. These techniques have shown, for example, that phosphorylation of three specific threonines in the myosin heavy chain modulates the assembly of myosin II into thick filaments and the contractile ring. Fully phosphorylated myosin cannot assemble into filaments and will not form a contractile ring, whereas cells with fully dephosphorylated myosins produce thick filaments in abundance but their cleavage furrow cannot disperse. Evidently, the cell must titrate very precisely the level of myosin phosphorylation as it enters division.

Genetic approaches have also been crucial in identifying novel cytoskeletal proteins needed to make a functioning contractile ring. A family of proteins called septins, for example, were discovered through mutants unable to make the septum in budding yeasts. Proteins closely related to yeast septins have since been found in other organisms, often localized close to their contractile ring. The assembly process is also influenced by specific small G proteins related to Rho, which as we saw in the previous chapter play a special role in controlling cortical actin.

## *Recombinant DNA techniques reveal a large family of unconventional myosins*

A staggering amount of genomic sequence information has been gathered since the early 1990s and has had a major impact on the study of cell movements. One direct consequence has been the identification of multiple genes related to myosin, recognized because they possess sequences similar to those encoding the head region of myosin II. Budding yeast has five myosin genes (in classes I, II, and V) whereas mice have at least 26 myosin genes, in seven classes. Any given vertebrate cell may contain more than a dozen different kinds of myosins.

To date, 14 structurally distinct classes of myosin heavy chains have been identified in addition to the well-characterized myosin-II's of muscle and nonmuscle cells (Table 7-1). These classes are empirically defined on the basis of sequence comparisons of their conserved motor or head domains, numbered in the order of their discovery, although their mechanochemical properties are not completely documented. The motor domains of characterized myosins, as already mentioned, bind actin in an ATP-sensitive manner and generate force through the hydrolysis of ATP. Almost all known myosins have an N-terminal motor domain linked to a C-terminal tail via a neck region that serves as a binding site for myosin light chains. All except one (myosin VI) move along actin filaments toward their barbed ends. The tail regions of the molecules are also characteristic of a given class (Figure 7-11). A subset of myosins have tail

[2] The first evidence for the participation of myosin II in cell division was actually obtained in 1977 from the injection of antibodies into a dividing sea urchin egg. There are other ways to tackle these questions than by genetics!

Table 7-1 Selected myosins and their function

| Class | Number of members | Heavy chains | IQ motifs | Structure | Potential function |
|---|---|---|---|---|---|
| I | 19 | 1 | 1–6 | | Drive membrane movements of many kinds: secretion, phagocytosis, pseudopod formation. Hair cell adaptation (hearing). |
| II | 21 | 2 | 2 | | Forms thick filaments and produces contraction: in stress fibers, contractile ring, and muscle. |
| III | ? | 1 | 1–2 | | Structural links and tension generation in microvilli and the cell body of photoreceptor cells (vision). |
| V | 5 | 2 | 6 | | Organelle transport and docking. |
| VI | 2 | 2 | 1 | | Stabilizes stereocilia (hearing). Organelle movement (toward the pointed end of actin filaments). |
| VII | 2 | 2 | 1 | | Membrane turnover in photoreceptors (human vision). |
| IX | ? | 2? | 4–6 | | Differentiation of leucocytes, signal transduction. |

Molecular structures are shown schematically with actin-binding heads to the left and tail regions to the right. IQ motifs are sites on the heavy chain to which one molecule of calmodulin (or one myosin light chain) binds. Where different functions are shown for a given class, they are typically performed by specific subtypes.

domains with predicted coiled-coil-forming α-helices suggesting that these myosins like myosin II are two-headed. Other myosins contain sequences recognized from other proteins, such as membrane-binding domains, kinases, or signaling motifs.

## *Myosin I's associate with membranes*

The first of these 'unconventional' myosins was isolated (in this case by biochemical rather than genetic techniques) by Tom Pollard and Ed Korn in 1972 from the small amoeba *Acanthamoeba*. Designated *myosin I*, this protein is now known to be just one of a large family of related myosins, with multiple members often in the same cell. Myosin I's are single-headed and have a globular tail region that is typically much smaller than the tail of myosin II. One or more IQ sequences each bind a molecule of calmodulin or a calmodulin-like light chain. Some myosin I's contain sequences rich in basic amino acids shown to promote binding to phospholipids, and these proteins are believed to operate with their tail attached to a membrane.

Recent studies suggest that myosin I's are extensively involved in the movements of membrane vesicles inside cells, especially in movements to and from the plasma membrane. Both *Acanthamoeba* and *Dictyostelium* have multiple myosin I genes, which gene knockout studies suggest act in concert or have overlapping functions. The uptake of small particles (phagocytosis) and uptake of fluid (pinocytosis) are impaired in cells lacking myosin I genes, and myosin I's are found near the Golgi in some vertebrate cells and may be involved in the production of secretory granules. Myosin I's are also thought to be involved in the cellular locomotion of *Dictyostelium*, since cells lacking specific myosin I's show defects in pseudopodial extension and in cortical stiffness. Development in *Dictyostelium* is also impaired by myosin I deletions, with defects in fruiting body formation and in chemotactic aggregation.

The involvement of a single-headed molecule in vesicle transport seems a little surprising, especially as kinetic studies show that in the mechanochemical cycle of this protein the head is detached for much of the time. How are myosin I molecules able to keep hold of an actin

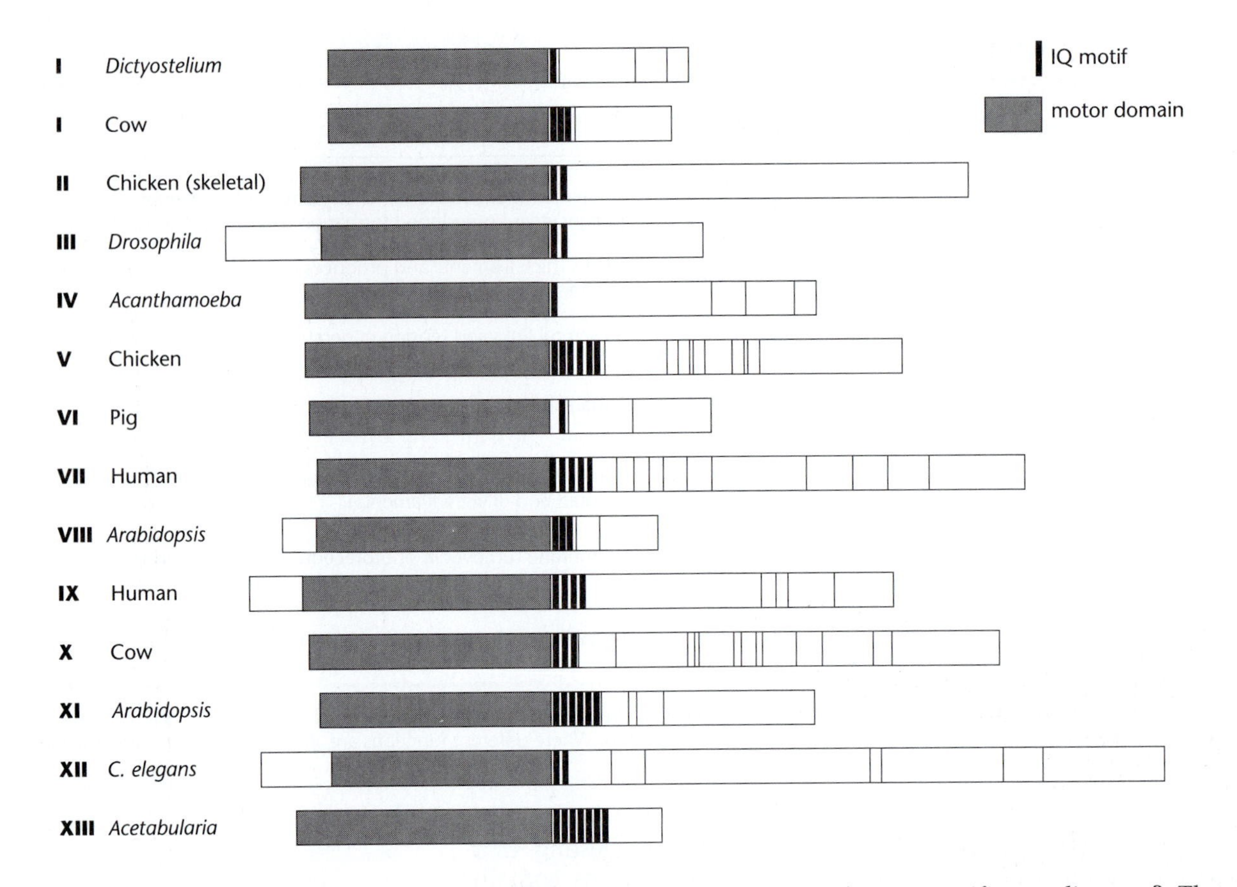


**Figure 7-11** Myosin sequences. Schematic view of the domain organization of 13 myosin classes, with N-terminal sequences to the left. Boxes indicate different functional domains predicted by sequence homology, such as the motor domains (shaded) and IQ motifs (black lines)—regions of light chain binding. Other domains are predicted to form coiled-coils, bind to other cytoskeletal proteins, and generate cytoplasmic signals.

filament long enough to travel any significant distance? The answer seems to be that the myosin I molecules do not work singly but in groups. Most vesicles, for example, probably carry multiple myosin I molecules on their surface, so that if one myosin head detaches the remaining heads can keep the vesicle attached.

## Myosin V attaches vesicles to actin filaments

Another type of myosin that seems certain to be involved in the movement and positioning of vesicles is myosin V. These are two-headed myosins with a high affinity for actin even in the presence of ATP, a pointer to the fact that they spend much of their time attached to a filament. The heads of myosin V are longer than those of conventional myosins and in the electron microscope the two heads can be seen to attach to equivalent positions on the helical actin filament—spanning a distance of about 36 nm. In nerve growth cones, myosin V associates with small vesicles on both microtubules and actin filaments, while, in species from yeasts to mice, mutations of myosin V cause a variety of defects associated with vesicle mobilization or movement inside cells.

The mouse mutants are especially interesting. Lack of the heavy chain of one isoform of myosin V produces mice with pale hair coloration, the genetic locus being called *dilute* for that reason. What this has to do with myosin becomes clearer when the melanocytes in the mouse skin—the cells that produce pigment granules—are examined in detail. Normally the long dendritic processes of these cells are full of vesicles containing black pigment. But in cells from a *dilute* mouse, the granules, although present in normal numbers, are clustered in its central region. Since, in a normal mouse, pigment granules are transferred from the dendritic ends of melanocytes to hair-forming cells, we can see why a *dilute* mouse has a paler coat.

Why then are the granules restricted to the cell body? Kinetic analysis of the movement of pigment granules shows, surprisingly, that mutant cells without myosin V are not impaired in this respect. Indeed, granules seem to move even more rapidly than normal, being carried in both directions along microtubules within the dendritic processes. What seems to be missing, however, is the ability of granules to leave the microtubule tracks and accumulate in the actin-rich regions at the tips of the processes. That is, the myosin V seems to act as a 'trap' for pigment granules, receiving them from the microtubule transport system and transferring them to a much slower mode of transport on actin filaments. In this way, accumulations of granules can build up in the dendritic tips.

## *Myosins are used in hearing and vision*

We know even less of the functions of other myosin classes, but evidence, once again from mutant studies, suggests an intriguingly wide range of activities. The myosin III found in *Drosophila* (product of the *ninaC* gene) seems from its amino acid sequence to have a protein kinase adjacent to its motor domain. Null mutations in *ninaC* cause visual defects in flies and degeneration of photoreceptor cells. The retinal cells in this species carry large numbers of microvilli, the membranes of which contain the light-sensitive pigment rhodopsin. Myosin III appears to be part of the structure of these microvilli and could be especially important for their rapid turnover that occurs during vision.

Hearing, no less than vision, is affected by the loss of specific myosins and both class VI and class VII myosins have been identified as deafness genes in mice and humans. Myosin VI is concentrated in the actin-rich cuticular plate that forms a base for the large stereocilia of auditory epithelial cells. As we saw in Chapter 6, these large microvilli move in response to sound vibrations and their movement generates the neural signals sent to the auditory region of the brain. Myosin VII also forms part of the structure of stereocilia, providing cross-links between actin filaments and the plasma membrane—defects in this protein have been correlated with several types of human deafness. Lastly, an isoform of myosin I may be a crucial part of the sensory transduction process itself. Auditory transduction is dependent on the action of a stretch-gated channel that is modulated or adapted to maintain sensitivity to changes in mechanical stimulation. The process of adaptation may be facilitated by a myosin motor restoring resting tension and myosin Iβ is present in the correct location to provide this restoring tension.

The possible functions of myosin do not end there. Evidence from sequence homologies suggests that some myosins are also involved in intracellular signaling—the process by which cascades of chemical changes carry information in cells. Myosins can be phosphorylated on multiple sites and some, like myosin III just mentioned, even have kinase domains so that they can actually catalyze the transfer of phosphate groups to other proteins. Myosin IX has a region of sequence homologous to a GAP domain—that is, it is potentially able to inactivate a G protein by stimulating the hydrolysis of bound GTP. Other myosins have binding motifs that implicate them in the formation of clusters of proteins involved in signaling.

The discovery of this plethora of motor proteins, largely thanks to genomic sequencing, is one of the astonishing findings of cell motility research. What these myosins do in the cell, and how their capacity for movement is harnessed, are likely to engage researchers for many years.

## Outstanding Questions

*What directs myosin II to specific locations of a cell? How does a cell build a contractile structure such as a contractile ring or a stress fiber? How does Rho affect myosin activity, and why? More generally, how are molecular motors switched on and off? How do different myosins interact with actin filaments? What supramolecular structures do they form, and what is their function in the cell? Why do some cells produce a contractile ring containing actin and myosin II and not others? Are unconventional myosins involved in cytokinesis in animal cells? If so, what do they do? What ubiquitous molecules or molecular processes are common to cytokinesis in all animal cells? Which myosins generate streaming in* Nitella *cells? How do single-headed myosins, such as myosin I, travel along actin filaments without detaching? How do myosins such as those of class IX participate in cell signals? Why do they have an actin-activated ATPase? What are the mechanochemical properties of these and other unconventional myosins?*

## Essential Concepts

- Myosins are a large family of motor proteins that produce movement along actin filaments.
- Head regions of myosin molecules engage in an ATP-driven cycle in which they attach to an actin filament and pull against it, and then detach. Sophisticated motility assays, utilizing laser tweezers and other electronic devices, allow the interaction of individual molecules to be studied.
- The rodlike tail region of the myosin II molecule allows it to aggregate into bipolar filaments with clusters of heads at either end. Interaction with actin filaments in antiparallel arrangement then leads to the shortening of the actin–myosin bundle.
- Contractile bundles of actin filaments and myosin II are found widely in animal cells: in muscle, in stress fibers, in adhesion belts, and in the contractile ring formed during cell division.
- Formation of contractile bundles is influenced by phosphorylation of myosin by specific kinases. These not only control assembly of myosin into filaments but also regulate myosin's actin-activated ATPase activity.
- The assembly of contractile bundles relies on other proteins to hold the interacting actin and myosin in the correct positions. The side-binding protein tropomyosin, in particular, often has a crucial influence on the type of bundle produced.
- The contractile ring, by which large animal cells divide their cytoplasm, is a bundle of actin and myosin II filaments that forms and disperses during the cell cycle.
- Genome analysis reveals that eucaryotic cells contain multiple different myosins. These all have head domains homologous to that of myosin II; their tail regions are highly divergent.
- Genetic techniques indicate that myosins have many different functions in cells. These include the movement of vesicles to and from the plasma membrane, the transport of organelles, and the production and stabilization of pseudopods, microvilli, and stereocilia.

## Further Reading


Adams, R.J., Pollard, T.D. Binding of myosin I to membrane lipids. *Nature* 340: 565–568, 1989.

Balasubramanian, M.K., et al. A new tropomyosin essential for cytokinesis in the fission yeast *S. pombe*. *Nature* 360: 84–87, 1992.

Block, S.M. Real engines of creation. *Nature* 386: 217–219, 1997.

Chisholm, W.L. Cytokinesis: a regulatory role for Ras-related proteins? *Curr. Biol.* 7: R648–R650, 1997.

Chu, Q., Fukui, Y. *In vivo* dynamics of myosin II in *Dictyostelium* by fluorescent analogue cytochemistry. *Cell Motil. Cytoskeleton* 35: 254–268, 1996.

Drechsel, D.N., et al. A requirement for Rho and Cdc42 during cytokinesis in *Xenopus* embryos. *Curr. Biol.* 7: 12–23, 1996.

Fincham, V.J., Frame, M.C. The catalytic activity of Src is dispensable for translocation to focal adhesions but controls the turnover of these structures during cell motility. *EMBO J.* 17: 81–92, 1998.

Finer, J.T., et al. Single myosin molecule mechanics: piconewton forces and nanometer steps. *Nature* 368: 113–120, 1994.

Funatsu, T., et al. Imaging of single fluorescent molecules and individual ATP turnovers by single myosin molecules in aqueous solution. *Nature* 374: 555–559, 1995.

Glotzer, M. The mechanism and control of cytokinesis. *Curr. Opin. Cell Biol.* 9: 815–823, 1997.

Holmes, K.C. The swinging lever-arm hypothesis of muscle contraction. *Curr. Biol.* 7: 112–118, 1997.

Jay, P.Y., et al. A mechanical function of myosin II in cell motility. *J. Cell Sci.* 108: 387–393, 1995.

Kimura, K., et al. Regulation of myosin phosphatase by Rho and Rho-associated kinase (Rho kinase). *Science* 273: 245–248, 1996.

Knight, A.E., Kendrick-Jones, J. A myosin-like protein from a higher plant. *J. Mol. Biol.* 231: 148–154, 1993.

Lin, J.J., et al. Tropomyosin isoforms in nonmuscle cells. *Int. Rev. Cytol.* 170: 1–38, 1997.

Mabuchi, L., Okuno, M. The effect of myosin antibody on the division of starfish blastomeres. *J. Cell Biol.* 74: 251–263, 1977.

Mehta, A.D., et al. Myosin-V is a processive actin-based motor. *Nature* 400: 590–593, 1999.

Mehta, A.D., et al. Single-molecule biomechanics with optical methods. *Science* 283: 1689–1695, 1999.

Mermall, V., et al. Unconventional myosins in cell movement, membrane traffic, and signal transduction. *Science* 279: 527–533, 1998.

Miller, K.G., Kiehart, D.P. Fly division. *J. Cell Biol.* 131: 1–5, 1995.

Mooseker, M.S., Cheney, R.E. Unconventional myosins. *Annu. Rev. Cell Dev. Biol.* 11: 633–675, 1995.

Pollard, T.D., Korn, E.D. *Acanthamoeba* myosin I. Isolation from *Acanthamoeba castellanii* of an enzyme similar to muscle myosin. *J. Biol. Chem.* 248: 4682–4690, 1973.

Rayment, I., et al. Structure of the actin-myosin complex and its implications for muscle contraction. *Science* 261: 58–63, 1993.

Rayment, I., et al. Three-dimensional structure of myosin subfragment-1: a molecular motor. *Science* 261: 50–65, 1993.

Robinson, D.N., Spudich, J.A. Towards a molecular understanding of cytokinesis. *Trends Cell Biol.* 10: 228–237, 2000.

Ruppel, K., Spudich, J.A. Structure–function analysis of the motor domain of myosin. *Annu. Rev. Cell Dev. Biol.* 12: 543–573, 1996.

Sato, M., et al. Dependence of the mechanical properties of actin/α-actinin gels on deformation rate. *Nature* 352: 828–830, 1987.

Simmons, R. Single-molecule mechanics. *Curr. Biol.* 6: 392–394, 1996.

Stites, J., et al. Phosphorylation of the *Dictyostelium* myosin II heavy chain is necessary for maintaining cellular polarity and suppressing turning during chemotaxis. *Cell Motil. Cytoskeleton* 39: 31–51, 1998.

Titus, M. Unconventional myosins: new frontiers in actin-based motors. *Trends Cell Biol.* 7: 119–123, 1997.

Titus, M.A. Myosin V: the multi-purpose transport motor. *Curr. Biol.* 7: R301–R304, 1997.

Walker, M.L., et al. Two-headed binding of a processive myosin to F-actin. *Nature* 405: 804–807, 2000.

Wang, Y., et al. Single particle tracking of surface receptor movement during cell division. *J. Cell Biol.* 127: 963–971, 1994.

Wessels, D., et al. A *Dictyostelium* myosin I plays a crucial role in regulating the frequency of pseudopods formed on the substratum. *Cell Motil. Cytoskeleton* 33: 64–79, 1996.

Wu, X., et al. Visualization of melanosome dynamics within wild-type and dilute melanocytes suggests a paradigm for myosin V functions *in vivo*. *J. Cell Biol.* 143: 1899–1918, 1998.

# CHAPTER 8

# Fibroblast Locomotion

Few forms of cell movement have been so prone to controversy as the migration of animal cells over surfaces. Ever since the first detailed observations of amoeboid locomotion were made in the early part of the nineteenth century, scientists have been speculating—and disagreeing—about its underlying mechanism. Are pseudopodia pushed out by hydrostatic pressure? By surface tension? Do fibroblasts crawl by means of actin and myosin? By membrane flow? The perennial fascination with the phenomenon of crawling cells as well as, in recent years, the recognition of its importance to the metastatic spread of cancer cells and the form-shaping episodes of development, all help to fuel the debate.

It has become evident, too, that the crawling movements of animal cells are extremely difficult to explore at the molecular level. Different parts of the crawling cell change at the same time and there is not an easily identifiable motile organelle such as a flagellum that can be studied in isolation. Although everyone now concurs that actin is a common element in most crawling movements, this protein undergoes many different transformations as the cell advances, assembling into filopodia and lamellipodia, associating with focal contacts, forming contractile bundles, and so on. A complete account would have to give a molecular explanation for these different transformations, explain how they are coordinated in time and space, and also relate them to biophysical parameters such as the viscosity of the cytoplasm and the adhesion of the cell to its substratum.

We are still a long way from such a comprehensive picture. But, as we describe in this chapter, recent discoveries allow us at least to identify the principal driving force for cell crawling and permit a broad-brush outline of how it is integrated with other components of the cell.

## *Crawling cells extend, adhere, contract, detach*

As we saw in Chapter 2, many different kinds of animal cells exhibit crawling movements. Free-living amoebae of all kinds crawl through moist soil or over the surfaces of aquatic plants in ponds in search of food. In the adult human body, white blood cells crawl through tissues in search of infecting bacteria, which they devour, and skin fibroblasts migrate into wounds as part of the repair process. In a developing embryo, orchestrated migrations of cells establish the foundations of tissues and organs. The crawling of nerve growth cones and other cell appendages is the basic mechanism by which detailed connections in the brain, spinal cord, and other complex organs are made.

Cells that crawl may be tiny, like white blood cells or the amoeboid form of *Dictyostelium* (about 10 μm in length), or enormous, like *Chaos*

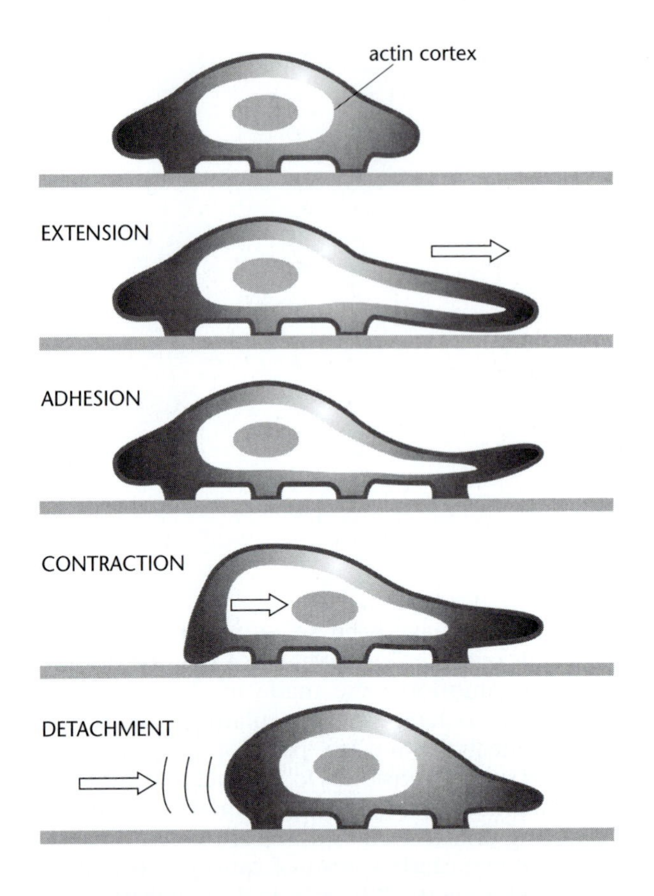


**Figure 8-1** Basic repertoire of movements shown by a crawling cell.

*carolinense*, which may reach several millimeters in linear dimension (and moves at tens of micrometers each second). Crawling cells extend blunt, stubby pseudopodia or fine filopodia, or they may migrate by means of large flat lamellipodia flowing out in flat sheets in advance of the cell. There are even, as mentioned in Chapter 2, protozoa that crawl by means of leglike bundles of cilia.

Regardless of its size and detailed shape, it seems that any crawling cell must, logically, perform a limited repertoire of movements. The cell must (1) *extend* its leading margin over the substratum; (2) *attach* to the substratum; (3) pull or *contract* using the newly formed points of adhesion as anchorage; and (4) release or *detach* its adhesions at the rear of the cell (Figure 8-1). In small simple cells, these movements may occur in a linear sequence, but in larger and more complex cells they usually occur simultaneously at different parts of the cell. Our task in this chapter is to describe the molecular machinery of these four elements of cell crawling and then to consider how they might be coordinated.

## Nematode sperm crawl without actin

Most of this chapter will be concerned with the role of actin filaments in crawling movements. However, there is one form of cell locomotion that, although it takes place without actin, illustrates the machinery of migration in a particularly basic form. The sperm cells of *Caenorhabditis elegans* and other nematodes lack flagella and migrate by a form of crawling similar to amoeboid locomotion (Figure 8-2). However, in contrast to that of amoebae, fibroblasts, and white blood cells, the migration of nematode sperm is driven by filaments made not of actin but from a totally unrelated 14 kDa protein termed *major sperm protein*, or *MSP*. During spermatogenesis, most of the actin and tubulin of the cell are sequestered into a compact body and discarded.

Activation of nematode sperm during copulation leads to the rapid assembly of MSP filaments. Within 6 minutes, the cells extend a filament-packed flattened pseudopodium, sometimes termed a villipodium, which

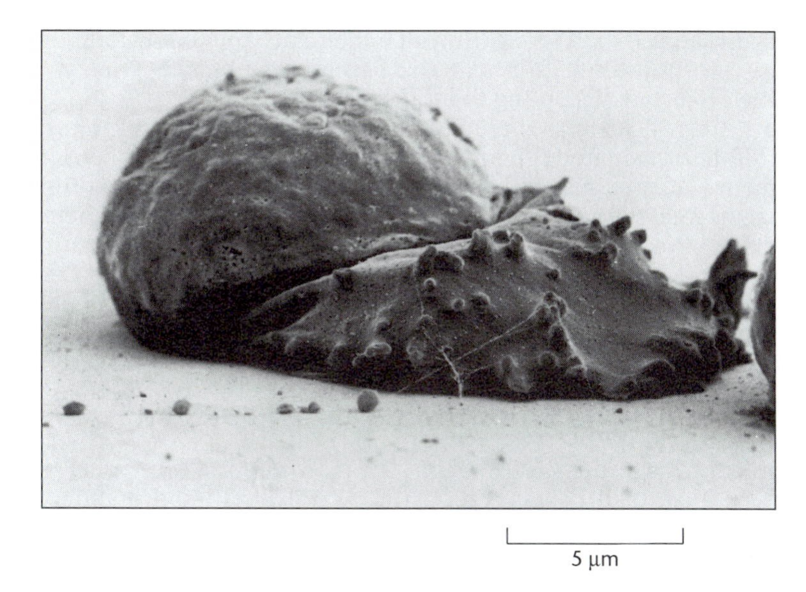


**Figure 8-2** Nematode sperm. (Scanning electron micrograph courtesy of Sol Sepsenwol, University of Wisconsin, Stevens Point.)

pulls the organelle-packed cell along at speeds up to 70 μm/min. Within the villipodium, MSP filaments are grouped into long, branched fiber complexes readily visible by light microscopy, which extend from the leading margin back to the cell body. Each fiber complex is constructed from a dense meshwork of 2 nm filaments that interdigitate with filaments from neighboring fiber complexes, to form a single interconnected unit. As a sperm cell crawls, the meshwork of MSP filaments remains stationary with respect to the substratum. Consequently, it is believed that filaments assemble at the leading margin of the pseudopodium and disassemble at its base in a continual treadmilling motion (Figure 8-3a,b).

MSP filaments are nucleated by specific structures associated with the plasma membrane, as may be demonstrated by examining cytoplasmic extracts derived from the sperm cells under a microscope. The extracts are initially free of filaments but they contain small vesicles derived from the sperm plasma membrane (but presumably inside out, with the cytoplasmic face of the membrane facing outward). Upon incubation in ATP, these vesicles nucleate a 'comet tail' of MSP fibers that grows from an attachment point on the coverslip, pushing the vesicle before it (Figure 8-3c). As in the motion of *Listeria* bacteria in actin-rich extracts described in Chapter 5, the sperm vesicles promote addition of subunits into filaments (in this case MSP filaments) proximal to the membrane and this polarized growth drives the vesicle through the cytoplasm.

In an intact sperm cell, presumably, MSP filament assembly drives the leading edge forward, triggered by the plasma membrane and ATP. How disassembly occurs at the trailing end of the villipodium is less clear, although a change in pH may be involved—a drop in intracellular pH below 6.0 has been shown to cause a rapid but reversible disassembly of filaments. Most puzzling of all is the fact that MSP filaments have no polarity, being built up of a dimeric subunit with a twofold rotational symmetry. This appears to rule out the operation of motor proteins and

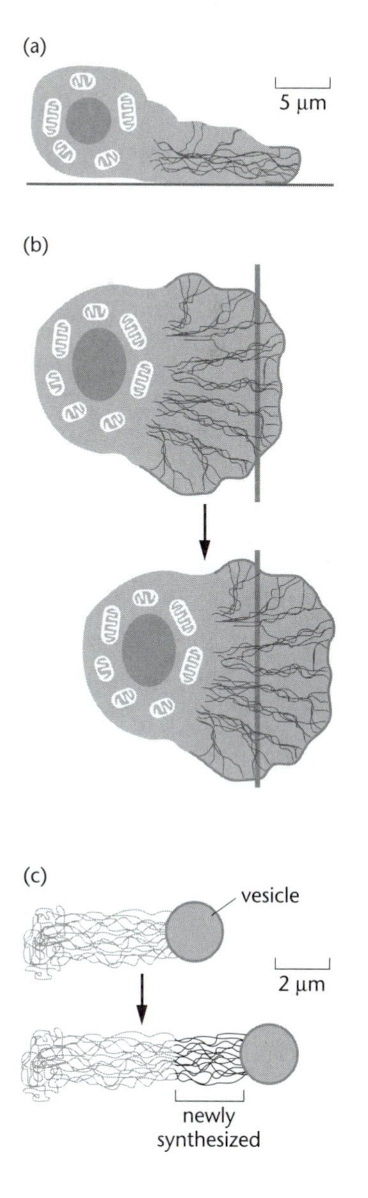


**Figure 8-3** MSP-based movement of nematode sperm and membrane vesicles. Side view (a) and top views (b) of a sperm cell crawling from left to right. A dense meshwork of MSP filaments fills the pseudopodium. Filaments assemble at the leading edge into a complex branched network. As the cell moves forward (b) new filament assembly occurs at the same rate as leading edge advancement. MSP filament disassembly occurs at the base of the pseudopodium near the cell body. Individual features of the branching network of fibers remain stationary with respect to the substratum as the cell translocates. (c) Inside-out vesicles derived from the plasma membrane can nucleate MSP filament assembly in cytosolic extracts. Newly associated filaments push the vesicle forward as they grow.

favors a push–pull model in which filaments anchored along their length to the substratum (through the plasma membrane), grow or shrink at their free ends according to local conditions.

We will see that the migration of other cells shows some similarities to that of nematode sperm (such as being driven by filaments that polymerize at one end and disassemble at the other) but in other respects seems much more complicated. But we would do well to remember, as we discuss the molecular minutiae of locomotion in fibroblasts and amoebae, that some of these interactions may be icing on the cake. Evidently, as we see in the case of nematode sperm, you *can* drive a cell with a much simpler engine.

## *The crawling of most cells is based on actin*

In contrast to nematode sperm, the crawling of all other cells that have been examined depends on actin. Actin fills the advancing margin of neutrophils, lymphocytes, fibroblasts, and amoebae and is the principal (sometimes the only) filamentous protein in filopodia, lamellipodia, and pseudopodia. The concentration of actin in these regions is remarkably high—estimated to be 12 mg/ml in lamellipodia and 40 mg/ml (that is, 1 millimolar!) in microvilli—so that actin-specific stains often display a distinctive cortical accumulation at the leading margin of the cell. Actin-binding drugs such as cytochalasin, phalloidin, and latrunculin cause a rapid, reversible cessation of crawling at low concentrations. By contrast, drugs such as colchicine and colcemid that act on microtubules have little or no effect on the forward progression of the cell itself, although they can have interesting effects on the direction taken by the cell, as we discuss later in the chapter.

A cell that displays actin-based crawling movements in a particularly pure form is the nonpigmented epidermal cell isolated from fish scales, sometimes termed a *keratocyte* (Figure 8-4). These small, flattened cells migrate unusually rapidly in culture, traveling at speeds of up to 10 μm/min or more. Electron microscopy and immunostaining with antibodies indicate that the predominant cytoskeletal element present in the flattened leading edge of the cells is actin filaments and that intermediate filaments and microtubules are largely confined to the trailing region around the cell nucleus. Furthermore, fragments of lamellipodia detached from the cell—produced either spontaneously as cells pull free of a tissue explant or deliberately by cutting with a glass knife—are able to migrate with equal rapidity to the intact cell. All indications are that in this simple cell, actin filaments, in association with myosins and other actin-binding proteins, not only move the cell over a surface but also maintain its distinctive shape and polarity.

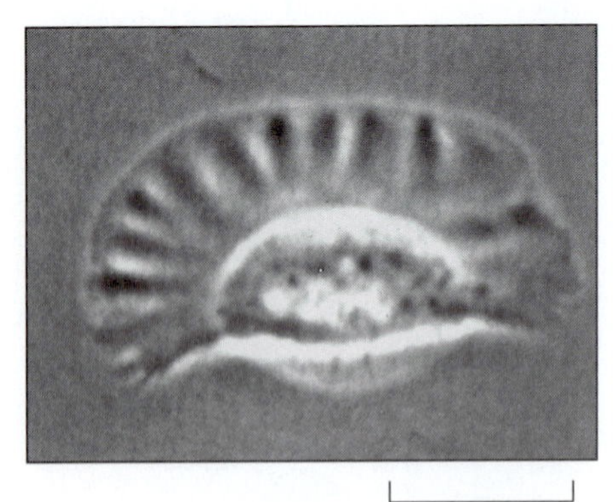


**Figure 8-4** Keratocyte in culture. The cell migrates with a fan-shaped lamellipodium at its front, or leading edge, which is uppermost in this figure. (Courtesy of Juliet Lee.)

## *A migrating cell advances by polymerizing actin*

There is good reason to believe that the leading margin of most migrating cells is driven forward by the growth of actin filaments. The action of cytochalasin, mentioned above, was an early indication of this fact, and so too was evidence from the growth of actin-rich cell extensions, such as the acrosomal process of *Thyone* sperm described in Chapter 6. Studies in which fluorescent actin is injected into cells in culture and then photobleached by a focused spot of light also provide support for an addition of actin to the leading margin of the membrane. The bleached spots in such experiments move steadily back from the leading edge, suggesting a flow of actin molecules analogous to the treadmilling motion described in Chapter 5.

However, the molecular mechanism is unlikely to be simple. Recall that the lamellipodium of a migrating fibroblast not only advances but also retracts—ebbing and flowing in fluid fashion, and even lifting off the

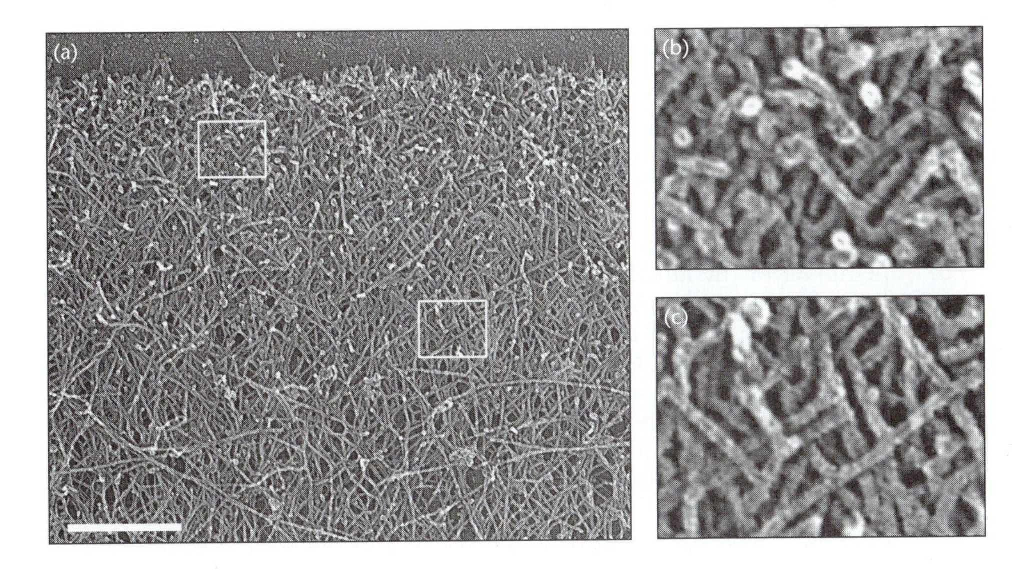


**Figure 8-5** Actin filament organization in a lamellipodium. Highly motile keratocytes from frog skin were fixed, dried, shadowed with platinum, and examined in an electron microscope. (a) Actin filaments form a dense network with filament fast-growing ends terminating at the leading margin of the lamellipodium (top of figure). Bar = 0.5 μm. (b) and (c) Boxed regions from (a) are enlarged to show Y-junctions between individual filaments. (Courtesy of Tatyana Svitkina and Gary Borisy.)

surface and moving over the dorsal surface (Chapter 2). If actin polymerization is responsible for these movements, then it is surely controlled in a very sensitive fashion. We surmise that there must be a 'polymerizing machine' carried at the leading margin of the cell that responds to external influences by adding or removing actin molecules from the ends of actin filaments. Deeper in the cytoplasm, other molecular machinery is needed to break or depolymerize the actin filaments (which do not grow for ever) and recycle their subunits to the front of the cell for further assembly.

What these molecular engines are and how they work are questions under active investigation. Some intriguing hints have already emerged from a detailed examination of the ultrastructure of lamellipodia. If cells migrating in culture are rapidly fixed and their plasma membranes are stripped by a brief exposure to detergent, then the underlying actin cortex becomes exposed and can be examined in an electron microscope. This reveals (following decoration with S1 fragments) that the filaments are, without exception, arranged with their barbed ends nearest to the leading edge, consistent with the idea that this is where actin molecules add. But, unexpectedly, filaments at the leading margin are also highly branched, like the surface of a bush or tree, with the shortest 'twigs' being found closest to the leading margin (Figure 8-5). Twigs apparently grow out from existing filaments, forming a series of 'Y'-shaped junctions every 20–50 nm or so, so that a dense brush of filaments fills this region of cytoplasm.

## *Arp2/3 and cofilin support treadmilling in the leading lamella*

The key to the production of the short twigs of actin filaments is the *Arp2/3 complex*, located at the Y-junctions. This complex of seven proteins, originally identified in studies of *Listeria* infection (Chapter 5), has an intriguing list of properties. First, it binds to the side of actin filaments; second, it adds to (caps) the pointed end of an actin filament; and third, it acts like the barbed end of a filament in being able to add new ATP-actin monomers. These properties, taken together with images of the leading edge of migrating cells such as those in Figure 8-5, led to the widespread belief that Arp2/3 is the principal agent promoting actin growth at the cell margin.

The general idea is as follows (Figure 8-6). Arp2/3 binds to the side of actin filaments close to the cell margin and is then activated by signals coming from the plasma membrane. Upon activation, the Arp2/3 complex nucleates a new filament, allowing actin molecules from the cytosol to add progressively to the barbed end. Because of the structure of the Arp2/3 complex, the new actin filament will be firmly based on the previous filament and grow from it at an angle close to 70°. Large numbers of relatively short filaments growing in concerted fashion around the perimeter of the lamellipodium push against the plasma membrane and cause it to advance. As the leading edge moves forward, barbed ends of existing twiglets are probably blocked by capping protein or gelsolin, thereby terminating their action and allowing new actin twigs to take over.

What happens next? Experiments in which the network of actin filaments is exposed to depolymerizing conditions reveal that loss of subunits occurs most readily from the *rear* of the actin brush (that is, the region closest to the cell center). There is also a graded distribution of molecules such that Arp2/3 complexes are most abundant at the leading edge, whereas the depolymerizing protein cofilin is most abundant at the rear. The distribution of these proteins is therefore consistent with a model in which net addition of actin monomers takes place at the front of the lamellipodial brush and net depolymerization (through loss of Arp2/3 and cofilin severing) at the rear. Overall, we expect actin subunits to move through the array, being added as ATP-actin at the front and lost as ADP-actin at the rear. The situation is therefore analogous to the treadmilling of an individual filament described in Chapter 5, except that it takes place over a larger area and entails the coordinated action of multiple actin filaments.

Clearly this bald account leaves much unsaid. We will return to questions of how the cell regulates the actin brush and how contractility enters the picture later in the chapter. For the present, however, we must continue with our account of a locomoting cell and progress to the next stage: adhesion to the substratum.

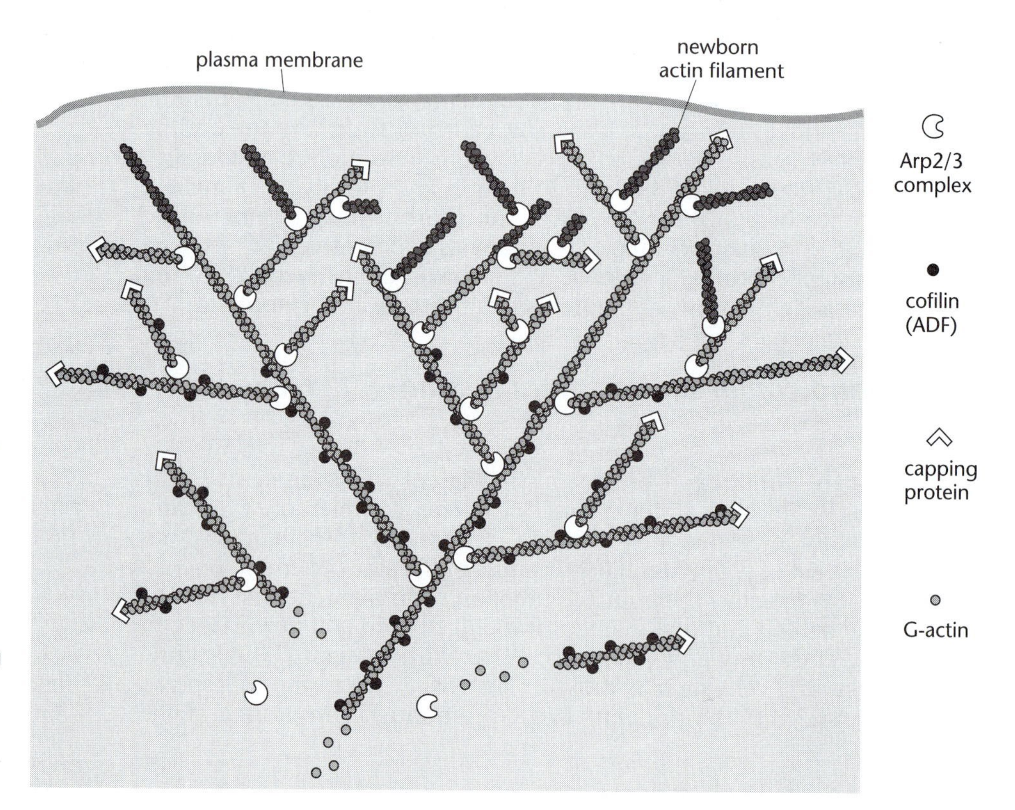


**Figure 8-6** Model for polymerization in a locomoting cell. New filaments at the leading edge grow as branches of preexisting filaments. Twigs are nucleated and protected from depolymerization by the Arp2/3 complex. Many barbed ends are believed also to be capped to prevent uncontrolled increase in filament mass. Release of the Arp2/3 complex from the Y-junctions at the base of the actin brush, followed by cofilin-mediated dissociation of actin subunits from pointed ends, leads to actin filament disassembly. Note that in this model, an individual filament does not treadmill but rather grows first at the barbed end and then later shrinks at the pointed end. However, the filament array as a whole does treadmill, reproducing itself at the front of the cell and dismantling itself at the lamellipodial rear. (Based on Svitikina and Borisy, 1999.)

## *Receptors on the underbelly of the cell attach it to the substratum*

A crawling cell must make contact with the surface over which it moves.[1] In the case of a skin fibroblast in culture, the contacts are relatively strong and these cells produce focal adhesions on surfaces coated with extracellular matrix components such as fibronectin or vitronectin. As we saw in Chapter 6, focal adhesions are sites to which actin filaments are anchored and bundled together into a stress fiber. These bundles, which contain myosin and have contractile properties, are especially abundant in cells such as the endothelial cells lining large arteries, which regularly have to exert tension. Fibroblasts also use stress fibers to pull on neighboring collagen fibrils, an important element in both wound healing and in the development of patterns of collagen in developing embryos.

Interestingly, stress fibers not only produce tension but are themselves generated in response to mechanical forces. We will return to the interplay between the cytoskeleton and mechanical forces in Chapter 18.

Rapidly migrating cells such as neutrophils or amoebae do not in general show well-defined points of adherence such as focal adhesions. Their affinity for the surface is weaker and less permanent (which, permits them to move more quickly, as discussed below). Although weaker and less well-defined, however, the attachments are nevertheless necessary for migration and, as in the case of focal adhesions, are probably mediated by integrins.

## *Crawling cells must pull themselves over the surface*

A crawling cell has not only to extend but also to contract. It needs to develop a force that pulls the bulk of the cell (including the nucleus and its intracellular organelles) behind its actively protruding margin. The origin of this force—usually termed *traction*—is presently unknown, although it must be coordinated, at least, with the protrusive activity at the front of the cell. A migrating fibroblast or keratocyte is typically wedge-shaped in outline, with a broad leading front and a narrow tail, or *uropod.* In time-lapse videos of migrating cells the uropod periodically snaps back into the cell, as though attached to the remainder of the cell by elastic. It seems that the contractile properties of the cell cortex, are employed in the uropod to 'gather' the contents of the cell behind the advancing front.

What is the mechanism of cortical contraction? One intriguing possibility that has been raised by experiments with *Dictyostelium* amoebae is that it could be driven by the localized disruption of actin filaments, perhaps catalyzed by myosin. Experiments in which actin filaments were anchored to a glass surface by a mixture of cortexillin (a protein that links actin filaments to membranes) and myosin heads, show that the filaments become increasingly tortuous in outline, and then break up into small fragments. The suggestion is that if myosin heads work against a rigidly anchored filament they can actually break it. In the cell, for example at the uropod of a migrating *Dictyostelium* amoeba, actin filaments anchored by cortexillin to the plasma membrane might be gathered together and then fragmented by myosin heads, perhaps those of the myosin I family.

Cortical contraction is an essential element in the phenomenon of cortical flow, discussed below, and important not only for the migration of cells but also for their division and shape determination.

## *Detachment is an essential part of cell migration*

Once the bulk of a cell has moved forward over a substratum, its trailing portions must free their adhesive contacts so they can be pulled along too. In some, possibly aberrant, situations, such as the migration of skin

[1] By 'contact' we mean here a proximity to the substratum that is sufficiently strong to allow the actin machinery to drive the cell. This is usually a specific molecular contact in which proteins on the cell surface bind to components of the extracellular matrix. Or it could conceivably be a contact through less specific viscous interactions, as discussed in Chapter 2.

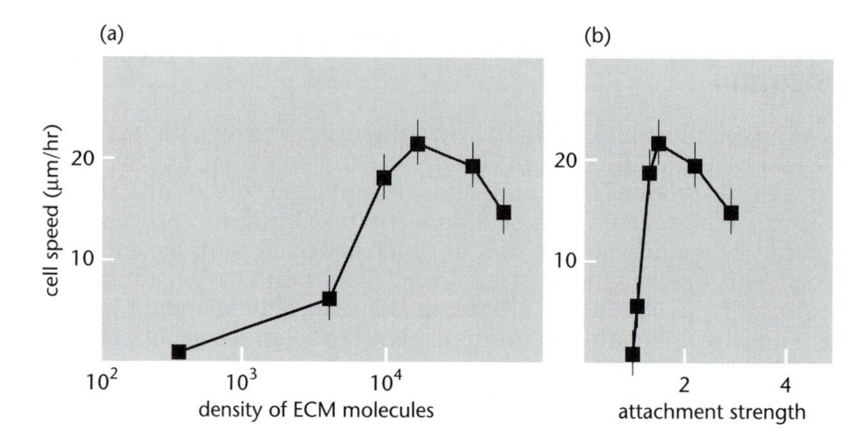


**Figure 8-7** Cell crawling depends on surface adhesion. The results of an experiment are shown in which cells were cultured on surfaces coated with different concentrations of extracellular matrix (ECM) molecules—the higher the concentration the more adhesive the surface. The speed of migration of cells on the dish showed a maximum at an intermediate level of adhesiveness.

fibroblasts on a tissue culture dish, the cell adopts the Draconian solution of leaving portions of its trailing cytoplasm behind. These fragments are apparently torn from the cell by the traction forces. Under other, perhaps more normal, conditions, focal adhesions have been shown to have a variable affinity for the substratum. In stationary cells, for example, focal adhesions labeled with fluorescent proteins can be seen moving over the substratum, whereas in migrating cells they are fixed in position. This suggests the existence of a clutchlike mechanism by which focal adhesions can be engaged or disengaged.[2]

You might think that if a cell sticks too firmly to the substratum it will crawl more slowly, and indeed this seems to be the case. Cells plated onto surfaces of different adhesiveness (produced, for example, by adding increasing quantities of fibronectin to the culture surface) migrate at the greatest speed at some intermediate level of adhesion. Both very weak adhesion and very strong adhesion slow down migration (Figure 8-7). The effect is most dramatic, as we shall see shortly, in cells that have lost their normal contractile machinery and are consequently unable to pull away from the surface. The retarding effects of strong adhesion may also explain why skin fibroblasts, which form distinct points of strong adhesion to the substratum, typically migrate more slowly than cells such as keratocytes or neutrophils that form only weak adhesions.

Can cells can contrive to break some adhesions and not others and in this way guide their direction of migration? No one knows.

## *Myosin II is needed for efficient, directed cell crawling*

Crawling cells contract, but how is this contraction produced? The first attempt to tackle this question experimentally employed mutants of the slime mold *Dictyostelium discoideum*. Thinking that the filament-forming myosin II (which together with actin is responsible for the contraction of muscle) would be a likely candidate, researchers prepared *Dictyostelium* mutants in which the gene for myosin II was removed or inactivated. Surprisingly, these cells were still able to adhere to surfaces, migrate over them, extend and retract filopodia and pseudopodia, ruffle, and engulf particles despite their lack of myosin II. The forces needed for these events are probably quite small, however, and, could be generated by unconventional myosins working within the network of actin and actin-cross-linking proteins.

Although myosin II is not absolutely essential for crawling, it does make a difference. Mutant *Dictyostelium* lacking myosin II advance at less than half the rate of wild-type cells: they are more flattened and less polarized in shape, and have a much diminished persistence of movement. These cells are also severely disadvantaged in motility when compared with normal cells when grown onto a highly adhesive surface, one coated with polylysine for example.

Exactly what part myosin II plays in a crawling cell is not known, but microscopy again offers some clues. In *Dictyostelium* amoebae, myosin II is concentrated at the rear of the cell, mainly in the uropod. In the

[2] Detachment of a cell from its substratum may also be helped by proteolytic enzymes that hydrolyze components of the extracellular matrix. Two proteolytic enzymes, urokinase and calpain, are associated with focal contacts of fibroblasts, for example, and may contribute to the normal migration of these cells on substrates such as fibronectin.

lamellipodia of keratocytes and fibroblasts, myosin molecules are most abundant at the base of the actin filament brush. They also occur as short bipolar filaments in bundles of actin filaments (arcs and stress fibers). This pattern suggests that myosin II helps develop tension between the leading margin of the cell and the cell body. Certainly, the ability of these mutant cells to develop 'cortical tension' is much less than that of wild-type cells (we will return to this topic in Chapter 18) and they are unable to 'cap' antibodies or lectins on their surface—that is, to collect these molecules into a patch. In the light of the preceding section, we might guess that their migration is paralyzed on adhesive surfaces because, lacking myosin II, they cannot pull away from the substratum.

## *Unconventional myosins may be even more important*

The role of unconventional myosins in cell crawling is even less clear. We saw in the previous chapters that animal cells contain a plethora of myosin-related proteins with a wide diversity of structures. Each is potentially able to develop force in conjunction with actin filaments, and in principle could contribute to the events in the locomotory cycle of, say, a fibroblast.

Myosin I's are especially interesting because of their abundance and association with membranes. Most kinds of myosin I have a stretch of basic residues in their tail that allows the molecule to bind to negatively charged phospholipids. We mentioned previously that this could permit myosin I's to transport membrane vesicles along actin filaments, but it could also link actin filaments to the plasma membrane. Might myosin I molecules, at the barbed ends of actin filaments, promote the addition of actin monomers (a similar function has been proposed for microtubule motors during mitosis, see Chapter 13)? Indeed, myosin I's could actually be part of the Arp2/3 complex that nucleates actin filaments, thereby explaining, among other things, the relatively high concentration of these nucleating structures at the leading margin of the cell.

Unconventional myosins could also be instrumental in developing tension within the actin cortex, and also perhaps in powering the depolymerization of actin filaments at the trailing end (uropod). Both functions are crucial for cell migration, as we now see.

## *Cell crawling is based on a cyclical flow of cortical actin*

At this point in the chapter we know enough to build a rudimentary crawling machine. Each of the four processes described in Figure 8-1—extension, adhesion, contraction, detachment—has a corresponding molecular machinery based on actin filaments. We have ideas how each might work in isolation, but what about the system as a whole? How are events at the front of the cell coordinated with those on its lower surface or tail over distances of 100 μm or more?

We will see below that, at least in some cells, calcium ions and other diffusible signals are able to synchronize motile activities at different regions of the cortex. Microtubules also may influence the direction of migration of some cells, but arguably the most fundamental level of integration is provided by the actin cortex itself. For not only is this meshwork of cross-linked actin filaments continuous in a structural sense, it is also able to undergo concerted movements. Indeed, a steady movement, or flow, of actin-containing structures takes place in almost all kinds of migrating animal cells (except nematode sperm) and is almost certainly a major factor in the integration of their locomotion.

The surface of an animal cell is pliable and easily deformed, so it is not surprising that the appearance of cortical flow varies according to the type of cell and its surroundings. Probably the simplest and most geometrically regular display of cortical flow is seen in fish epidermal cells. Immunofluorescent labeling and particle marking of migrating keratocytes reveals a regular 'fountain flow' in the highly flattened lamellipodium of these cells in which cortical elements add (through actin polymerization) to the center of the semicircular leading margin, sweep laterally to either side, and then pull inward (perhaps by the action of myosin II) at the rear of the cell body (Figure 8-8). A similar movement can also occur in cell fragments lacking microtubules, showing that cell-wide coordination can be achieved by actin alone.

Skin fibroblasts in culture have a more irregular form and are less flattened, and their cortical flow shows itself mainly in the phenomenon described in Chapter 2 of ruffling, in which lamellipodia sporadically rear up on the upper surface and sweep rearward. Fibroblasts also produce bands of actin filaments, of unknown function, known as *actin arcs*, orthogonal to the direction of movements. Actin arcs, which contain myosin II, form sporadically at the leading edge and sweep backward as the cell migrates. As these structures move, they carry with them markers on the outside of the cell, such as aggregates of lectins.

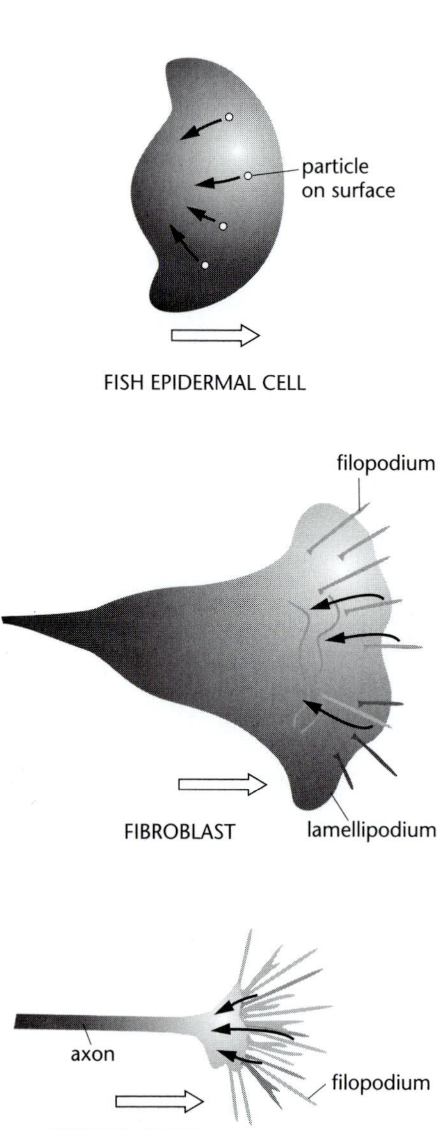


**Figure 8-8** Actin flow in three kinds of migrating cell. In fish epidermal cells (*keratocytes*) the lamellipodia are flattened and smooth. Movement of the underlying cytoskeleton is revealed by particles sitting on the surface, which are carried back continuously from the leading edge. Note that the particles move not only in relation to the cell but relative to the substratum as well. The surface of a *fibroblast* is covered with highly irregular projections such as lamellipodia and filopodia, all of which show a concerted movement backward unless attached to the surface. The leading tip of a growing axon (*growth cone*) shows similar rearward movements that carry both surface projections and adventitious particles on the surface.

## *There are interesting parallels between cell migration and cell division*

There are interesting parallels between the crawling movements that are the subject of this chapter and the movements of cytokinesis described in Chapter 7. Both movements are driven by the cortex and in both actin-containing structures move from the perimeter of the cell toward its center. Some migrating cells, as just mentioned, sporadically produce bundles of actin filaments and myosin II molecules at right angles to the direction of movement. In neutrophils, actin arcs typically extend around the entire perimeter of the cell, often forming a waistlike constriction, which in structure and composition resembles the contractile rings formed during the division of animal cells (Figure 8-9). Another point of similarity is in the distribution of leukosialin, a highly-charged surface glycoprotein, which accumulates in the uropod of migrating neutrophils and in the cleavage furrow of dividing lymphocytes.

In both migrating cells and dividing cells, a meshwork of actin-containing structures beneath the plasma membrane moves as a continuous unit. Events at one region of the cell influence events at other regions by modulating the addition or removal of components to or from the cortex, or by locally modifying tension in the cortical sheet. In a migrating cell, the cortex grows at the front end by the continual polymerization of actin filaments and their incorporation, together with associated proteins, into the cortical meshwork. Farther back in the cell, the cortex shrinks by fragmentation and depolymerization of actin filaments and dispersal of cortical components into the cytosol. These two activities are linked mechanically through the meshwork of the cortex and by cortical contractions in the uropod, which make this region of the cell round up and detach from the substratum. In a dividing cell there is an accumulation of actin filaments in a ring around the waist of the cell. As these contract (with the aid of myosin II), they pull on neighboring regions of the actin cortex, causing them to move toward the center.

## *Cytoplasmic streaming and hydrostatic pressure contribute to amoeboid locomotion*

All of the movements we have so far described in a locomoting cell occur in the same direction. An unending stream of actin filaments, actin-binding proteins, transmembrane proteins, and lipids moves from the leading margin of the cell to its rear. Experiments show very clearly that these movements can continue for many hours in the absence of protein or lipid synthesis, so they cannot depend on these molecules being made at one end and degraded at the other. In other words, most molecules carried in this flow must travel in a cycle, being used over and over again. But how, if this is so, do they return to the front of the cell?

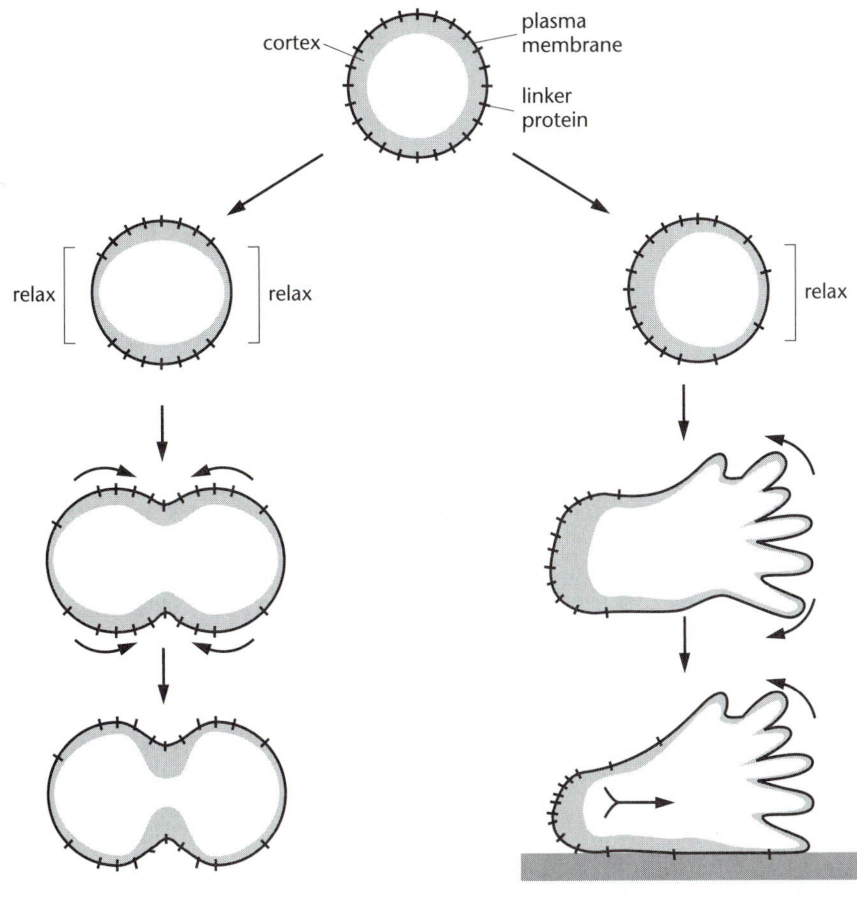


**Figure 8-9** Cortical movements during cytokinesis and cell migration. During cytokinesis, contraction at the equator leads to the accumulation of actin, myosin, and associated proteins in a band around the equator of the cell—the contractile ring. In a migrating cell, the trailing end contracts and cortical components are pulled in a unidirectional flow toward the region of greatest tension. Since migration continues for long periods of time in the absence of protein synthesis, components of the cortex must be recycled through the cytoplasm to form again at the leading edge.

Part of the return journey, especially over short distances, could be by diffusion. For example, the immediate source of actin for assembly at the front of the cell is thought to be the pool of monomeric actin, complexed to monomer-binding proteins such as thymosin. This actin pool is probably continually replenished by actin molecules released from the base of the cortical brush, perhaps through the action of fragmenting proteins such as cofilin. If these actin molecules are freely diffusing in the cytosol, then why should they not travel by diffusion to the leading margin of the cell? The same may be said of other components of the cortex, such as actin-binding proteins.

However, transport by diffusion becomes less and less effective as the distances become greater, for reasons given in Chapter 1, and it seems that many large cells employ a more deliberate and active mechanism. When giant amoebae such as *Amoeba proteus* migrate, they show dramatic and vigorous internal streaming in which components of the cytoplasm flow into each advancing pseudopodium. These streaming movements were among the first examples of cell movements to be studied under the light microscope and, historically, were the source of much speculation and debate.

One of the earliest and most durable hypotheses relating to amoeboid movements—advanced by such authors as Dujardin, Eckert, Schulze, and Wallich in the period 1835–1875—is that cytoplasmic streaming (and hence pseudopodial extension) is driven by pressure. The proposition was that the amoeboid cell contracts its outer, cortical layer of cytoplasm selectively in certain regions and thereby squeezes a stream of more fluid cytoplasm into the advancing pseudopodium. It was further supposed that at the tip of the pseudopodium the fluid stream is converted to more rigid cortex while at the same time, farther back in the body of the cell, the converse activity takes place. Since, as we now know, both the cortical layer and the inner cytoplasm of an *Amoeba* are full of actin, we see in this century-old formulation an anticipation of the present view of molecular events in a migrating cell.

Many experiments have been performed to test the idea of pressure-driven pseudopodial expansion. Pseudopodia have been cut and the exposed ends subjected to positive or negative pressure. Intracellular electrodes capable of recording pressure have been inserted into migrating amoebae. Although such experiments are often difficult to perform and interpret, the overall conclusion is that giant amoebae do indeed develop positive internal pressures, and that these pressures are necessary (although not, by themselves, sufficient) to drive pseudopodial expansion. Further support for this idea comes from the cytoplasmic streaming of syncytial acellular slime molds such as *Physarum*. Because these multinucleated 'cells' are so large, it is possible to use intracellular pressure-measuring devices to show unequivocally that streaming is the direct consequence of periodic contractions of the outer cortical layer of the cytoplasm (although, in this case the streaming is not always directly related to forward migration).

The movements of a giant amoeba seem at first glance to be unrelated to those of a vertebrate cell such as a neutrophil. But the differences may be superficial and attributable to the sizes of the two cells. Thus, although it is probably the case that actin filaments continually polymerize at the tip of an advancing amoeboid pseudopodium, the relatively enormous size of these structures makes it difficult to perform the labeling experiments necessary to demonstrate this phenomenon. Conversely, the relatively minute volumes of cytoplasm in a migrating neutrophil or fibroblast will make it difficult to detect any streams of cytoplasm that might occur.[3]

## How does the membrane recycle?

The front of a migrating animal cell undergoes a continual, local, expansion of its surface area as new filopodia or lamellipodia are added, each enclosed in a plasma membrane. The question thus arises: where does this new membrane come from? One possibility is that it is recruited locally, from neighboring regions of the cell (like a sheet pulled from a bed). Another is that new plasma membrane is created locally by the insertion of membrane components from the cytoplasm. These two possibilities are not mutually exclusive, since we know that cell membranes have a complex composition made of many different kinds of lipid and protein molecules, each of which could travel by a different path.

It was suggested some years ago that the lipid molecules in a cell undergo a continual movement somewhat similar to the cyclic movement of the actin cortex just described. This hypothetical 'lipid flow' was postulated to sweep rearward with the plasma membrane and then return to the front of the cell in the form of vesicles moving through the cytoplasm. Furthermore, the continual flow of lipid was proposed to sweep with it, as if in a wind, proteins inserted in the membrane. Since this motion included membrane proteins attached to the cytoskeleton, lipid flow was thought to be responsible for driving cell locomotion.

To test this interesting idea, various experiments were performed in which labeled lipid molecules and extremely small (less than 50 nm diameter) gold particles were observed on the surface of migrating cells. These experiments revealed that lipid molecules and membrane proteins do not all move steadily in the same direction (Figure 8-10). Except for those that are attached to the cytoskeleton, molecules in, or on, the membrane diffuse freely and move as rapidly forward as to the rear of the cell. Thus there is no evidence for an unending cyclical flow of all the lipid molecules of the cell and certainly no reason to think that it drives cell migration.

A corollary of the experiments just mentioned is that membrane constituents needed for protrusion of the cortex such as lamellipodia and

[3] A happy intermediate may be provided by the eosinophils of newts, which are both large enough to allow streaming to be seen but small and flat enough for individual actin filaments to be labeled and identified.

filopodia could be supplied from local sources. As the actin-containing extension grows, the plasma membrane that surrounds it could be provided by the free diffusion of lipid molecules and membrane proteins in the plane of the plasma membrane.

Although a universal 'lipid flow' does not exist, there is in some cells a directed stream of selected membrane constituents through the cytoplasm. Movements of this kind are especially conspicuous in differentiated cells that are structurally polarized and in which some regions of the plasma membrane differ in composition from others. Most epithelial cells, for example, have one domain of plasma membrane facing the lumen and another facing the lateral and basal regions. These two domains differ in their composition of both membrane proteins and lipids—differences that are created and maintained by targeted delivery of membrane vesicles containing the appropriate membrane constituents. An extreme case of such targeted movement of membrane components is seen in nerve cells, where the cell body, dendrites, and axon each receive specialized deliveries of membrane vesicles.

Whether targeted membrane movements make any contribution to the machinery of cell migration is unresolved. One way to approach this question is to ask what happens if we inhibit the machinery that carries membrane vesicles—which brings us to the connection between microtubules and cell migration.

(a) freely diffusing

leading edge of cell

track of particle

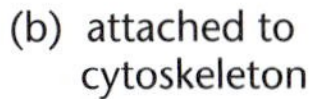


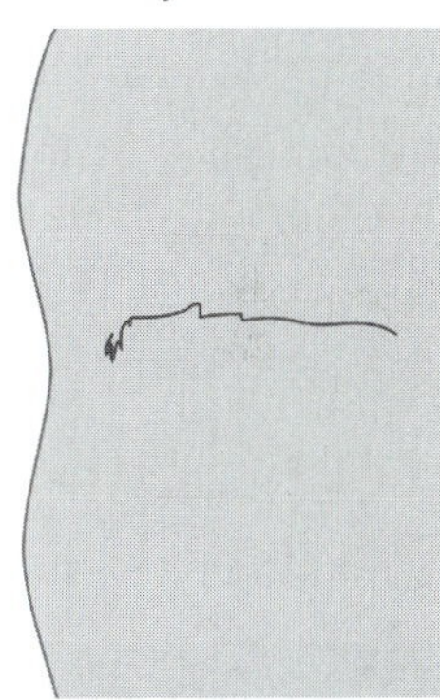

**Figure 8-10** Particle movements on the cell surface. Gold beads, 50 nm in diameter were observed on the surface of a fibroblast in tissue culture. In (a) the bead has not attached to the underlying cytoskeleton and diffuses freely over the cell surface. In (b) the bead has attached to the cytoskeleton via integrin and in this case shows a slow persistent movement rearward from the leading edge. The duration of both tracks was approximately 60 seconds. (Based on Felsenfeld et al., 1996.)

## *Microtubules influence the migration of fibroblasts*

So far in this chapter we have emphasized the actin-based cytoskeleton. Actin is certainly the major cytoskeletal player in cell locomotion and there are examples of cells, or fragments of cells, that can crawl over a surface and contain actin filaments but not microtubules or intermediate filaments (see Figure 4-1). However, observations of this nature do not teach us that other filaments of the cytoskeleton *never* influence cell migration. On the contrary, there is good reason to think that microtubules, especially, often exert a controlling influence on the location of the actin-based machinery and its activity.

The earliest observations implicating a role for microtubules in cell migration were made in 1970 by Vasiliev and coworkers, who treated fibroblasts in culture with colchicine and other mitotic poisons. These treatments destroy most if not all of the microtubules in the cell but leave the fibroblasts still able to crawl. However, the cells are subtly changed by this treatment, since their form is less polarized and ruffling occurs around most of the cell perimeter instead of in just one region. Coincident with this loss of structural polarity, the trajectory taken by the cells over the substratum is more disordered than in untreated cells and they turn more frequently than before.

A more dramatic effect has been observed in certain chemotactic cells. The eosinophils of newts are large cells that migrate by a form of actin-based ruffling movement that closely resembles that of skin fibroblasts. Their movement is stimulated and directed by factors in the serum, and when these contact the surface of the cell, they induce lamellipodial expansion and subsequent migration in the direction of the stimulus. Eosinophils have a well-developed array of interphase microtubules and when treated with antimitotic agents, they start to bleb and stop migration (in contrast to fibroblasts).

How microtubules affect cell migration is not known, but several hypotheses have been put forward. One is that the array of microtubules directs a stream of membrane vesicles to one region of the cell surface, thereby reinforcing migration in a particular direction (this is indeed what happens in a growing nerve axon). Another possibility is that microtubules might influence migration by acting directly on the cell cortex. A vivid illustration of the latter effect is that removal of microtubules leads to an increase in stress fibers and a severalfold increase in the traction forces produced by the cell. It has also been observed that growing microtubules stimulate Rac GTPase, which drives actin polymerization and the protrusion of lamellipodia. We will return to the interactions between actin filaments and microtubules in Chapter 19.

## Chemoattractants polarize the motile machinery

The primary purpose of cell crawling is to carry the cell to a specific location, and this often takes place in response to *chemotaxis*—movement toward a distant source of diffusing attractant molecules (Chapter 3). From what we have said about the machinery of motion, it is clear that the concentration of chemoattractant molecules in the environment must in some way modify the distribution of actin in the cell cortex so that the cell moves in a specific direction.

The earliest cytoskeletal event in chemotaxis is probably a local stimulation of actin polymerization. In the chemotactic response of *Dictyostelium* amoebae to cyclic AMP, for example, the level of polymerized actin in the cell as a whole rises within 10 seconds. Newly formed actin filaments provide the cytoskeletal framework for pseudopodial extensions, generating a sudden flurry of filopodia and ruffles on the cell surface. These quickly become restricted to one end as the cell becomes polarized and begins to migrate. In a gradient of chemoattractant, ruffling is at the front end of the cell, which is the end that points in the direction of the highest concentration (Figure 8-11).

A similar response is seen in neutrophils, the white blood cells that hunt down and kill infecting bacteria. As we mentioned in Chapter 3, these cells are attracted by the formylated peptides produced by bacteria and will crawl toward these molecules by extending actin-rich pseudopodia. Experiments with permeabilized neutrophils show that one of the earliest effects of chemoattractants is a redistribution of Arp2/3 complex into discrete foci on the side of the cell facing the chemoattractant. The foci nucleate actin-rich structures that grow out of the membrane sites, rather like the rockets of actin filaments forming behind a *Listeria* tail.

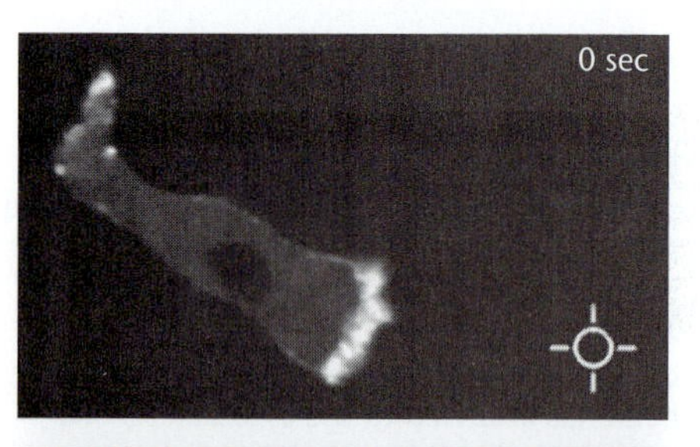


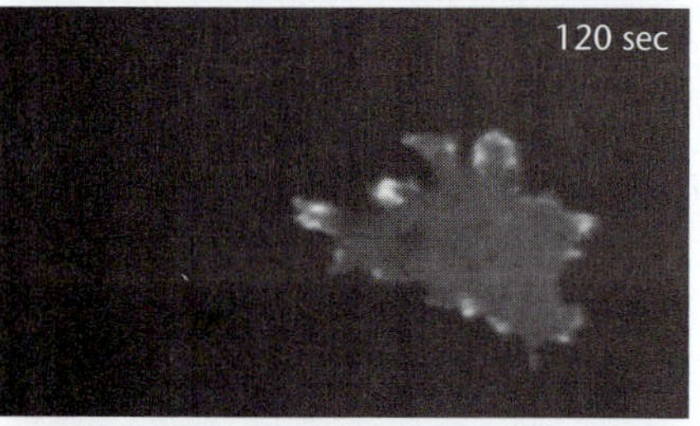


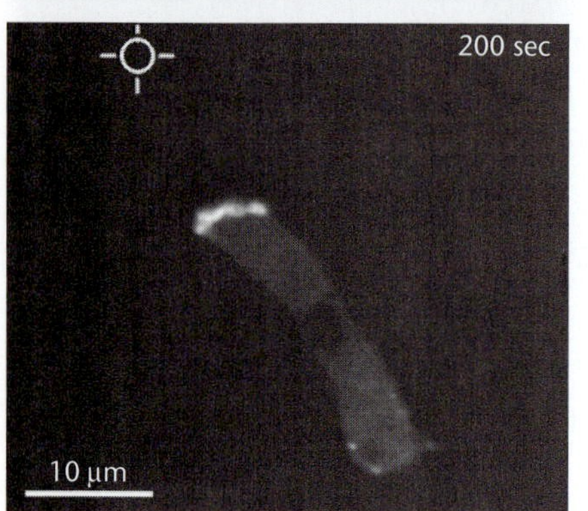


**Figure 8-11** Chemotaxis of *Dictyostelium*. A single amoeba is shown migrating toward the mouth of a pipette (position indicated by cross hairs) containing the chemoattractant cAMP. The cell has been stained with antibodies to coronin, an actin-binding protein enriched in the leading lamellipodia (Courtesy of Mary Ecke and Günther Gersich.)

## Networks of signals regulate the motile machinery

Although the cytoskeleton reorganizes rapidly in response to a chemoattractant, this is not the earliest response of the cell. In *Dictyostelium* amoebae and leucocytes, the receptors for chemoattractant molecules such as cAMP or formylated peptides are coupled to trimeric G proteins. Occupation of the receptor triggers a rapid dissociation of the G protein components, which diffuse away from the receptor and initiate a variety of local responses. Each of the many steps in this cascade of reactions follows its own time course and each has a characteristic distribution and action on the motile machinery.

The receptors themselves are uniformly distributed over the surface of the cell, even in the presence of a chemoattractant gradient. However, the products of their stimulation are often highly localized. One of the earliest detectable responses, for example, is the conversion of the signaling lipid $PIP_2$ to its phosphorylated derivative, $PIP_3$, which accumulates transiently in the region of the cell facing up the gradient. $PIP_2$ has important effects on the binding of the actin cytoskeleton to the plasma membrane and a fall in local concentration of this lipid can promote such events as the extension of pseudopodia. $PIP_3$ has been shown to be the site of anchorage to the membrane of proteins that activate kinases, such as myosin light chain kinase, and the small G proteins Rac, Rho, and Cdc42. The

**Table 8-1 Proline-rich regulators of actin polymerization**

| Protein | Function |
|---|---|
| VASP | Binds to ActA (*Listeria*) and to adherens junction proteins such as vinculin. Promotes actin filament assembly. Regulated by phosphorylation. |
| Ena | Homologue of VASP and product of the *enabled* gene in *Drosophila*. |
| Mena | Mammalian form of Ena, highly enriched in nerve growth cones and vital for neuronal development. |
| WASP | Adapter protein that links the Arp2/3 complex. Contains a GTPase domain and interacts with the G protein Cdc42. |
| Scar | Adapter protein related to WASP that binds to Arp2/3 and enhances its nucleating capacity. Thought to couple G-protein receptors to actin cytoskeleton. |

crucial role of the latter in coordinating actin-based responses was discussed in Chapter 6.

Changes in intracellular $Ca^{2+}$ also follow chemoattractant stimulation. In some cells (such as newt eosinophils), a transient rise in $Ca^{2+}$ in the region close to the chemotactic stimulus is followed by a slower and more persistent elevation in regions farthest from the chemoattractant source. Calcium ions affect a myriad processes in a living cell, but in this instance the most relevant action may to be to activate myosin molecules in the cortex, thereby causing contraction. This may help to establish a cycle of cortical actin, which as we saw above consolidates the polarity of the cell.

The last players to be mentioned are members of a large and heterogeneous family of proteins that act as essential agents in the nucleation of actin filaments, and which in some cases are activated by G protein-coupled receptors (Table 8-1). These proteins (which include the VASP mentioned in connection with *Listeria* motility in Chapter 5) are characterized by having a proline-rich domain that binds to profilin. Different members also contain domains that link them to sites engaged in the active growth of actin filaments, such as *Listeria* bacteria, adherens junctions, or the Arp2/3 complex, or to other agents that regulate the assembly process, such as Rho or Cdc42.

The 'wiring' of this biochemical circuitry—that is, the binding strengths between individual components and the rates of individual reactions—has been shaped by evolution so as to produce movement in the correct direction. Many other links in the web of biochemical reactions controlling the cell's signaling machinery remain to be found, and the larger question of how they all cooperate and work together is still to be answered. These are among the most exciting challenges facing present-day investigators of cell motility.

## Outstanding Questions

*Why do nematode sperm use ATP to produce movement? How does disassembly of MSP filaments pull the cell body and how are assembly and disassembly coupled? How, and in what form, do actin molecules move to the leading edge of a migrating fibroblast? What limits the addition of actin molecules at the front of the cell (is it the supply of actin monomers; the availability of uncapped barbed ends; the activation of Arp2/3 complexes)? What part do myosin molecules play in fibroblast migration—especially unconventional myosins? Are changes in intracellular pH important for migration (as they appear to be in nematode sperm)? Do actin filaments polymerize at the tip of an amoeba pseudopod? Is gel osmotic pressure a significant force in cells? Do vertebrate tissue cells maintain a significant positive hydrostatic pressure? Can cells contrive to break some adhesions and not others and in this way guide their direction of migration? How do microtubules influence cell migration? What is the biochemical circuitry that controls cell migration? How does this operate in a chemotactic response?*

## *Essential Concepts*

- When an animal cell crawls over a surface it performs a repeated cycle of movements in which it (1) extends over the substratum, (2) forms an attachment to the surface, (3) pulls the remainder of the cell to the point of anchorage, and (4) detaches anchorage points at the rear of the cell.
- Nematode sperm crawl by a variant of the normal process that does not require actin but does reveal the importance of assembly and disassembly of protein filaments.
- The crawling of amoebae, white blood cells, and cells from skin and other tissues, is driven by a cortical layer of actin filaments in association with multiple other proteins.
- Extension of the leading margin of the cell is caused by the polymerization of actin filaments. In lamellipodia, a branched network of actin filaments, growing at their barbed ends, pushes the plasma membrane forward.
- Crucial factors in this leading edge extension are the Arp2/3 complex, gelsolin, and capping protein. At the base of the cortical actin meshwork, cofilin promotes the disassembly of filaments.
- Transmembrane proteins (integrins) anchor the cell to its substratum by binding both to molecules of the extracellular matrix on the outside of the cell and to the actin cytoskeleton on the inside.
- Cortical contractions due to myosin molecules pull structures toward the center of the cell, causing the uropod to retract and unattached structures on the dorsal surface to move backward.
- The cycle is completed by a forward movement of actin and other constituents through the cytoplasm to the leading margin of the cell.
- In very large cells, such *Amoeba proteus*, the forward extension of pseudopodia is driven by streams of cytoplasm caused by the squeezing action of the cortex.
- The direction taken by a migrating cell can be influenced by the distribution of microtubules in its cytoplasm. Microtubules interact directly with cortical actin and also direct streams of membrane vesicles to the leading margin.
- The direction taken by a migrating cell is responsive to both physical and chemical features of its substratum. Cells can also respond to gradients of diffusing substances, as shown by the chemotaxis of *Dictyostelium* amoebae and neutrophils.
- Cell locomotion is regulated by a web of reactions that convey signals from receptors in the plasma membrane to every part of the motile machinery. Key components in the regulatory network are calcium ions, inositol phospholipids, the Rho family of G proteins, and proteins related to WASP.

## Further Reading


Benink, H.A., et al. Analysis of cortical flow modles *in vivo*. *Mol. Biol. Cell* 11: 2553–2563, 2000.

Bray, D., White, J. Cortical flow in animal cells. *Science* 239: 883–888, 1988.

Bresnick, A.R. Molecular mechanisms of nonmuscle myosin-II regulation. *Curr. Biol.* 11: 26–33, 1999.

Butcher, E.C., Picker, L.J. Lymphocyte homing and homeostasis. *Science* 272: 60–66, 1996.

Cox, D., et al. Genetic deletion of ABP-120 alters the three-dimensional organization of actin filaments in *Dictyostelium* pseudopods. *J. Cell Biol.* 128: 819–835, 1995.

De Bruyn, P.P.H. Theories of amoeboid movement. *Q. Rev. Biol.* 22: 1–24, 1947.

DiMilla, P.A., et al. Maximal migration of human smooth muscle cells on fibronectin and type II collagen occurs at an intermediate attachment strength. *J. Cell Biol.* 122: 729–737, 1993.

Felsenfeld, D.P., et al. Ligand binding regulates the directed movement of β1 integrins on fibroblasts. *Nature* 383: 438–440, 1996.

Gerisch, G., et al. Genetic alteration of proteins in actin-based motility systems. *Annu. Rev. Physiol.* 53: 607–628, 1991.

Hallett, M.B. Controlling the molecular motor of neutrophil chemotaxis. *BioEssays* 19: 615–621, 1997.

Hotary, K., et al. Regulation of cell invasion and morphogenesis in a three-dimensional type I collagen matrix by membrane-type matrix metalloproteinases 1, 2, and 3. *J. Cell Biol.* 149: 1309–1323, 2000.

Italiano, E., et al. Reconstitution *in vitro* of the motile apparatus from the amoeboid sperm of *Ascaris*: direct evidence that filament assembly and bundling move membranes. *Cell* 84: 105–114, 1996.

Jay, P.Y., et al. A mechanical function of myosin II in cell motility. *J. Cell Sci.* 108: 387–393, 1995.

Jin, T., et al. Localization of the G protein β2 complex in living cells during chemotaxis. *Science* 287: 1034–1036, 2000.

Lasky, L.A. Selectins: interpreters of cell-specific carbohydrate information during inflammation. *Science* 258: 964–969, 1992.

Lee, J., et al. Traction forces generated by locomoting keratocytes. *J. Cell Biol.* 127: 1957–1964, 1994.

Machesky, L.M. Complex dynamics at the leading edge. *Curr. Biol.* 7: 164–167, 1997.

Machesky, L.M., Insall, R.H. Signaling to actin dynamics. *J. Cell Biol.* 146: 267–272, 1999.

Mitchison, T.J., Cramer, L.P. Actin-based cell motility and cell locomotion. *Cell* 84: 371–379, 1996.

Mullins, R.D., et al. The interaction of Arp2/3 complex with actin: nucleation, high affinity pointed end capping, and formation of branching networks of actin filaments. *Proc. Natl. Acad. Sci. USA* 95: 6181–6186, 1998.

Nobes, C.D., Hall, A. Rho, Rac and Cdc42 GTPases regulate the assembly of multi-molecular focal complexes associated with actin stress fibers, lamellipodia, and filopodia. *Cell* 81: 53–62, 1995.

Pollard, T.D., et al. Molecular mechanisms controlling actin filament dynamics in nonmuscle cells. *Annu. Rev. Biophys. Biomol. Struct.* 29: 545–576, 2000.

Ridley, A.J. Membrane ruffling and signal transduction. *BioEssays* 16: 321–327, 1994.

Ridley, A.J. Rho: theme and variations. *Curr. Biol.* 6: 1256–1264, 1996.

Roberts, T.M., Stewart, M. Nematode sperm locomotion. *Curr. Biol.* 7: 13–17, 1995.

Roberts, T.M., Stewart, M. Nematode sperm: amoeboid movement without actin. *Trends Cell Biol.* 7: 368–373, 1997.

Schafer, D.A., Schroer, T.A. Actin-related proteins. *Annu. Rev. Cell Biol.* 15: 341–363, 1999.

Seveau, S., et al. Leukosialin (CD43, sialophorin) redistribution in uropods of polarized neutrophils is induced by CD43 cross-linking by antibodies, by colchicine or by chemotactic responses. *J. Cell Sci.* 110: 1465–1475, 1997.

Small, J.V. Lamellipodia architecture: actin filament turnover and the lateral flow of actin filaments during motility. *Semin. Cell Biol.* 5: 157–163, 1994.

Small, J.V., et al. Actin filament organization in the fish keratocyte lamellipodium. *J. Cell Biol.* 129: 1275–1286, 1995.

Smilenov, L.B., et al. Focal adhesion motility revealed in stationary fibroblasts. *Science* 286: 1172–1177, 1999.

Soll, D.R. The use of computers in understanding how animal cells crawl. *Int. Rev. Cytol.* 163: 43–104, 1995.

Stefansson, S., Lawrence, D.A. The serpin PAI-1 inhibits cell migration by blocking integrin αvβ3 binding to vitronectin. *Nature* 383: 441–443, 1996.

Svitkina, T.M., et al. Analysis of the actin-myosin II system in fish epidermal keratocytes: mechanism of cell body translocation. *J. Cell Biol* 139: 397–415, 1997.

Svitkina, T.M., Borisy, G.G. Arp2/3 complex and actin depolymerizing factor/cofilin in dendritic organization and treadmilling of actin filament array in lamellipodia. *J. Cell Biol.* 145: 1009–1025, 1999.

Weiner, O.D., et al. Spatial control of actin polymerization during neutrophil chemotaxis. *Nat. Cell Biol.* 1: 75–81, 1999.

Welch, M.D. The world according to ARP: regulation of nucleation by the Arp2/3 complex. *Trends Cell Biol.* 9: 423–427, 1999.

Yanai, M., et al. Intracellular pressure is a motive force for cell motion in *Amoeba proteus*. *Cell Motil. Cytoskeleton* 33: 22–29, 1995.

Zheng, J.Q., et al. Turning of nerve growth cones induced by neurotransmitters. *Nature* 368: 140–101, 1994.

Zigmond, S.H., et al. Mechanism of Cdc42-induced actin polymerization in neutrophil extracts. *J. Cell Biol.* 142: 1001–1012, 1998.

# CHAPTER 9

# The Molecular Basis of Muscle Contraction

It comes as something of a relief, after considering the complex variety of molecular events in a migrating cell, to turn to a movement that is as regular and precise as a piece of clockwork. Vertebrate skeletal muscle contracts by means of enormous multinucleated cells that evolved specifically to generate extremely rapid, repetitive, and forceful movements. The cytoplasm of these giant cells is crammed full of a highly organized, almost crystalline, array of cytoskeletal filaments whose only function is to produce contractile force. Muscle contraction is the most obvious form of cell movement and it is the one that has been studied for the longest time.[1] Because muscle cells are so large and have such a high degree of internal order they are ideal specimens for the diffraction of electromagnetic radiation. Studies using light microscopy, electron microscopy, and x-ray diffraction provide structural data to complement information from biochemical and physiological data. It is now possible to trace muscle contraction in a series of logical steps, from the contraction of the whole muscle down to events occurring at the molecular and atomic level: a wonderful correspondence between structure and function!

## *Tension measurements define the performance of a muscle*

Muscles can only pull; they cannot push. The complex choreography of the body operates on the lever principle, with each muscle being typically attached to two or more bones either directly or through tendons. Muscles often act in antagonistic pairs, with one relaxing as the other contracts: as you raise your forearm toward your shoulder the biceps contracts and the triceps relaxes; the opposite happens as you move your forearm down again. There are also muscles that are coupled through tissue rigidified by hydrostatic pressure rather than bone, such as muscles in the elephant's trunk and the octopus's tentacle, and many muscles in invertebrates.

Every skeletal muscle in a vertebrate body receives a nerve supply from the central nervous system, so that its contraction is under voluntary control. A single muscle dissected from the animal and mounted in a suitable bathing fluid remains active for days (Figure 9-1). Stimulated

[1] The Roman Galen, in the second century AD, made a detailed anatomical examination of the mode of action of muscles and recognized the heart as a muscle.

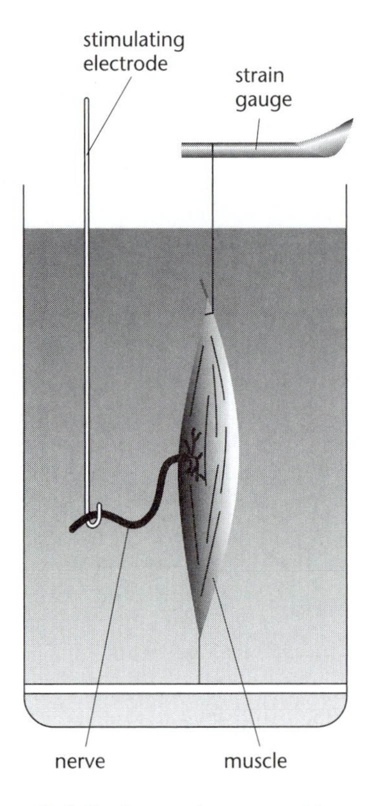


**Figure 9-1** Single muscle mounted in preparation for tension measurements.

electrically, by an electrode placed directly on its surface or through a remaining length of motor nerve, the isolated muscle contracts. A single shock causes a twitch contraction, whereas repeated shocks above a certain frequency fuse to produce a steady tetanic contraction (Figure 9-2).

The maximum possible rate of contraction is shown by a muscle unattached to any load. Conversely, if the muscle is weighted with a sufficiently large load, then no shortening will occur, although tension still develops under this *isometric* condition. At intermediate loadings, the muscle shortens at intermediate rates and the *power output*—given by the velocity of shortening multiplied by the load—is a maximum.[2]

A weight attached to an isolated muscle causes the muscle to stretch slightly. Tension develops as passive elastic elements, such as the titin filaments described later in this chapter, resist the stretching. If the stretched muscle is then stimulated to contract, it develops a greater tension than before—the difference between the unstimulated and stimulated tension (the *developed tension*) being due to the contractile machinery of the muscle. The developed tension shows a maximum of about 3 $kg/cm^2$ when the muscle is held at a length close to its length in the body. The existence of this maximum has a simple explanation in the molecular mechanism of muscle contraction, as we will explain below.

## Muscles can work aerobically or anaerobically

Muscles are, traditionally, one of the principal tissues biochemists use to study the enzymology of energy production. Where the ATP comes from to drive muscle contraction; how energy consumption varies with the work done; what waste products are produced and how they are excreted—these are all understood in considerable detail. Although it may seem that this large body of knowledge is peripheral to our subject of cell movements, we should remember that muscles are close cousins of the contractile portions of an amoeba or a fibroblast. A solitary myosin molecule moving in, say, a nerve cell consumes ATP molecules that, as in muscle, must be generated by a coordinated set of metabolic reactions. Are these reactions aerobic or anaerobic? Are they localized to the site of movement? Do nerve cells store energy-rich intermediates for use by their motor molecules? Is the build up of waste products a limiting factor (does a fibroblast get muscle cramps)? We have not the slightest idea how to answer these questions, so it makes sense to look to muscle contraction for guidance.

Vertebrate muscles receive a blood supply, which ensures that during exercise they are supplied with oxygen from the lungs and that a rapid exchange of glucose, lactate, and other metabolites between contracting muscles and the liver can occur. Oxygen is needed for oxidative phosphorylation, which takes place, in muscle as in all eucaryotic cells, in mitochondria. In this process, food stores such as carbohydrates and fats are

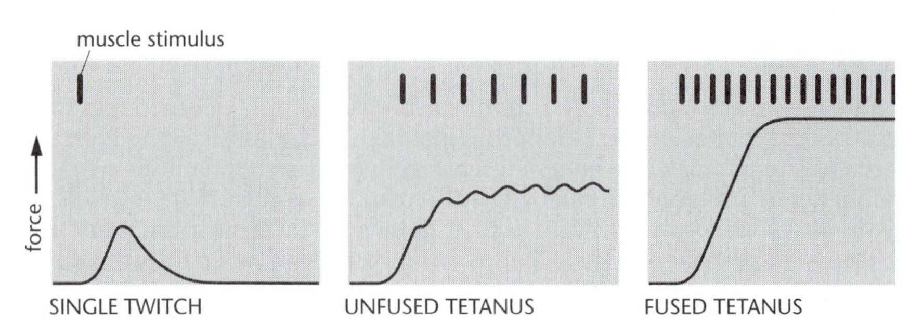


**Figure 9-2** Muscle tetanus. Increasing the rate of stimulation of a muscle causes the individual twitches to fuse into a continuous contraction or tetanus. This is reached at about 50 cycles per second in mammalian muscle (about the frequency of most domestic electricity supplies).

[2] The trade-off between force exerted and velocity of movements is familiar to cyclists and one of the reasons why bicycles have gears.

converted into energy-rich intermediates, notably adenosine triphosphate (ATP), through a controlled oxidation process that releases gaseous $CO_2$.

However, oxygen is not needed for the contraction itself. A skeletal muscle held in a bath saturated with $N_2$ rather than $O_2$, or a bath that contains an inhibitor of oxidative phosphorylation, can still contract, although only for a short period. Metabolic energy is obtained under these conditions by *glycolysis*: a series of coupled enzymatic reactions in which glycogen is degraded anaerobically to generate ATP.

In the body, skeletal muscles work anaerobically when a sudden burst of work is required. Anaerobic contractions occurring under these conditions are not limited by the supply of oxygen and are the most powerful and rapid contractions of which the body is capable. They are profligate of fuel, however, and limited in duration by the build-up of lactic acid (Figure 9-3). Eventually, the accumulation of lactic acid prevents further contraction.

Some muscles, such as the slow type of skeletal muscle or the nonstriated smooth muscle are more heavily dependent on oxygen. Although slower to act, such muscles do not accumulate lactic acid to the same degree and so can continue working for longer periods of time (Chapter 10).

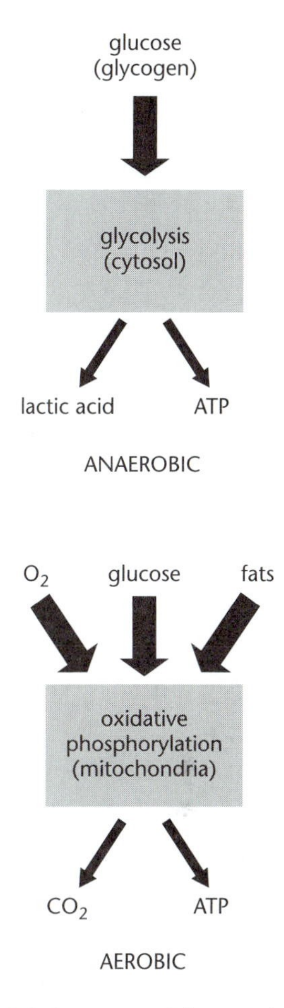


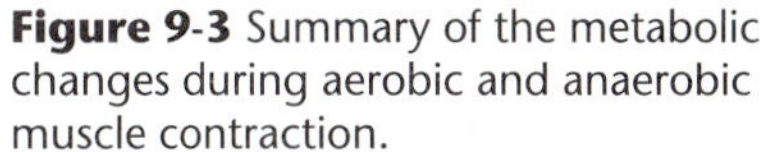

**Figure 9-3** Summary of the metabolic changes during aerobic and anaerobic muscle contraction.

## *Creatine phosphate is an energy carrier for muscle contraction*

The basis of muscle contraction is the interaction between myosin II and actin, which utilizes ATP as a source of chemical energy. However, it is surprisingly difficult to demonstrate this fact experimentally by making measurements on intact whole muscle. The quantity of free ATP in a muscle does not change significantly even after prolonged contraction and falls only with the onset of death.[3] The reason for this apparent contradiction is that an extremely effective back-up source is able to maintain ATP levels at an essentially constant level. The source of this energy-rich reserve of energy, in vertebrate muscle, is the small molecule creatine phosphate.

*Creatine phosphate* is one of a number of so-called high-energy phosphate compounds used by cells in their energy conversions. The common feature of these compounds is that the hydrolysis of one of their phosphate groups under the conditions existing in the cell is highly favorable: that is, it has a large free energy change. Hydrolysis may therefore be used to drive a second, less favorable reaction in the cell provided there is a suitable enzyme present that can couple the two reactions together. For creatine phosphate, hydrolysis is so favorable that it can be used to drive the formation of ATP and the enzyme creatine kinase catalyses the reversible reaction

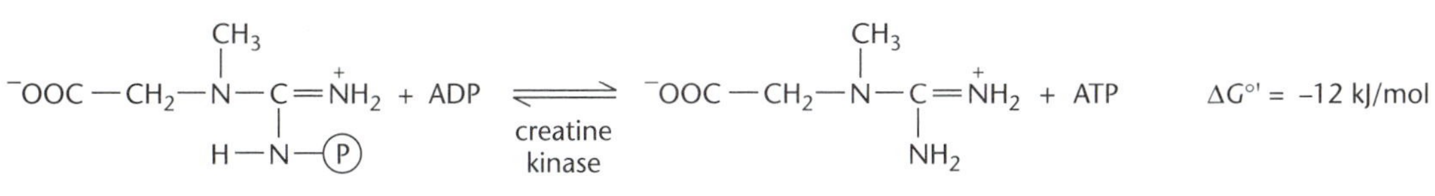


$\Delta G^{\circ\prime} = -12$ kJ/mol

The concentration of creatine phosphate in the cytosol bathing a myofibril is so high (about 25 mM) that it drives this reaction to the right, maintaining a constant high level of ATP.

Regeneration of creatine phosphate depends on oxygen and takes place in mitochondria. Mitochondria possess a different isoform of creatine kinase, localized between the inner and outer mitochondrial membranes. This enzyme is positioned so that it captures newly made ATP molecules as they leave the mitochondrion and catalyzes the above reaction in the opposite direction, with net synthesis of creatine phosphate. Newly made creatine phosphate molecules then diffuse from

[3] The rigid conformation of skeletal muscle following death is due to depletion of ATP. It is termed rigor, since it the cause of rigor mortis.

Figure 9-4 Creatine phosphate shuttle. A cyclic metabolic reaction that replenishes ATP depleted during muscle contraction.

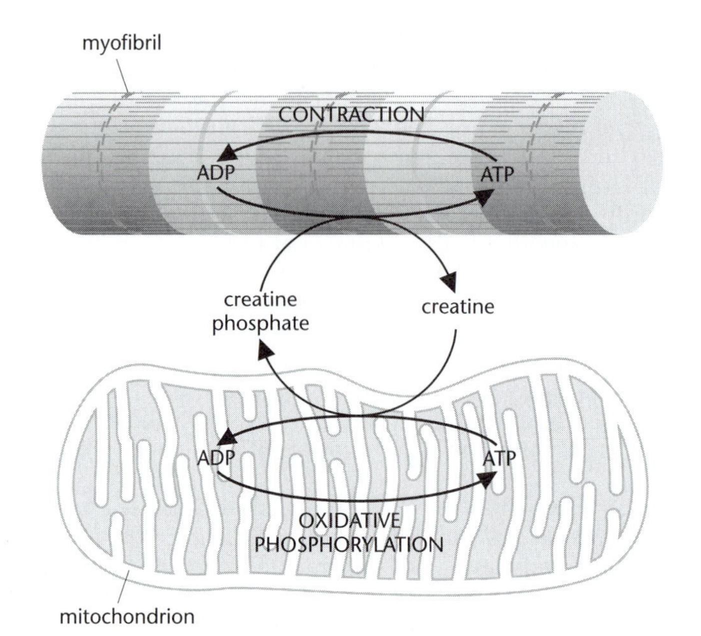


mitochondria to the contractile machinery, where they replenish ATP. Creatine phosphate is therefore an energy carrier that ferries high-energy phosphate groups between mitochondria and the contractile machinery (Figure 9-4).

## *Each muscle fiber is a large multinucleated cell*

Individual skeletal muscles are subdivided into a large number of long, thin compartments each aligned with the axis of contraction (Figure 9-5). The largest compartments, visible to the naked eye, are *fascicles*, which are associated in various patterns to form the several types of muscle recognized by anatomists. Fascicles are themselves composed of numerous threadlike *muscle fibers*, which can be seen in the light microscope.

Muscle fibers are typically 10–100 μm in diameter and range from less than a millimeter to a centimeter or so in length. Despite their relatively enormous size, each fiber is one cell. It is enclosed in a plasma membrane, which is tightly associated with the basal lamina to form a distinct muscle

**Figure 9-5** Levels of organization of skeletal muscle.

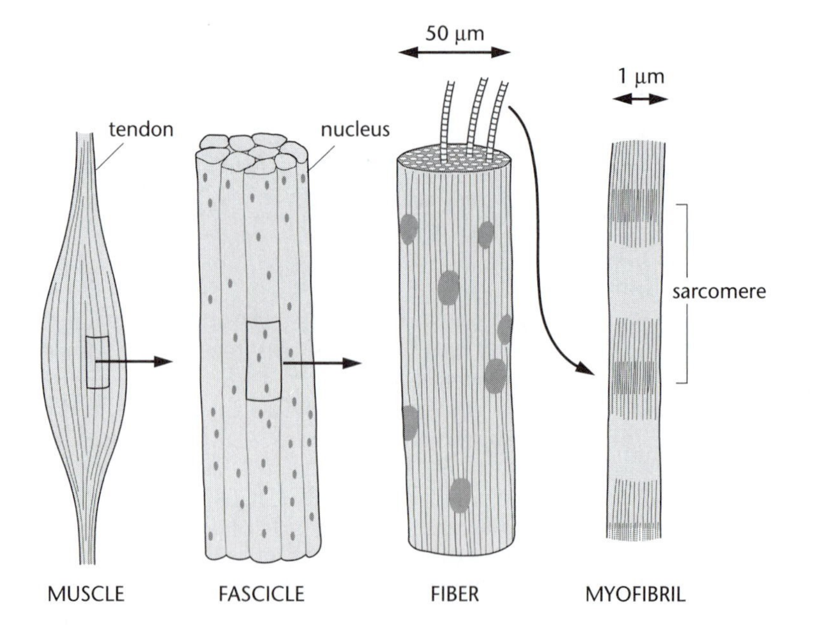

'skin' or *sarcolemma*. Muscle fibers are produced during development by the fusion of many mononucleated precursor cells and the nuclei of these precursor cells are retained in the adult muscle cell.

Although in some small vertebrates, such as mouse, fibers extend the entire length of the muscle, this is rarely true in larger muscles (longer than 2–3 cm). In the large muscles of humans, for example, most fibers are substantially shorter than the fascicles in which they run and are attached at their ends to other fibers rather than to tendons.

## Myofibrils are the contractile units of a muscle cell

Some 80% of the cytoplasm of a skeletal muscle fiber is occupied by cylindrical rods of protein known as myofibrils (see Figure 9-5). Many thousands of myofibrils, each about 1 μm in diameter, are contained within the cross-section of a single muscle fiber. Fibers also contain mitochondria, sandwiched between the myofibrils, and nuclei, which are squeezed to a peripheral location, just beneath the plasma membrane. Each fiber is enclosed by a sheath of connective tissue, known as a *basal lamina*.

*Myofibrils* are the structures responsible for muscle contraction, as can be demonstrated by isolating fragments of myofibrils from homogenized muscle and observing the rodlike fragments under the microscope. If ATP is added, the myofibril fragments snap into crumpled knots, like stretched rubber bands that have been cut. Using new laser tweezers techniques and sophisticated electronic measuring devices, myofibrils can be mounted and their mechanical performance (how the tension changes with degree of stretch and rate of shortening) can be measured with enormous accuracy.

The most distinctive feature of myofibrils is their banded appearance, due to the highly regular, almost crystalline, arrangement of their contractile apparatus. We discuss below how changes in these bands during contraction, led to the discovery of the molecular mechanism by which muscle contracts. Before describing the contractile process itself, however, it is useful to ask how signals delivered by a nerve axon reach myofibrils. This is not a trivial question since many myofibrils are embedded in the cytoplasm a distance of 50 μm or more away from the plasma membrane. Recall also that the time it takes a signal to reach them may be a matter of life or death for the animal.

## Internal membranes relay nerve stimulation to myofibrils

*Excitation–contraction coupling*, as it is known, depends on a series of highly specialized signal-relay devices. First in the chain comes the neuromuscular synapse. Here an action potential, reaching the end of the nerve axon, jumps to the plasma membrane of the muscle cell (sometimes termed the sarcolemma) by means of a chemical transmitter, acetylcholine, which diffuses across the small gap between the muscle and nerve (Figure 9-6).

The muscle cell plasma membrane is excitable so that action potentials spread rapidly and without decrement from the site of the neuromuscular junction (Figure 9-6b). A highly developed system of tubular

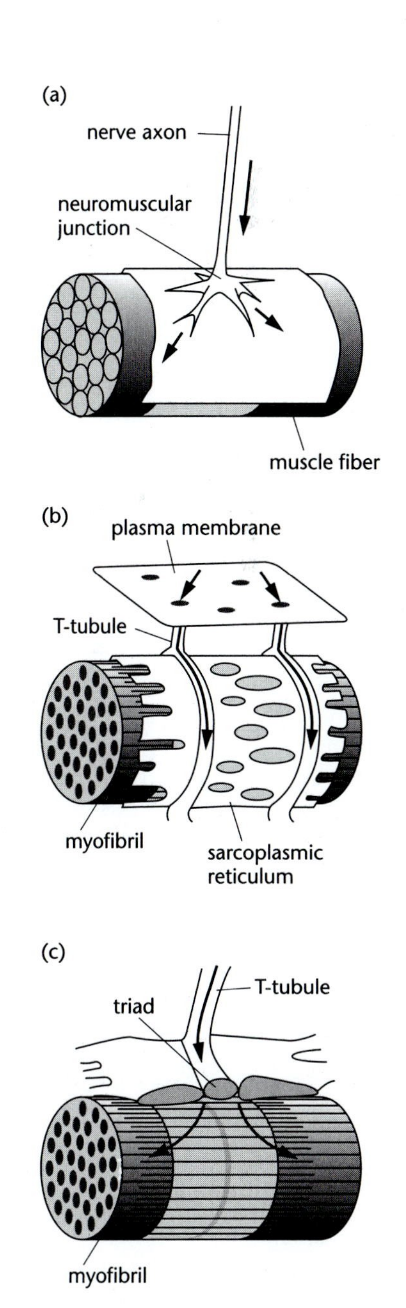


**Figure 9-6** Signal relay during muscle stimulation. (a) An action potential in the nerve axon is relayed to the plasma membrane of the muscle cell via the neuromuscular junction, a special form of chemical synapse. (b) The action potential spreads in the muscle cell plasma membrane and enters the interior of this large cell via tubular invaginations, the T-tubules to reach individual myofibrils. (c) Junctions between the T-tubules and the sarcoplasmic reticulum, a set of internal membranes surrounding each myofibril, couple the electrical signal to release of calcium ions into the myofibril to trigger contraction.

invaginations of the plasma membrane known as transverse tubules, or *T-tubules*, then carries the excitation to the interior of the muscle cell, where they make contact with a ragged sheath of membrane sacs known as the *sarcoplasmic reticulum* surrounding each myofibril. The sarcoplasmic reticulum is a specialized compartment of the endoplasmic reticulum containing high concentrations of $Ca^{2+}$. Calcium ions, released from this source when excitation spreads inward from the T-tubules, trigger muscle contraction.

The site of contact between the T-tubule and sarcoplasmic reticulum is known as a *triad junction* (Figure 9-6c) and it is here that an action potential in the T-tubule causes release of calcium ions from the sarcoplasmic reticulum. In the electron microscope, the two apposing membranes of a triad junction, 10–20 nm apart, are linked at regular intervals by protein 'feet.' These form the calcium-release channel: the port through which calcium ions flood to activate muscle contraction.

On the other side of a triad, the calcium release channel makes contact with a matching array of intrinsic proteins in the T-tubule membrane. Although the mechanism is still debated, evidence grows that the signal for contraction is relayed by a direct mechanical coupling between the two sets of membrane proteins. In this view, a voltage-dependent conformational change in the T-tubule protein pushes open the calcium-release channel in the sarcoplasmic reticulum, allowing calcium ions to flood onto the contractile apparatus.

The entire chain of chemical and electrical changes, from the delivery of an action potential to the muscle cell to the receipt of calcium ions by the contractile machinery, takes less than 15 milliseconds. It occurs with every movement we make, however slight.

## Muscle contraction is caused by the sliding of actin and myosin filaments

A myofibril shows alternating light and dark bands with a periodicity in the resting state of 2.3–2.6 μm, the dark bands being those with the greater amount of protein (Figure 9-7). When illuminated with polarized light, the dark bands are also found to be more *anisotropic*—that is, their refractive index depends more on the plane of polarization of the incident light. For this reason they are usually called A-bands (for anisotropic), in contrast to the lighter and relatively isotropic I-bands. Each I-band, has a thin line of high refractive index called a Z-disc (also Z-band or Z-line), whereas each A-band has a band, less dense than the Z–line, called the H-zone, and a central M-line. The entire repeating structure, from one Z-disc to the next, is known as a *sarcomere*.

The banded appearance of the sarcomere is produced by hundreds of protein filaments bundled together in a highly ordered arrangement. There are two principal types of filament in these bundles: thick filaments about 15 nm in diameter and thin filaments about 7 nm in diameter. Thick filaments occupy the central region of the sarcomere and thin filaments extend on either side of the Z-disc of the sarcomere. Selective extraction of myofibrils with solutions of high or low ionic strength, followed by biochemical analysis, demonstrates that the thick filaments are made mostly of myosin and the thin filaments mostly of actin. Both types of filament contain other proteins that help to hold them in the correct steric arrangement and regulate the contractile process.

Arrays of actin and myosin filaments overlap in the sarcomere, the individual filaments interdigitating rather like two stiff bristle brushes pushed together. The arrangement can be seen clearly in cross-sections made at different positions along the sarcomere and examined by electron microscopy (Figure 9-7b). Comparison with the sarcomeric patterns seen by light microscopy shows that the A-bands correspond to the position of the thick, myosin-containing filaments. These filaments contain a greater

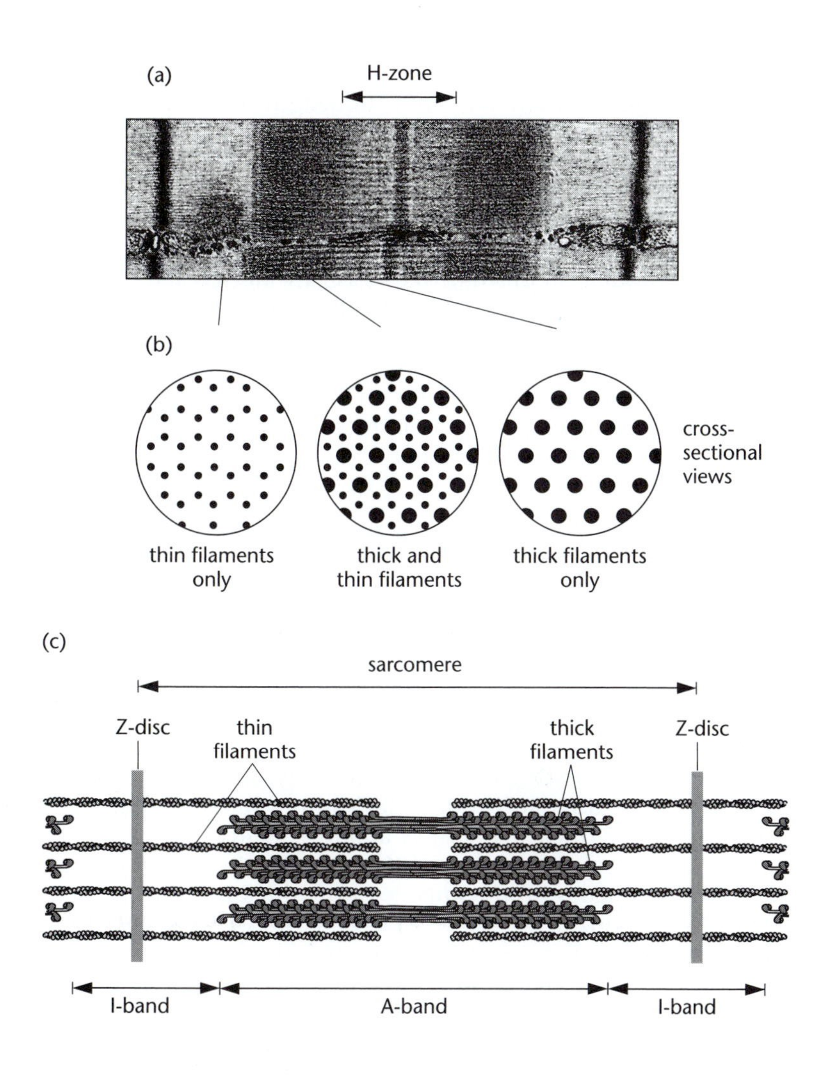


**Figure 9-7** Arrangement of filaments in a sarcomere. (a) Electron micrograph of section of single sarcomere. (b) Pattern of filaments in cross-sections made at different positions along sarcomere. (c) Schematic longitudinal section through a sarcomere showing thick and thin filaments. (Micrograph courtesy of Roger Craig.)

mass of protein and in consequence they have a higher refractive index (and hence a greater ability to change the plane of polarized light). The I-bands contain the actin-rich thin filaments.

Changes in the banding pattern during muscle contraction provided the vital clue to the mechanism of contraction. Although each sarcomere is reduced in length as the muscle contracts—typically from a length of 2.5 μm down to about 1.5 μm—the A-band does not shorten. The I-band, by contrast, contracts in unison with the muscle. In a muscle that has contracted to about 65% of its rest length (the usual limit of shortening under physiological conditions), the I-bands disappear. Conversely, in a muscle that has been passively stretched, the length of the I-band increases proportionately.

The changes in sarcomeric spacing were explained in two papers published in 1954, one by Hugh Huxley and Jean Hanson and the other by Andrew Huxley and Ralph Niedergerke. These authors suggested that when muscle shortens or lengthens, the two sets of protein filaments, thick and thin, slide past one another, thereby generating the observed changes in sarcomeric patterns (Figure 9-8). A corollary of this sliding filament hypothesis was that the force for contraction is developed by the interaction between the two sets of filaments in the region of their overlap.

The sliding filament hypothesis, now universally accepted, readily explains why maximum tension is developed by a muscle held at its normal rest length. As shown in Figure 9-8 (contracted), the resting length of muscle corresponds to the maximum overlap of thick and thin

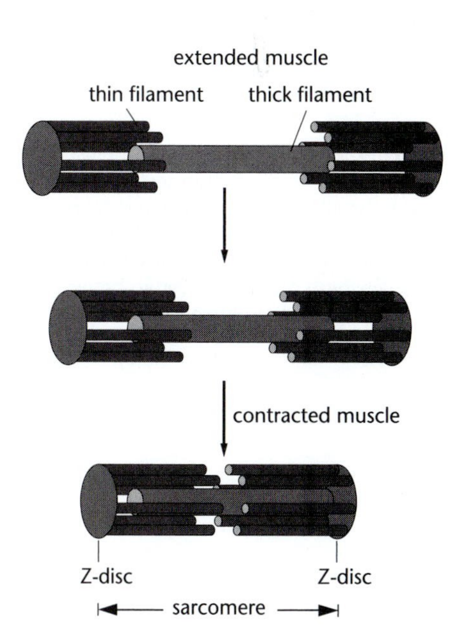


**Figure 9-8** The sliding filament hypothesis. Schematic diagram showing the relation between filament overlap and contraction. Three stages of contraction are shown.

filaments within each sarcomere and hence would be expected to develop the greatest tension. Furthermore, since the length of A-bands is remarkably constant in different muscles (close to 1.55 μm in vertebrates), the maximum length of overlap possible is also constant.

It has been suggested that the value of this maximum tension—about 3 kg/cm$^2$ (30 N/cm$^2$)—may be limited by the breakage of the muscle. The force necessary to break actin–actin bonds in an actin filament has been measured by manipulation of single fluorescently labeled actin filaments. The value obtained, close to 120 pN, may be compared to the maximum tension sustained by a single thin filament in muscle of around 150 pN (divide 30 N by the number of actin filaments per square centimeter—about $2 \times 10^{11}$, close to the Z-disc). Why then do the filaments not break? Presumably because they are strengthened by proteins such as tropomyosin that bind along their length.

## *Actin and myosin filaments are held in precise positions in the myofibril*

Skeletal muscle myosin has a high-molecular-weight heavy chain, of the 'conventional' or myosin II family. As we saw in Chapter 7, myosin II has the distinctive property of forming large filamentous aggregates in solutions of isotonic ionic strength—the basis for the formation of muscle thick filaments. Naturally occurring thick filaments have a precisely defined structure, consisting of about 300 myosin molecules neatly stacked by means of their stiff rodlike tails into a bipolar structure. The overall length of the thick filament is about 1.55 μm, with the central bare zone occupying about a tenth of this length. Myosin molecules are arranged with their heads on the outside of the thick filament and pointing away from the midzone, toward either end of the filament.

Within the myofibril, as already mentioned, the two sets of filaments, one based on actin the other on myosin, are held in precisely interdigitating arrays. Thick filaments come within about 13 nm of the adjacent thin filaments, close enough for myosin heads, projecting from the thick filaments on thin flexible necks, to reach across and bind to actin molecules. In this way, the myosin heads form cross-bridges visible in electron micrographs (Figure 9-9).

Not only are the two sets of filaments held at an optimum distance from each other to allow them to interact, but their relative polarities in

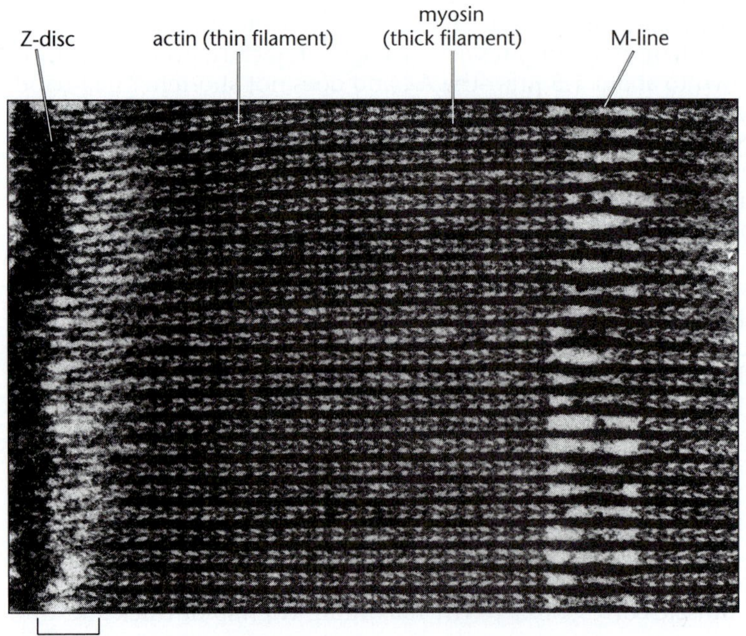


**Figure 9-9** Electron micrograph of insect flight muscle. This very thin section shows clearly the alternating myosin and actin filaments and the cross-bridges that link the two. Note that insect flight muscle has an unusually high degree of overlap between the myosin and actin filaments. (Courtesy of Mary C. Reedy.)

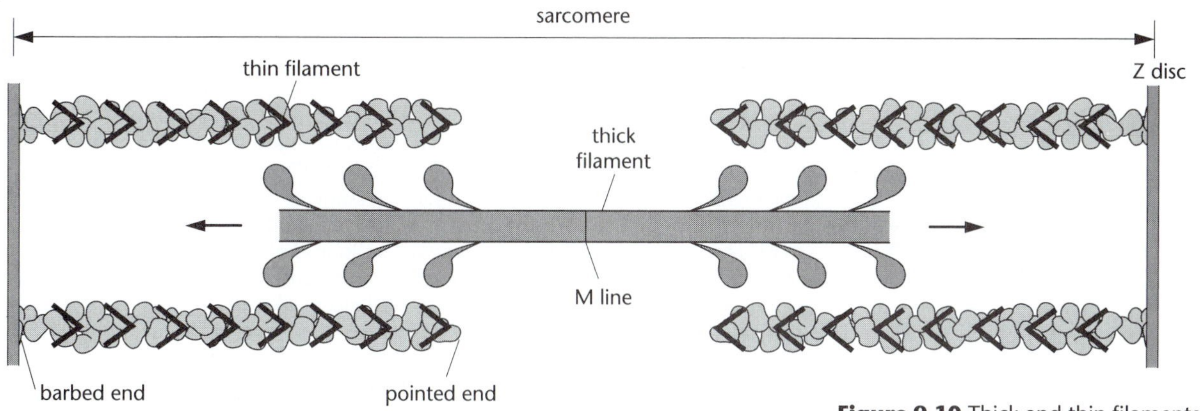


**Figure 9-10** Thick and thin filaments both reverse polarity midway between the Z lines.

the two halves of the sarcomere are the same. As illustrated schematically in Figure 9-10, actin filaments are anchored by their barbed or plus ends to the Z-disc and their pointed or minus ends point toward the middle of the sarcomere; the bipolar thick filaments are built from myosin molecules arrayed with their heads pointing outward from the center of the sarcomere. Consequently, when the two sets of filaments slide against each other during contraction, myosin heads on either side of the midline of the sarcomere undergo a sterically equivalent interaction with adjacent actin filaments—moving in both cases toward the barbed ends of the actin filaments.

In a sense, the overlapping filaments of muscle provide platforms that enable myosin heads to interact with their binding sites on actin with optimal spacing and orientation.

## *Cross-bridges made of myosin heads interact cyclically with actin filaments*

We mentioned earlier, in the discussion of contraction in nonmuscle cells in Chapter 7, that the heads of myosin interact cyclically with actin. The details of this repeated sequence, on which the conversion from chemical energy to mechanical energy depends, are best understood in muscle. Here they form the basis of the *cross-bridge cycle* in which a change in shape, or angle of the myosin head generates the force for contraction.

Information on the cross-bridge cycle comes from a convergence of techniques. Chemical kinetic measurements, using stopped flow and other methods, reveal short-lived intermediates in the 'chemical cycle'—the sequence of intermediates that form during the hydrolysis of ATP on the myosin head. Mechanical measurements on whole muscle fibers as well as single myosin molecules using laser tweezers, provide estimates of the physical movement of the myosin head (its step size) and of the force generated at each stage.

Determination of the three-dimensional structure of the myosin head by x-ray diffraction analysis, in 1993, was a major advance. This not only provided insight into the static anatomy of the myosin molecule, showing the position of sites of binding to actin, ATP and so on, but also threw a light on its movements. Thus, if a muscle fiber is mounted in such a way that tension measurements and x-ray diffraction analysis are performed at the same time, then rapid molecular-scale changes can be deduced. This applies not only to the relatively large movements of the contractile cycle but also the smaller elastic deformations undergone by the myosin molecule.

As indicated in Figure 9-11, ATP binds to the myosin head, causing the latter to detach from the actin filament. The myosin head, attached to its tail (and hence to the thick filament) by a flexible 'neck' of polypeptide chain, is then free to move by thermal motion. The head is thought

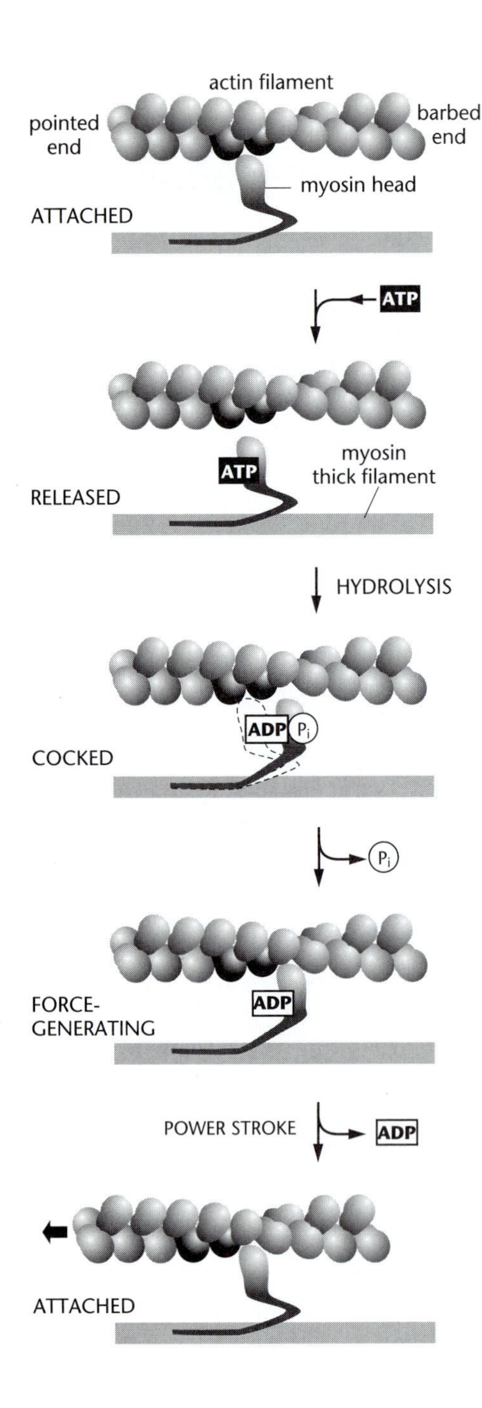


**Figure 9-11** Summary of the cross-bridge cycle. (1) Attached—at the start of the cycle shown in this figure, a myosin head lacking a bound nucleotide is locked tightly onto an actin filament in a *rigor* configuration. In an actively contracting muscle this state is very short-lived, being rapidly terminated by the binding of a molecule of ATP. (2) Released—a molecule of ATP binds to a narrow hole on the 'back' of the head (that is, on the side farthest from the actin filament) and immediately causes a slight change in the conformation of the domains that make up the actin-binding site. This reduces the affinity of the head for actin and allows it to slide along the filament, powered by other cross-bridges. (3) Cocked—the cleft on the myosin head closes like a clam shell around the ATP molecule, triggering a large shape change that causes the head to be displaced along the filament by a distance of about 5 nm. Hydrolysis of ATP occurs, but the ADP and $P_i$ produced remain tightly bound to the protein. In fact the head probably alternates between uncocked (ATP) and cocked ($ADP/P_i$) in a rapid equilibrium at this stage. (4) Force-generating—the weak binding of the myosin head to a new site on the actin filament causes release of the inorganic phosphate produced by ATP hydrolysis, concomitantly with the tight binding of the head to actin. This release triggers the 'power stroke'—the force-generating change in shape during which the head regains its original conformation. In the course of the power stroke, the head loses its bound ADP, thereby returning to the start of a new cycle.

to rock, or change its angle reversibly, at this stage, changes in shape being accompanied by the interconversion of ATP and a partially hydrolyzed intermediate ($ADP.P_i$). At the next step, the head binds back again to an actin filament and releases the products of ATP hydrolysis. Release of inorganic phosphate ($P_i$), which is normally the slowest step in the chemical cycle, is powerfully stimulated by contact with actin, and closely followed by release of ADP. A large drop in free energy takes place at this stage of the cycle—this is the 'power stroke' in which a change in form or orientation of the myosin head develops tension.

Many of the detailed atomic-level changes occurring in the myosin head during the cross-bridge cycle have been deduced. Thus, the head contains a *gamma phosphate sensor*, consisting of two amino loops of the polypeptide backbone, which changes position when the head binds ATP. The motor domain of the microtubule-based motor kinesin, discussed in Chapter 12, has a similar pair of loops, suggesting a common evolutionary origin for these two proteins. The gamma phosphate sensor loops switch back to their original position when inorganic phosphate is released from the head. We will now see that this small movement is relayed through the molecule to a rigid α-helix that drives the power stroke.

## A rigid α-helix in the myosin head acts like a lever

Estimates of the size of the 'mechanical cycle'—the physical movements and force production of each head—come not only from measurements of the contraction of whole muscles but also from analysis of the motion of single myosin molecules using laser tweezers. In one technique, already described in Chapter 7, single actin filaments are held in position above a lawn of myosin molecules and then lowered so that a single myosin attaches. The investigator then measures the displacement of the tethered actin filament with each cycle of the myosin head, or uses a servo loop to balance the movement and thus measure the force exerted by the head. These are delicate measurements of movements occurring close to the background of thermal, or Brownian, motion but the step size is probably close to 10 nm and the force generated per step is between 2 and 4 piconewtons (pN).

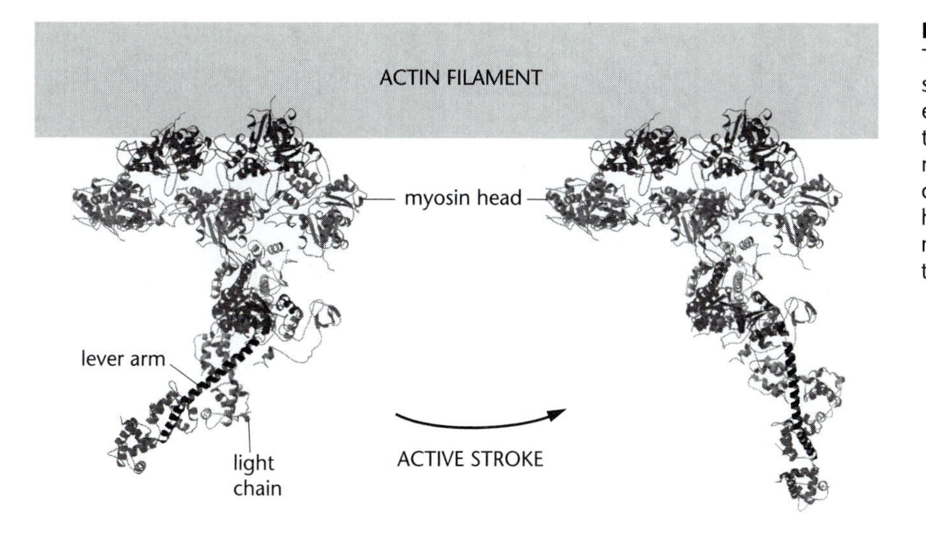


**Figure 9-12** Bending of the myosin head. The head region of the myosin II molecule is shown with α-helices as ribbons. An extended length of α-helix extending from the head is thought to act as a lever during muscle contraction. Small changes in conformation of the head due to the hydrolysis of ATP are magnified by movements of the lever, which sweeps through an angle of about 30°.

The three-dimensional atomic structure of the head region of myosin reveals an extended stretch of α-helix of over 70 amino acids and about 8 nm long. This α-helix extends from the ATP-binding pocket and is enveloped along most of its length by the light chains, which presumably serve to strengthen or rigidify it (Figure 9-12). The α-helix is believed to serve as a lever that magnifies the small conformational changes associated with ATP hydrolysis; movement of this lever arm could produce a step of a suitable size.

Support for this idea comes from an ingenious experiment in which three variant *Dictyostelium* myosins were genetically engineered to produce α-helical rods of different length. The first myosin mutant was deleted for both light chain binding sites, the second was deleted for just one of the two light chain binding sites, and the third myosin was engineered so that it carried an additional light chain binding site and could consequently bind to three instead of the normal two light chains. The three mutant constructs together with the wild type thus comprised a series of levers of increasing length, able to accomodate 0, 1, 2, or 3 light chains, respectively. The four proteins were expressed in cells, purified, and scored for motility in a cell-free assay and for ATPase activity. They moved actin at average sliding velocities that increased monotonically with length. Not only did the short lever arm constructs move more slowly but, importantly, the one with longer than normal α-helix moved even faster than the wild type—an improvement on nature! (Figure 9-13).

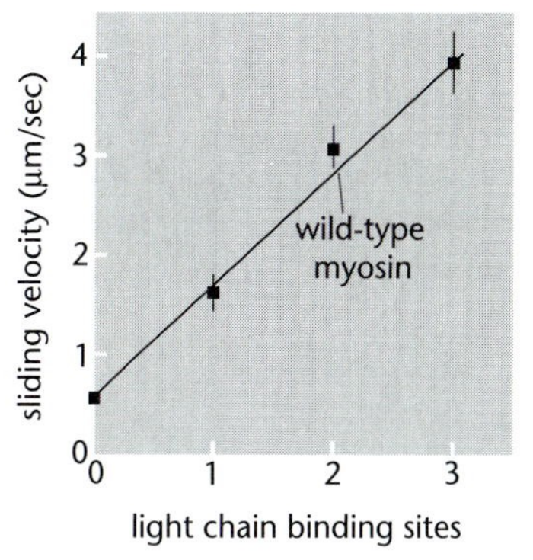


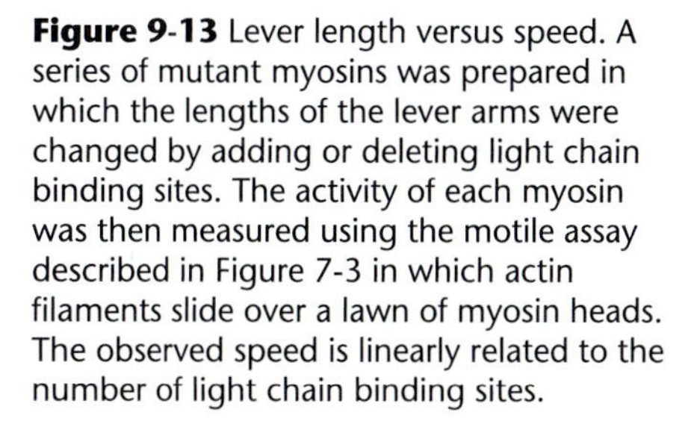

**Figure 9-13** Lever length versus speed. A series of mutant myosins was prepared in which the lengths of the lever arms were changed by adding or deleting light chain binding sites. The activity of each myosin was then measured using the motile assay described in Figure 7-3 in which actin filaments slide over a lawn of myosin heads. The observed speed is linearly related to the number of light chain binding sites.

## *Troponin and tropomyosin make muscle contraction $Ca^{2+}$-sensitive*

Returning to the myofibril, we know that its contraction must be under neural control and we have described the membrane systems of the muscle fiber that carry excitatory signals from the nerve axon to the interior of the muscle cell. The end result of this chain of signals, it may be recalled, is a rise in the local concentration of calcium ions from about $10^{-7}$ M to about $10^{-5}$ M. We now have to examine how this local increase of calcium ions triggers the myofibril to contract.

In vertebrate striated muscle, the interaction of actin and myosin is regulated by two accessory proteins, tropomyosin and troponin. These are associated with actin in such a way that they inhibit the cross-bridge cycle unless calcium ions are present at sufficiently high concentration. Tropomyosin, which was introduced in Chapter 7, is a rigid rod-shaped molecule with an unusually high content of α-helix, making it extremely stable to heat and organic solvents. The isoform of tropomyosin found in

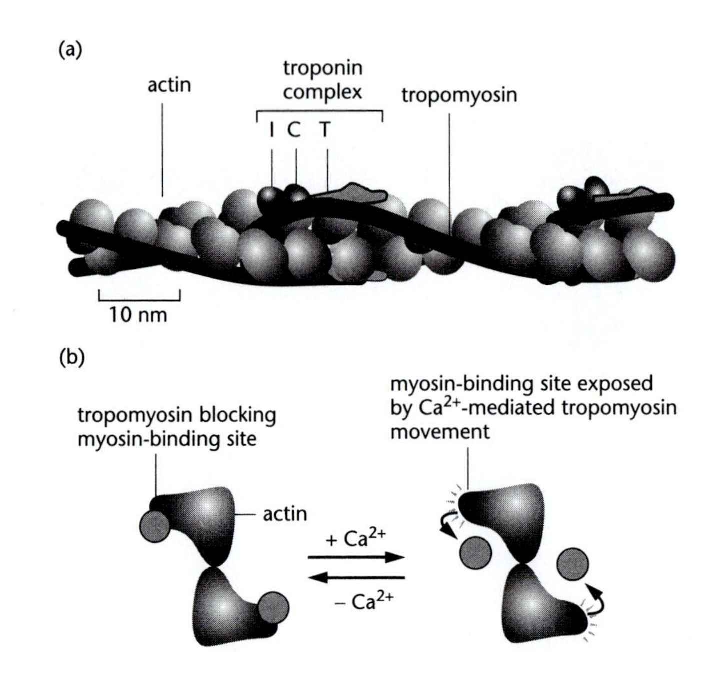


**Figure 9-14** Control of skeletal muscle contraction by troponin. (a) Thin filament in striated muscle, showing the position of tropomyosin and the troponin complex along the actin filament. Each tropomyosin molecule has seven evenly spaced regions of homologous sequence, each of which is thought to bind to an actin monomer, as shown. (b) A thin filament shown end-on, illustrating the slight movement of tropomyosin, caused by $Ca^{2+}$ binding to troponin. This movement allows the myosin head to interact with actin.

skeletal muscle has a molecular weight of 70,000 and consists of two almost identical subunits, α-Tm and β-Tm, which wind around each other in a coiled-coil structure similar to that in the myosin tail. Each tropomyosin molecule is about 40 nm long and extends along the length of seven actin monomers in an actin filament. Sequential tropomyosin molecules overlap, forming paired threads that wind around the actin filament, adding to its strength and rigidity (Figure 9-14).

Troponin, unlike tropomyosin, is present only in striated muscle cells. It is a complex of three polypeptide chains, known as troponin C (Tn-C) (calcium binding), troponin I (Tn-I) (inhibitory) and troponin T (Tn-T) (tropomyosin binding). Troponin binds to a specific site on tropomyosin and is thereby positioned at regular intervals of seven actin monomers on each side of the actin filament (see Figure 9-14). The combination of troponin and tropomyosin, added to pure actin filaments, produces a complex that binds in a $Ca^{2+}$-sensitive fashion to myosin. When cytosolic calcium ions are low in concentration, actin activation of the myosin ATPase is abolished and the ATPase activity approaches that of myosin alone. In the presence of calcium concentrations around $10^{-5}$ M, the full ATPase activity is obtained.

A 'myosin on' switch is controlled by troponin C, which undergoes a large conformational change when it binds a calcium ion (Figure 9-14). Atomic structures show that a crucial glutamic acid in troponin C, which is normally attracted to a lysine elsewhere in the structure, is pulled away when a calcium ion binds to the protein, thereby causing a large shape change. This in turn shifts tropomyosin in the groove of the actin helix and probably relieves a steric block—rather like removing a small wedge that prevents the binding of myosin heads. In the absence of calcium ions the tropomyosin/troponin complex therefore blocks the progression of myosin through the cross-bridge cycle.

## *Troponin C belongs to a family of calcium-binding proteins*

Troponin C, which confers $Ca^{2+}$-sensitivity on muscle contraction, belongs to the same protein family as calmodulin—a protein that mediates the effects of calcium ions on many cytoplasmic enzymes. *Calmodulin* is a small (molecular weight 16,500), acidic protein which, like troponin C, is able to bind four calcium ions per molecule. Association with calcium ions produces a conformational change in calmodulin and this in turn changes the activity of enzymes that associate with calmodulin. In some enzymes calmodulin is a permanent regulatory subunit, in others, calmodulin binds to its target enzyme only when it is itself complexed with calcium ions.

The two types of light chains associated with the head region of skeletal muscle myosin—the so-called 'regulatory' and 'essential' light chains—also belong to this family of proteins (Chapter 7).

The common basis on which proteins of this family bind calcium ions is seen clearly in the case of parvalbumin, a 12 kDa, $Ca^{2+}$-carrying protein from muscle cytoplasm. This protein acts as a calcium buffer when muscle relaxes: accepting calcium ions as they dissociate from troponin C and carrying them to the $Ca^{2+}$ pumps in the sarcoplasmic reticulum. Parvalbumin has two similar $Ca^{2+}$-binding sites, each about 30 amino acids. Each site is formed by a small loop containing aspartate and glutamate (which bind to the metal ion) held in rigid position by two α-helices. One of the calcium-binding sites in parvalbumin is formed by helices E and F of this protein, positioned like the thumb and forefinger of the right hand, hence giving rise to the term *EF hand*. More than a hundred proteins are now known to have domains closely related to the parvalubumin EF hand.

Parvalbumin has two EF hands whereas troponin, calmodulin, and the myosin light chains each have four. Their similarity in amino acid sequence indicates that each has evolved from a common ancestral protein but they are not identical. In fact, some sequences that can be recognized by their amino acid sequence have changed so much that they no longer bind calcium ions with appreciable affinity. Thus, in the myosin light chains, all but one of the EF hands (that in the regulatory light chain) have lost their ability to bind calcium.

## *Accessory proteins maintain the precise architecture of a myofibril*

The contractile apparatus of skeletal muscle is not only remarkably fast and powerful, but it is also beautifully ordered at the molecular level. Presumably this regular pattern of molecules evolved as a consequence of the selective pressures to optimize performance. The speed and efficiency of vertebrate striated muscle depends on its myosin molecules being held at just the correct orientation and distance from actin filaments. A dozen or more 'accessory' proteins, each with a different, characteristic, location in the myofibril—in the Z-disc or M-line, for example, or along the length of a thick filament—serve to hold different parts of the final assembly in place (Table 9-1).

**Table 9-1 Protein components of vertebrate skeletal myofibrils**

| Protein | Location | % total protein | Molecular weight |
|---|---|---|---|
| myosin | thick filaments (A-band) | 44 | 520,000 (2 heavy + 4 light chains) |
| actin | thin filaments (I-band) | 22 | 42,000 |
| titin | from M line to Z-disc | 10 | 3,000,000 |
| tropomyosin | thin filaments (I-band) | 5 | 66,000 (dimer) |
| troponin | thin filaments (I-band) | 4 | 70,000 (troponin C, I, and T) |
| nebulin | I-band | 5 | 800,000 |
| M protein | M-line | 2 | 165,000 |
| C protein | A-band | 2 | 135,000 |
| α-actinin | Z-disc | 2 | 190,000 |

Other proteins found in the vertebrate striated myofibril include phosphofructokinase, H protein, I protein, myomesin, creatine kinase, Cap Z, tropomodulin, filamin, desmin, vimentin, synemin, and many others.

The Z-discs, which form the division between one sarcomere and the next, are sites to which actin filaments are anchored. They are held in a precise square lattice with the barbed ends of the filaments of one sarcomere held in one orientation and the barbed ends of filaments belonging to the adjoining sarcomere held in the other. This precise structural arrangement is generated by proteins in the Z-disc, the most familiar being the large, flexible cross-linking protein α-actinin, already encountered as a cross-linking protein of the actin cortex (Chapter 6). The skeletal muscle isoform of this protein is insensitive to calcium ions: sequence analysis shows that the EF hands present in nonmuscle α-actinin have mutated to a nonfunctional form. Skeletal muscle α-actinin binds only to the terminal regions of actin filaments (near their barbed ends) embedded in the Z-disc, being excluded from other locations by tropomyosin. By linking pairs of neighboring actin filaments together, α-actinin creates a square lattice of actin filaments, which may be seen in cross-sections of the myofibril at this level.

The Z-disc is also the location of a special capping protein, Cap Z (related to the capping protein of nonmuscle cells, see Chapter 6), which associates with the barbed end of each actin filament. The pointed end of these filaments is capped by tropomodulin, a protein that binds to tropomyosin. These two proteins help to keep the actin filaments in place and contribute to the stabilization of their length.

Myosin filaments also have a precisely determined length and regular packing and are held squarely in the center of each sarcomere. The M-line (the midpoint of the thick filaments, where the polarity of myosin filaments reverses) contains several unique proteins, including M-protein, and myomesin which serve to hold the myosin filaments, in this case in a regular hexagonal array. Each myosin-containing A-band of skeletal muscle is also divided along its length into distinct zones, three of which can be uniquely identified by electron microscopy and by antibody staining. The central region is further subdivided by a series of 11 transverse stripes or lines at regular intervals. The stripes are composed of C, I, and X proteins which are, as we will see shortly, themselves positioned by a molecular template of much larger dimensions.

The M-line is the location of high concentrations of the muscle isozyme of creatine phosphokinase that catalyzes the phosphorylation of creatine, as described above. Other enzymes such as those involved in glycolysis are also found in specific locations of the myofibril. Organization of the cytoplasm in and around the myofibrils is clearly of importance in channeling metabolites to the contractile machinery. The crucial role of the cytoskeleton in maintaining the internal structure of the cytoplasm will be taken up again in Chapter 18.

## *Titin and nebulin act as molecular rulers for the sarcomere*

One of the most astonishing discoveries about skeletal muscle in recent years is that it contains a second system of protein filaments. Made of the proteins titin and nebulin, these filaments are left behind with the insoluble residue after other muscle proteins have been extracted. They comprise a substantial 13% of the total protein of the myofibril, titin being the third most abundant protein after myosin and actin (see Table 9-1). *Titin* (named for its titanic molecular weight of 3,000,000) and *nebulin* (also large, but named for its association with lateral 'nebulous stripes' in the I-bands of the myofibril) are closely associated with myosin and actin filaments, respectively. They provide elastic strength to the myofibril, serving to hold Z-discs and M-lines together, and conferring a restoring force if the myofibril is passively stretched. They also have an

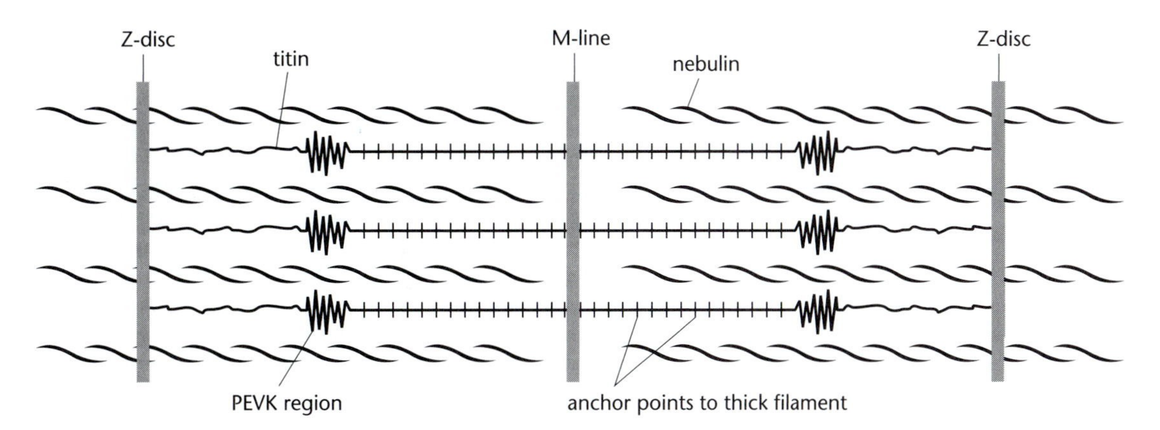


**Figure 9-15** Titin and nebulin. Schematic view of the position of these proteins in a myofibril. Each giant titin molecule extends from the Z-disc to the M-line—a distance of over 1 μm. Part of the titin molecule is closely associated with myosin molecules in the thick filament; the rest of the molecule is elastic and changes length as the muscle contract and relaxes. At weak or moderate stretch, such as the muscle usually encounters, elasticity is provided by the unfolding and straightening of a specific 'PEVK' domain that acts as an entropic spring. Each nebulin molecule extends from the Z-disc along the length of one thin actin filament and probably determines the thin filament length.

essential role in the formation of muscle by providing a molecular template for the assembly of the sarcomeric structures.[4]

Each titin molecule is a long string of about 300 globular domains, each approximately 2.5 nm in diameter and 4 nm long. The amino acid sequences of the domains show that they belong to two large superfamilies of proteins: 132 domains are of the fibronectin family and up to 166 of the immunoglobulin family. But each domain is unique in detailed sequence and has specific properties, so that the molecule as a whole serves like a molecular ruler (Figure 9-15). At the C-terminal end, normally embedded in the M-line of muscle, a globular head binds to other M-line proteins and is the site of regulation by phosphorylation. At the other end, the N-terminus of titin is buried in the Z-disc. Between these two, the repeating domains have different properties depending on position. Close to the myosin-containing A-band, most domains have binding sites for myosin and perhaps accessory proteins such as C protein. Interestingly, the sequence of domains in this region shows an 11-fold super repeat that precisely matches a corresponding 43 nm repeat in the helical array of myosin molecules in the thick filaments. In the I-band (that is, the actin-only region), the titin molecule forms a separate filament distinct from actin filaments and contains a region of amino acid sequence thought to unfold under tension, thereby conferring a springlike elasticity to the filament.

Nebulin is also a protein ruler, which in this case marks positions along the actin filament. One end is embedded in the Z-disc, whereas the other is close to the end of the thin filament and appears to limit the length of the filament (how it does this is not known). The nebulin amino acid sequence consists almost entirely of a series of 35-amino-acid motifs, with a super repeats every seven motifs. The number seven is significant here because, as mentioned previously, troponin molecules are positioned every seven actin molecules along the filament. Indeed, present evidence suggests that the nebulin molecule associates so closely with an actin filament that each small motif is bound to an actin molecule and every seventh motif has a binding site for troponin. If this proves to be the case, then nebulin may specify the structure of an actin filament with single-molecule precision!

If titin and nebulin are in fact molecular rulers for the sarcomere, then we might expect them to vary from one type of muscle to the next. Different kinds of muscle show characteristic and highly reproducible sarcomeric spacings and filaments lengths. In the following chapter we describe the assembly process of muscle and we will see that titin and nebulin do indeed vary in the expected way.

[4] Titin has also been found in smooth muscle cells and blood platelets and, most surprisingly, in the nucleus of tissue cells. In these locations it appears to act as a molecular shock absorber that helps keep large structures (like pairs of chromosomes) intact.

## Outstanding Questions

*How much of the enzymology of energy production in a muscle is carried over to nonmuscle movements? Is cell crawling or swimming aerobic or anaerobic? Do motor molecules use energy-rich intermediates, such as creatine? Is the build-up of waste products a problem? What are the actual atomic-level events accompanying the cross-bridge cycle in skeletal muscle? How many distinct structural states are there, and what is their significance for the overall contraction? Why are some motile events, such as muscle contraction, triggered by changes in $Ca^{2+}$ ions and not others? How are the multiple heads in a thick filament coordinated? What is the effect of tension transients on the cross-bridge cycle? How is the length of actin filaments in muscle determined so that it is precisely 1 μm? How is myosin assembly controlled so that it produces bipolar filaments with regular domains along its length? What keeps Z-discs and M-lines in their correct place?*

## Essential Concepts

- Skeletal muscles are built up of long, thin fibers, each of which is a single, large multinucleated cell.
- The cytoplasm of each muscle fiber is filled with myofibrils, the contractile units of the muscle. These have regular striations along their length, indicative of a highly organized internal structure.
- Myofibrils are stimulated to contract through a system of internal membranes that carry excitatory signals from the plasma membrane.
- ATP is the immediate source of energy for contraction but it is rapidly replenished, following hydrolysis, by a large reservoir of creatine phosphate.
- Muscle contraction is produced by the sliding of actin filaments against myosin filaments.
- The head regions of myosin molecules, projecting from myosin filaments, undergo an ATP-driven cycle in which they attach to adjacent actin filaments, undergo a change in shape that pulls one filament against the next, and then detach.
- Two accessory proteins—troponin and tropomyosin—allow the contraction of skeletal muscle to be regulated by $Ca^{2+}$. In the absence of calcium they bind to the actin filaments in such a way that they inhibit the binding of myosin heads.
- The cycle is facilitated by accessory muscle proteins that hold the actin and myosin filaments in parallel overlapping arrays with the correct orientation and spacing for sliding to occur.
- Two extremely long stringlike molecules, titin and nebulin, act like molecular rulers for myofibril assembly, determining the precise length of actin and myosin filaments and their location in the sarcomere.

## Further Reading


Anson, M., et al. Myosin motors with artificial lever arms. *EMBO J.* 15: 6069–6074, 1996.

Bagshaw, C.R. Muscle Contraction. London: Chapman and Hall, 1993.

Block, S.M. Fifty ways to love your lever: myosin motors. *Cell* 87: 152–257, 1996.

Cooke, R., et al. A model of the release of myosin heads from actin in rapidly contracting muscle fibers. *Biophys. J.* 66: 778–788, 1994.

Dobbie, I., et al. Elastic bending and active tilting of myosin heads during muscle contraction. *Nature* 396: 383–387, 1999.

Ebashi, S. Calcium binding and relaxation in the actomyosin systems. *J. Biochem.* 48: 150–159, 1963.

Erickson, H. Stretching single protein molecules: titin is a weird spring. *Science* 276: 1090–1092, 1997.

Flucher, B.E., Franzini-Armstrong, C. Formation of junctions involved in excitation–contraction coupling in skeletal and cardiac muscle. *Proc. Natl. Acad. Sci. USA* 93: 8101–8106, 1996.

Funatsu, T., et al. Imaging of single fluorescent molecules and individual ATP turnovers by single myosin molecules in aqueous solution. *Nature* 374: 555–559, 1995.

Goldman, Y.E. Wag the tail: structural dynamics of actomyosin. *Cell* 93: 1–4, 1998.

Gregorio, C.G., et al. Muscle assembly: a titanic achievement? *Curr. Opin. Cell Biol.* 11: 18–25, 1999.

Holmes, K.C. The swinging lever-arm hypothesis of muscle contraction. *Curr. Biol.* 7: 112–118, 1997.

Huxley, A.F., Niedergerke, R. Interference microscopy of living muscle fibres. *Nature* 173: 971–972, 1954.

Huxley, H.E. A personal view of muscle and motility mechanisms. *Annu. Rev. Physiol.* 58: 1–19, 1996.

Huxley, H.E., Hanson, J. Changes in the cross-striations of muscle contraction and their structural interpretation. *Nature* 173: 973–976, 1954.

Kitamura, K., et al. A single myosin head moves along an actin filament with regular steps of 5.3 nanometers. *Nature* 397: 129–134, 1999.

Labeit, S., Kolmerer, B. Titins: giant proteins in charge of muscle ultrastructure and elasticity. *Science* 270: 293–296, 1995.

Molloy, J.E., et al. Movement and force generated by a single myosin head. *Nature* 378: 209–212, 1995.

Roberts, T.J., et al. Muscular force in running turkeys: the economy of minimizing work. *Science* 275: 1113–1115, 1997.

Ruppel, K., Spudich, J.A. Structure-function analysis of the motor domain of myosin. *Annu. Rev. Cell Dev. Biol.* 12: 543–573, 1996.

Schröder, R.R., et al. Three-dimensional atomic model of F-actin decorated with *Dictyostelium* myosin S1. *Nature* 364: 171–174, 1993.

Schroeter, J.P., et al. Three-dimensional structure of the Z band in a normal mammalian skeletal muscle. *J. Cell Biol.* 133: 571–583, 1996.

Service, R. Flexing muscle with just one amino acid. *Science* 271: 31, 1996.

Taylor, M.V. A myogenic switch. *Curr. Biol.* 6: 924–926, 1996.

Trinick, J., Tskhovrebova, L. Titin: a molecular control freak. *Trends Cell Biol.* 9: 377–380, 1999.

Tskhovrebova, L., et al. Elasticity and unfolding of single molecules of the giant muscle protein titin. *Nature* 387: 308–312, 1997.

Uyeda, T.Q.P., et al. The neck region of the myosin motor domain acts as a lever arm to generate movements. *Proc. Natl. Acad. Sci. USA* 91: 4459–4464, 1996.

Whittaker, M., et al. Smooth muscle myosin moves 35A upon ADP release. *Nature* 378: 748–751, 1995.

Wojtas, K., et al. Flight muscle function in *Drosophila* requires colocalization of glycolytic enzymes. *Mol. Biol. Cell* 8: 1665–1675, 1997.

# CHAPTER 10

# Muscle Development

Anyone who has cooked chicken knows that there are different types of muscle. White breast and red thigh are both skeletal muscles, but they differ in color, anatomy, function (and flavor). Further dissection reveals other types of muscle in the same animal, such as smooth muscle in the digestive, respiratory, urinary, and genital tracts, and cardiac muscle of the heart. Nonmammalian species have even more diverse and highly specialized muscles. The flight muscle of some insects, for example, can contract at frequencies up to 600 times per second. The muscle of a clam can hold shut for many hours, maintaining enormous tensions with minimal expenditure of energy.

All of these different types of muscle are driven by the same basic molecular machinery described in the previous chapter, but they have been fine-tuned over the course of evolution to match the precise mechanical task they perform. How they achieve this fine-tuning is the subject of this chapter. We will see that the largest differences arise during the early development of the muscle, in the way that embryonic precursor cells build the muscle. Then there is the use of different genes—different forms of actin, myosin, and so on—that are switched on for particular muscle types, and confer on their muscles specific mechanical properties. Lastly, there are many subtle modifications occurring during the lifetime of the animal, in response to exercise, for example, or hormonal changes.

The mechanisms of fine-tuning are of great interest to muscle builders and to doctors (since many wasting diseases are caused by a failure of muscle to form properly). They are relevant to a book on cell movements because they show us some of the general strategies a cell can use to build and modify large cytoskeletal assemblies in the cell with precision and subtlety.

**Figure 10-1** Muscle somites. Scanning electron micrograph of the trunk region of a swordtail embryo showing the regular array of somites. (Courtesy of B. Sadaghiani and J.R. Vielkind.)

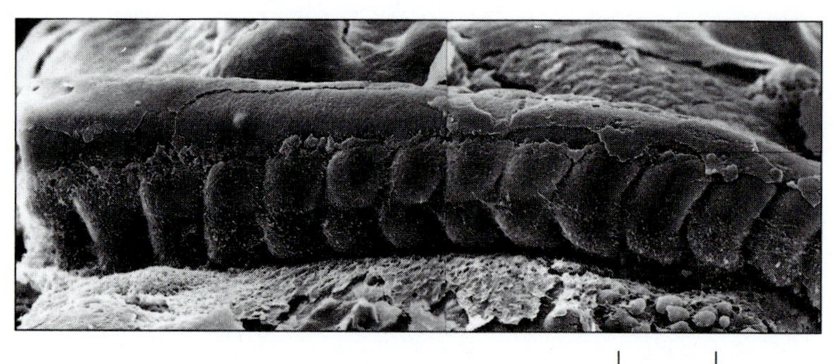


## *Skeletal muscle forms by fusion of mononucleated myoblasts*

In vertebrate animals, muscle precursor cells, or *myoblasts*, come from the *somites*, a series of tissue masses segmentally arranged on either side of the body (Figure 10-1). Initially the myoblasts, specified by signals coming from other regions of the embryo, are scarcely distinguishable from their neighboring cells. As development proceeds, however, the myoblasts migrate from the somite and take up positions in the body, such as in the region of a prospective limb (Figure 10-2a–c). Here they remain, inconspicuously mixed with connective-tissue cells, until the time comes for them to differentiate. The migration of muscle precursors can be traced by marking them using either a nontoxic dye or a heritable genetic label. Studies in which cells were grafted from quail embryos into chick embryos, for example, provided an important source of information.

At the appropriate time, the myoblasts begin to make muscle-specific proteins, such as the muscle isoforms of actin and myosin and the acetylcholine receptor, and reorganize their cytoskeleton. The cells become spindle-shaped and they fuse with each other into long multinucleated *myotubes* (Figure 10-2d)—a process that evidently involves specific recognition between myoblasts since they do not fuse with other cells.[1] As thick and thin filaments accumulate in the cytoplasm of the myotubes, they align into myofibrils and the myotube differentiates into a striated muscle fiber.

The entire series of events leading to muscle formation, including cell fusion, is under the control of master genes. Transcription-regulating genes in the *myoD* family are expressed only in myoblasts and mature muscle. If they are introduced experimentally into fibroblasts, they cause the cells to fuse with each other and to synthesize musclelike proteins.

Myoblasts taken from an embryo and put into culture usually go through a small number of cell divisions before they fuse into myotubes. However, cell lines have been isolated in which division continues indefinitely in the presence of fibroblast growth factor (FGF). Removal of FGF from such cultures triggers irreversible withdrawal of the cells from the cell cycle and, if other conditions are suitable, the cells differentiate into muscle.

## *Stem cells restore damaged muscle*

Once a fully differentiated fiber is formed, it does not divide further and in most cases it is destined to survive for the life of the animal. However, if the muscle fiber is damaged or even destroyed completely, new muscle

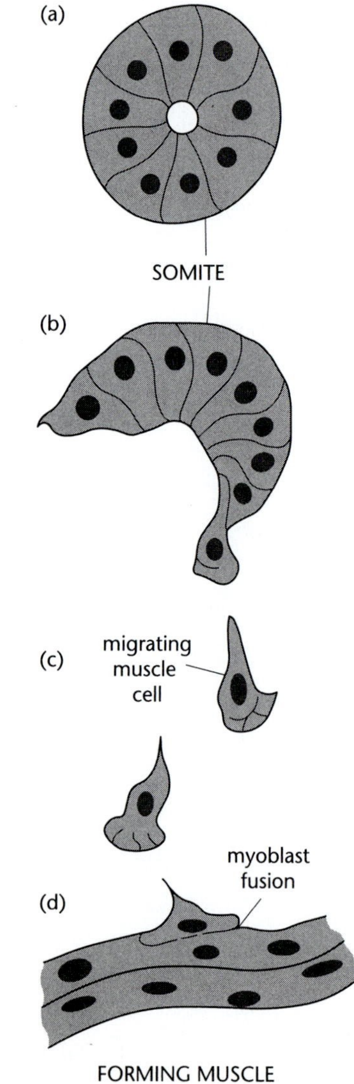


**Figure 10-2** Migration of limb muscle cells. (a) Muscle cells originate in the somites, specified by specific signals from neighboring cells. (b) As development proceeds, the muscle precursors (myoblasts) break free of the somites and migrate as individual cells into the limb bud. (c) and (d) Eventually the myoblasts fuse to form long spindle-shaped myotubes—the precursors of muscle fibers.

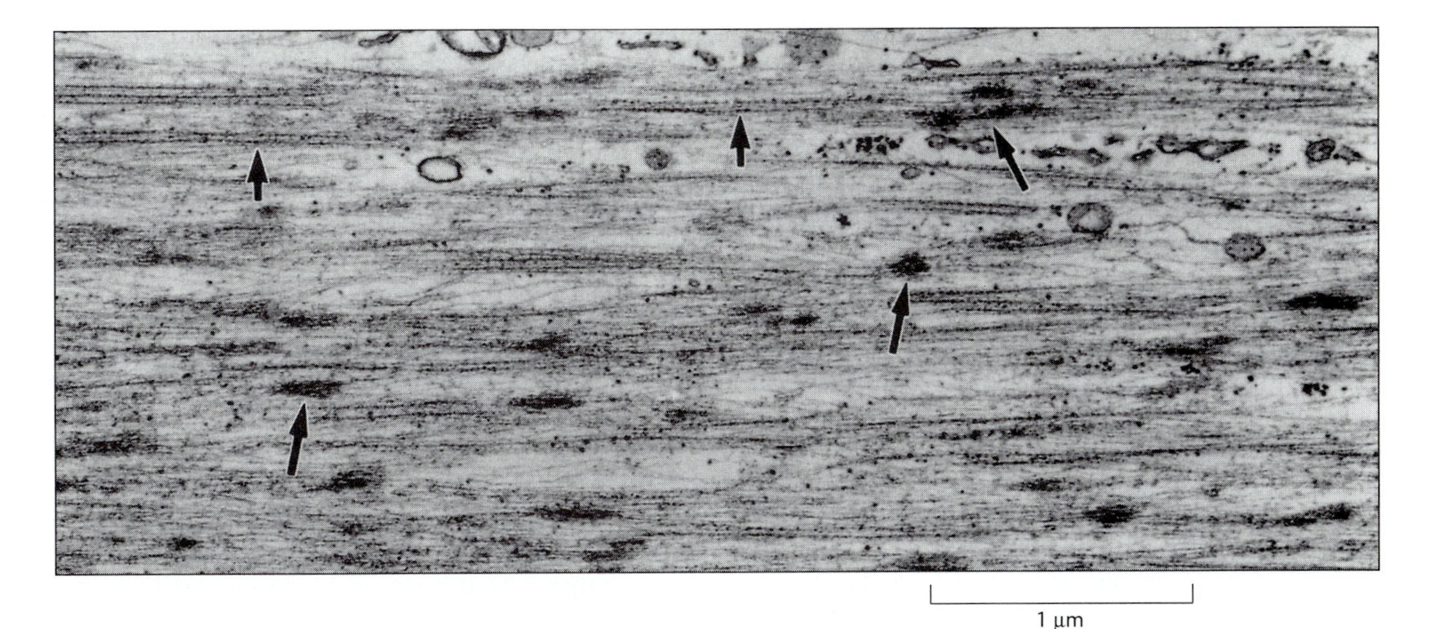


**Figure 10-3** Developing chick muscle. In a two-day embryo, thick and thin filaments appear to be separately organized. Short arrows indicate myosin-containing thick filaments, long arrows indicate actin filaments attached to forming Z-discs (Z-bodies). (From Almenar-Querait et al., 1999.)

tissue can take its place. Regeneration is especially efficient if the wound leaves intact part of the extracellular matrix, which plays an instructive role in both muscle and nerve regeneration. New muscle fibers assemble from a population of undifferentiated myoblast precursors present in the mature muscle. These appear in histological sections like normal muscle nuclei, but on closer examination are seen to be separate cells, totally enclosed in an inpocketing of the muscle plasma membrane. In response to tissue damage these quiescent cells are activated and then multiply and fuse, recapitulating the events that gave rise to muscle fibers in the embryo.

## *Titin plays a crucial role in myofibril assembly*

The most important task confronting a differentiating myotube is to build *myofibrils*, the complex arrays of thick and thin filaments that produce muscle contraction. This is achieved in a carefully controlled assembly process initiated by the synthesis of specific embryonic isoforms of muscle proteins (Figure 10-3). Once immature myofibrils have been made, their constituent proteins are replaced by other adult isoforms, expressed in a developmental sequence leading to one of the various specific types of mature, fully-differentiated muscle.

We still do not know how the precise, complicated architecture of a muscle sarcomere is built up. One of the earliest stages, however, appears to be formation of Z-discs, which it will be recalled are the sites in muscle at which actin filaments from adjacent sarcomeres meet. Filament barbed ends are held in the Z-disc in a multiprotein complex together with α-actinin and titin. Complexes of these three proteins, sometimes called I-Z-I complexes, can be found in the cytoplasm of differentiating myotubes at early stages.

In skeletal muscle (but apparently not in cardiac muscle) the actin filaments in I-Z-I complexes have a precise length of 1 µm. This is believed to be due to the presence of long nebulin molecules, which as mentioned previously act as tape measures for actin assembly. The two ends of the actin filaments are also capped at this stage—by tropomodulin at their pointed ends and by Cap Z at their barbed ends (buried in the Z-disc).

Thick filaments must now assemble. We know that, in a test tube, myosin II molecules can make bipolar filaments by themselves (Chapter 7). But the *in vitro* products are variable in length, diameter, and internal arrangement. In the cell, it is likely that this process is regulated by, among other things, titin. The C-terminal end of titin has regularly repeating domains that match the periodic repeats in a myosin filament (Chapter 9). Titin molecules are thought to extend out from a Z-disc to

[1] It is more accurate to say that myoblasts do not fuse with each other but prefer to fuse with existing myotubes. Consequently, the number of muscle fibers is developmentally regulated by special initiating events and each particular muscle in the adult has a characteristic number of fibers.

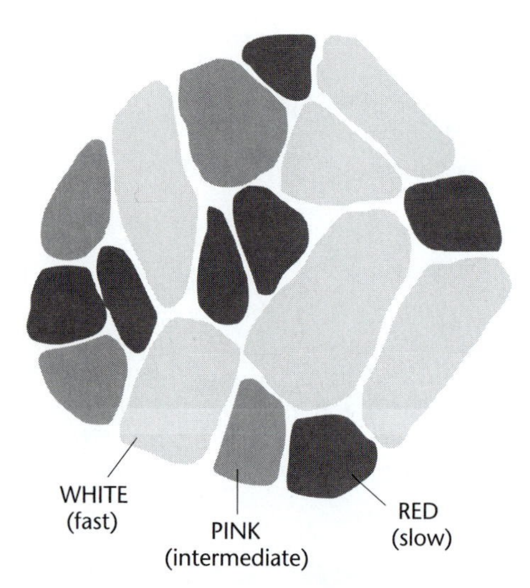


**Figure 10-4** Fiber types. Schematic cross-section of a typical vertebrate skeletal muscle showing the different types of fiber it contains. Fast, slow, and intermediate fibers can be identified in actual cross-sections by antibody staining for different isoforms of myosin or histological staining for different levels of lactate dehydrogenase.

make contact with thick filaments and to regulate their precise length and packing. How they do this is presently unclear, although the M-line protein myomesin and C protein, both of which bind to titin and myosin, are probably important linkers. There is also likely to be signaling control over the process, and it is interesting that titin has sequences at either end resembling those of signaling molecules in other pathways (a kinase and a SH3 domain).

Subsequent events include the assembly of troponin, tropomyosin, and all the other proteins in a myofibril into the structure. Given the complexity of the final product, it seems inescapable that its assembly will take place in a precisely defined sequence, with different proteins being added at specific locations and times. But we know very little of this.

## *Microtubules align forming myoblasts*

Microtubules are prominent features of myoblasts and early myotubes, extending parallel to the long axis of the cell. From what we know of microtubule function in other cells, it is reasonable to suppose that they might help to organize myofibril assembly. Consistent with this view, treatment of myotubes with drugs such as colchicine that lead to microtubule depolymerization result in flattened amorphous cells in which the muscle filaments, although present in normal quantities, are in disarray. The effects can be reversed, so that washing out the drug leads to the spontaneous reassembly of microtubules and the recovery of the aligned form of the developing myotubes. Shapeless myotubes ('myosacs') are also produced in myoblasts that lack a specific kinase that phosphorylates microtubules, another indication that a series of events involving microtubules is required to initiate their formation.

The role of microtubules is especially clear in the development of the indirect flight muscle of *Drosophila*, described later in the chapter. As this muscle forms, uniformly striated myofibrils appear first within 'sleeves' of microtubules. Later in development, the sleeves disappear, consistent with a scaffolding role.

As skeletal muscle myofibrils grow, other changes take place. Microtubules disappear from the cytoplasm and nuclei move from their central location to the periphery. Interestingly, although the many nuclei belonging to each muscle fiber share the same cytoplasm, they retain a fair degree of autonomy of function. For example, in the muscles of human–mouse heterokaryons, which have both human and mouse nuclei within the same muscle fibers, mouse myosin can be shown, by staining with specific antibodies, to be localized close to mouse nuclei.

## *Skeletal muscle contains fast and slow fibers*

The skeletal muscles of vertebrates range in color from white to red depending on the types of muscle fiber they contain. Most muscles contain a mixture of fast and slow fibers, as well hybrid slow/fast fibers with intermediate properties (Figure 10-4, Table 10-1).

**Table 10-1 Vertebrate muscle types**

| Fiber type | Characteristics |
|---|---|
| white, FG (fast glycolytic) | Large-diameter fibers. Rapid contraction. Low density of mitochondria. Fewer blood vessels. Principally anaerobic. |
| pink, FOG (fast oxidative-glycolytic) | Intermediate size. Rapid contraction. More mitochondria and oxidative enzymes. |
| red, SO (slow oxidative) | Slow twitch. Relatively small diameter. Many mitochondria. Oxidative metabolism. Resistant to fatigue. |

**Figure 10-5** Cross-reinnervation experiment. (a) The nerves to two different muscles are transposed; in this case the nerve to the slow twitch soleus muscle of rat has been made to innervate the fast-twitch extensor digitorum longus muscle, and vice versa. (b) The contractile responses observed several months later indicate that crossed innervation tends to increase the speed of contraction of the slow muscle and to slow the contraction of the fast muscle.

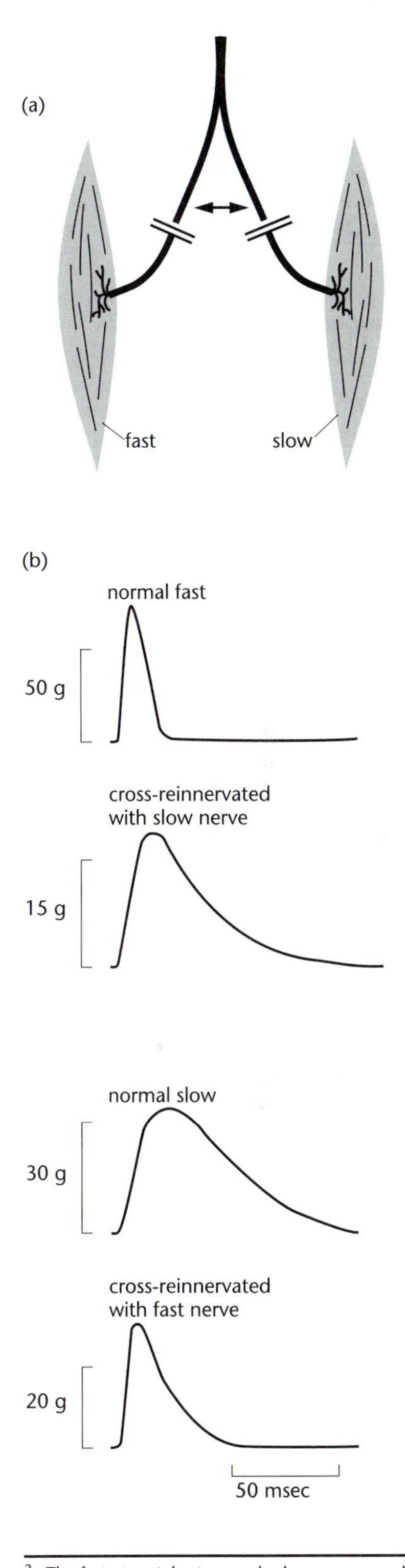


*Fast* muscle fibers (also called white fibers), work mainly by anaerobic metabolism. Fibers of this kind tend to be reserved for short bursts of intense activity such as the escape reflex of a fish, rather than sustained contraction. Fast muscles are distinguished by a dense accumulation of sarcoplasmic reticulum and its associated calcium pumps and a high concentration of the calcium-sequestering protein parvalbumin. The myosin of fast muscles is a distinct isoform with a higher rate of shortening than for slower muscles.[2] Since they use anaerobic rather than aerobic metabolism, fast muscle fibers tend to be large and pale in color and to have few blood vessels. Their cytoplasm contains many glycogen granules, used as an immediate source of glucose, but relatively few mitochondria. Fast muscle contraction is powerful but limited in duration since the reaction pathway by which they make ATP leads to the accumulation of lactic acid, which eventually prevents further contraction.

*Slow*, or red, muscle fibers rely on aerobic metabolism for their energy supply. For this reason they are heavily vascularized and contain large quantities of the oxygen-carrying protein myoglobin in their cytoplasm. *Myoglobin* is a red, heme-containing protein, related in structure and sequence to hemoglobin, which is able to combine with oxygen with a high affinity. Myoglobin is used by slow muscles to supply oxygen to their mitochondria. Slow muscles tend to be smaller, to have many large mitochondria, and to use lipid droplets rather than glycogen granules to store energy.

## Muscle type is influenced by innervation

In an adult vertebrate animal, each muscle and the motor neuron that innervates it act together as a unit. Their development and differentiation are closely linked, so that if the motor axon is cut then the specific muscle it innervates will wither. The interdependence between nerve and muscle is also evident in the normal physiological properties of the muscle fiber, which match the firing pattern of the motor neuron that supplies it. Red muscles, which are slow to contract and slow to relax, are activated by low rates of firing. White muscle fibers, which contract and relax rapidly, are normally activated and develop their maximal tension at high rates of firing.

This remarkable matching of electrical signals to force generation is revealed in experiments in which nerves to two muscles in the leg of a rat, one fast and the other slow, are cut and transposed (Figure 10-5). Muscles connected to the 'wrong' axon change their properties so that fast fibers become slower in character and previously slow fibers became more like fast fibers. Such transformations, which entail the modulation of muscle protein isoforms as well as a cohort of other cytoplasmic changes, are not complete. Authentic slow or fast fibers arise only as the result of a developmentally regulated pattern of gene expression.

How does a nerve affect the state of differentiation of its target muscle? Evidence suggests that the electrical activity supplied by the axon to the muscle is itself an important factor, since it can be mimicked by cutting the nerve and putting an electrode in its place. A fast muscle stimulated artificially in this way with a slow pattern of shocks acquires, over the course of a week or so, characteristics of a slow muscle. A slow muscle

[2] The fastest vertebrate muscles known are used to make sounds, such as the swimbladder muscle of a toadfish, which produces the 'boatwhistle' mating call of this species, and the muscle responsible for the eponymous 'rattle' of the rattlesnake (which in some species sounds more like an industrial buzzer). These muscles are capable of contractions exceeding 100 Hz (compared to 0.5–5 Hz for locomotory muscle) thanks to a variety of molecular specializations, including an ability to make very rapid calcium transients.

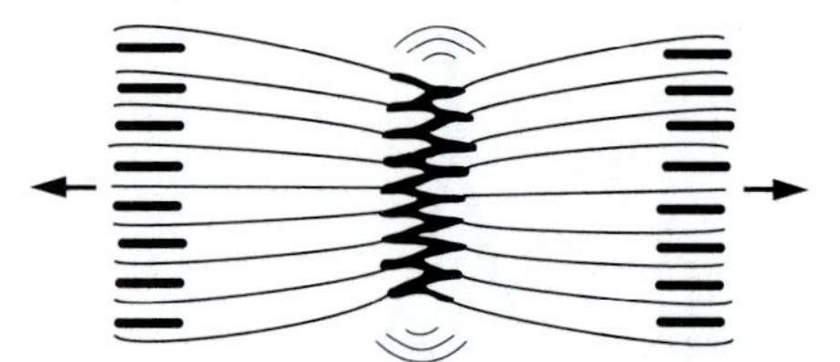


splitting of Z-disc relieves tension

**Figure 10-6** Possible mechanism of Z-disc splitting. Actin filaments are closer together at the Z-disc than at the region of overlap with thick filaments. Consequently, when muscle contracts, a lateral tension will develop in the Z-disc (top), which can be relieved by the splitting of the Z-disc (bottom).

stimulated with a fast pattern of shocks becomes faster. Remarkably, therefore, the ultrastructure of the cell and the types of genes it expresses, are affected (within limits) by the frequency of electrical stimulation it receives on its plasma membrane.

## Exercise affects muscle development and gene expression

Exercise is a related issue. The harder a muscle works, the larger it becomes: an adaptive change, since the force developed by a muscle is proportional to its cross-sectional area. However, muscles that grow through exercise do not make more muscle cells (fibers). Instead, it is the myofibrils inside the fibers that become more numerous, increasing from a few myofibrils at the myotube stage to more than 1000 in the mature muscle fiber following exercise. The increase in myofibrils arises by the muscle fibers splitting along their length once they have reached a certain diameter.

What mechanism might allow myofibrils to multiply in response to exercise? An ingenious explanation has been proposed that depends on the arrangement of filaments in the myofibril. At the Z-disc, it may be recalled, the thin filaments are packed in a regular square lattice while in the region of overlap with thick filaments packing is hexagonal. Because of this geometrical incongruence, it has been suggested that, as the muscle contracts, lateral mechanical tension will develop at each Z-disc. This will increase with the diameter of the myofibril and eventually cause the Z-disc to split into two (Figure 10-6).

Hypertrophy can be simulated under culture conditions. If cultured myotubules are subjected to a regime of regular electrical stimulation, or to enforced mechanical stretching (for example, by growing muscle cells on a flexible plastic substratum that is stretched periodically), the treated cells grow larger and make more actin and myosin. The contractions of myotubes, which appear spontaneously after several days in culture, have also been shown to be essential for the appearance of the neonatal isoform of myosin—which is synthesized at the same time as the myotubes increase in size. Other examples of the influence of mechanical tension on cytoskeletal assemblies are given in Chapter 18.

A second form of muscle hypertrophy occurs when a muscle is experimentally stretched to a new rest length (a similar change occurs naturally as the bone length of a young animal increases). Muscles adapt readily to the increase in length by adding further sarcomeres to their myofibrils. Autoradiography indicates that the new sarcomeres come from the recruitment of more myoblasts, which fuse at the growing end of the muscle fiber.

## Different genes are expressed at different stages of muscle development

Each of the different types of skeletal muscle so far described, as well as cardiac and smooth muscle, contains a different complement of sarcomeric proteins. Further diversification occurs during muscle development, as additional isoforms of proteins are expressed in a defined sequence.

A major source of variation lies in the expression of different genes for muscle proteins, many of which belong to multigene families. Thus, vertebrates possess at least 10 genes for the myosin heavy chain, four of which are expressed in substantial amounts in particular types of adult skeletal muscle fiber. During the development of chicken pectoralis muscle, for example, three isoforms of muscle heavy chain (known as *embryonic, neonatal,* and *adult* myosin heavy chain) and three of the thick filament-associated C protein are expressed in sequence, with further changes occurring as slow and fast muscle fibers differentiate. These names are slightly misleading since some of the 'embryonic' and 'neonatal' forms of myosin heavy chains are retained in adult muscle. When an adult muscle grows through exercise, for example, there is a transient expression of embryonic myosin, especially close to the tendons at each end.

Other sarcomeric proteins change. Actin, which exists in myoblasts as the βγ nonmuscle isoforms, finishes in the mature skeletal muscle as the muscle α form (Chapter 5). Curiously, and for reasons that are presently unexplained, embryonic skeletal muscles transiently express both the cardiac-specific isoform of actin and the cardiac-specific forms of myomesin and troponin C. This may be a vestigial regulatory mechanism due to the common ancestry of these two muscle types; or it could be a specific developmental requirement met by the cardiac gene products.

## *Myofilaments can add and lose subunits in the cell*

The above discussion ignores the question of how the isoform composition of myofibrils can change without disrupting the myofibrillar apparatus. In fact, a similar process must occur during the life of the animal to allow normal turnover of the molecules in the myofibril. In an adult rat, for example, the half-life of myosin heavy chains in heart muscle has been estimated to be 5 days, that of actin 10 days, and that of troponin I about 3 days. These protein exchanges presumably occur while the heart continues to beat 600–700 times per minute.

The distinct half-lives of different myofibrillar proteins are most simply explained if molecules free in the cytoplasm can exchange with those in the assembled structure. This interpretation is supported by experiments in which isolated myofibrils are mixed with labeled contractile proteins, when the myofibrils are found to incorporate subunits along their length. An even more rapid exchange of contractile proteins is observed in nonmuscle cells. Microinjection of fluorescently labeled α-actinin and actin into fibroblasts shows that these proteins readily exchange with α-actinin and actin in focal contacts and stress fibers.

Studies such as these reveal that even the most stable cytoskeletal assemblies can exchange subunits with those free in the cytoplasm. Presumably this ability has been selected for during evolution so that these proteins might even have structural features that allow them to worm their way into the myofibril.

## *Muscle proteins are further diversified by alternative RNA splicing*

The variations of muscle type so far encountered depend on the selective expression of different genes, but an even richer source of variation is provided by tissue-specific RNA splicing. The coding sequences for most eucaryotic proteins are present in discontinuous segments (exons) interspersed among sequences that do not form part of the mature mRNA (introns). Primary transcripts of these genes contain sequences corresponding to both exons and introns, but the intron sequences are removed by a nuclear multistep process known as mRNA splicing.

In most instances, each and every one of the exons present is incorporated into the mature mRNA. Thus, all 41 exons of the mammalian myosin heavy chain gene are, so far as we know, invariably assembled into the same messenger RNA. However, there are also cases in which exons are joined together in some but not all transcripts from a gene. This *alternative mRNA splicing* yields mRNAs with different primary structures and therefore different protein sequences from the same gene. For reasons discussed below, muscle proteins are particularly prone to this form of diversification. Variant forms of tropomyosin, troponin, titin, and (in some species) myosin heavy and light chains are known to be generated by this mechanism.

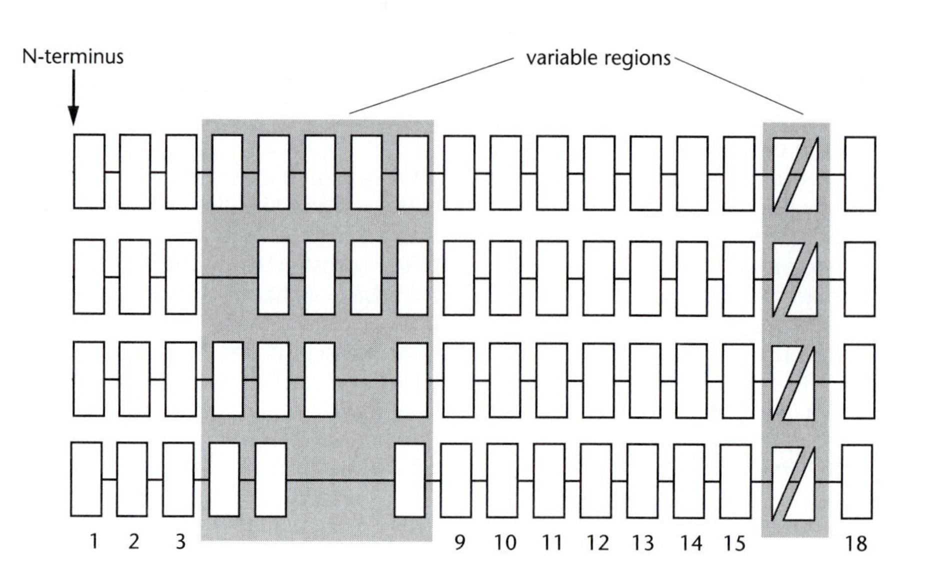


**Figure 10-7** Alternatively spliced variants. Exons in the major forms of troponin T of adult rabbit muscle are shown. Exons 16 and 17 are spliced in alternative fashion to generate α and β isoforms. An additional exon is present in fetal rabbit muscle, between adult exons 8 and 9.

A remarkable illustration of this process is provided by troponin T, where a single gene is modified by alternative RNA splicing to give at least 10 distinct protein sequences. The isoforms produced in this way differ in size, in amino acid sequence, and in the calcium sensitivity they confer on the contractile machinery. Analysis of the products of RNA splicing reveal that variable recombination is confined to one set of exons, while a second set remains invariant (Figure 10-7). The protein therefore exhibits variable and constant regions somewhat reminiscent of immunoglobulins. In the immune system however, diversity is generated in large part by rearrangements of DNA: an irreversible process well-suited to terminally differentiated cells with a short life span.

Muscle cells, in contrast, have a very long life span and so require a plastic mechanism capable of generating diversity in a reversible fashion. Alternative mRNA splicing is especially suited to this purpose because it allows multiple isoforms to be generated without affecting the levels of transcription. Thus, a muscle could synthesize the same overall quantity of troponin T mRNA while changing the precise composition of isoforms of this protein in the muscle.

## The molecular lesion in muscular dystrophy is known

The gene product defective in two kinds of common muscular dystrophy, Duchenne and the less severe Becker dystrophy, has been shown to be an actin-binding protein *dystrophin*. This protein was identified first by mapping and sequencing DNA, which showed it to be an extremely large protein (molecular weight 427,000) with a family resemblance to α-actinin and spectrin. The molecule has four domains, one of which consists of a long chain of 25 repeating triple-helical segments similar to those present in spectrin. The N-terminal domain of dystrophin shares sequence similarity to α-actinin and binds to actin filaments. Unlike spectrin and α-actinin, however, dystrophin is probably monomeric in the cell rather than a dimer, so it is unlikely to cross-link actin filaments into a network. Rather, it appears to attach the actin cytoskeleton to a complex of membrane glyoproteins through its C-terminal domain. When dystrophin is absent or defective, the cell surface has a reduced stiffness and the plasma membrane tears easily when the muscle contracts. This is believed to lead to progressive degeneration of the muscle, although many aspects of this tragic condition remain unexplained.

The dystrophin gene is normally carried on the X gene and the disease is most commonly found in males; Duchenne dystrophy typically causes progressive degeneration in young boys, leading to paralysis and eventual death in the third decade. Other forms of muscular dystrophy are produced by defects in other parts of the membrane complex, such as the integral membrane protein dystroglycan, and these are not confined to males.

A number of dystrophic animal models of the disease are known, including dogs, cats, and mice. Methods have been developed by which the native dystrophin gene can be introduced into the muscles of such animals, where it causes a significant level of recovery. Research workers are now searching for suitable ways to introduce the dystrophin gene into human muscles.

## *Cardiac muscle fibers consist of chains of mononucleated cells coupled together*

Cardiac muscle is responsible for the beating of the heart, carrying out this rhythmic contraction autonomously for the lifetime of the animal. It consists of long, striated fibers with a sarcomeric pattern closely similar to that described for skeletal muscle. Superficially it is a form of red muscle similar to slow skeletal muscle and uses aerobic metabolism almost exclusively. However, in this case the red color arises not from the presence of myoglobin but from the extensive vascularization of the heart muscle.

Cardiac muscle also has a unique cellular organization. Its fibers are branched rather than single and are interrupted at intervals of about 100 μm or so by irregular transverse divisions (Figure 10-8). Known as *intercalated discs*, these are regions of contact between the individual mononucleated cells that make up the cardiac muscle fiber. In one part of the intercalated disc the two plasma membranes come into close proximity to form electrically communicating gap junctions; elsewhere, regions similar to desmosomes form mechanically robust attachment between the cells. There are also points of insertion of actin filaments somewhat similar to the focal adhesions of cultured fibroblasts described previously and containing some of the same proteins, such as vinculin.

Each cardiac muscle cell has a single nucleus and many large mitochondria scattered throughout the cytoplasm, often in close association with droplets of lipid that provide fuel for their contraction. Intracellular

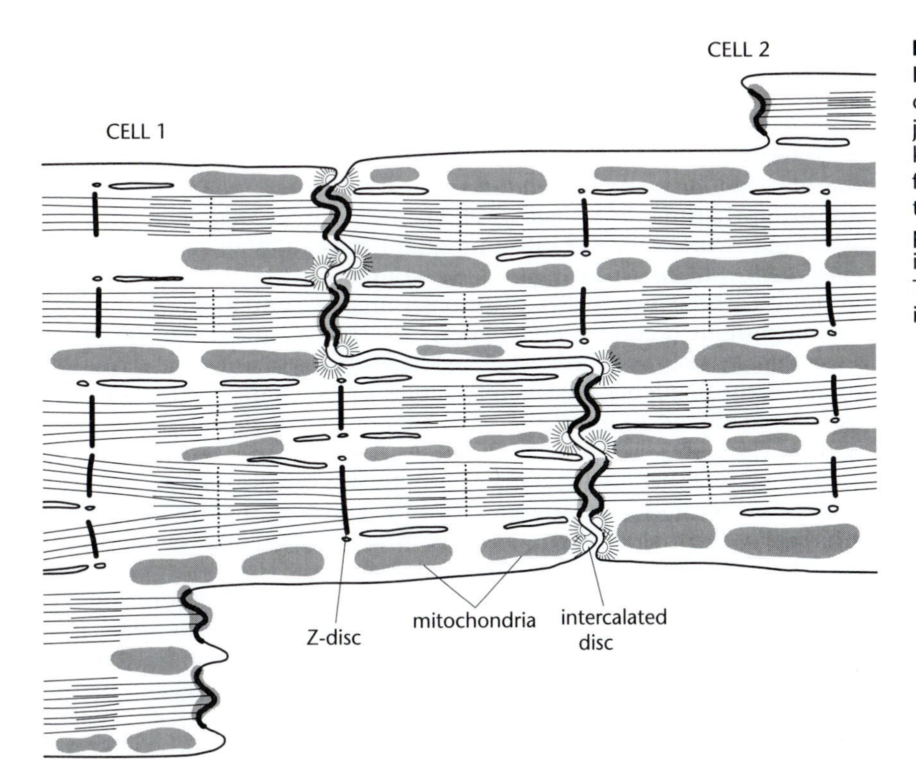


**Figure 10-8** Structure of heart muscle. Heart muscle is composed of many discrete cells, each with its own nucleus. The cells are joined end-to-end by specialized junctions known as intercalated discs. Actin filaments from sarcomeres in adjacent cells insert into the dense material associated with the plasma membrane in the region of each intercalated disc as though they were Z-discs. Thus myofibrils continue across the muscle, ignoring cell boundaries.

membranes, the analogues of the T-system and sarcoplasmic reticulum are also present, although in simpler and sparser form. As just mentioned, the cells are also coupled electrically through low-resistance gap junctions, so that contraction spreads as a wave from cell to cell without need for neural control. Although the heart is extensively innervated by sympathetic and parasympathetic axons, these have a regulatory function: speeding up or slowing the heart beat.

The myofibrillar banding pattern in cardiac muscle is closely similar to that described for skeletal muscle. All of the common myofibrillar proteins are found, although many are present in a distinctive cardiac isoform. Thus there is a distinct cardiac actin and a distinctive cardiac myosin (although the latter is restricted to the ventricular region of the heart and elsewhere the slow adult skeletal myosin is found) and specific cardiac forms of troponin I and T.

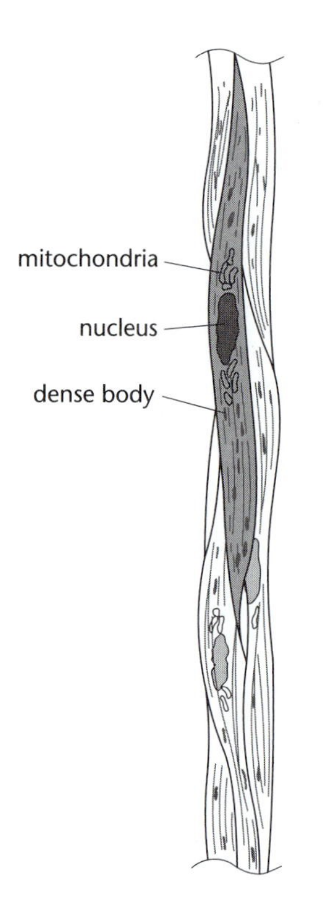


**Figure 10-9** The morphology of smooth-muscle cells.

## *Smooth muscle contains a network of myofibrils linked by dense bodies*

Smooth muscle is found in the digestive, respiratory, urinary, and genital tracts, in the walls of arteries, veins, and larger lymphatic vessels, in the uterus, and in the ducts of liver and spleen. It consists of long spindle-shaped cells, typically 200 μm long by 5 μm wide in the human intestine, containing a single centrally positioned elongated nucleus (Figure 10-9). Individual smooth-muscle cells are surrounded by extracellular matrix and disposed with different orientations in different tissues. In the blood vessel wall, for example, the cells are arranged circumferentially, whereas in the wall of the uterus and bladder they are arranged in criss-cross fashion. In most types of smooth muscle, electrically communicating gap junctions exist between smooth muscle cells, ensuring that contractions spread from cell to cell.

Smooth muscle is so called because its cytoplasm does not contain striations, in contrast to that of skeletal and cardiac muscle. Indeed, the spatial organization of the contractile machinery within a smooth-muscle cell is still poorly understood. Filaments of actin and myosin are present in large numbers, aligned with the long axis of the cell. They are loosely associated into thin myofibrils consisting of a centrally located myosin filament surrounded by multiple actin filaments but seem to lack the precise spatial order of skeletal myofibrils (Figure 10-10). The molar ratio of actin to myosin in smooth muscle is about 25 to 1, which is several times that in skeletal muscle.

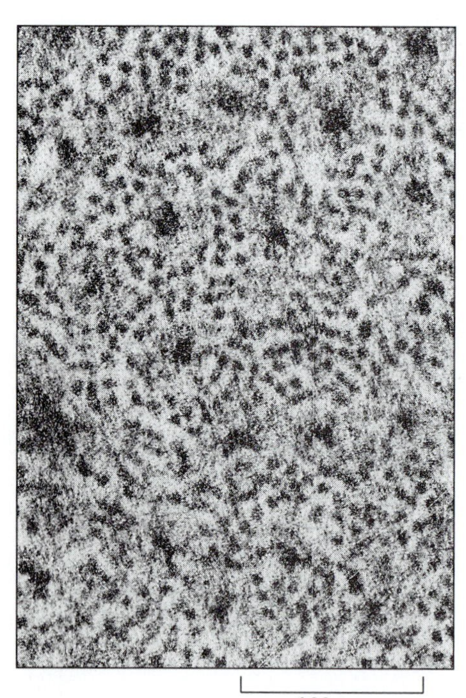
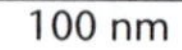


**Figure 10-10** Smooth muscle cytoplasm. Micrograph of a vascular smooth-muscle cell in cross-section, showing thick and thin filaments. (Courtesy of Andrew Somlyo.)

Interspersed with the myofibrils just mentioned are regions rich in intermediate filaments (usually desmin or vimentin) containing numerous ovoid-shaped, *dense bodies*, about 100 nm in diameter. A typical smooth-muscle cell of the intestine might have 3000 or so dense bodies scattered throughout its cytoplasm. In common with the Z-discs of skeletal muscle, the dense bodies of smooth muscle contain the actin cross-linking protein α-actinin and appear to serve as anchorage points for actin filaments of myofibrils. Their association with the system of internal intermediate filaments presumably serves to integrate contractions over the entire cell and to allow the very high degree of shortening achieved by these cells.

The plasma membrane of a smooth-muscle cell also shows two structurally distinct domains, arranged like the staves of a barrel around the cell. One domain is characterized by a smooth plasma membrane and a dense feltlike arrangement of dense material on the inner face known as a *dense plaque*. This region of the cell is another site of insertion of actin filaments of the myofibrils and has many similarities to the focal adhesions already described for fibroblastic cells, containing many of the same proteins (Table 10-2). When the smooth-muscle cell contracts, tension is exerted on the dense plaque regions, causing the membrane in these regions to pull inward. Because the dense plaques are linked through

**Table 10-2 Major cytoskeletal components of smooth muscle**

| Cellular location | Components |
|---|---|
| cytoplasm: associated with myofibrils | actin (smooth muscle), myosin (muscle type II), tropomyosin, caldesmon, calponin, myosin light chain kinase, myosin light chain phosphatase, calmodulin |
| cytoplasm: dense bodies | actin (nonmuscle), α-actinin, calponin |
| cytoplasm: associated with intermediate filaments | actin (nonmuscle), desmin (or vimentin), filamin, calponin |
| membrane: dense plaques | actin (nonmuscle), filamin, calponin, vinculin, talin, paxillin, tensin, integrins, plectin |
| membrane: caveolar domain | dystrophin, caveolin, $IP_3$ receptor, $Na^+$/ $Ca^{2+}$ exchanger, $Na^+$/ $K^+$ pump |

integrins to components of the extracellular matrix, tension generated by individual cells is transmitted from cell to cell throughout the smooth muscle.

The second plasma membrane domain lacks dense material but is rich in *caveolae*—small invaginations rich in cell signaling components. The function of these structures in smooth-muscle cells is not fully understood. One possibility, based on their activity in other situations, is that they may serve as stretch receptors to monitor the degree of contraction exerted by the cell and to relay information relating to this contraction through signaling pathways to the contractile machinery.

## *Smooth-muscle contraction is regulated by multiple enzyme cascades*

Contraction of smooth muscle is slow. Typically, it takes about 5 seconds to reach maximum tension, in contrast to less than 0.1 second for skeletal muscle. The switching mechanism is also different, and a plethora of different effectors stimulate smooth-muscle cells in various tissues. For instance, gut smooth-muscle contraction is stimulated by cholinergic neurons and inhibited by adrenergic neurons and is also affected by ATP, serotonin, prostaglandin, enkephalins, somatostatin, and substance P. Aorta smooth muscle is stimulated by norepinephrine. Many smooth muscles also show a mechanically-induced response in which they contract in response to the application of stretch.

The slow response of smooth muscle is inherent in the slower rate of ATP hydrolysis of smooth as compared to striated muscle myosin, and in the lower state of organization of its contractile apparatus and sarcoplasmic reticulum. It also reflects the different regulatory mechanisms present, a variety of which have been identified in smooth muscle.

The dominant form of regulation in smooth muscle is probably at the level of myosin. Myosin light chain kinase (MLCK) mediates phosphorylation of myosin light chains and hence affects the activity of the protein, as previously described for nonmuscle myosin II molecules (Chapter 7). The kinase is itself activated by a rise in calcium ions, through its interaction with $Ca^{2+}$-calmodulin; increased phosphorylation by this enzyme leads to the cycling of myosin heads along actin filaments and a consequent development of force or contraction in the smooth-muscle cell.

In a test tube, phosphorylation of smooth-muscle myosin changes both its conformation and its ability to aggregate and can lead to the depolymerization of myosin filaments. In the living cell, however, myosin filaments are not dissembled into myosin molecules when the muscle relaxes. They are probably maintained in filamentous form both by the high concentrations existing in the cell and by other proteins that associate with myosin.

In many smooth muscles, especially those that maintain tension for long periods of time, the level of myosin phosphorylation is not closely correlated with the tension developed, suggesting that other mechanisms of regulation are present. Three actin-binding proteins are present in sufficiently large quantities in smooth-muscle cells to be used in regulation—filamin, caldesmon, and calponin The latter two can block myosin's interaction with actin in a calcium-sensitive manner. Whether this is their role in the cell remains unclear: they could also contribute to the structure of the loose myofibrils by holding actin filaments and myosin filaments at the correct distance from each other. To further complicate the situation, there is evidence that proteins similar to troponin are also present in smooth muscle, so that a mechanism of actin-linked regulation closer to that in skeletal muscle may also exist.

Why does contractile machinery of smooth muscle have such a confusing variety of control mechanisms? Perhaps it is because this tissue performs such a wide variety of functions and is stimulated in so many different ways. Perhaps the apparent disorder of the contractile machinery of smooth muscle, compared to the highly regular skeletal myofibril, actually reflects a highly diverse and complicated molecular anatomy.

### *Insect flight muscles and molluscan catch muscles are highly specialized*

An extremely high degree of specialization is displayed in some invertebrate muscles that have to perform tasks that have no parallel in the vertebrate body. These include insect flight muscle—the most rapidly contracting muscle known—and molluscan muscle, which can maintain enormous tensions for days with negligible expenditure of energy.

The *flight muscles* of many insects contract at an amazingly rapid rate. Those of midges, for example, contract 600–900 times every second (the degree of contraction with each cycle, it should be noted, is only about 1–2% of the rest length). These muscles also consume energy at a rate far higher than that found in vertebrate muscles: bee flight muscle has a maximum consumption rate of 2400 (kcal/kg)/hr, which may be compared to the maximum consumption of leg muscles of man of 50–60 (kcal/kg)/hr.

Tests with single muscles from unusually large insects show that they can carry out many contractions in response to a single nerve stimulus. This ability seems to depend on a form of stretch-induced activation: an experimentally produced increase in length is followed after a short delay by an increase in tension. Thus, the muscles in an antagonistic pair of flight muscles vibrate for some time in response to a single stimulus, like a tuning fork. In the electron microscope, insect flight muscles have a typical sarcomeric arrangement of striated muscles, distinguished by unusually short I-bands and an extremely high content of mitochondria and lipid droplets. They also have conspicuous filaments stretching from the ends of the I-bands to the Z-discs, rather like the titin and nebulin filaments in vertebrate striated muscle. It is possible that the elasticity of this third system of filaments plays a major role in the rapid contraction.

Mussels and scallops can keep their shells tightly closed for days against all efforts to open them; oysters, scarcely breathing, have been shown to remain closed against a force of 500 g for 20–30 days. Such performances would require enormous expenditure of energy by vertebrate skeletal muscle and clearly call for a special mechanism.

The muscle involved in the case of scallop is a cord of uniform diameter that is partly striated and partly not. The striated portion is able to contract rapidly, whereas the contraction of the other portion is slower but is responsible for the holding ability. The latter muscle, sometimes called a *catch muscle*, maintains tension without prolonged nervous stimulation. Various explanations have been offered for this remarkable ability. For example, it has been suggested that the protein paramyosin, which forms the core of the thick filaments of these and many other invertebrate muscles, provides a 'catch,' like the latch of a door, enabling tension to be maintained for long periods.

## *Genetic analysis enables muscle development to be dissected*

The isolation of muscle mutants has been particularly instructive for the fruit fly *Drosophila melanogaster* and the nematode *Caenorhabditis elegans*. These organisms are well suited for genetic analysis and both offer mutations that affect muscle structure and function but not viability. In *Drosophila*, the functions of the indirect flight muscles and the jump muscle are not essential to survival in the laboratory or for fertility and therefore can be mutated at will. Likewise, the body wall muscle of *C. elegans* is not essential, since nonmotile organisms can feed and digest.

*Drosophila* mutants that fail to make actin or fail to make myosin provide information on the normal path of myofibrillar assembly. Null mutants that fail to express the special isoform of actin present in the indirect flight muscles lack thin filaments but do synthesize both thick filaments and Z-bodies (structures that contain the same set of proteins found in the mature Z-disc, see Figure 10-3). Conversely, a mutant unable to express the isoform of myosin heavy chain special to the indirect flight muscle (its specificity may be due to tissue-specific RNA splicing of the myosin gene) lacks thick filaments but has both thin filaments and Z-discs in a quasi-sarcomeric arrangement. These observations lend support to the view that thick filaments, thin filaments, and Z-bodies can assemble independently.

An interesting sidelight of the myosin mutant just mentioned is that not only myosin but also other thick filament-associated proteins such as myosin light chains are absent from the flight muscles. It seems that these are synthesized but degraded because there is no place for them to go.

The nematode *C. elegans* has two primary muscle types: body wall muscle for locomotion and pharyngeal muscle for feeding. A combination of genetic and molecular approaches has identified over 80 genes involved in muscle development and function in this organism. Included in this set of genes are those encoding the structural components of nematode thick filaments, myosin and paramyosin, as well as components of thin filaments, actin and tropomyosin. Interaction of these two filament types is a complicated process that involves both thick and thin filament regulatory networks. For the thin filaments, regulation is through the troponin/tropomyosin complex, whereas regulation of thick filaments is mediated by *twitchin*, a protein related to mammalian titin. Mutations in regulatory components have been described. Those affecting thin filaments invariably lead to late embryonic or early larval lethality, whereas mutations affecting thick filament regulation lead to unregulated spontaneous contractions and uncoordinated behavior.

## Outstanding Questions

*How is a skeletal muscle myofibril built? Do the structural proteins achieve this by spontaneous self-assembly, or do other proteins (such as scaffolding proteins or chaperones) play an essential role? What mechanism allows myofibrils to multiply in response to exercise? Why does a developing skeletal muscle transiently express cardiac isoforms of actin and troponin? How does a nerve affect the state of differentiation of its target muscle? How are the proteins in a myofibril degraded and replaced without disrupting muscle integrity or its ability to develop tension? Have sarcomeric proteins evolved the ability to worm their way into an assembled myofibril? How is the contractile apparatus of a smooth muscle organized? What is the function of caveolae and dense plaques? Why does smooth muscle have such a confusing variety of control mechanisms?*

## *Essential Concepts*

- During embryonic development in vertebrate animals, mononucleated myoblasts fuse into long myotubes and progressively differentiate into mature skeletal muscle fibers.
- A population of myoblasts remains dormant in adult muscle but can be activated if the muscle is damaged, when they proliferate, fuse, and form new fibers.
- Myoblast fusion and muscle differentiation are under the control of a small number of master genes.
- Myofibrils form by a regulated process in which thick and thin filaments assemble first and are then linked together. Titin plays a crucial role in this process, helping to form the Z-disc and acting as a template for thick filament assembly.
- Microtubules have a role in early myoblast development serving to align myofibrils with the long axis of the cell.
- Changes in myofibrillar structure and protein components can take place even in an adult muscle. Exercise, hormonal changes, and the pattern of electrical stimulation received from motor neurons all activate different genes and modify the paths of mRNA maturation, leading to changes in the composition and structure of myofibrils.
- Different forms of muscle contain subtly different forms of the same basic contractile machinery. In vertebrates, these include fast, slow, and mixed skeletal muscles, smooth muscle and cardiac muscle, all of which evolved to perform specific functions in the body.
- Vertebrate smooth muscle consists of long, thin mononucleated cells embedded in an extracellular matrix. Within the cytoplasm, filaments of actin and myosin associate into loose myofibrils that are anchored to the plasma membrane and to a system of intermediate filaments.
- Contraction of smooth muscle is triggered by a plethora of stimuli and mediated by changes in $Ca^{2+}$ through a variety of regulatory mechanisms. These include phosphorylation of myosin light chains and $Ca^{2+}$-dependent changes in proteins associated with actin filaments.
- Invertebrates have many highly specialized forms of muscle, such as insect flight muscle, which is able to contract at very high frequencies, and molluscan catch muscle, which can maintain tension for very long periods.
- Genetic analysis allows the function of individual molecules in muscle development and contraction to be identified.

## Further Reading


Allen, E.R., Pepe, F.A. Ultrastructure of developing muscle cells in the chick embryo. *Am. J. Anat.* 116: 115–148, 1965.

Almenar-Queralt, A., et al. Tropomodulin assembles early in myofibrillogenesis in chick skeletal muscle: evidence that thin filamants rearrange to form striated myofibrils. *J. Cell Sci.* 112: 1111–1123, 1999.

Barrat, J.M., Epstein, H.F. Protein machines and self assembly in muscle organization. *BioEssays* 21: 813–823, 1999.

Blake, D.J., et al. The emerging family of dystrophin-related proteins. *Trends Biol. Sci.* 4: 19–23, 1994.

Breitbart, R.E., et al. Alternative splicing: a ubiquitous mechanism for the generation of multiple protein isoforms from single genes. *Annu. Rev. Biochem.* 56: 467–495, 1987.

Campbell, K.P. Three muscular dystrophies: loss of cytoskeleton–extracellular matrix linkage. *Cell* 80: 675–679, 1995.

Cerny, L.C., Bandman, E. Contractile activity is required for the expression of neonatal myosin heavy chain in embryonic chick-pectoral muscle cultures. *J. Cell Biol.* 103: 2153–2161, 1986.

Cross, R.A. Smooth operators. The molecular mechanics of smooth muscle contraction. *BioEssays* 11: 18–21, 1989.

Danowski, B.A., et al. Costameres are sites of force transmission to the substratum in adult rat cardiomyocytes. *J. Cell Biol* 118: 1411–1420, 1992.

Duxson, M.J., Sheard, P.W. Formation of new myotubes occurs exclusively at the multiple innervation zones of an embryonic large muscle. *Dev. Dyn.* 204: 391–405, 1995.

Ehler, E., et al. Myofibrillogenesis in the developing chicken heart: assembly of Z-disk, M-line and the thick filaments. *J. Cell Sci.* 112: 1529–1539, 1999.

Franzini-Armstrong, C., Fischman, D.A. Morphogenesis of skeletal muscle fibers. In Myology (A.G. Engel, C. Franzini-Armstrong, eds.) pp. 74–96. New York: McGraw-Hill, 1996.

Fulton, A.B., L'Ecuyer, T.J. Cotranslational assembly of some cytoskeletal proteins: implications and propects. *J. Cell Sci.* 105: 867–871, 1993.

Goldspink, G. The proliferation of myofibrils during muscle fibre growth. *J.Cell Sci.* 6: 593–603, 1970.

Horowitz, A., et al. Antibodies probe for folded monomeric myosin in relaxed and contracted smooth muscle. *J. Cell Biol.* 126: 1195–1200, 1994.

Kislauskis, E.H., et al. Isoform-specific 3′-untranslated sequences sort α-cardiac and β-cytoplasmic actin messenger RNAs to different cytoplasmic compartments. *J. Cell Biol.* 123: 165–172, 1993.

L'Ecuyer, T.J., et al. Transdifferentiation of chicken embryonic cells into muscle cells by the 3′ untranslated region of muscle tropomyosins. *Proc. Natl. Acad. Sci. USA* 92: 7520–7524, 1995.

Moessier, H., et al. The SM 22-promoter directs tissue-specific expression in arterial but not in venous or visceral smooth muscle cells in transgenic mice. *Development* 122: 2415–2425, 1996.

Pasternak, C., et al. Mechanical function of dystrophin in muscle cells. *J. Cell Biol.* 128: 355–361, 1995.

Perry, S.V. Troponin T: genetics, properties and function. *J. Muscle Res. Cell Motil.* 19: 575–602, 1998.

Reedy, M.C., Beall, C., Ultrastructure of developing flight muscle in *Drosophila*. I. Assembly of myofibrils. *Dev. Biol.* 160: 443–465, 1993.

Rome, L.C., et al. The whistle and the rattle: the design of sound producing muscles. *Proc. Natl. Acad. Sci. USA* 93: 8095–8100, 1996.

Schafer, D.A., et al. Inhibition of CapZ during myofibrillogenesis alters assembly of actin filaments. *J. Cell Biol.* 128: 61–69, 1995.

Small, J.V. Structure–function relationships in smooth muscle: the missing links. *BioEssays* 17: 785–792, 1995.

Somlyo, A.P., et al. Filament organization in vertebrate smooth. muscle. *Phil. Trans. R. Soc. Lond. B* 265: 223–229, 1975.

Somlyo, A.P., Somlyo, A.V. Signal transduction and regulation in smooth muscle. *Nature* 372: 231–236, 1994.

Stockdale, F.E., Miller, J.B. The cellular basis of myosin heavy-chain isoform expression during development of avian skeletal muscle. *Dev. Biol.* 123: 1–9, 1987.

Trinick, J. Titin as a scaffold and spring. *Curr. Biol.* 6: 258–260, 1996.

Xie, X., et al. Structure of the regulatory domain of scallop myosin at 2.8 Å resolution. *Nature* 368: 306–312, 1994.

Yagami-Hiromasa, Y., et al. A metalloprotease-disintegrin participating in myoblast fusion. *Nature* 377: 652–656, 1995.

# CHAPTER 11

# Microtubules

There is something queenly about microtubules. Although typically anchored near the cell center, close to the nucleus, they can reach out across the cell to influence events at the cell cortex and plasma membrane. In contrast to actin filaments, which are far more numerous in most cells and typically work in interconnected bundles and meshworks, microtubules are often solitary structures, providing distinct tracks through the cell along which vesicles, organelles, and other cell components move. It is this capacity to direct traffic, above all, that enables microtubules to impart polarity to the cell, to regulate cell shape and cell movements, and to dictate the plane of cell division.

In this chapter we introduce the protein tubulin and examine how it assembles into microtubules. As we will see, this seemingly prosaic process is surprisingly subtle at the molecular level. Nucleotide hydrolysis and conformational changes make the assembly of tubulin into polymer a sensitive, hesitant process responsive to its surroundings and easily switched from net growth to net disassembly. Specific proteins and other molecules in the cytoplasm modulate the process, generating often complex dynamic behavior, such as treadmilling and dynamic instability, and creating polarized arrays of microtubules with specific geometry.

## *Tubulin is present in most eucaryotic cells but is usually purified from brain*

Microtubules have a highly distinctive structure and are present in virtually all animal and plant cells. However, they were not seen reliably in the earliest electron micrographs of eucaryotic cells because the harsh fixatives used at the time caused these delicate structures to disintegrate.[1] Not until glutaraldehyde was introduced into electron microscopy in 1962 were microtubules consistently preserved and visualized. Only then could cell biologists explore the extensive and varied display of microtubules in the cytoplasm of plant cells, animal cells, and protozoa.

1 Some large arrays of microtubules can be seen by light microscopy. The marginal bundles of microtubules in amphibian and avian red blood cells, for example, were described by Retzius in 1875.

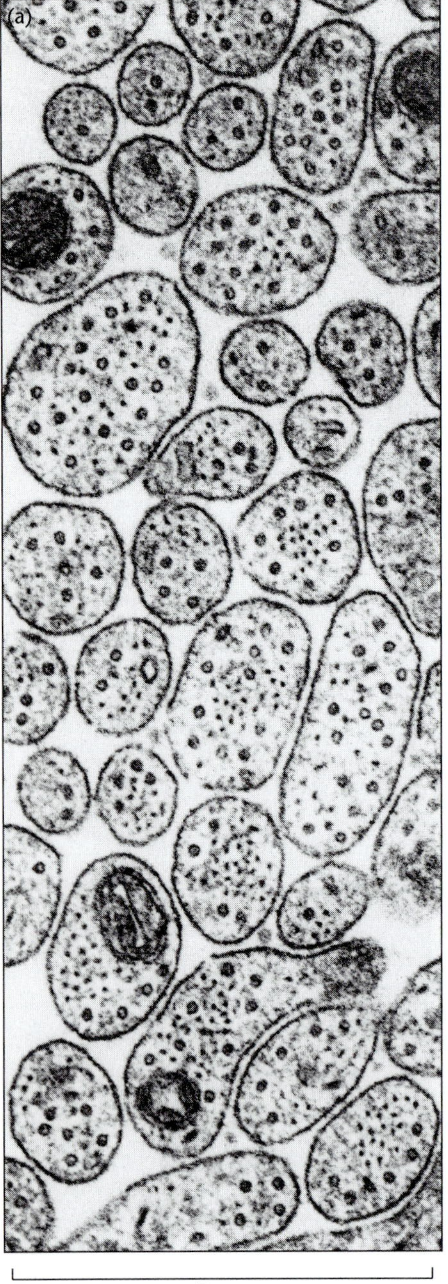



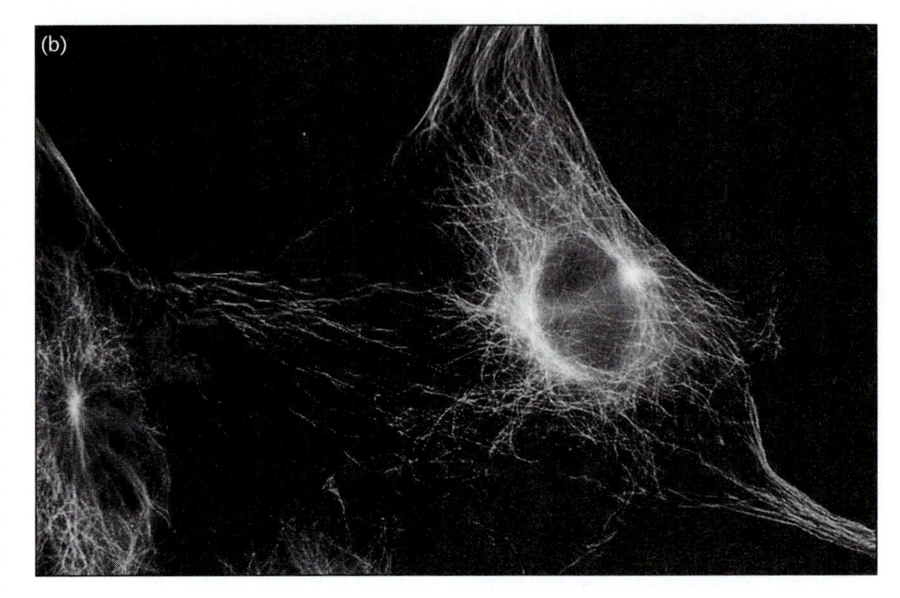



**Figure 11-1** Microtubules in cells.
(a) Section through a culture of rat neurons in tissue culture, viewed by electron microscopy, reveals multiple ringlike cross-sections of microtubules within the axons.
(b) Fibroblast in tissue culture stained with fluorescent antibodies to tubulin. Microtubules radiate from positions close to the nucleus throughout the cytoplasm.
(a, courtesy of V.J. Obremski and M.B. Bunge, The Miami Project to Cure Paralysis, University of Miami, School of Medicine; b, courtesy of Mary Osborn, MPI Goettingen.)

A few years later, the chemical composition of microtubules was established. One of the richest sources was found to be brain, due to the large number of microtubules in nerve axons (Figure 11-1a). In 1966, Ed Taylor and his colleagues in Chicago fractionated the subunits of brain microtubules free of other proteins using binding to the antimitotic drug colchicine to monitor the purification. The protein they isolated, known as *tubulin*, was later shown to reassemble into microtubules under defined conditions. In 1975, Klaus Weber and colleagues prepared antibodies to tubulin, coupled them to fluorescent antibodies, and used them as a specific stain to reveal microtubules in the light microscope (Figure 11-1b).

The colchicine-binding protein tubulin isolated from cow or pig brain has a molecular weight of about 100 kDa. It is a globular, slightly acidic protein, with an α-helical content of about 25% and an isoelectric point of about pH 5.4. On SDS acrylamide gel electrophoresis the subunit molecular weight of tubulin is 50 kDa, indicating that the stable species isolated in solution is a dimer. Further fractionation, for example by two-dimensional gel electrophoresis, resolves brain tubulin into two isoforms, termed α- and β-tubulin, present in equal amounts. Chemical cross-linking experiments demonstrate that each dimer contains one α- and one β-tubulin.

The many similarities between tubulin and actin led investigators to ask whether tubulin is also associated with a molecule of ATP. While a tightly bound nucleotide was indeed found, it proved unexpectedly to be GTP not ATP. Each dimer of tubulin in solution is now known to contain two molecules of GTP, one of which is hydrolyzed in the course of polymerization (see Table 11-1).

## *Tubulin is a GTPase distantly related to Ras*

Because of difficulties in preparing crystals of tubulin for x-ray diffraction analysis, it was not until 1998 that the atomic structure was obtained. The successful study employed electron crystallography, in which a beam of electrons is directed onto a two-dimensional sheet of protein and a diffraction pattern is obtained. In this case the specimen was a sheet of regularly arranged tubulin protofilaments formed in the presence of $Zn^{2+}$ ions. The resulting analysis revealed a globular, dimeric molecule composed of α- and β-tubulin. The two isoforms, α- and β-tubulin, almost indistinguishable in structure, are linked in such a way that the GTP molecule on the β subunit is exposed to water whereas that on the α subunit is buried in the dimer (Figure 11-2). The structure explains a long-standing

**Table 11-1 Comparison of actin and tubulin**

| Property | Actin | Tubulin |
|---|---|---|
| polypeptide molecular weight | 43,000 | α 50,000; β 50,000 |
| unpolymerized form | globular monomer | globular αβ dimer |
| bound nucleotide (unpolymerized) | ATP, one per monomer | GTP, two per dimer |
| form of polymer | two-stranded filament | hollow tube made of 13 protofilaments |
| diameter of polymer | 8 nm | 25 nm |
| polypeptides per μm of polymer | 370 | 3200 |
| persistence length* | 10–20 μm | 2–6 mm |

* Persistence length is a measure of the stiffness of a filament (Chapter 18).

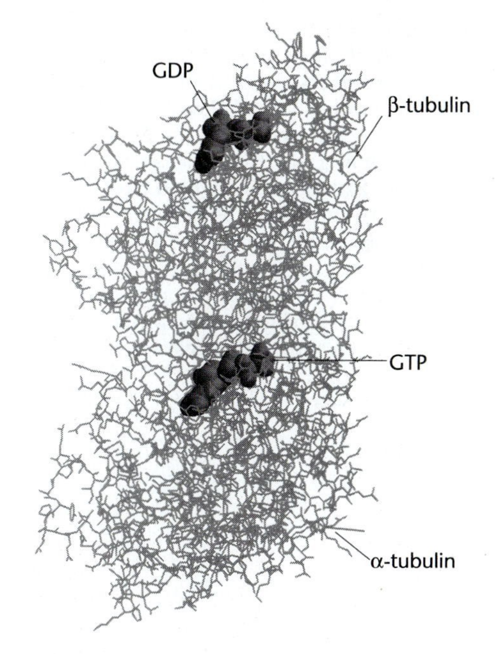


**Figure 11-2** Structure of the tubulin molecule. Polypeptide chains of α- and β-tubulin subunits are shown as a wire frame with their bound nucleotides as space-filling structures.

observation that only the GTP associated with β subunits is hydrolyzed to GDP as dimers add to the polymer (the enzymatic steps by which hydrolysis is achieved, however, are still not known).

When the detailed three-dimensional folds of a protein molecule are examined for the first time, they sometimes reveal unexpected similarities to other proteins. In the case of tubulin, its structure was found to resemble that of the Ras superfamily of G proteins, despite the fact that tubulin and Ras have markedly different amino acid sequences. As we mentioned in Chapter 6, G proteins act as self-inactivating switches in intracellular signaling cascades. The similarity of their strands and helices to tubulin is so close (especially in the GTP-binding domains) that some biologists now believe that the two proteins must have evolved from a common ancestral protein. Since tubulin also hydrolyzes GTP and undergoes a conformational change, the mechanism of this movement might also be conserved between the two kinds of proteins.[2]

At the same time as the structure of tubulin was elucidated, other scientists solved (in this case by x-ray diffraction) the structure of a protein called FtsZ. This bacterial protein, involved in the formation of septum during cell division (Chapter 13), shows a strong similarity in sequence to tubulin and the structures of the two are also very similar. Because of the ease of mutation of bacterial proteins, studies with FtsZ have, by extension, given useful information on the tubulin molecule. They have shown, for example, that the GTPase activity depends both on amino acids close to the GTP binding site of β-tubulin (as one might expect) and on a loop of amino acids in α-tubulin that is opposite the GTP-binding site. We will see below that this second site comes into action when a new tubulin dimer docks onto the end of a growing microtubule.

## *Vertebrates have multiple tubulin genes*

Like actin, myosin, tropomyosin, and many other cytoskeletal proteins, tubulin is produced by a family of closely related genes. The slime mold *Dictyostelium* has one α-tubulin and one β-tubulin; the plant *Arabadopsis* has 4 α and 8 β; the mammal *Homo sapiens* has at least 3 α-tubulins and 8 β-tubulins. This multiplicity of tubulins is further increased by a large number of posttranslational modifications. Tubulin molecules can, in different cells and at different times, be phosphorylated, acetylated, tyrosinated, and polyglutamylated. The function of most of these modifications is still not understood, although they usually occur on microtubules that are stable rather than dynamic.[3]

Tubulin amino acid sequences, determined by sequencing cDNA and genomic clones, reveal a highly conserved polypeptide backbone in which individual sequences diverge from one another in 2–17% of the approximately 450 residues. However, the 15 or so carboxy-terminal residues constitute a major variable region for β-tubulin and, to a lesser extent, for α-tubulin. In the case of β-tubulin, the C-terminal variable domain defines specific isoforms of tubulin that can be recognized from species to species, possibly indicating that they have been maintained by

2 The ATP-binding sites of myosin and the microtubule motor protein kinesin are also Ras-like, making it possible that they too are distantly related in an evolutionary sense. All of these proteins show a nucleotide-induced conformational change, the machinery of which could conceivably have arisen just once in evolution. One wonders what the function of this Ur-protein might have been!

3 Other more distant relatives of tubulin also exist, such as γ-tubulin, mentioned below, present in centrosomes, and δ-tubulin found in centrioles.

**Figure 11-3** Carboxy-terminal amino acid sequences of tubulin isoforms. Three of the six isoforms of chicken β-tubulin are shown, including two variants of class IV. The sequences are depicted with the C-terminus to the right and with the acidic amino acids, glutamate and aspartate, in bold type. Each isoform has a distinctive tissue distribution: for example class II is abundant in brain but is also found in smaller amounts in other tissues; class IVb is a major component of testes. Class VI tubulin (not shown) is found only in erythrocytes and platelets.

| | |
|---|---|
| I | **Glu Glu Glu Glu Asp** Phe Gly **Glu Glu** Ala **Glu Glu Glu** Ala |
| II | **Asp Glu** Gin Gly **Glu** Phe **Glu Glu Glu** Gly **Glu Glu Asp Glu** Ala |
| IVa | **Glu Glu** Gly **Glu** Phe **Glu Glu Glu** Ala **Glu Glu Glu** Val Ala |
| IVb | **Glu Glu Glu** Gly **Glu** Phe **Glu Glu Glu** Ala **Glu Glu Glu** Ala **Glu** |

positive selective pressures (Figure 11-3). This region of the tubulin molecule, which is rich in acidic amino acids, is the site to which a number of microtubule-associated proteins bind.

The difference between α-tubulin and β-tubulin is important for formation of the heterodimer and hence of the microtubule, but why there should be families of slightly different α- and β-tubulins is not understood. In general, particular isoforms of tubulin are not restricted to specific microtubule-containing structures, such as ciliary axonemes or mitotic spindles. For example, microtubules in the testis of the fruit fly *Drosophila melanogaster*, including those in flagellar axonemes and mitotic spindles as well as the cytoplasm, are made of a tubulin heterodimer composed of a particular α isoform, and a β isoform specific to testis. However, this should not be taken to indicate that these isoforms do not fulfill stringent structural requirements. Replacement of these tubulins by even closely similar isoforms (such as another α differing by as little as 2% in sequence) causes defects in the sperm and sterility.

Similar experiments reveal that microtubule architecture can depend on tubulin isoforms. If the testis specific β isoform just mentioned is replaced by a homologous β-tubulin from a moth, then the diameter of the resulting microtubules increases. The large size is due to an increase in the number of protofilaments in the microtubule wall, from 13 (which is typical of many organisms, as we see below) to 16, which is a distinctive feature of moth microtubules. Another example occurs in the nematode worm *Caenorhabditis elegans*. Most microtubules in this organism have (unusually) 11 protofilaments, but one class of touch-sensitive neurons in the nose of this animal contains arrays of larger microtubules each with 15 protofilaments. These large microtubules are composed of specific α- and β-tubulins, loss of which by mutation leads to the simultaneous loss of both 15-protofilament microtubules and the animal's sensitivity to touch. These and other examples indicate that individual isoforms of tubulin probably evolved in concert with specific microtubule-binding proteins, thereby creating structures or performing functions that cannot be replaced by other isoforms.

## Tubulin polypeptides need molecular chaperones to fold correctly

A widely accepted tenet of molecular biology is that the amino acid sequence of a protein contains all the information needed to specify its three-dimensional structure. However, it is an experimental fact that many proteins, including actin and tubulin, have *never* been recovered in a functional form following denaturation with agents such as urea. This paradox has been largely resolved by the discovery of a class of proteins known as *molecular chaperones* that function by promoting the proper folding of other proteins. Both actin and tubulins undergo facilitated folding through interaction with specific cytosolic chaperones.

Exactly how molecular chaperones work is not understood, but they appear to have an affinity for exposed hydrophobic patches on incompletely folded protein molecules. They bind and release their protein, hydrolyzing an ATP molecule with each cycle, in an interaction that has been likened to a 'protein massage.' In the course of this massage, those regions of the client protein likely to have misfolded are given another chance to fold. The chaperone that interacts with folded actin or tubulin molecules, for example, is a toroidal structure, termed the *TCP1 complex*, made of two stacked rings of subunits that encloses the incompletely folded polypeptide in a central cavity. By undergoing an ATP-dependent change in conformation, TCP1 facilitates the folding of the entrapped actin or tubulin into a correct state. Folding is a stochastic event and client polypeptides interact repeatedly with a chaperone complex until they achieve the correct folded structure, and are then released.

In the case of vertebrate tubulin, the folding program continues with the cooperative action of a series of five or so additional protein cofactors. These bind reversibly, in sequence, to the tubulin polypeptides released from the TCP1 treatment, bringing the α and β subunits together in a supercomplex in which they can achieve a native conformation. Final release of the functional heterodimer requires an input of energy in the form of hydrolysis of a GTP molecule. Interestingly, TCP1 and other chaperones are often located in places in the cell associated with the nucleation of microtubules, such as a centrosome, described below. We could imagine functional tubulin dimers being assembled 'to order,' as and when microtubules need to be made.

Evidently this complicated machinery coevolved with tubulin and has been maintained through hundreds of millions of years. One might therefore ask whether the tubulin polypeptide found in present-day cells does in fact contain all of the information required to form its folded structure? It could be that the tubulin gene has functioned in the environment of its molecular chaperones and protein cofactors for so long that it is now completely dependent on them to achieve a correctly folded state.

## *Specific drugs bind to tubulin and affect its polymerization*

The essential role of microtubules in cell division, discussed in Chapter 13, makes them a target for many naturally produced antimitotic poisons (Table 11-2). *Colchicine*, the best known, is obtained from the autumn crocus (*Colchicum autumnale*). One of the oldest drugs in the pharmacopeia, it has been used as a drug of choice in the relief of acute gout for over a thousand years. Colchicine is a heterocyclic alkaloid (Figure 11-4) that binds strongly to tubulin dimers and, as mentioned above, was the basis for the original isolation of tubulin from brain. Association with the drug causes tubulin dimers to change in conformation and inhibits the assembly of dimers into microtubules. Effects on polymerization in cell-free extracts, and on sensitive processes in the cell such as axonal growth and mitosis, can be seen at concentrations as low as 0.01 μg/ml.

**Table 11-2 Microtubule-acting drugs**

| Drug | Source | Action |
|---|---|---|
| colchicine | meadow saffron (Autumn crocus, *Colchicum autumnale*) | Inhibits polymerization. |
| colcemid (desacetyl-*N*-methylcolchicine) | chemically modified colchicine | Inhibits polymerization. More easily reversed than colchicine. |
| podophyllotoxin | May apple (*Podophyllum peltatum*) | Inhibits polymerization. |
| griseofulvin | from the bacterium *Penicillium griseofulvin* | Inhibits polymerization. |
| vinblastine, vincristine | periwinkle (*Vinca minor* and *major*) | Inhibit polymerization. Produce precipitates of tubulin in cytoplasm. |
| nocadazole | synthetic benzimidazole | Inhibits polymerization. Very easily reversed. |
| thiabendazole (TBZ) | synthetic benzimidazole | Disassembles microtubules. A potent antimitoic agent widely used as a fungicide. |
| taxol | yew tree (*Taxus brevifolia*) | Favors polymerization. Stabilizes microtubules. Used as an anti-cancer drug. |

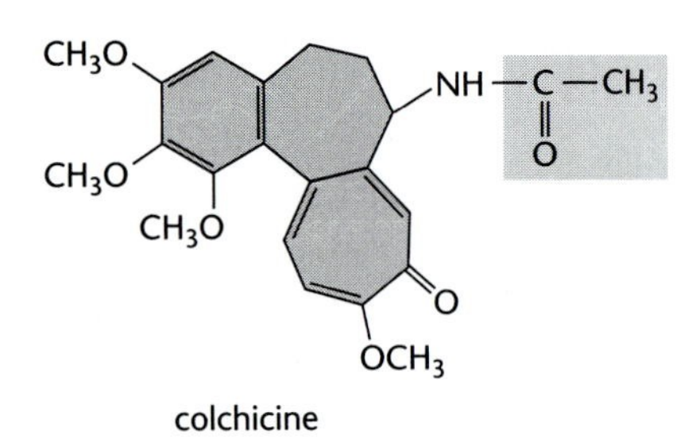


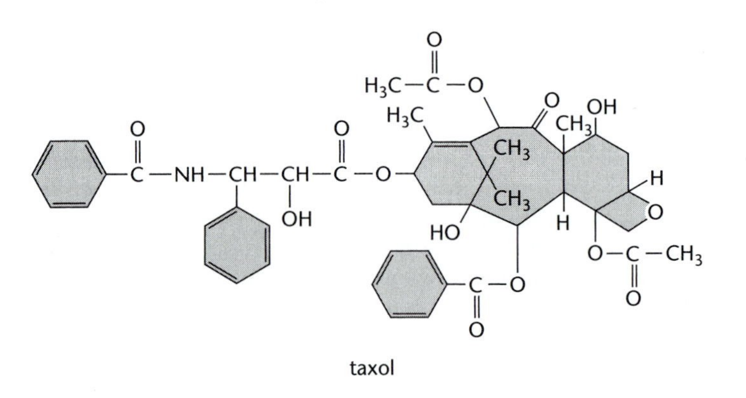


**Figure 11-4** Chemical structures of colchicine and taxol. Colchicine is a tricyclic water-soluble drug that prevents the polymerization of tubulin. Taxol has the opposite effect and stabilizes tubulin in its assembled, microtubular form.

A number of other plant alkaloids have a similar action on microtubules. These include *podophyllotoxin*, from the root of the May apple (*Podophyllum peltatum*), which shares the same site on tubulin as colchicine, and *vinblastine* from the periwinkle (*Vinca major*, or *minor*), with a different site of action. Besides inhibiting polymerization, vinblastine also causes tubulin to form crystal-like aggregates, which can sometimes be found in treated cells.

The plant alkaloid *taxol* from the yew tree (genus *Taxus*) (see Figure 11-4) is unique in that it stabilizes microtubules against breakdown rather than stimulating breakdown. Cells treated with taxol develop large aggregates of microtubules in their cytoplasm, an effect that inhibits many cellular processes as surely as microtubule depolymerization.

Besides these naturally occurring compounds, many synthetic drugs have been made that bind to tubulin. Some of these, such as *colcemid*, were obtained by chemical modification of existing drugs, whereas others were identified in drug screening programs because of their inhibitory effects on cell division. The latter category includes *nocodazole*, which is a potent inhibitor of tubulin polymerization, and *TBZ* (thiabendazole), which destabilizes existing microtubules.

## Purified tubulin assembles into microtubules

It was some time before the colchicine-binding protein from mammalian brain could be coaxed to produce microtubules. Conditions adequate to promote actin polymerization were tried without success, until it was found by chance that microtubules are unstable in the presence of $Ca^{2+}$ above 10 μM. Successful polymerization was achieved by warming a tubulin-rich solution with added GTP and the $Ca^{2+}$-chelator EGTA at 37°C for 30 minutes. Centrifugation then produced a glassy pellet of assembled microtubules, which could be redissolved, clarified by centrifugation, and again converted to the polymerized form.

The polymerization of tubulin under defined conditions shows broadly similar characteristics to that of actin described in Chapter 5. After an initial lag phase, proteins assemble rapidly into microtubules, eventually attaining a plateau in which microtubules are at equilibrium with a critical concentration of tubulin dimer. Kinetic analysis of the rates of growth are broadly consistent with a simple addition and loss of tubulin dimers to and from the microtubule ends, although the details are complicated, as we will see.

## Microtubules are hollow tubes made of protofilaments

Tubulin assembly produces microtubules of variable length (typically between 1 and 10 μm, but often much longer) but with a precisely determined diameter close to 25 nm. The cylindrical walls are about 5–6 nm thick and the central channel, or *lumen*, is about 14 nm in diameter. The lumen appears empty in synthetic microtubules, but those in nerve cells often contain plugs of amorphous material.[4]

[4] Could microtubules act as pipelike conduits for the transport of molecules or ions? Bacterial flagella evidently carry flagellin subunits from the cell body to the tip of the flagella (Chapter 16) and calculations indicate that a tubulin dimer should diffuse to the center of a 40 μm microtubule in a few minutes. But whether this pathway is used, and if so why, and how, has not been decided.

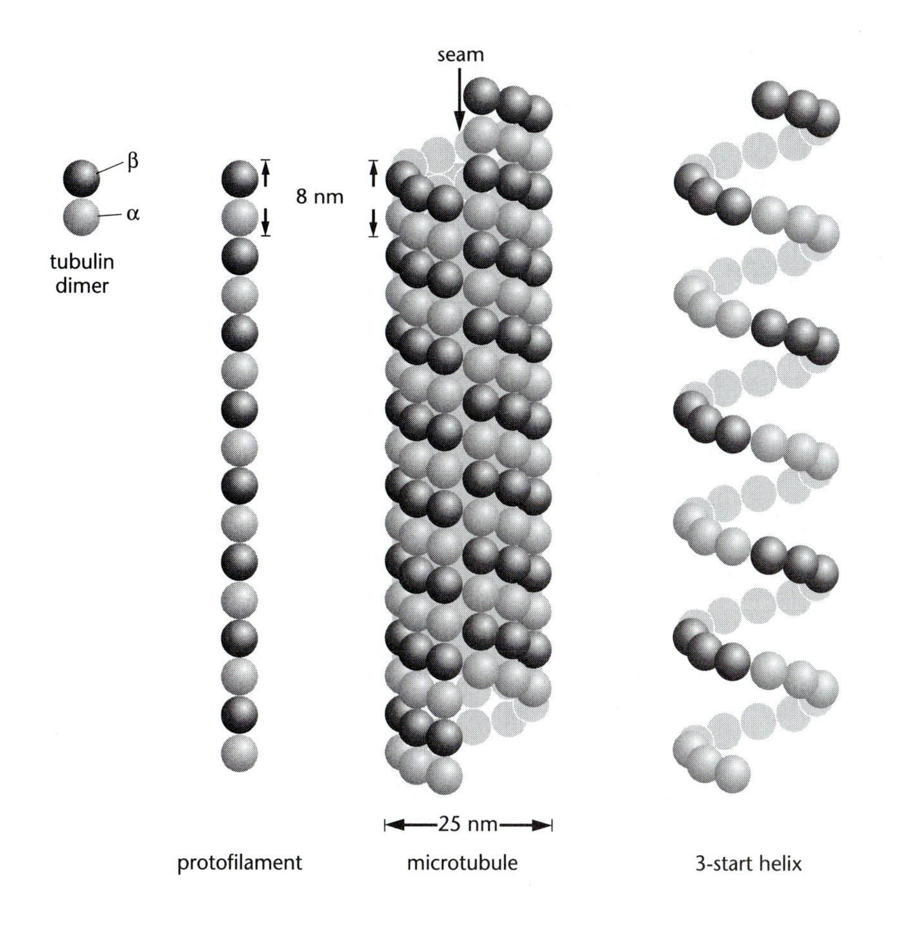


**Figure 11-5** Lattice structure of a microtubule. The cylindrical wall is built from dimers of α- and β-tubulin in a lattice that features a discontinuity or seam. The lattice can be represented as a circular array of protofilaments, each a linear sequence of tubulin dimers. Alternatively, it can be thought of as being made of three parallel helices that deviate from the horizontal with a 10° pitch, thereby forming a path that travels up the microtubule lattice (this path is called a 3-start helix because, if you follow the path of adjacent monomers for one complete turn, you end up three monomers above where you started).

The wall of a microtubule may be thought of as being made of parallel protofilaments, each a linear chain of tubulin subunits with α- and β-tubulins alternating along its length (Figure 11-5). The number of protofilaments in microtubules assembled from brain tubulin varies between 10 and 15, with most having 13 protofilaments. Although there are many exceptions, most microtubules in living cells, and most microtubules nucleated from centrosomes and axonemes in cell-free extracts, have 13 protofilaments (Figure 11-6).

The detailed spatial arrangement of neighboring protofilaments in the wall of a microtubule has been the source of some disagreement. Originally it was postulated that tubulin molecules packed in a helically symmetric structure called an A-lattice. However, evidence from microtubules decorated with the motor protein kinesin established that the actual structure is a B-lattice, which for a 13-protofilament microtubule is not truly helical. As shown in Figure 11-5, this molecular lattice has a discontinuity, or seam, at one point on its perimeter. It is not know whether this 'defect' has any functional role, but it could be related to the way microtubules grow.

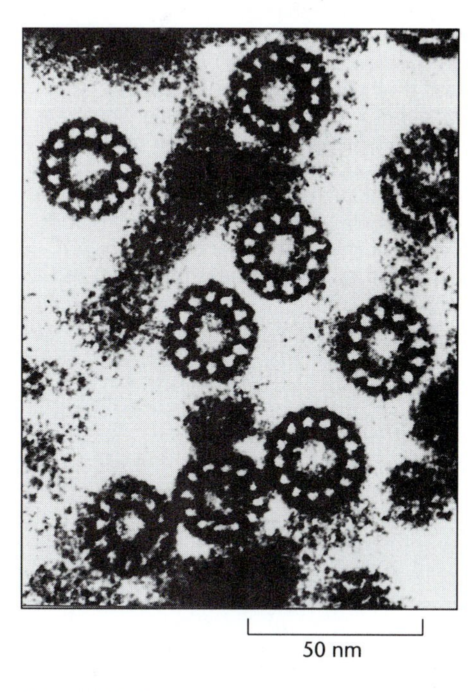


**Figure 11-6** Microtubules in cross-section. These microtubules, in the developing wing of a fruit fly, with diameters in the 20–24 nm range, are composed of 12 or 13 protofilaments. (Courtesy of John Tucker.)

## *Microtubules are polarized structures*

Each protofilament in a microtubule has a structural polarity, with β-tubulin exposed at one end and α-tubulin exposed at the other; this 'structural arrow' is the same for all 13 protofilaments, giving the microtubule as a whole structural polarity. The β-tubulin end, which is called the *plus end,* carries a rapidly exchangeable GTP molecule. The α-tubulin end, which is called the *minus end,* has an exposed domain thought to catalyze hydrolysis of GTP. Tubulin dimers add more rapidly to the plus end than to the minus end (which is why the ends were originally named in this way).

The fact that the structure has a definite polarity, with two different ends, is of pivotal importance to the organization of a cell's contents. First of all, as we will see shortly, polarity forms the basis of the dynamic instability of microtubules growth, by which cells control *where* microtubules are formed. And once they are formed, the *function* of microtubules also depends on their having a unique orientation—the many organelles that move along microtubules, for example, do so in a unique direction, dictated by the microtubule polarity. Thus we find that the formation and function of nerve axons, the segregation of chromosomes at mitosis, the establishment of morphogenetic axes in a egg, and many other processes, all rely on microtubules being not only in the right place in the cell, but also having the correct polarity.

How can we find out what polarity microtubules have, especially when they are in the middle of a cell? One way is by electron microscopy. The surface of microtubules can be 'decorated' by adding extra side-arms of tubulin or dynein in such a way that the underlying polarity of the protofilaments is revealed. Another, more recent, technique is to detect microtubule polarity by means of organelle transport. For example, a motor protein coupled to a histological marker such as β-galactosidase might be introduced into a cell. An array of microtubules with the same polarity will cause an accumulation of β-galactosidase in one region, and if the direction of the motor is known, the polarity of the microtubules can be inferred.

## *Dynamic instability is driven by hydrolysis of GTP*

Although a population of microtubules exhibits a bulk steady-state length, individual microtubules persist in prolonged states of polymerization or depolymerization and only occasionally switch between the two. Free ends are seen to alternate, in an abrupt, stochastic manner, between periods of steady growth and phases of disassembly, a phenomenon called *dynamic instability* (Figure 11-7). It was originally used to explain

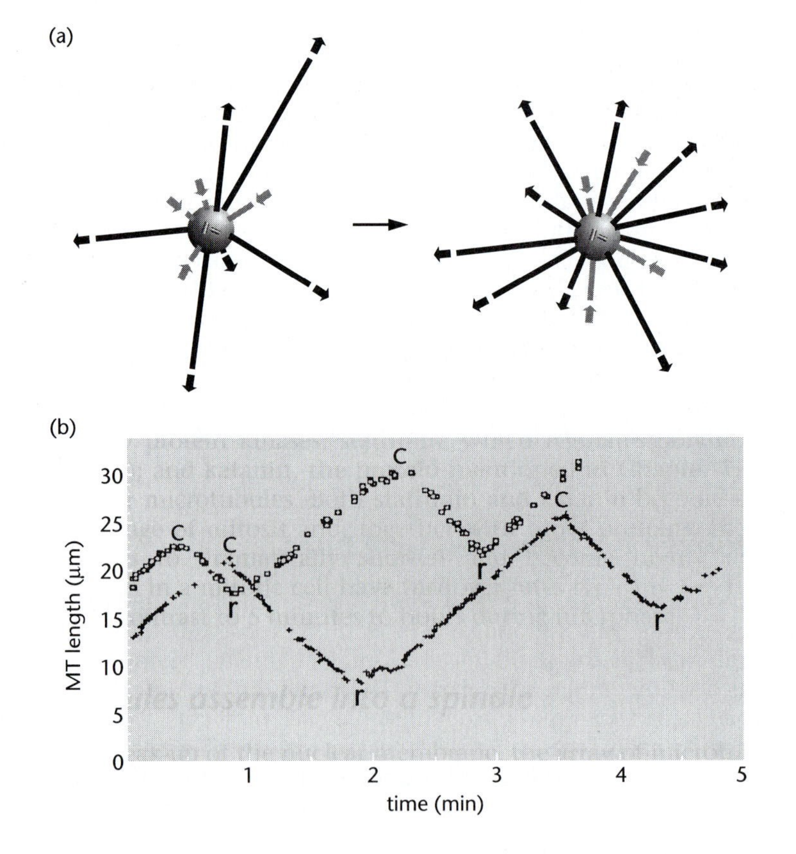


**Figure 11-7** Dynamic instability. (a) Schematic view of microtubules anchored by their minus ends to an organizing center and their plus ends exposed to the cytoplasm. The array changes continually as microtubules alternate between phases of growth and catastrophic disassembly. (b) Actual fluctuations at the plus end of microtubules undergoing dynamic instability. Phases of linear growth and shrinkage are separated by random catastrophes (c) and rescues (r).

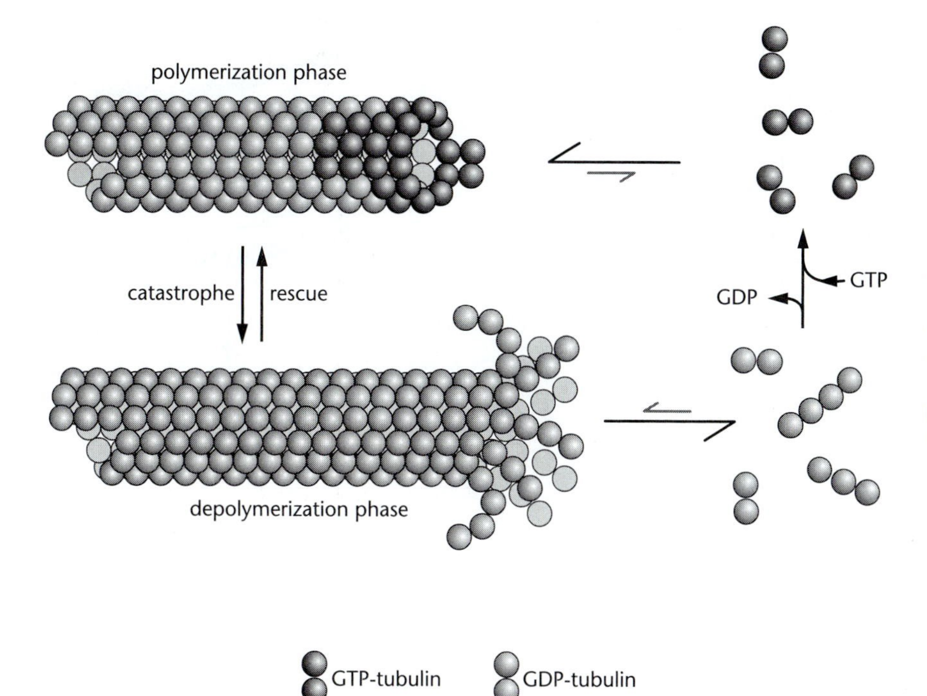


**Figure 11-8** Schematic view of the plus end of a microtubule, showing it in a growing and a shrinking phase. During rapid growth, GTP-containing tubulin dimers (shaded) add to the growing end faster than the GTP they carry can be hydrolyzed. A GTP cap is thereby created that facilitates further growth. Following the loss of this cap, the inner segment of GDP-containing subunits undergoes a catastrophic disassembly. Under typical conditions and with physiological concentrations of tubulin, microtubules assemble at 1–2 μm/min and shorten at 20–30 μm/min, switching between the two phases stochastically less than once a minute.

the distribution of lengths of microtubules in cell-free extracts, and has been confirmed by watching single microtubules using dark-field or video microscopy. Dynamic instability is now generally accepted as a dominant mechanism governing polymerization of microtubules in living cells.

Dynamic instability can be measured by four parameters: (1) the rate of polymerization; (2) the rate of depolymerization; (3) the frequency of *catastrophe* (the transition from polymerization to depolymerization); and (4) the frequency of *rescue* (the transition from depolymerization to polymerization) (Figure 11-8). In the polymerization phase, GTP-tubulin subunits add to the end of a microtubule and then, shortly afterwards, hydrolyze their bound GTP and subsequently release the hydrolyzed phosphate ($P_i$). In the depolymerization phase, GDP-tubulin subunits are released from the microtubule at a very rapid rate. Values of the above four parameters are known for microtubules under a wide variety of conditions.

Dynamic instability is a nonequilibrium behavior and thus requires an energy supply, the obvious source being GTP hydrolysis by β-tubulin during polymerization, which yields ~12.5 kcal/mol under the conditions of the cell. Polymerization itself does not require GTP hydrolysis and, interestingly, a microtubule lattice with bound GTP is actually more stable than one with GDP. As we will now see, the principal role of GTP hydrolysis is probably to *weaken* the microtubule lattice and, in effect, store mechanical strain energy in its tubulin subunits.

## *Growing and shrinking ends are structurally different*

Dynamic instability has a structural correlate. The ends of a microtubule have a molecular anatomy that depends on whether they are growing or shrinking. Growing microtubules typically have a curved sheet of protofilaments at their ends, a terminal region in which the microtubule wall opens up, it is believed along the seam. At a depolymerizing end, by contrast, protofilaments splay apart to form characteristic curved 'ram's horns' that break off as curved oligomers and depolymerize. There may

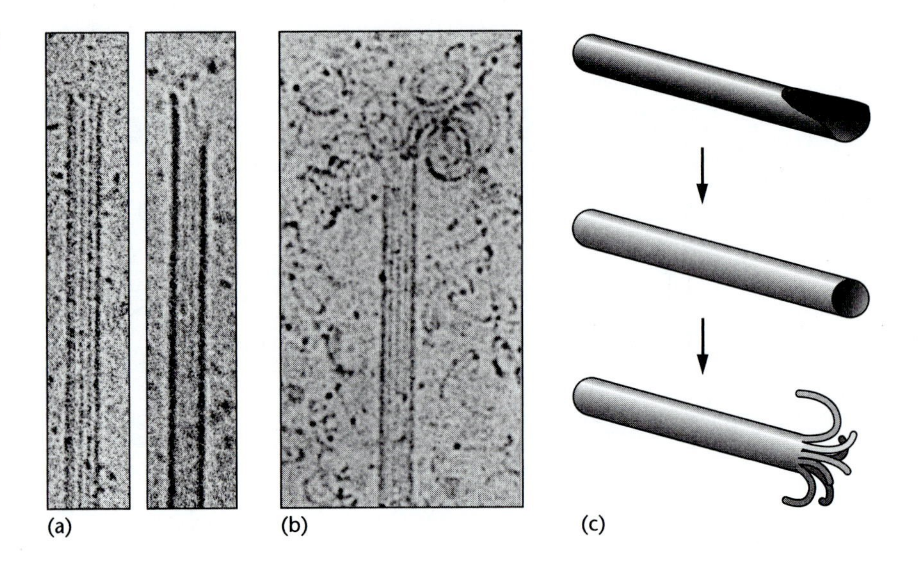


**Figure 11-9** Growing and shrinking ends of microtubules. The electron micrograph in (a) was made of rapidly growing microtubules; the microtubule in (b) was undergoing disassembly. In (c) the different states are shown schematically, together with a postulated third, intermediate, state. (a, and b, courtesy of Eva Mandelkow.)

also be a third, intermediate state in which the microtubule is a simple closed tube (Figure 11-9).

A simple explanation links these morphological changes to GTP hydrolysis. According to this hypothesis, GTP-tubulin assembles into protofilaments that are straight, but GDP-tubulin, because it has a different conformation, produces protofilaments that are curved. Since only straight protofilaments can readily pack side by side to form the wall of a microtubule, the GDP-tubulin subunits will be forced into a conformation that is not the most stable. They will be held in place by interactions with their neighbors in the lattice and by a short region of GTP tubulin at the plus end of the microtubule, but the wall structure will be, in a sense, under tension. The mechanical strain energy produced in the tubulin dimers by hydrolysis of GTP is stored in the microtubule lattice like energy stored in a compressed spring.[5] However, if the cap of GTP tubulin is removed, then the remaining protofilaments spring into their more stable curved conformation, producing a spray of protofilament ends.

According to this interpretation, the end of a microtubule showing dynamic instability is on a knife edge. Any momentary local increase in concentration of GTP-tubulin will make its GTP cap larger and further consolidate subunit addition. Conversely, if for any reason a transient fall in concentration occurs (for example, if the microtubule grows into a solid obstacle), the GTP cap will shrink. The unstable GDP-containing subunits of the microtubule wall may then be exposed, leading to a prolonged period of disassembly. Thus, in a solution containing nothing but tubulin and GTP, a nucleated microtubule will alternate in stochastic fashion between these two modes.

In a cell, of course, we expect things to be more complicated. Indeed, as we will see, numerous factors have been discovered that either suppress dynamic instability, making the microtubule more stable, or promote instability, leading to an even more rapid breakdown.

## *The centrosome is the major microtubule-organizing center in animal cells*

Dynamic instability in living cells can be visualized in real time by labeling the microtubules with fluorescent tags. In the flattened leading edge of a fibroblast in culture, for example, microtubules grow and shrink continually, reflecting the extreme sensitivity of this region of the cell to external influences. Most of these microtubules (but not all, as we will see later)

[5] Measurements of flexural rigidity of microtubules indicate that the energy stored in the lattice of a GDP microtubule is comparable to the energy available from GTP hydrolysis. A single microtubule can therefore store a remarkably large amount of energy—a microtubule 50 μm long stores the energy released from 80,000 GTP molecules for many minutes.

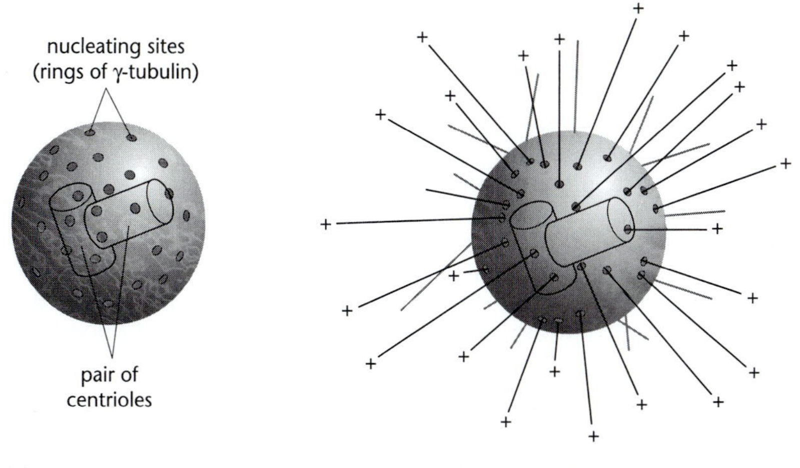


**Figure 11-10** Polymerization of tubulin on a centrosome. (a) Schematic drawing showing that a centrosome consists of an amorphous matrix of protein containing the γ-tubulin rings that nucleate microtubule growth. In animal cells, the centrosome contains a pair of centrioles each made of a cylindrical array of short microtubules. (b) A centrosome with attached microtubules. The minus end of each microtubule is embedded in the centrosome, having grown from a nucleating ring, whereas the plus end of each microtubule is free in the cytoplasm.

appear to have their minus ends attached to an organizing center near the nucleus. This center, known as the *centrosome*, has an important controlling influence on the location, and orientation of microtubules in the cytoplasm.

In animal cells, the centrosome is typically present on one side of the cell nucleus, at the center of an array of microtubules that extends outward into the cytoplasm (Figure 11-10). Centrosomes are formed by an amorphous matrix of filaments made of several α-helical coiled-coil proteins, such as *pericentrin*, together with hundreds of ring-shaped structures containing another isoform of tubulin, *γ-tubulin*. Each γ-tubulin ring serves as the starting point or *nucleation site* for the growth of one microtubule. Tubulin dimers add to the γ-tubulin ring by their α-tubulin ends, with the result that the minus end of each microtubule is embedded in the centrosome, and growth occurs only at the plus end—that is, the outward-facing end.

The γ-tubulin rings in the centrosome should not be confused with centrioles, which are curious structures made of a cylindrical array of short microtubules. The centrioles have no direct role in the nucleation of microtubules in the centrosome (the γ-tubulin rings alone are sufficient), but they are found associated with the centrosome in most animal cells. Centrioles and the closely similar structures called basal bodies associated with cilia and flagella, are discussed in Chapter 15.

Microtubules need nucleating sites such as those provided by the γ-tubulin rings in the centrosome. Under most conditions, it is much harder to start a new microtubule from scratch, by first assembling a ring of αβ dimers, than to add tubulin subunits to a preexisting microtubule structure. Solutions containing αβ-tubulin can polymerize spontaneously when at a high concentration, but at low concentrations, similar to those encountered in the living cell, polymerization requires a nucleating center (Figure 11-11). By providing organizing centers containing nucleation sites, and keeping the concentration of free αβ-tubulin subunits low, cells can control where microtubules form.

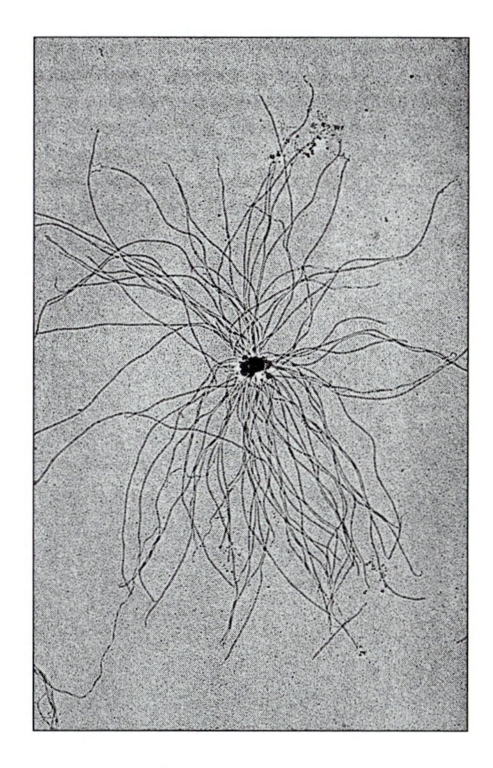

**Figure 11-11** Nucleated growth of microtubules. An isolated centrosome forms the center of an array of microtubules growing under cell-free conditions. (Courtesy of T. Mitchison and M. Kirschner, *Nature* 312: 232–237, 1984. © Macmillan Magazines Ltd.)

## *Asymmetric microtubule arrays can be generated by dynamic instability*

We have just said that cytoplasmic microtubules in animal cells radiate out from the centrosome where their minus ends are anchored. But in most cells this array is not perfectly symmetric, and has a preferred orientation. Long, narrow cells with a polarized shape typically have sets of microtubules that run longitudinally, along their long axis (an extreme example being nerve axons). Mitotic and meiotic spindles are often precisely positioned in the cytoplasm with reference to structures on the cortex and the plane of division. In these and many other situations the assembly of microtubules must be controlled so that many microtubules extend toward specific regions of the cell periphery.

Dynamic instability in conjunction with the selective stabilization of plus ends provides one important mechanism by which this polarity is achieved. The average lifetime of microtubules attached to the centrosome of a typical animal cell is quite short—of the order of several minutes. Thus the array is continually changing in form as new microtubules grow and replace others that have depolymerized. Any individual microtubule that grows from such a structure can be stabilized if its plus end is somehow capped so as to prevent its depolymerization. The location of the capping structure will consequently become linked to the centrosome by relatively stable microtubules.

This simple 'search–capture' mechanism provides a starting point for the generation of microtubule arrays of specific geometry (Figure 11-12). For example, it is implicated in the attachment of the meiotic spindle to a specific cortical site in marine eggs prior to polar body formation. The rotation of the nucleus prior to asymmetric division in a nematode embryo (Chapter 20) and the positioning of the budding yeast spindle to the mother-bud neck prior to anaphase (Chapter 19) are two other examples.

## *Microtubules undergo a slow maturation*

The continual formation and loss of microtubules is characteristic of cells undergoing a major internal reorganization, such as cells that are dividing or crawling over a substratum. However, in cells that have become part of

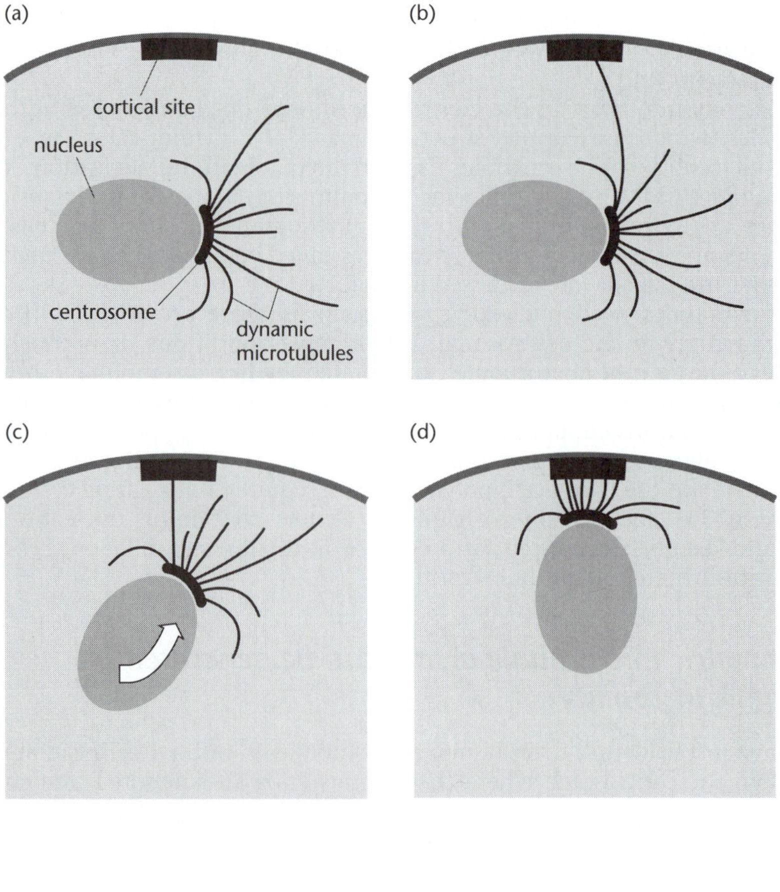

**Figure 11-12** Generation of cellular asymmetry by microtubule search–capture. In this schematic example, the nucleus is pulled toward a specific cortical site by a microtubule that has by chance grown into, and become stabilized by, a protein complex associated with the plasma membrane. Subsequent shortening of the anchored microtubule then pulls the nucleus toward the cortical location, thereby creating cellular asymmetry.

an established tissue, microtubules can become relatively permanent features. Consider nerve cells that can no longer divide after they differentiate, or cells that possess cilia or other structures built of microtubules. This microtubule 'maturation' depends partly on subtle chemical modifications of the tubulin molecules by enzymes in the cytoplasm.

A number of enzymes modify selected amino acids in tubulin. One of these is *tubulin acetyltransferase*, which acetylates a specific lysine of the α-tubulin subunit. In *Chlamydomonas*, for example, this enzyme is located principally in the flagellar axoneme and acetylates tubulin molecules after they assemble at the distal tip of the axoneme. A specific *deacetylase* enzyme is located in the cytoplasm and removes acetyl groups from unpolymerized tubulin. As a consequence of the localization of these two enzymes in *Chlamydomonas*, tubulin molecules in the axonemal microtubules, which are relatively long-lived, are acetylated, whereas those in cytoplasmic microtubules, which turnover rapidly, are mainly nonacetylated.

A second, more unusual, form of posttranslational modification is the removal of the C-terminal tyrosine residue from α-tubulin molecules that have become incorporated into a microtubule. This modification is catalyzed by a detyrosinating enzyme in the cytoplasm of many vertebrate cells, and, as in the case of acetylation, there is an oppositely acting enzyme that restores a tyrosine to unpolymerized tubulin. In cells with highly unstable microtubules, any given tubulin molecule is generally not present in a microtubule long enough to be detyrosinated. However, detyrosinated tubulin becomes enriched in older microtubules that survive the normal rapid turnover of newly formed microtubules.

Detyrosination and acetylation thus mark the conversion of transiently stabilized microtubules into a much more permanent form. When cultured fibroblasts are treated with drugs that depolymerize microtubules, a small population of microtubules that is spared during the drug treatment can be selectively labeled with antibodies that recognize either acetylated or detyrosinated tubulin.

Somewhat surprisingly, microtubules formed from acetylated or detyrosinated tubulin are not detectably more stable than microtubules from unmodified tubulin. It seems that these modifications exert their action indirectly, by triggering the binding of specific microtubule-associated proteins that stabilize microtubules and make them less likely to depolymerize. As we will see shortly, microtubule-associated proteins have important, although often subtle, effects on the dynamic properties of microtubules.

## *Microtubules treadmill through the cytoplasm*

In our discussion of actin polymerization in Chapter 5 we introduced the dynamic behavior known as treadmilling. In this phenomenon, it may be recalled, subunits move through a polymer driven by differences in binding affinity at the two ends. Energy for treadmilling comes from the free energy of hydrolysis of nucleotide bound to the polymer, which in the case of actin is ATP. Microtubules are also capable of treadmilling and tubulin dimers can move through the structure driven by hydrolysis of their bound GTP. Microtubule treadmilling was demonstrated over twenty years ago in the test tube and has subsequently been observed in a variety of conditions in living cells.

One of the best-studied examples occurs during mitosis. We will see in Chapter 13 that a steady stream of tubulin dimers moves from the plus ends of the microtubules of the mitotic spindle to their minus ends, buried in the spindle poles. Treadmilling can also occur in interphase cells. Depending on the cell type, there is usually a significant proportion of microtubules that have both ends free in the cytoplasm. Microtubules in the flattened lamella of a fibroblastic cell in culture can be seen to flex,

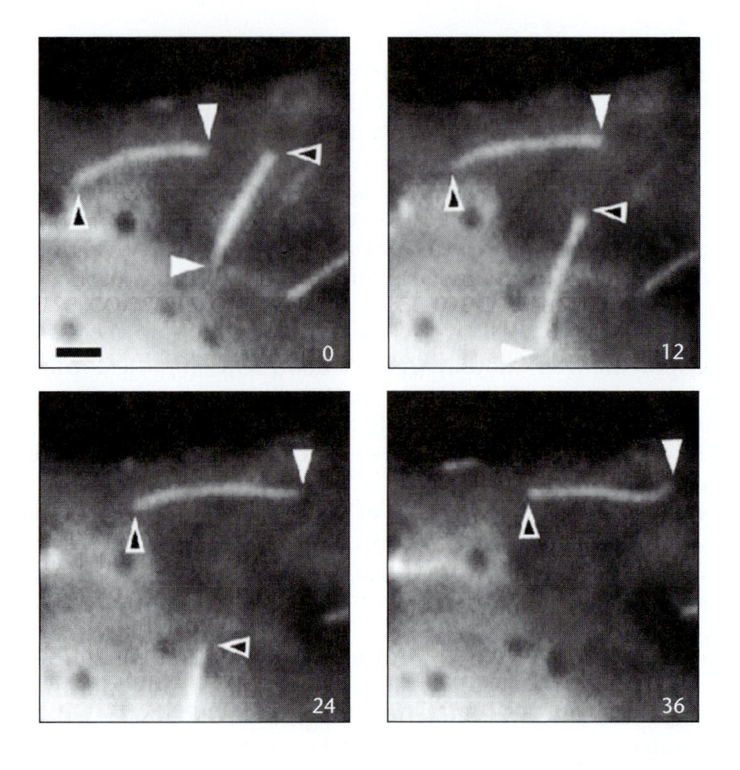


**Figure 11-13** Microtubules moving through cytoplasm. Fluorescently labeled microtubules in centrosome-free cell fragments persistently grow at one end and shorten at the other end. Numbers indicate time in seconds. Bar = 1 μm. (Courtesy of Vladimir Rodionov.)

bend, and occasionally break. Microtubules may also be released from their attachment to the centrosome or severed by the action of specific enzymes (which occurs during mitosis—see Chapter 13). Once these 'orphan' microtubules are produced, they can treadmill through the cytoplasm, with net growth occurring at the plus end and added subunits eventually being released by minus end shortening (Figure 11-13).

It is interesting to ask what is the relation between treadmilling and dynamic instability? Why does the plus end of a microtubule in one case fluctuate vigorously and in the other grow more-or-less uniformly? If we consider a population of microtubules in the cell, then the answer may be an automatic consequence of whether their minus ends are capped. If minus ends are exposed—as in a microtubule free in the cytoplasm—then they will lose subunits rapidly. The concentration of free tubulin in the cytoplasm will therefore rise until it reaches a level at which growth at the plus end balances the loss at the minus end—the condition of treadmilling. But if the minus ends are buried in a centrosome they cannot release subunits. The concentration of free tubulin in the cytoplasm will therefore be lower and the exposed plus ends will reach their own equilibrium, showing the alternating growth and shrinkage known as dynamic instability.

## *Many proteins bind to microtubules and modify their function in the cell*

The length, formation, rate of treadmilling, stability, and other properties of microtubules in the cytoplasm of a living cell are controlled by accessory proteins (Table 11-3). These are sometimes referred to collectively as *microtubule-associated proteins*, or *MAPs*, although the acronym MAP has other meanings, such as 'mitosis (or M-phase)-associated protein,' 'maturation-associated protein' or 'mitogen-activated protein,' and so is probably best avoided. The first proteins of this kind, found by copurifying them with tubulin through repeated cycles of polymerization and depolymerization, were from brain. Proteins such as MAP2 and tau (described in

**Table 11-3 Microtubule-binding proteins**

| Protein | Function |
|---|---|
| dynein | Giant motor protein that undergoes ATP-dependent movement along microtubules. Axonemal dyneins, located in cilia and flagella, generate bending movements. Cytoplasmic dyneins drive organelle transport toward the minus end. |
| kinesin | Motor protein that uses the energy of ATP hydrolysis to move along a microtubule. Conventional kinesin drives organelles along microtubules toward their plus end. |
| MAP2 | Heterogeneous group of related proteins that promote assembly and stabilize microtubules in nerve cells. |
| CLIP-170 | Cytoplasmic protein that binds to microtubules in a phosphorylation-dependent manner and is usually found at the plus ends of microtubules. |
| tektin | Family of fibrous proteins that form specialized protofilaments in microtubules of cilia, flagella, and centrioles. |
| stathmin/Op18 | Small protein that controls the dynamics of microtubule polymerization in cells. |
| STOPs | Family of calcium-regulated proteins that associate with microtubules and stabilize them against disassembly. |
| katanin | Disrupts tubulin–tubulin bonds, leading to eventual severing of a microtubule. |
| tau protein | Promotes assembly, stability, and bundling of microtubules. Most prominent in nerve axons. |
| BPAG1 (bullous pemphigoid antigen) | Large linker protein that links neurofilaments, actin filaments, and microtubules in nerve cells. |

Many other proteins associate with microtubules in special structures, such as mitotic spindles (Chapter 13), cilia and flagella (Chapter 14), and centrosomes (Chapter 15).

the following chapter) associate along the length of microtubules in axons and dendrites, stabilize them, and prevent their dynamic instability. Another source of microtubule-binding proteins is the ciliary axoneme, which as we will find in Chapter 14 contains many proteins stably associated with microtubules. Motor proteins that move along microtubules have also been identified, including those belonging to the dynein and kinesin families.

From the standpoint of dynamic instability, a protein of great interest is *stathmin/Op18*, a protein found widely in proliferating cells and in neurons. Its expression level changes during development, tissue differentiation and tumor formation. Stathmin is a small protein (molecular weight 19 kDa) with a C-terminal α-helical domain and a N-terminal regulatory domain containing four serine phosphorylation sites. These sites are targets of protein kinases that are regulated both during the cell cycle and by signal transduction cascades. Stathmin inhibits microtubule polymerization in a dose-dependent fashion, as shown both in test tube experiments and in living cells. Phosphorylation of stathmin on all four sites completely blocks its microtubule-destabilizing effect. Thus, unphosphorylated stathmin promotes microtubule disassembly and this activity is turned off by phosphorylation.

It is likely that many microtubule-associated proteins remain to be discovered. Differentiated cells of many kinds contain highly stable microtubules that fail to be solubilized, and are consequently discarded, during conventional biochemical extractions. The stability of these microtubules probably arises from microtubule-associated proteins that are for the most part unknown and probably cell-specific. The recently discovered family of cross-linker proteins (Chapter 17), includes molecules, such as BPAG1, that bind to microtubules and link them to other components of the cytoskeleton. The identity of all these microtubule-binding proteins, and how they work together to regulate microtubules in the cell, is an important area of future research.

## Outstanding Questions

*How are microtubule nuclei created and put into position? What is the structure of the pericentriolar matrix? What specifies the number of protofilaments in a microtubule? How is microtubule growth controlled in the cell so that it occurs only in specific locations? Is it by dynamic instability, or are there gradients of signaling molecules in the cytoplasm? What detailed, molecular-level steps occur as a tubulin dimer adds to a growing microtubule? How is the loss of the GTP cap related to the closure of the flattened sheet of protofilaments? How does a microtubule stop growing? What is the origin of 'orphan' microtubules unattached to organizing centers? How are microtubules released from the centrosome? What conditions allow a microtubule to treadmill through the cytoplasm? How and where did microtubules first appear during evolution? Why do plants such as yew, crocus, and periwinkle make alkaloids that block microtubules, and how do they protect their own microtubules? What, if anything, moves down the lumen of a microtubule? How is it propelled? What proteins associate with microtubules in a cell (especially cold-stable microtubules)? How does tyrosination stabilize a microtubule? What is the function of glycosylation of microtubules?*

## Essential Concepts

- Microtubules are stiff, hollow tubes built from tubulin dimers. They are polarized structures that grow faster at one end (the plus end) than at the other (minus) end.
- Tubulin molecules bind and hydrolyze GTP and undergo a conformational change when they do so. Their atomic structure shows a similarity to that of G proteins of the Ras family.
- Newly made tubulin polypeptides require chaperones and other protein cofactors to fold correctly into assembly-competent dimers.
- Purified tubulin dimers self-assemble to form microtubules. Assembly proceeds by addition of dimers to the free ends of microtubules, with one end growing faster than the other under most conditions.
- Hydrolysis of GTP accompanies, but lags behind, assembly of dimers, resulting in the formation of a cap of GTP-containing tubulin. This cap stabilizes the microtubule and promotes its further growth.
- Each tubulin dimer has a tightly bound GTP molecule that is hydrolyzed to GDP after the tubulin assembles into a microtubule. Hydrolysis reduces the affinity of the dimer for its neighbors in the microtubule wall, decreasing the stability of the polymer.
- Microtubule-organizing centers, such as centrosomes, protect the minus end of microtubules and nucleate new microtubules.
- Slowly growing microtubules are especially unstable and liable to rapid disassembly. They show a phenomenon called dynamic instability in which they alternate in stochastic fashion between periods of growth and periods of rapid shrinkage.
- Dynamic instability can be increased, decreased, or even suppressed entirely by proteins that associate with the microtubule. For example, a protein called stathmin accelerates the breakdown of microtubules, an effect that is regulated by phosphorylation.
- Conversely, the fluctuating end of a microtubule can be stabilized by protein complexes that capture the plus end—a process that helps specify the position of microtubules in a cell.

## Further Reading


Belmont, L., Mitchison, T. Catastrophic revelations about Op18/stathmin. *Trends Biol. Sci.* 21: 197–198, 1996.

Drechsel, D.N., Kirschner, M.W. The minimum GTP cap required to stabilize microtubules. *Curr. Biol.* 4: 1053–1061, 1994.

Erickson, H.P. Atomic structures of tubulin and FtsZ. *Trends Cell Biol.* 8: 133–137, 1998.

Hutchens, J.A., et al. Structurally similar *Drosophila* alpha-tubulins are functionally distinct *in vivo*. *Mol. Biol. Cell* 8: 481–500, 1997.

Hyams, J.S., Lloyd, C.W. Microtubules. New York: Wiley Liss, 1994.

Inoué, S., Salmon, E.D. Force generation by microtubule assembly/disassembly in mitosis and related movements. *J. Cell Biol.* 6: 1619–1640, 1995.

Kikkawa, M., et al. Direct visualization of the microtubule lattice seam both *in vitro* and *in vivo*. *J. Cell Biol.* 127: 1965–1971, 1994.

Lawler, S. Microtubule dynamics: if you need a shrink try stathmin/Op18. *Curr. Biol.* 8: R212–R214, 1998.

Ledbetter, M.C., Porter, K.R. A 'microtubule' in plant cell structure. *J. Cell Biol.* 19: 239–250, 1963.

Lewis, S.A., et al. The α- and β-tubulin folding pathways. *Trends Cell Biol.* 7: 479–484, 1997.

Liang, P., MacRae, T.H. Molecular chaperones and the cytoskeleton. *J. Cell Sci.* 110: 1431–1440, 1997.

Lowe, J., Amos, L.A. Crystal structure of the bacterial cell-division protein FtsZ. *Nature* 391: 203–206, 1998.

Ludueña, R.F. Multiple forms of tubulin. Different gene products and covalent modifications. *Int. Rev. Cytol.* 178: 207–275, 1998.

Mandelkow, E.M., et al. Microtubule dynamics and microtubule caps: a time-resolved cryo-electron microscopy study. *J. Cell Biol.* 114: 977–991, 1991.

Margolis, R.L., Wilson, L. Opposite end assembly and disassembly of microtubules at steady state *in vitro*. *Cell* 13: 1–8, 1978.

McNally, F.J. Modulation of microtubule dynamics during the cell cycle. *Curr. Biol.* 8: 23–29, 1996.

McNally, F.J., Vale, R.D. Identification of katanin, an ATPase that severs and disassembles stable microtubules. *Cell* 75: 419–429, 1993.

Mitchison, T.J. Localization of an exchangeable GTP binding site at the plus end of microtubules. *Science* 261: 1044–1047, 1993.

Mogensen, M.M., et al. Centrosomal deployment of γ-tubulin and pericentrin. *Cell Motil. Cytoskeleton* 36: 276–290, 1997.

Moritz, M., et al. Microtubule nucleation by γ-tubulin-containing rings in the centrosome. *Nature* 378: 638–640, 1995.

Müller-Reichert, T., et al. Structural changes at microtubule ends accompanying GTP hydrolysis: information from slowly hydrolyzable analogue of GTP, guanyl(αβ)methylenediphosphonate. *Proc. Natl. Acad. Sci. USA* 95: 3661–3666, 1998.

Nogales, E., et al. High-resolution model of the microtubule. *Cell* 96: 79–88, 1999.

Nogales, E., et al. Structure of the αβ tubulin dimer by electron crystallography. *Nature* 391: 199–203, 1998.

Odde, D. Diffusion inside microtubules. *Eur. Biophys. J.* 27: 514–520, 1998.

Raff, E.C., et al. Microtubule architecture specified by a β-tubulin isoform. *Science* 275: 70–73, 1997.

Rodionov, V.I., Borisy, G.G. Microtubule treadmilling *in vivo*. *Science* 275: 215–218, 1997.

Rodionov, V. et al. Centrosomal control of microtubule dynamics. *Proc. Natl. Acad. Sci. USA* 96: 115–120, 1999.

Roobol, A., et al. Cytoplasmic chaperonin complexes enter neurites developing *in vitro* and differ in subunit composition within single cells. *J. Cell Sci.* 106: 1477–1488, 1995.

Sabatini, D.D., et al. Preservation of ultrastructure and enzymatic activity of aldehyde fixation. *J. Histochem. Cytochem.* 10: 652–662, 1962.

Shelanski, M.L., Taylor, E.W. Isolation of a protein subunit from microtubules. *J. Cell. Biol.* 34: 549–555, 1967.

Wade, R.H., Chrétien, D. Cryoelectron microscopy of microtubules. *J. Struct. Biol.* 110: 1–27, 1993.

Wade, R.H., et al. Organisation and structure of microtubules and microtubule–motor protein complexes. *Eur. Biophys. J.* 27: 446–454, 1998.

Walker, R.A., et al. Dynamic instability of individual microtubules analyzed by video light microscopy: rate constants and transition probabllities. *J. Cell Biol.* 107: 1437–1448, 1988.

Waterman-Storer, C.M., Salmon, E.D. Actomyosin-based retrograde flow of microtubules in the lamella of migrating epithelial cells influences microtubule dynamic instability and turnover and is associated with microtubule breakage and treadmilling. *J. Cell Biol.* 139: 417–434, 1997.

Waterman-Storer, C.M., Salmon, E.D. Microtubule dynamics: treadmilling comes around again. *Curr. Biol.* 7: R369–R372, 1997.

Weber, K., et al. The specific visualisation of cytoplasmic microtubules in tissue culture cells. *Proc. Natl. Acad. Sci. USA* 72: 459–463, 1975.

Weisenberg, R.C. Microtubule formation *in vitro* in solutions containing low calcium concentrations. *Science* 177: 1104–1107, 1972.

Weisenberg, R.C., et al. The colchicine-binding protein of mammalian brain and its relation to microtubules. *Biochemistry* 7: 4466–4472, 1968.

Wilson, P.G., Borisy, G.G. Evolution of the multi-tubulin hypothesis. *BioEssays* 19: 451–454, 1997.

Zheng, Y., et al. Nucleation of microtubule assembly by a γ-tubulin-containing ring complex. *Nature* 378: 578–583, 1995.

# CHAPTER 12

# Organelle Transport

How do microtubules organize the contents of a cell? We saw in the previous chapter that these hollow rods extend throughout the cytoplasm, often radiating outward from the cell center or lying parallel to the long axis of a polarized cell. So one might suppose that they act as stiffeners, like the ribs of an umbrella, holding the less rigid portions of the cytoplasm in position. However, although microtubules do in fact have a static mechanical function, as we will see in Chapter 18, their most potent influence is on the *dynamics* of the cell—by providing tracks for organelle transport.

To a molecule or small vesicle in the cytoplasm, a long, straight microtubule often offers the fastest and the most direct highway across the cell. Thus we find that nuclei, mitochondria, synaptic vesicles, vesicles and tubules of endoplasmic reticulum, lysosomes, protein complexes, and RNA molecules all have the capacity to move deliberately in one direction

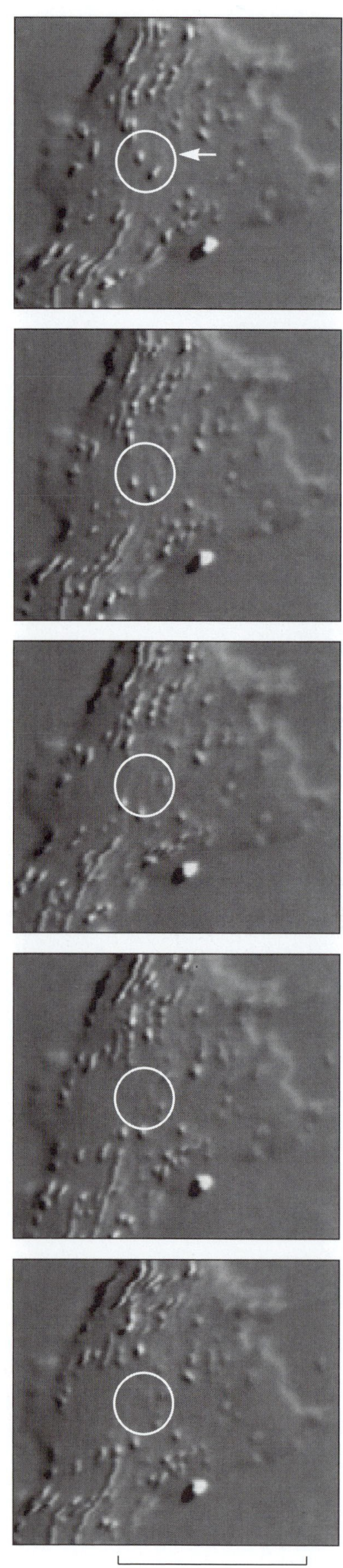


or the other along microtubules, destined for locations such as the cell body, the nucleus, the plasma membrane, or the Golgi apparatus. These ubiquitous movements, driven by molecular motors, constitute perhaps the most important organizing principle of eucaryotic cells. Thanks to organelle movements, animal and plant cells can maintain their highly differentiated cytoplasm, become enormously large, as for muscle cells, or adopt highly asymmetric shapes like a nerve cell.

Organelle transport has been observed, measured, and speculated about since the middle of the nineteenth century. But it was not until the 1980s that any real understanding of its molecular basis was achieved. The breakthrough came in 1983 with the observation, at the Marine Biology Laboratory at Wood's Hole, Massachusetts, that extracts of squid giant axons show movements in the absence of a plasma membrane. The extracts provided an assay by which the proteins responsible for movement could be identified and led to the discovery of two families of molecular motors—kinesin and cytoplasmic dynein—that operate on microtubules. How these molecules move, what they carry, and how their speed and direction are controlled are central issues of the present chapter. We will also try to present a broader picture of the role of these motors in the context of the whole cell, focusing especially on nerve cells. Historically important, the nerve axon retains its position today at the center of studies of the fascinating phenomenon of organelle transport.

## *Membrane-bound organelles move rapidly along microtubules*

Mitochondria, vesicles, lysosomes, lipid droplets, and ingested particles can be watched in the light microscope as they move within epithelial cells or fibroblasts flattened onto the surface of a tissue culture. These organelles typically show what is termed *saltatory movement*, which is rapid, though sporadic, and interspersed with long periods of quiescence.[1] In well-spread cells the movements tend to be directed toward or away from the cell center. Velocities range between 0.5 and 5 μm/sec and individual saltations (jumps) do not usually exceed 1–2 μm (Figure 12-1). Many organelle movements take place in close proximity to microtubules, and when electron microscopy is used, this often reveals molecular links between the surface of the moving organelle and a microtubule (Figure 12-2).

Microtubules play a vital part in positioning membrane-bounded organelles within a eucaryotic cell. The membrane tubules of the endoplasmic reticulum (ER), for example, align with microtubules and extend almost to the edge of the cell, whereas the Golgi apparatus is located near the centrosome. When cells are treated with drugs that depolymerize microtubules, both types of organelle change their location: the ER collapses to the center of the cell, while the Golgi apparatus fragments into small vesicles that disperse throughout the cytoplasm. When the drug is removed, the organelles return to their original positions, dragged by motor proteins moving along the re-formed microtubules. The normal position of each of these organelles is thought to be determined by their association with specific microtubule-dependent motors—a kinesin for the ER and a dynein for the Golgi apparatus.

**Figure 12-1** Organelle transport. Video-enhanced images of a flattened region of an invertebrate nerve cell recorded at intervals of 400 msec. Numerous membrane vesicles and mitochondria are present, many of which can be seen on inspection to move in this sequence (see arrow). (Courtesy of Paul Forscher.)

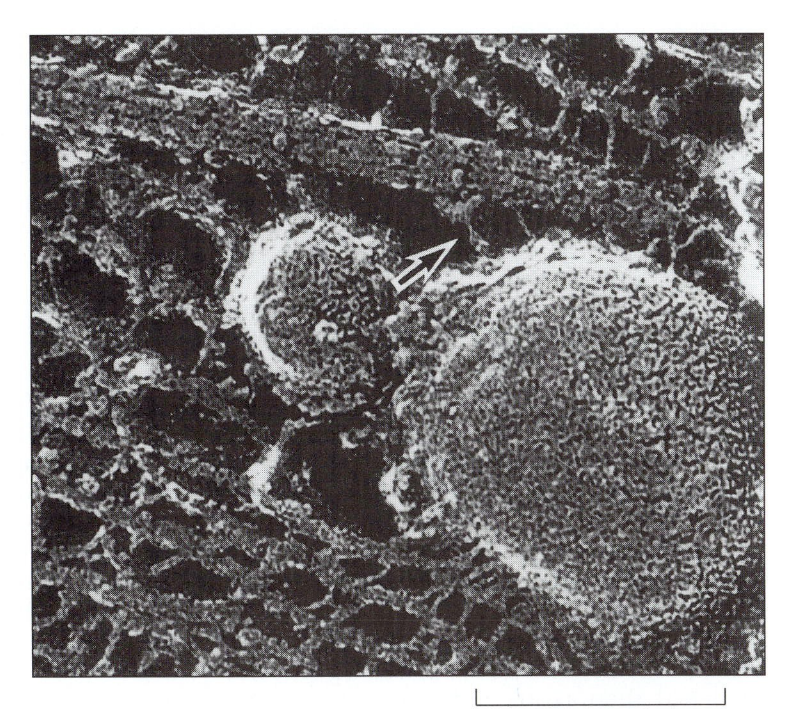


**Figure 12-2** Membrane organelle undergoing fast transport in a nerve axon. The axon was rapidly frozen and water was removed by sublimation before examination in the electron microscope. A short cross-bridge (arrow) could be a motor molecule linking the organelle to a microtubule. (Courtesy of N. Hirokawa, University of Tokyo.)

## *Nerve cells have an exaggerated dependence on organelle transport*

In our description of microtubule-based organelle movements in this chapter, we will focus on nerve cells. These highly elongated cells carry electrical signals in the brain and spinal cord and are the primary substrate for our sensations, thoughts, and emotions. Nerve cells are also (from the more prosaic standpoint of cell movements) notable for their highly asymmetric form and for their dependency on organelle transport. Each nerve cell has a long, thin microtubule-rich process, termed an axon, that can extend millimeters or even centimeters from the cell body (Figure 12-3). Since the nucleus and most biosynthetic organelles are located in the cell body, and since the axon itself lacks the machinery of protein synthesis, most proteins in the axon have to be delivered by organelle transport (usually termed *axonal transport*). Mitochondria, synaptic vesicles, and components of the cytoskeleton all move in an unceasing stream from the cell body out to the distant terminals of the axon, while endocytic vesicles, mitochondria (again), and phagocytic vesicles containing materials taken up at the nerve terminals or produced by the normal turnover of membranes and protein structures, make the return journey back to the cell body.

Axonal transport has been studied over many years, both in the animal and in explants, and is conventionally divided into two distinct rates. Fast axonal transport carries membrane vesicles and organelles in both directions at speeds of around 100–400 mm/day (1–5 μm/sec). As we will see, this movement is driven by motor proteins and is closely related to organelle transport in other cell types. Slow axonal transport, which is harder to see, carries structural proteins of the axonal core, such as tubulin and actin, unidirectionally away from the cell body at rates of 1–8 mm/day. Although this slower form of transport is still poorly understood, recent evidence suggests that it too may be driven by motor proteins.

**Figure 12-3** Typical human motor neuron drawn to scale. The axon is 30 cm long and has a diameter of 2 μm. Omitted from the diagram are the numerous branches formed by the axon at its terminal region, as it enters the muscle.

[1] The movements of cytoplasmic organelles are purposeful and directed and distinct from Brownian movement. In fact, one of the diagnostic signs of a cell's demise is the onset of the unrestrained manic dance of its cytoplasmic constituents due to thermal motion.

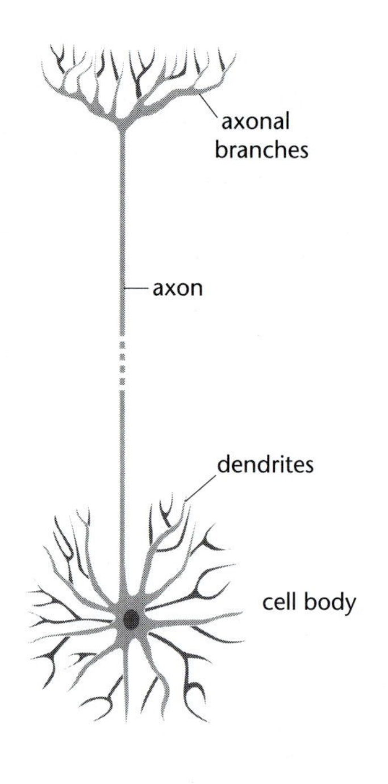


**Figure 12-4** Nerve axons and dendrites. Form of a typical vertebrate nerve cell. Detailed morphology varies enormously from cell to cell, especially in the number and length of the processes and the extent of their branching. Axons in humans, for example, range from less than a millimeter to more than a meter in length (see Figure 12-3).

## *Axons and dendrites have different arrays of microtubules*

Before describing the molecular motors that move organelles in nerve cells, we need to consider the position and polarity of the tracks along which they move. During embryonic development, nerve cells produce long branching processes that grow out and make contact with their synaptic targets, usually other nerve cells. These processes are of two types (Figure 12-4). *Axons*, as just mentioned, are long (sometimes very long), uniform in diameter, unbranched except at their synaptic terminals, and carry signals from the nerve cell body to the distant synaptic target. *Dendrites* are shorter and fatter, and taper in diameter; they are more highly branched than axons and receive signals from surrounding cells. The differentiation of axons and dendrites confers a morphological polarity to the nerve cell essential for its processing of neuronal signals.

The distinct structure and function of dendrites are reflected in their different ultrastructure and macromolecular constitution. The axonal cytoplasm is relatively simple and more clearly differentiated, lacking Golgi elements and ribosomes. Dendrites are rather like extensions of the cell body and contain most of the same components—ribosomes, smooth and rough endoplasmic reticulum, and Golgi elements. There are other differences; for example, the abundant intermediate filaments in axons (neurofilaments) are more highly phosphorylated than those in dendrites (Chapter 17).

The arrangement and composition of the microtubules in axons differ from those in dendrites (Figure 12-5). In both axons and dendrites the microtubules are discontinuous and have their plus and minus ends exposed to the nerve cell cytoplasm (rather than being firmly anchored to the centrosome or other large microtubule-organizing center). In an axon, the microtubules (which might be 100 µm or longer in a mature rat axon) are arranged with their plus ends farthest from the cell body. In a dendrite, microtubules have mixed polarity, with approximately equal numbers pointing in the two directions.

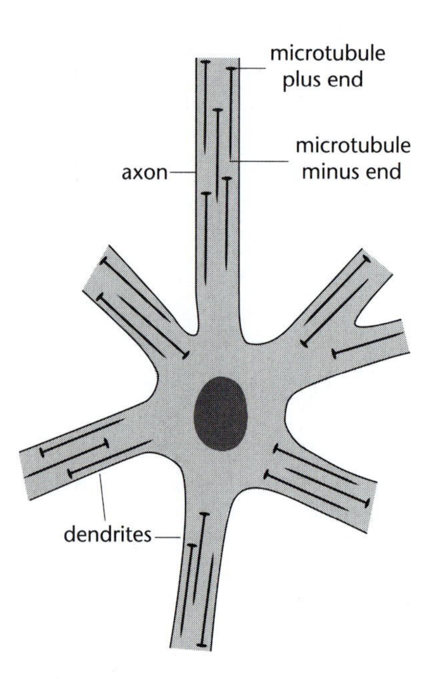


**Figure 12-5** Microtubules in axons and dendrites. In this figure, each microtubule is represented as a 'pin' with its plus end as the head and its minus end the point.

The differences in microtubule polarity may contribute to the accumulation of dendrite-specific molecules. This is because specific organelles move along microtubules toward either their plus end or their minus end. In many cells, secretory vesicles and fragments of smooth endoplasmic reticulum are carried to the plus ends of microtubules, so that we would expect to find them in both axons and dendrites, as indeed is the case. Golgi elements and (at least in some specialized insect cells) polyribosomes are carried toward the minus end of microtubules which could explain why they are present only in dendrites. It even appears that specific mRNA molecules are transported into dendrites, such as the mRNA for MAP2.

Axons and dendrites thus differ fundamentally in the molecules they contain, even though they are bathed in the same cytosol. The basis for this segregation of molecules—a phenomenon sometimes referred to as *neuronal polarity*—is still not completely understood. We do not know, for example, how a nerve cell selects one process, and no more than one, to become an axon. But some at least of the mechanisms responsible for polarity are becoming clearer, and they all involve motor proteins.

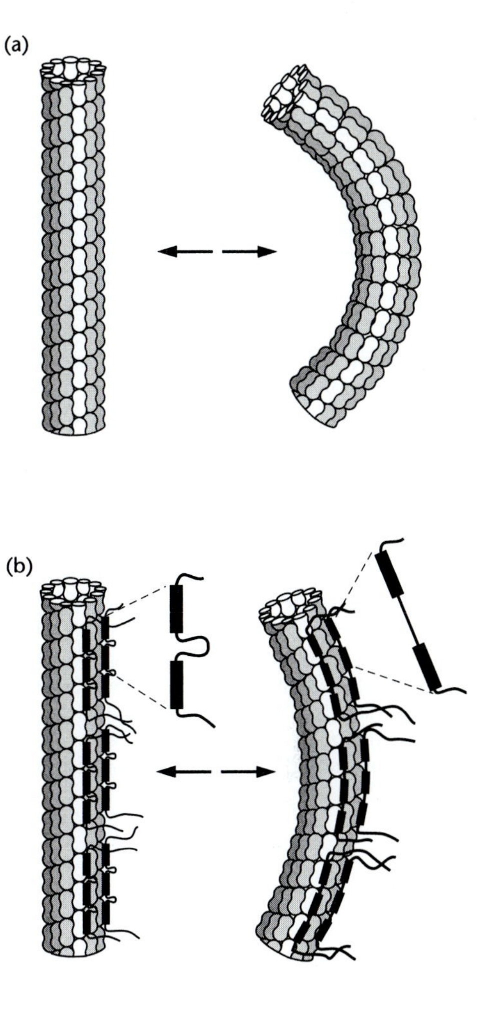


**Figure 12-6** Binding of MAP2 and tau to microtubules. (a) In the absence of these proteins (as in many nonneuronal cells) the flexibility of a microtubule is limited only by the interactions between adjacent tubulin dimers in the microtubule wall. (b) In nerve cells, MAP2 or tau bind to adjacent tubulin subunits through their repeated tubulin-binding motifs, coupling them together closely and restricting their flexibility. This not only makes the microtubule stiffer, but also promotes its stability and hence favors polymerization.

## MAP2 and tau stabilize neuronal microtubules

Nerve microtubules are associated with two related proteins, *MAP2* and *tau*. These are abundant, heat-stable proteins that bind along the length of microtubules; in the test tube, both accelerate the polymerization of microtubules, reducing the lag phase (see Figure 5-5). When MAP2 is expressed in nonneuronal cells the microtubules appear to become stiffer than those in control cells not expressing MAP2. Furthermore, the cells frequently form processes that are long and cylindrical, resembling those produced by nerve cells. MAP2 and tau contain homologous microtubule-binding domains consisting of three or four repeats of an 18-amino-acid sequence. These repeats are thought to bind to tubulin subunits in the wall of a microtubule, tethering them together and reducing their ability to move relative to each other. The ability of MAP2 and tau to 'stitch' subunits together could explain why they are able to promote polymerization and to stiffen microtubules (Figure 12-6).

Tau, or at least certain phosphorylated variants of tau, is restricted to axons and is found as a family of related polypeptides generated by alternative RNA splicing. One form, which is found predominantly in fetal brain, has three tandem repeats, whereas an adult form has four. The latter protein is especially important from a medical standpoint since it accumulates in many neurodegenerative disorders. Various forms of dementia, including presenile dementia (Alzheimer's disease), are characterized by the loss of microtubules from axons and their replacement by tangled filaments composed in large part of highly phosphorylated tau.

MAP2 and tau have multiple sites for protein kinases, both in their microtubule-binding domain and in their side-arms. High levels of phosphorylation reduce their affinity for tubulin and their ability to interact with other components in the cell. Phosphorylation of tau, in particular, has a marked effect on its shape and properties, making the molecule longer and stiffer. In view of the high density of tau cross-links in the axonal cytoplasm, these changes may be expected to cause a profound modulation of the cytoplasmic properties, affecting such processes as the transport of organelles within the axon.

## Kinesin carries organelles toward the plus end of microtubules

In the studies mentioned in the introduction to this chapter, cytoplasm extruded from a squid giant axon was fractionated to yield progressively simpler motile preparations. It had been noted previously that movements in the intact axon were arrested upon addition of the nonhydrolyzable ATP analogue AMP-PNP. This observation prompted investigators to isolate proteins that bound to microtubules in the presence of AMP-PNP and resulted in the discovery of *kinesin*, now known to move membrane-bound organelles in animal cells of many kinds.

When added to a mixture of purified microtubules and vesicles isolated from squid brain, or even to microtubules and polystyrene beads, kinesin produces rapid movement. Vesicles or beads are carried smoothly from one end of a microtubule to another, without the characteristic hesitations

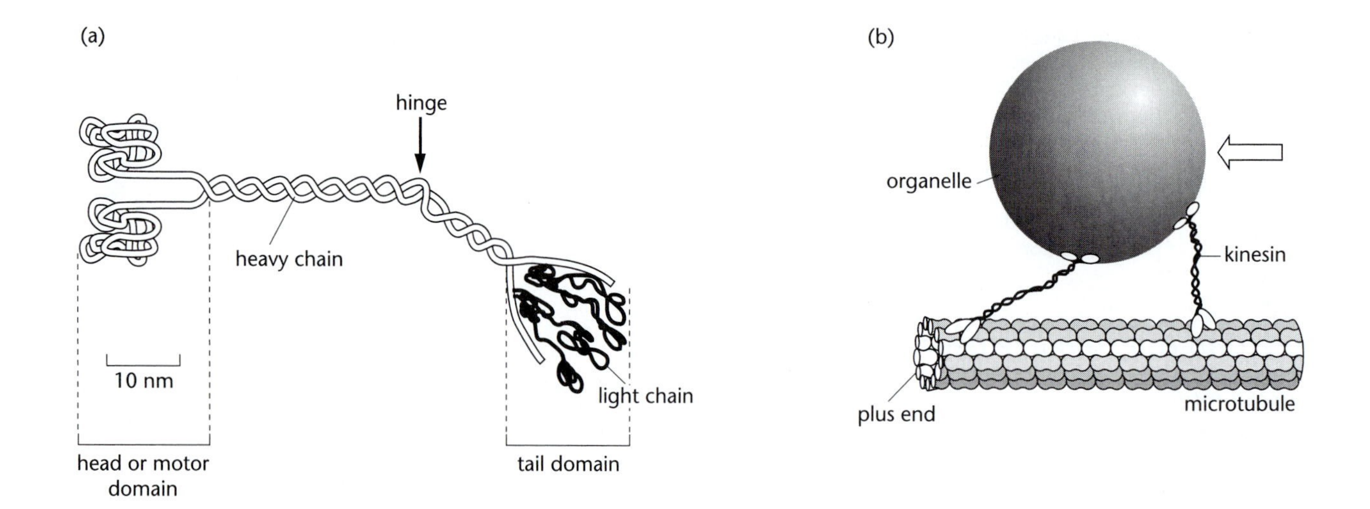


**Figure 12-7** Conventional kinesin (from brain). (a) The molecule, with overall length about 80 nm, has two globular heads that bind to microtubules and a fan-shaped tail responsible for attachment to membrane organelles. The long coiled-coil tail has a hinge region near its middle that allows the molecule to fold. (b) Kinesins can transport organelles along microtubules.

seen in intact cells. Kinesin is even capable of moving microtubules by themselves, so that if this protein is bound to a glass slide, it causes added microtubules to glide like snakes over the surface of the glass. We will describe in Chapter 19 the interesting arrays of microtubules that can be produced using kinesin alone.

The direction of movement produced by kinesin was identified by adding it together with polystyrene beads to arrays of microtubules polymerized onto centrosomes. Whereas crude extracts of axonal cytoplasm generate transport in both directions, purified kinesin produces only movements away from the center of the array—that is, toward the plus ends of the microtubules. Since microtubules in nerve axons have their plus ends farthest from the cell body, movement in this direction corresponds to fast transport away from the cell body.

## *Cells contain a superfamily of kinesinlike proteins*

The kinesin obtained from squid, cow, or mouse brain and now termed *conventional kinesin*, consists of two heavy chains each of 120 kDa, and two 64 kDa light chains. The heavy chains form three structurally and functionally distinct regions: a pair of globular heads at the N-terminus, connected to a long stalk of predominantly α-helical coiled-coil configuration and a fan-shaped tail at the C-terminus (Figure 12-7a). As for myosin, the globular heads can be detached by proteolytic action and retain the ability to cleave ATP and move. The overall length of the molecule is about 80 nm and its shape is reminiscent of the larger myosin molecule although the two proteins have very little similarity in sequence. Another important difference is that the light chains of kinesin are associated with the tail rather than with the head as in myosin.

Conventional kinesins carry small membrane organelles or protein complexes toward the plus end of microtubules in nerve cells and other cells (Figure 12-7b). Several different subtypes have been found, and it is thought that the light chains play an important part in selecting the type of membrane cargo carried, although the full story has yet to be told. The light chains may also be important for the regulation of kinesin's ATPase activity. In the absence of microtubules, the kinesin molecule appears to fold up so as to bring the heavy and light chains together, and this may suppress the ATPase activity.[2]

When the amino acid sequence of conventional kinesin was identified, it was realized that a number of products of cloned genes that were essential for mitosis also possessed a kinesinlike head or 'motor' domains. This discovery led to a systematic search for proteins with related sequences, eventually revealing a large superfamily of kinesins all with

[2] It may be recalled from Chapter 7 that some conventional myosins also show 'foldback inhibition.' The advantage in either case is presumably that the motor proteins do not consume ATP needlessly when they are not attached to their cognate protein filament.

**Table 12-1 Selected kinesins and their function**

| Class | Heavy chain | Motor domain | Structure | Function |
|---|---|---|---|---|
| conventional kinesin (brain) | homodimer | N | | anterograde transport in axons |
| N-monomeric | monomer | N | | carries mitochondria and synaptic vesicles |
| internal | homodimer | I | | destabilizes microtubules, axonal transport, mitosis |
| N-heteromeric | heterodimer | N | | intraflagellar transport, axonal transport, pigment granule movement |
| N-bipolar | homotetramer | N | | organelle transport, mitosis |
| C-mitotic | homodimer | C | | organelle transport, mitosis |

Selection of kinesins known to be important in organelle transport. Many have other functions, for example during mitosis or meiosis. (Nomenclature is from Kreis and Vale, 1999.) The location of the motor domain in the kinesin sequence is given as N = N-terminus; C = C-terminus; I = internal location.

the same highly conserved ATP-binding region (catalytic core) in their motor domain (Table 12-1). The second part of the motor domain, termed the neck region, was found to be conserved only within certain kinesin classes.

Other anatomical features of the kinesin molecules are highly variable. Thus, the motor domain might be at the N-terminal end of the molecule, as in conventional kinesin, at the C-terminal end, as in the Ncd protein from *Drosophila*, or in the middle of the polypeptide chain, as in the mouse protein KIF2. Most kinesins possess two identical motor domains, but the list includes heterodimers, tetramers, and even monomeric species. Some, but not all, kinesins possess a long α-helical coiled-coil domain termed the stalk. Some but not all have an additional globular domain, or tail, at the end of the stalk.

All kinesin family members appear to bind to microtubules and move along them, but their detailed activities are varied, and many are not completely understood. Some kinesins have specific functions in mitosis or axonal transport, others play a crucial role in the assembly of cilia. One interesting subtype of kinesin, termed Kin I, which has a centrally located catalytic domain, acts as a microtubule-destabilizing agent both in mitotic cells and in growing nerve cells. It moves to the end of a microtubule and there acts as an enzyme to catalytically destabilize the terminal tubulin dimers. ATP binding then dissociates the kinesin–tubulin complex, allowing Kin I to recycle.

Other kinesins, such as Ncd from *Drosophila* and KIF2 from mouse, move in the opposite direction to conventional kinesins—toward the minus end of microtubules. How this change in polarity is achieved is something of a puzzle, since the motor domains, which provide both ATP and microtubule binding, are very similar in both amino acid sequence and three-dimensional structure. Experiments in which selected regions from different kinesins are spliced together indicate that the Ncd motor domain does not by itself confer minus-end directionality, but that splicing the neck and stalk of Ncd onto a conventional kinesin motor

**Figure 12-8** Duty ratio. The head of a motor protein makes a working stroke of duration $t_{ON}$ while it is attached to its filament. It makes a recovery stroke of duration $t_{OFF}$ while detached. The duty ratio is the fraction of time the motor is bound and developing force, that is
$t_{ON} / (t_{ON} + t_{OFF})$.

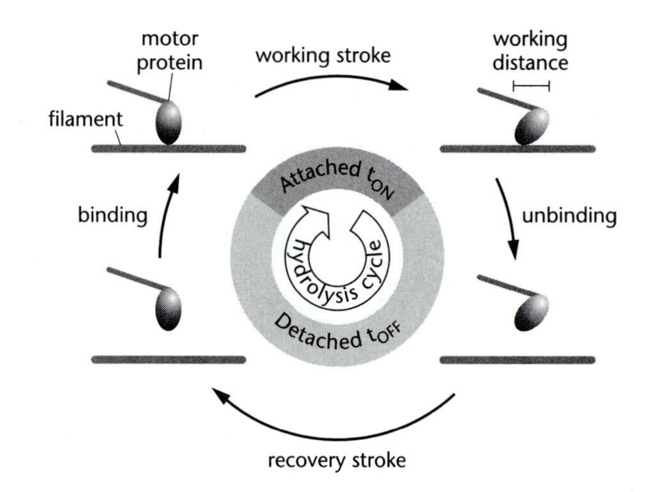


domain does reverse its direction of movement. The 'reversed' kinesin motor appears to have evolved by the appearance of new mechanical elements in the protein structure that alter the coupling between the ATP binding pocket and the rest of the head.

## *Kinesin is a processive motor*

Kinesins, like myosins, use the energy of ATP hydrolysis to power their movement along a protein polymer. We might anticipate, therefore, that the two molecular motors would operate in a broadly similar fashion, despite being physically different in size (kinesin is about half the size of myosin II) and having almost no homology in amino acid sequence. In support of this idea, conventional kinesin is strikingly similar in molecular shape to muscle myosin, with two globular heads and a fibrous tail. In both myosin and kinesin, the globular heads contain the region that hydrolyzes ATP and generate force, whereas the tail is attached to the cargo to be moved.

Detailed measurements using motility assays, however, reveal interesting differences between the motion of the two kinds of motors. These differences can be understood in terms of the *duty ratio*, which is the fraction of time that a motor spends attached to its filament (Figure 12-8). Differences in the duty ratio have important consequences for the function of the motors in the cell.

In the cross-bridge cycle of muscle myosin (as we saw in Chapter 9) the head makes a power stroke while it is attached to the actin filament and a recovery stroke while it is detached. By recovering its initial conformation while detached, the motor avoids stepping backward and so progresses a distance equal to the power stroke with each cycle. However, there is a discrepancy between the length of the power stroke, which is about 10 nm, and the distance between attachment sites on an actin filament, which occur every 36 nm on the helical filament. Because of this, each individual myosin head (the two heads work independently) can only spend at most about 10/36 of the time attached—that is, its duty ratio is about 0.35 or less. In a myofibril with thousands of myosin heads along the thick filament, whenever any individual head loses contact with the actin filament it will be carried along by the action of the others. But, if myosin molecules are extracted from muscle and scattered at low density over a cover slip, then their low duty ratio means that they cannot keep hold of a filament. Filaments attach, move, detach, and diffuse away.

Conventional kinesin, by contrast, moves in linear fashion along a protofilament of the microtubule. Tubulin dimers are spaced at intervals of 8 nm (see Figure 11-5), and the distance between one attachment point

**Figure 12-9** Dynein-driven movements. (a) In the cytoplasm, conventional dynein molecules carry organelles such as mitochondria or vesicles toward the minus end of microtubules. (b) If the dynein tail is anchored to a large permanent structure such as the actin cortex or the surface of a coverslip, then the cyclical action of the dynein head will drive the microtubule with its plus end leading. As one might expect from its function, cytoplasmic dynein, like kinesin, is a processive motor that can move for long distances along a microtubule without detaching.

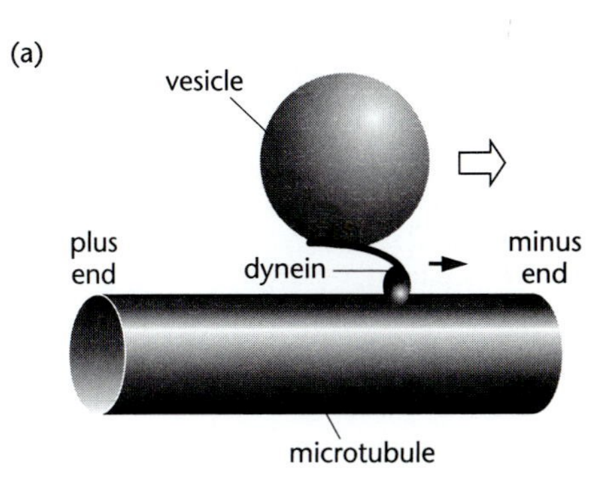


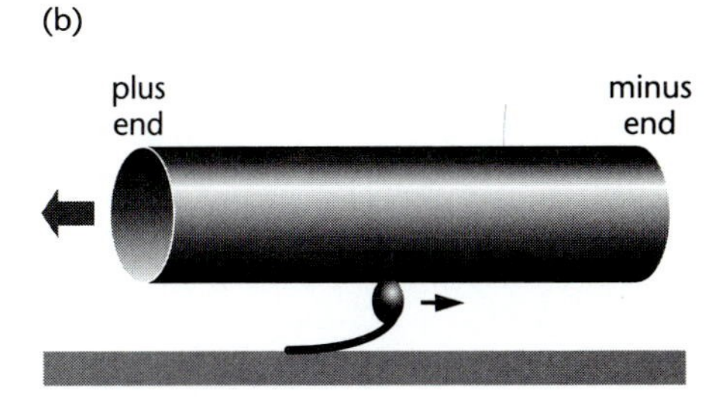


and the next, for any given head, will be 16 nm. The working stroke for each head is about 8 nm, giving a duty ratio close to 0.5, but since the two kinesin heads work in coordinated fashion, the duty ratio for the whole molecule is close to 1.0. In other words, the molecules can progress continuously along the microtubule for long distances without detaching.[3] For this reason, kinesin is often referred to as a 'processive' motor, by analogy with processive enzymes such as DNA polymerase that perform many repeated catalytic cycles without detaching from their particular molecules. One further consequence of the difference in duty ratio is that, because myosin can 'skip' over long distances with each cycle, it typically moves at a much faster rate than kinesin, at least when its load is light.

## *Cytoplasmic dynein carries organelles toward the minus end of microtubules*

Most kinesins, including the most abundant form present in nerve cells, move toward the plus ends of microtubules. Since nerve axons have their microtubules arrayed with plus ends farthest from the cell body, conventional kinesin and most of its relatives will carry organelles toward the synaptic terminals. However, direct observation shows that movements occur in both directions along an axon, and in fact the largest organelles travel in the retrograde direction, back toward the cell body. Thus, it was widely anticipated that a second class of motor molecules must exist that operates in the minus-end direction. The required motor, known as *cytoplasmic dynein* was identified in 1987, as a high-molecular-weight, microtubule-associated protein. When added to a preparation of purified microtubules on a glass coverslip, cytoplasmic dynein causes them to glide in an ATP-dependent fashion as already described for kinesin. In contrast to kinesin, however, dynein causes microtubules to glide with their plus ends leading, indicating that the glass-attached dynein molecules advance along the microtubules toward their minus ends (Figure 12-9).

Cytoplasmic dynein belongs to a family of related proteins, the best-studied members of which are the ciliary dyneins responsible for microtubule sliding in eucaryotic cilia and flagella (Chapter 14). All dyneins are extremely large proteins. Brain dynein includes two heavy chains of around 530 kDa each, three intermediate chains (74 kDa) and four light chains (~55 kDa). When viewed by electron microscopy, cytoplasmic dynein is seen to be a two-headed particle closely similar in morphology to the dynein from *Chlamydomonas* flagella. As we will see in Chapter 14, where we discuss ciliary dynein, each head of the dynein molecule is an ATPase that binds to the wall of a microtubule. In the absence of ATP, isolated dynein heads form a series of side-arms inclined toward the minus end of the microtubule, thereby indicating the polarity of the microtubule (rather like myosin heads on actin filaments).

Antibody injection experiments, motility assays, and subcellular fractionation all point to an essential role of cytoplasmic dynein in the mobilization and positioning of intracellular membranes. In nerve axons,

[3] Although the processivity of conventional kinesin is widely held to depend on its having two heads, this view is challenged by a recent observation that a single-headed kinesin can move more than a micrometer along a microtubule without detaching.

as already indicated, cytoplasmic dynein acts as a motor for retrograde transport, carrying membrane-associated structures back from the nerve terminals to the cell body. In intestinal epithelial cells, dynein has been shown to maintain the position of the Golgi apparatus, endosomes, and lysosomes close to the nucleus. In addition to membrane movements, cytoplasmic dynein moves microtubules, positions the nucleus, and changes the form of the mitotic spindle. As in the case of the kinesins, it is likely that different forms of organelle transport are mediated by specific subtypes of cytoplasmic dynein, several distinct forms of which have been identified.

## *Dynactin couples dynein to its cargo*

Although dynein drives organelle transport, it does not work unaided. Experimentally, salt-washed membrane vesicles mixed with purified cytoplasmic dynein reveal no movement whatever. Other components need to be added, especially a large protein complex called *dynactin* (Figure 12-10). Dynactin is built from 10 subunits, which include a short filament composed of the actin-related protein Arp-1, also called *centractin*. This short filament is believed to serve as an attachment site to membrane vesicles, perhaps through spectrin, which as we saw in Chapter 6 is often located on the cytosolic surface of membranes. The dynactin complex also contains components that bind passively to microtubules, perhaps maintaining an attachment while the dynein heads are detached.

Because it forms an essential link between the motor (dynein) and its cargo (a vesicle) dynactin is a natural target for treatments designed to interfere with transport. For example, antibodies to dynactin result in the slowing and stopping of vesicle movement along microtubules in squid axoplasm. They also cause dispersal of the Golgi complex, redistribution of early and late endosomes toward the cell periphery, and the suppression of the transport of intermediate components from the ER to the Golgi. A different method to obtain a very potent inhibition is the genetically controlled overexpression of one of the proteins in the dynactin complex, termed p50 or dynamitin. Presumably because of the position of p50 in the dynactin structure, large quantities of soluble p50 cause the complex to break up and the function of dynactin to be lost.

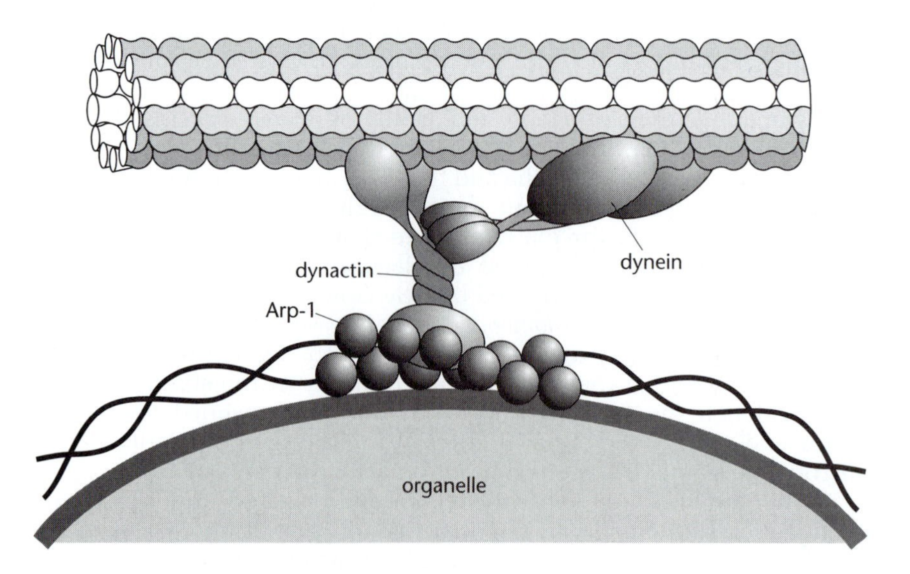


**Figure 12-10** The dynein/dynactin complex. Cytoplasmic dynein is linked to the membrane of a cargo organelle by dynactin.

It may be highly significant that dynactin contains an actin-related subunit. Indeed, there are numerous instances, some of them mentioned later in this book, in which this protein appears to be attached to actin filaments in a cell. With an anchorage on the cell cortex and motor domains that bind to microtubules, dynactin is potentially able to link the two major cytoskeletal systems of a eucaryotic cell.

## *Dynein can cause microtubules to slide*

By analogy with the family of myosins discussed in Chapter 7, it is likely that the dynein/dynactin combination moves cargoes other than membrane vesicles. In fact, evidence is accumulating that these proteins contribute to a wide range of movements, including (i) the orientation of the spindle and the migration of nuclei in budding yeasts; (ii) the migration of nuclei in filamentous fungi; (iii) the orientation of centrosomes in early embryos of the nematode worm; and (iv) the organization of astral microtubules in the mitotic spindle. In Chapter 14 we will relate that a member of the dynein family (the first discovered, in fact) is the primary source of movement in eucaryotic cilia and flagella. In this ciliary dynein, both head and tail regions are associated with microtubules. The head produces ATP-driven movement, whereas the tail is anchored to a neighboring microtubule—with the result that the two microtubules slide against each other.

Microtubule sliding is the basis of many cell movements and changes in shape. It occurs during the closing stages of mitosis and forms the mechanistic basis of ciliary and flagellar bending. Many protozoa change their body shape through the active contraction of huge bundles of microtubules in their cytoplasm (Chapter 15).

One of the best-documented examples in higher animals is the change in length of elongated fish photoreceptor cells. In the dark, cones elongate and rods contract; in the light cones contract and rods elongate. These movements, which compensate for the absence of a pupil, ensure that the appropriate receptor type is positioned first for optimal photodetection. Studies with isolated cells permeabilized by treatment with detergents indicate that the *contraction* of photoreceptors is an actin-based movement, probably mediated by myosin, whereas *extension* depends on both microtubule polymerization and microtubule sliding.

## *Motor proteins are responsible for neuronal polarity*

Returning to nerve cells, we have still to explain how axons and dendrites are produced and by what mechanisms these two types of process acquire different compositions (and hence functions). Although we are still far from a complete answer to these questions, it has become clearer over the past few years that microtubule motors have a powerful organizing influence.

As already mentioned, a mature nerve axon contains within its length many complete microtubules, all oriented with their plus ends distal. Where are these microtubules made? In most cells, microtubules are nucleated on rings of γ-tubulin associated with the centrosome, and the same could be true of nerve cells. Antibody staining reveals γ-tubulin only in the cell body close to the nucleus, and experiments in which microtubules are depolymerized and then allowed to regrow support the view that these form exclusively at the centrosome, even in a nerve cell. But if this is so, how are the microtubules released and how do they travel into the axon?

It is relevant to note here that animal cells contain a protein that is able to cut microtubules. Called 'katanin' after the long single-edged sword (*katana*) of the Japanese samurai, this pentameric protein associates with the side of a microtubule and uses the energy of ATP hydrolysis to

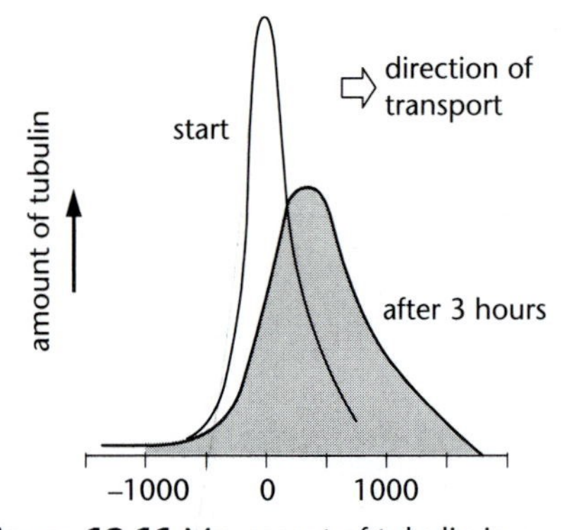


**Figure 12-11** Movement of tubulin in a large axon. Fluorescent tubulin injected into the giant axon of a squid undergoes both diffusive spread and active transport. The machinery driving the latter is presently not known. (Based on Galbraith et al, 1999.)

develop first a kink and then a break in the microtubule. Katanin is present in neurons, especially close to the centrosome and there is evidence that it functions to release microtubules from their nucleation site on the centrosome.

How then do microtubules move into the axon? This question has been the subject of active debate for many years and is still not fully resolved. One idea is that microtubules assembled in the nerve cell body move into the axon like logs in a sluice, driven by dynein. One could imagine, for example, dynactin complexes positioned on the actin cortex of the axon driving microtubules down the axon, a possibility supported by the observation that disruption of the dynactin complex blocks transport of microtubules in an axon. On the other hand, direct examination of microtubules in axons, even rapidly growing ones, fails to show forward movement of these structures. Another possibility is that microtubules could advance down the axon by adding subunits at their plus ends and either depolymerizing at the minus end or fragmenting. In this second scenario the tubulin molecules themselves would have to move in some unassembled (although possibly oligomeric) form in the axon and then add where necessary to the exposed ends of microtubules. Recent studies of the movement of labeled proteins in axons support this second mechanism (Figure 12-11).

Dendrites present yet another set of questions and possibilities. Recall that in these shorter processes the microtubules are arranged with dual polarity—some with their plus ends pointing outward and others pointing the other way. Although direct evidence for the assembly of these structures is lacking, it is intriguing to note that dendrites contain a species of tetrameric kinesin that has two pairs of heads, each able to walk along a different microtubule toward its minus end. A motor protein with this structure would be able to slide microtubules of opposite polarity against one another.

## *Vesicles can carry more than one type of motor*

We see that at any instant of time in any nerve axon, one population of organelles is traveling out from the cell body driven by kinesins while, at the same time, another population of organelles is making the return journey back to the cell body, driven by cytoplasmic dynein. But this instantaneous picture raises a number of important logistic questions. How do dynein molecules get to the nerve terminal in the first place, to begin their journey? What happens to kinesin molecules after they reach the nerve terminal and discharge their cargo? An additional complication is that some organelles, such as mitochondria, can and do move in *either* direction, moving for a period away from the cell body and then switching to move toward the cell body. How is the movement of such organelles controlled?

To begin with, there is evidence that some membrane organelles carry both kinesin and cytoplasmic dynein. Studies in which long axons have been tied off so the traffic accumulates on either side show that cytoplasmic dynein is ported to the cell periphery on membrane organelles. Furthermore, biochemical studies frequently show both kinds of motor present with a given membrane preparation.[4]

In fact, there are even indications that vesicles can carry not only kinesin and dynein but also myosin molecules as well. Vesicles in squid giant axons and mitochondria in vertebrate axons have both been shown to move even when microtubules are absent, traveling in this case along oriented actin filaments. In the melanophores of *Xenopus* discussed below, pigment-containing granules move on both microtubules and actin filaments, and have been shown to carry myosin V on their surface. In the cytoplasm, it seems, actin filaments and microtubules provide alternative sources of directional traffic, a topic we return to in Chapter 19.

[4] The characteristically hesitant, saltatory movement of vesicles mentioned earlier in this chapter could be partly due to the switching from one type of motor to another on the same organelle.

A partial answer to the questions posed at the start of this section is that the different types of motor molecule travel continually both up and down the axon on membrane organelles. Sometimes they move passively, as passengers on an organelle, and at other times they move actively by means of their ATP-driven movements. Moreover, different motor proteins can and do occupy the same organelle. Thus, dynein could be carried to the axon terminal by a kinesin, and kinesins could be carried back to the cell body by dynein. If we accept this view then the searchlight falls onto the switches and the manner in which one type of motor molecule is turned on while the others are turned off. Evidence is accumulating from organelle transport in other cell types, and especially pigment cells, that changes in protein phosphorylation are crucial.

## Pigment granule movement may be regulated by phosphorylation

The ability to change skin color rapidly is a characteristic of many lower vertebrates, particularly fish and amphibia. Thus cichlid fish, while usually appearing mottled in appearance, can also become almost pure black or pure white during the establishment of sexual dominance or for the purposes of camouflage. These same fish can also display red or yellow pigmentation or reflect light in an iridescent pattern. These changes are mediated by the transport of organelles called pigment granules within a specialized class of dermal cells, *chromatophores*. Their color is derived from the coordinated transport of pigment granules to and from the cell center. When the pigment is dispersed throughout the cytoplasm, the pigment absorbs light and the cell appears black (or colored). When the pigment granules are transported to the cell center, most of the cytoplasm is unpigmented, and the cell appears essentially colorless (Figure 12-12)

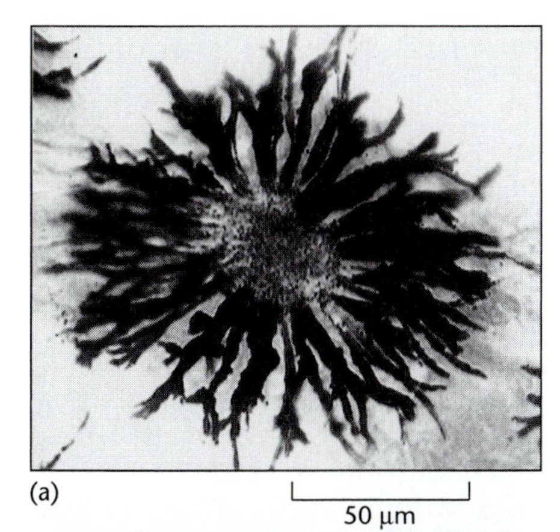


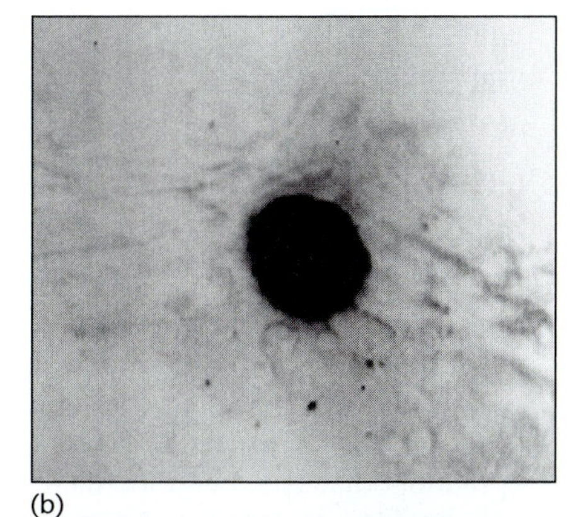

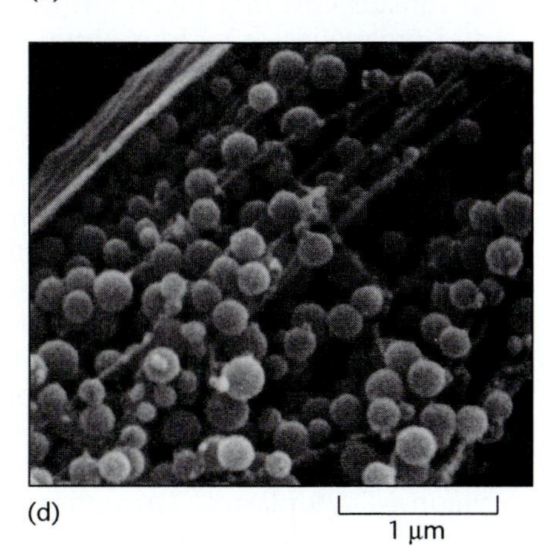


**Figure 12-12** Fish pigment cells. These pigment-containing cells are responsible for changes in skin coloration in several species of fish. The dark brown pigment is contained in large pigment granules, which can change their location in the cell in response to a neuronal or hormonal stimulus. (a and b) Bright-field images of the same cell in a scale of an African cichlid fish, showing its pigment granules either dispersed throughout the cytoplasm or aggregated in the center of the cell. (c) An immunofluorescent image of another cell from the same fish stained with antibodies to tubulin, showing huge bundles of parallel microtubules extending from the centrosome to the periphery of the cell. (d) Scanning electron micrograph of a pigment cell following a brief exposure to detergent. The plasma membrane and soluble contents of the cytoplasm have been removed, exposing the array of microtubules and associated pigment granules. (a, b, and c, courtesy of Leah Haimo; d, from M.A. McNiven and K.R. Porter, *J. Cell Biol.* 103: 1547–1555, 1986. © Rockefeller Univeristy Press.)

In the animal these changes are controlled by neuronal or hormonal stimuli but they can also be observed in chromatophores isolated in culture. Pigment cells removed from scales of an angel fish, squirrel fish, or black tetra, attach to the bottom of a culture dish and carry out spontaneous cycles of aggregation and dispersion during which the pigment granules move radially toward or away from the center of the cell. Even when they are in the dispersed state, pigment granules are not stationary but shuttle back and forth over distances of a few micrometers.

Movement of pigment granules takes place mainly along microtubules, although actin filaments are also involved. Radial arrays of microtubules in chromatophores parallel the paths taken by pigment granules and the disruption of microtubules with mitotic inhibitors abolishes the rapid transport of granules to the cell perimeter. Treatment with an actin inhibitor such as latrunculin, by contrast, allows transport to the periphery but prevents the slower dispersal to an even distribution throughout the cytoplasm.

In model systems obtained by selectively extracting the plasma membranes from cells with nonionic detergent, the movement of pigment-containing granules is dependent on the presence of ATP. Furthermore, the direction of movement can be controlled by manipulating the concentration of $Ca^{2+}$. Present evidence suggests that dispersion of the granules (movement to the plus end of microtubules) is driven by a kinesin, whereas aggregation appears to be driven in large part by dynein.

Pigment granule dispersion can be induced in live chromatophores by agents that elevate cyclic AMP, such as caffeine or dibutyryl cyclic AMP. A spontaneous rise in cyclic AMP also occurs naturally as pigment granules disperse. Cyclic AMP has its effect through the activation of a specific protein kinase, PKA, which phosphorylates either kinesin itself or a protein associated with kinesin. Pigment granule aggregation can be induced by agents that lower cyclic AMP levels and/or raise $Ca^{2+}$ levels. Calcium ions appear to operate through the activation of a $Ca^{2+}$-regulated protein phosphatase calcineurin. Because the kinase and the phosphatase are regulated by cyclic AMP and $Ca^{2+}$ respectively, cells can control the direction of transport by regulating the level of one or other of these second messengers.

## *Microtubules form the basis of feeding tentacles of protozoa*

Microtubule-based transport of organelles is seen in many cells other than nerve cells and many curious motile mechanisms remain to be uncovered. One of the most remarkable examples occurs in protozoa known as *Foraminifera*, which have cell bodies covered with a hard shell (test) through the apertures of which emerges a feathery growth of many slender pseudopodia. The pseudopodia unite and ramify into an extensive network of threads covering neighboring surfaces, within which particles of food and organelles of various kinds are carried in streams toward and away from the cell body.

The pseudopodia are remarkably similar in form and ultrastructure to small axons and the movements have many of the features described above for fast axonal transport. A membrane-permeabilized preparation has been prepared from a closely related organism, *Reticulomyxa*, in which organelles move along microtubules. As in the axon, bidirectional transport occurs on a single microtubule, indicating that the polarity of the microtubule is not the sole determinant of directionality.

In another type of feeding tentacle, microtubules may perform a structural rather than a motile role. The stiff, needlelike pseudopodia of heliozoa contain huge bundles of parallel microtubules. In addition to their rapid collapse in contact with prey, axopodia also show a form of vigorous cytoplasmic streaming that carries trains of food particles and large organelles to and from the cell body. The moving particles are sandwiched between the bundle of microtubules in the axopodial core and the plasma membrane, often distorting the latter like a small animal swallowed by a boa constrictor.

## Ribosomes, viruses, and nuclei also move along microtubules

One of the most remarkable examples of microtubule-based transport occurs in the ovaries of some insect species where the oocytes are connected to nutritive cells by way of channels known as *nutritive tubes*. In the water boatman *Notonecta*, for example, the nutritive tubes are approximately 20 μm in diameter and may be up to 5 mm in length. These tubes act as conduits for ribosomes and other cytoplasmic constituents, which pass from the nurse cells and accumulate in the oocytes. The tubes are packed with perhaps 30,000 longitudinally oriented microtubules each of which is surrounded by a clear zone about 50 nm in diameter. The microtubules have their minus ends in the oocyte, so that the organelle movement corresponds to dynein-based retrograde movement in axons, although the actual motors responsible have not yet been identified. As in axons, movements occur at both slow (ribosomes) and fast (mitochondria and mRNA) rates.

Viruses may also utilize microtubules in order to spread through cells and tissues. For example, retroviruses, the envelope of which derives more or less directly from host cell membranes, are often associated with spindle microtubules. Adenoviruses and herpes virus, enclosed in protein capsids, appear to migrate towards the nucleus along microtubules, and their infectious cycle can be delayed by treatment with the tubulin-binding drug vinblastine.

There are even cases in which the nucleus, the largest 'organelle' of all, undergoes directed movement. This occurs in filamentous fungi, where nuclei can travel often long distances toward the advancing hyphal tip. Yeast cells also show nuclear migration during mating, when nuclei of two cells move together and fuse. In Chapter 20 we will see that the nuclei in a *Drosophila* blastoderm undergo a phase of migration to a cortical position, and mention a possible role for nuclear migration in the development of the mammalian brain.

In these and countless other situations, organelles move within cells over predetermined routes by mechanisms that are still largely unexplored. Each organelle, whether a vesicle, a filament, or a nucleus, presents its own challenges to the motile machinery. To judge from the few systems that have been explored in detail, each form of organelle transport will have led to the evolution of a novel set of fascinating molecular devices.

## Outstanding Questions

*How are microtubules initiated in cells such as neurons? What mechanism allows a nerve cell to make just one axon and multiple dendrites? How are tau and MAP2 routed into the appropriate type of extension from the cell body? How do the heads of conventional kinesin alternate during movement along a microtubule? How is a single-headed kinesin able to move more than a micrometer along a microtubule without detaching? How is the movement of organelles along microtubules controlled and how are motor proteins moved to their starting point? What happens to kinesin molecules when they reach the nerve terminal? What are the origins of the saltations seen in organelles moving in a living cell? What is the function of the different types of tail domain of microtubule motors? What are their cargoes and how are they recognized? What are the consequences for human health of disorders in motor function? Are there attachment sites for microtubules on the plasma membrane?*

Nutritive tubes

## *Essential Concepts*

- Mitochondria, membrane vesicles, protein complexes, RNA molecules, and other organelles move continually from one location to another in the eucaryotic cytoplasm along tracks provided by microtubules.
- By means of microtubule-based transport, cells establish and maintain regions of different function and are able to become large and asymmetric in shape.
- Nerve axons have a particularly acute dependence on organelle transport. Most protein synthesis occurs in the cell body and the products have to be transported over long distances (often many millimeters) to the axon terminals.
- In nerve axons, microtubules are longitudinally aligned with their plus ends pointing away from the cell body. Microtubule-associated proteins such as tau bind along the length of microtubules, promoting their stability and increasing their mechanical stiffness.
- Microtubules in dendrites are also arranged in parallel bundles, but in this case both polarities are present. Dendrites also have a distinct set of microtubule-associated proteins, including MAP2, and carry different organelles.
- Movement away from the nerve cell body is driven by a family of kinesins. Most of these have two heads and a tail and move in a similar fashion to conventional myosin, although with special features that adapt them to organelle transport.
- A large superfamily of proteins related to kinesins has been identified by recombinant DNA techniques. These have similar motor domains but differ markedly in their molecular anatomies and in their motile function in the cell.
- Movement toward the cell body is driven by a microtubule motor called cytoplasmic dynein. This is a large protein related to the dynein found in eucaryotic cilia and flagella.
- Attachment of cytoplasmic dynein to membrane cargoes occurs through a protein complex called dynactin. The dynein/dynactin complex drives many movements inside cells, including the movement of microtubules themselves.
- Some membrane vesicles carry both kinesin and dynein, and perhaps also myosin molecules on their surface. Control of the direction of the organelles is then exerted by switching motors on and off at the appropriate time.
- An enormous variety of cell movements, such as color changes in fish scales, diurnal changes in shape of retinal cells, tentacle feeding in protozoa, and transport of ribosomes, nuclei, cytoskeletal filaments, and virus particles, depend on organelle transport. The molecular basis for most of these movements remains to be discovered.

## Further Reading


Ahmad, F.J., et al. Cytoplasmic dynein and dynactin are required for the transport of microtubules into the axon. *J. Cell Biol.* 140: 391–401, 1998.

Allan, V.J., Schroer, T.A. Membrane motors. *Curr. Opin. Cell Biol.* 11: 476–482, 1999.

Block, S. Nanometres and piconewtons: the macromolecules and mechanics of kinesin. *Trends Cell Biol.* 5: 169–175, 1995.

Burkhardt, J.K., et al. Overexpression of the dynamitin (p50) subunit of the dynactin complex disrupts dynein-dependent maintenance of membrane organelle distribution. *J. Cell Biol.* 139: 469–484, 1997.

Chang, S., et al. Speckle microscopic evaluation of microtubule transport in growing nerve processes. *Nat. Cell Biol.* 1: 399–403, 1999.

Craig, A.M., et al. Neuronal polarity. *Curr. Biol.* 2: 602–606, 1992.

Delacourte, A., Buée, L. Normal and pathological tau proteins as factors for microtubule assembly. *Int. Rev. Cytol.* 171: 167–224, 1997.

Desai, A., Mitchison, T.J. A new role for motor proteins as couplers to depolymerizing microtubules. *J. Cell Biol.* 128: 1–4, 1995.

Endow, S.A., Waligora, K.W. Determinants of kinesin motor polarity. *Science* 281: 1200–1202, 1998.

Galbraith, J.A., et al. Slow transport of unpolymerized tubulin and polymerized neurofilament in the squid giant axon. *Proc. Natl. Acad. Sci. USA* 96: 11589–11594, 1999.

Gee, M., Vallee, R. The role of the dynein stalk in cytoplasmic and flagellar motility. *Eur. Biophys. J.* 27: 466–473, 1998.

Gilbert, S.P., et al. Pathway of processive ATP hydrolysis by kinesin. *Nature* 373: 671–676, 1995.

Goldstein, L.S.B., Philp, A.V. The road less traveled: emerging principles of kinesin motor utilization. *Annu. Rev. Cell Dev. Biol.* 15: 141–183, 1999.

Haimo, L.T., Thaler, C.D. Regulation of organelle transport: lessons from color change in fish. *BioEssays* 16: 727–733, 1994.

Henningsen, U., Schliwa, M. Reversal in the direction of a molecular motor. *Nature* 389: 93–95, 1997.

Hirokawa, N. Kinesin and dynein superfamily proteins and the mechanism of organelle transport. *Science* 279: 519–526, 1998.

Hirokawa, N., et al. Brain dynein (MAP1C) localizes on both anterogradely and retrogradely transported membranous organelles *in vivo*. *J. Cell Biol.* 111: 1027–1037, 1990.

Holleran, E.A., et al. The role of the dynactin complex in intracellular motility. *Int. Rev. Cytol.* 182: 69–109, 1998.

Howard, J. Molecular motors: structural adaptations to cellular functions. *Nature* 389: 561–567, 1997.

Kashina, A.S., et al. A bipolar kinesin. *Nature* 379: 270–272, 1996.

Kikkawa, M., et al. Three-dimensional structure of the kinesin head-microtubule complex. *Nature* 376: 274–277, 1996.

Kreis, T., Vale, R. Guidebook to the Cytoskeletal and Motor Proteins. Oxford, UK: Oxford University Press, 1999.

Kumar, J., et al. Kinectin, an essential anchor for kinesin-driven vesicle motility. *Science* 267: 1834–1837, 1995.

Lee, K.D., Hollenbeck, P.J. Phosphorylation of kinesin *in vivo* correlates with organelle association and neurite outgrowth. *J. Biol. Chem.* 279: 5600–5605, 1995.

Leibler, S., Huse, D.A. Porters versus rowers: a unified stochastic model of model proteins. *J. Cell Biol.* 121: 1357–1368, 1993.

Matus, A. Stiff microtubules and neuronal morphology. *Trends Neurosci.* 17: 19–23, 1994.

McNiven, M.A., Ward, J.B. Calcium regulation of pigment transport *in vitro*. *J. Cell Biol.* 106: 111–125, 1988.

Meyhöfer, E., Howard, J. The force generated by a single kinesin molecule against an elastic load. *Proc. Natl. Acad. Sci. USA* 92: 574–578, 1995.

Michele, D.E., et al. Thin filament dynamics in fully differentiated adult cardiac myocytes: toward a model of sarcomere maintenance. *J. Cell Biol.* 145: 1483–1495, 1999.

Moore, J.D., Endow, S.A. Kinesin proteins: a phylum of motors for microtubule-based motility. *BioEssays* 18: 207–219, 1996.

Morris, N.R. Nuclear migration from fungi to mammalian brain. *J. Cell Biol.* 148: 1097–1102, 2000.

Overly, C.O., et al. Organelle motility and metabolism in axons vs dendrites of cultured hippocampal neurons. *J. Cell Sci.* 109: 971–980, 1996.

Panda, D. et al. Rapid treadmilling of brain microtubules free of microtubule-associated proteins *in vitro* and its suppression by tau. *Proc. Natl. Acad. Sci. USA* 96: 12459–12464, 1999.

Quintyne, N.J., et al. Dynactin is required for microtubule anchoring at centrosomes. *J. Cell Biol.* 147: 321–334, 1999.

Rodionov, V.I., et al. Functional coordination of microtubule-based and actin-based motility in melanophores. *Curr. Biol.* 8: 165–168, 1998.

Rogers, S.L., Gelfand, V.I. Myosin cooperates with microtubule motors during organelle transport in melanophores. *Curr. Biol.* 8: 162–264, 1998.

Stephen, S., et al. Poly(A) mRNA is attached to insect ovarian microtubules *in vivo* in a nucleotide-sensitive manner. *Cell Motil. Cytoskeleton* 43: 159–166, 1999.

Suzaki, T., et al. Structure and function of the cytoskeleton in heliozoa. *Eur. J. Protistol.* 30: 404–413, 1994.

Thormählen, M., et al. Interaction of monomeric and dimeric kinesin with microtubules. *J. Mol. Biol.* 275: 795–809, 1998.

Vale, R.D., Milligan, R.A. The way things move: looking under the hood of molecular motor proteins. *Science* 288: 88–95, 2000.

Vale, R.D., et al. Identification of a novel force generating protein (kinesin) involved in microtubule-based motility. *Cell* 41: 39–50, 1985.

Vallee, R.B., Sheetz, M.P. Targeting of motor proteins. *Science* 271: 1539–1544, 1996.

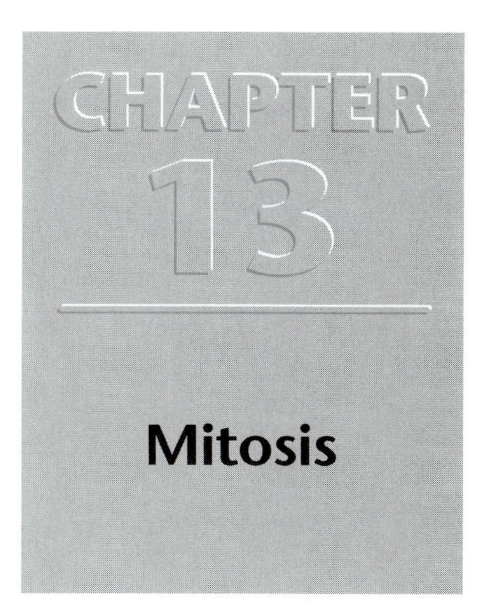


# Mitosis

The dynamic features of microtubule assembly discussed in previous chapters form the basis of the most intricate set of cytoplasmic movements known. During mitosis in plant and animal cells, populations of microtubules grow, attach to chromosomes, and maneuver them with amazing precision into two newly created nuclei. In a dividing human cell, for example, 46 newly replicated chromosomes are handled separately. Each of the 46 will split into two daughter chromosomes, each with an equal chance to go into one or other daughter cell, as though their fate is determined by the spin of a coin. And yet this capricious coin-tossing is controlled with such accuracy that the new nuclei receive one complete set of chromosomes and no more than a complete set. Analysis of yeast cells shows that the segregation of chromosomes has a normal error rate of less than one in every 100,000 divisions.

Most of this chapter is concerned with nuclear division (mitosis) in a typical vertebrate cell. As we will see, this engages the full repertoire of microtubule machinery, including nucleation, polymerization, depolymerization, fragmentation, and movements driven by molecular motors. In the final part of the chapter we briefly discuss the corresponding events that occur in plant cells and algal cells, and in meiosis, the specialized type of cell division by which the sex cells of most eucaryotes are produced. To set the scene for our discussion, however, we begin with the simplest division of all—that of cells that do not possess a nucleus.

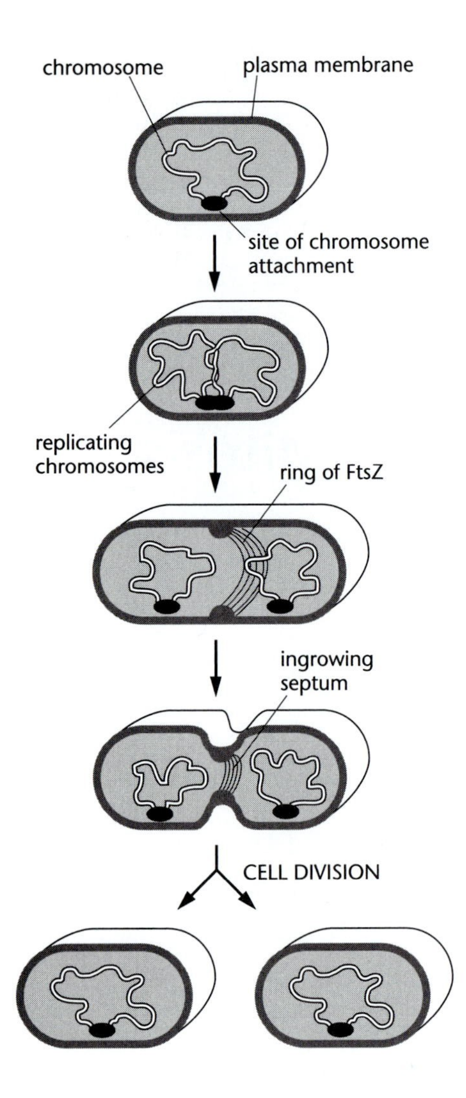


**Figure 13-1** Cell division in bacteria. The bacterial chromosome containing a single circular DNA molecule is attached to the plasma membrane and remains attached both during and after replication. The two DNA molecules become separated by cell growth, and division is initiated by formation of a ring of FtsZ molecules around the waist of the cell. A complex of proteins associated with FtsZ then catalyzes the formation of the cell wall and plasma membrane, resulting in the inward growth of a septum and eventual division of the cell.

## *Bacteria use FtsZ, a distant relative of tubulin, to divide*

Cell division is at its simplest and most rapid in bacteria, which do not have a nucleus and contain a single chromosome. In *Escherichia coli*, for example, the whole cell cycle can take as little as 20 minutes in favorable growth conditions. The single circular chromosome of this organism, containing a single DNA molecule, is anchored to the plasma membrane and remains attached as it replicates, the two chromosomes separating as the cell grows. When the cell has approximately doubled in size, cell division occurs by simple fission as a new cell wall and plasma membrane are laid down between the two chromosomes.

Bacteria do not possess a cytoskeleton in the usual sense and there is no evidence for an organized structure resembling the mitotic spindle of eucaryotic cells. But proteins have been identified that are necessary for septation (that is, division by the formation of a wall) and one of these appears to be related to tubulin. Septation proteins are recognized because their inactivation results in nondividing filamentous bacteria. Known as "Fts" proteins (for *f*ilamenting *t*emperature *s*ensitive), these are the products of a cluster of cell division genes physically close to those involved in cell wall synthesis and designated by letters such as *FtsL*, *FtsM*, and so on. *FtsZ* originally attracted attention because it had a limited sequence similarity to tubulin. It was subsequently found to be extremely similar in three-dimensional structure to tubulin, to bind and hydrolyze GTP, and to produce linear polymers (although probably not tubular). Between cell divisions, *FtsZ* is apparently distributed throughout the cell, perhaps in an unpolymerized form. As division begins FtsZ collects in a ring around the waist of the cell, possibly serving to nucleate a complex of other cell division proteins, including FtsA (which may be a distant relative of actin). In some fashion not yet understood, these protein cooperate to organize extracellular peptidoglycan and additional membrane, resulting in the inward growth and eventual division of the bacterial cytoplasm (Figure 13-1).

How FtsZ operates, whether it contracts, and whether it is associated with motorlike proteins are among the fascinating questions awaiting future research.

## *Eucaryotic cells go through a fixed sequence of events as they divide*

Division in eucaryotic cells is significantly more complicated than in bacteria. The cells are 10–20 times larger in linear dimension and contain many organelles and other cytoplasmic structures, all of which have to be duplicated and then segregated into one of the two daughter cells. The DNA of eucaryotic cells, in particular, is enormously long and requires an intricate intracellular apparatus for its maintenance and replication. For instance, all 46 chromosomes of a human tissue cell have to replicate individually and then physically move into one or another newly formed daughter cells. These elaborate rearrangements require many coordinated movements and major structural changes in the cytoskeleton.

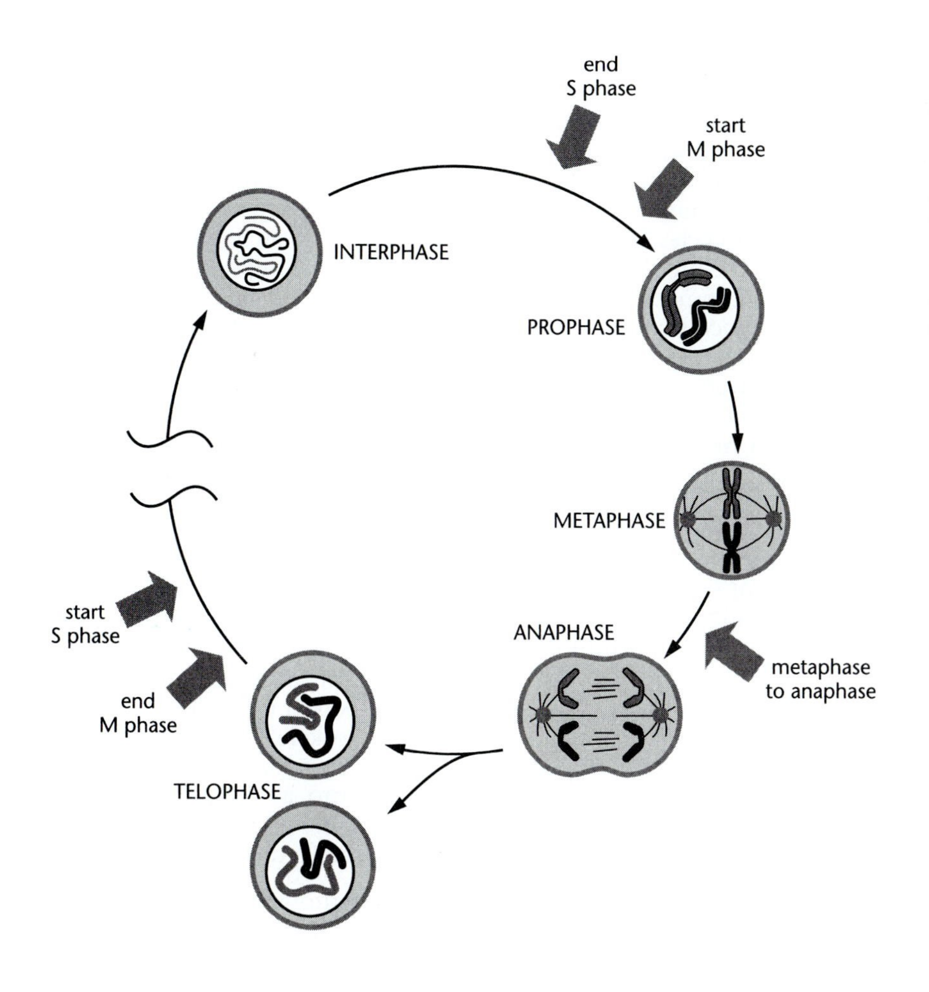


**Figure 13-2** Schematic view of the eucaryotic cell cycle showing principal points of control. Biochemical switches involving cyclin-dependent kinases and proteases operate at the beginning and end of S phase (synthetic phase) and at the beginning and end of M phase (mitosis). In this way events are driven unidirectionally around the cycle. Note that the durations of the different stages are not shown to scale. In particular the M phase usually occupies only a very small fraction of the total.

As it grows and divides, a eucaryotic cell passes through a number of stages characterized by distinct biochemical events affecting both DNA and the cytoskeleton (Figure 13-2). The sequence of stages is referred to as the *cell cycle*, since it repeats over and over for as long as cell division continues. Nuclear division, or *mitosis*, is actually a relatively short episode in the cell cycle. It is preceded by a prolonged *interphase*, during which the cell duplicates its contents, both nuclear and cytoplasmic, and closely followed by *cytokinesis*—the process by which the two newly produced cells are physically separated (Chapter 7). Mitosis and cytokinesis are closely synchronized both in time and space and together constitute the *M-phase* of the cell cycle.

## Phosphorylation and proteolysis control the eucaryotic cell cycle

Progression from one cell cycle stage to the next is controlled, as if by clockwork, by a cascade of biochemical reactions. These occur in a single direction and can if necessary be halted at critical steps, sometimes called *checkpoints*. One point of no return, as we will see, is when the nuclear membrane breaks down and the cell becomes committed to mitosis. A second irreversible step is the separation of chromosomes, which marks the exit from mitosis. This ability to arrest the cycle is crucial for the cell, since it allows division to respond to external conditions, such as the presence of growth factors. It is also necessary to synchronize nuclear events with cytoplasmic events—ensuring, for example, that mitosis does not begin before all of the DNA has been replicated.

Summarizing a great deal of complicated biochemistry, it can be said that progression through the cell cycle is driven by the episodic synthesis

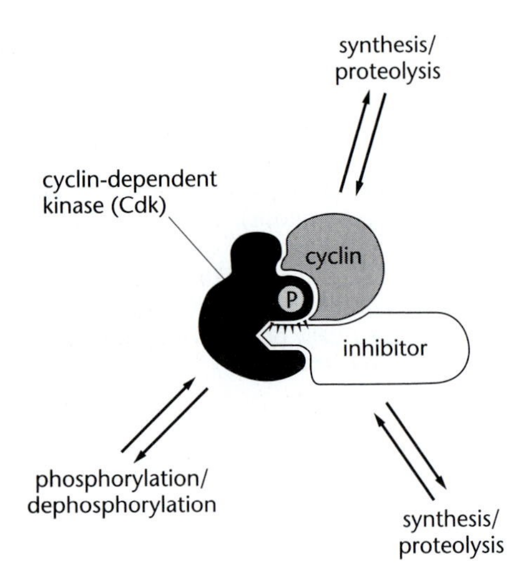


**Figure 13-3** Control reactions of the cell cycle. The cell cycle is regulated by a series of protein kinases called Cdk's (cyclin-dependent kinases) which are themselves part of a network of biochemical signals. The latter include (i) association of the Cdk with a specific cyclin; (ii) addition or removal of phosphate groups from the Cdk molecule; (iii) association with specific inhibitor molecules. Both cyclins and Cdk inhibitors can be rapidly degraded by targeted proteolysis.

and precipitous breakdown of small proteins called *cyclins*. These operate by activating specific protein kinases, called cyclin-dependent kinases, or *Cdk's,* which phosphorylate many different target proteins in the cytoskeleton and elsewhere, and in this way control events such DNA synthesis and entry into mitosis (Figure 13-3). The 'precipitous breakdown' of cyclin just mentioned is due to targeted proteolysis at crucial stages of the cycle. Enzymes covalently attach a small protein called ubiquitin to the protein to be degraded, thereby tagging it for destruction by a protein complex called a proteasome. The irreversible nature of this destructive process confers a unidirectional character to the cell cycle.

Eucaryotic cells have multiple cyclins and Cdk's, which rise and fall in concentration (and activity) at different times in the cell cycle. As well as triggering events outside the cycle, cyclin-Cdks also phosphorylate components of the cycle machinery itself, thereby forming a complex network of control reactions. For example, one cyclin-Cdk phosphorylates itself in autocatalytic fashion to generate an almost explosive increase in activity at the beginning of M-phase. This same cyclin-Cdk later in the cycle activates a specific proteolytic cascade that leads to the destruction of its own cyclin, thereby terminating the action of its Cdk. Another cyclin-Cdk is actually activated by proteolysis, through the targeted degradation of an inhibitory protein. This network of reactions is a dramatic illustration of the close interplay between cell signals and the cytoskeleton. Cyclins, kinases, and proteases, bound up in intricate reaction pathways, interact with proteins of the cytoskeleton at specific times of the cycle to direct the physical events of cell division.

Mitosis itself is traditionally subdivided into a sequence of stages according to the behavior of the chromosomes, the principal stages being *prophase, metaphase, anaphase,* and *telophase* (see Figure 13-2). These will be discussed in turn, together with the cytoskeletal mechanisms that drive them.

## Breakdown of the nuclear membrane commits the cell to mitosis

One of the earliest indications that a cell is about to enter prophase is that its chromosomes condense in the nucleus. Condensation involves short-range movements of DNA and chromatin, the mechanism of which is presently unknown. Shortly after this the nuclear envelope disassembles and breaks up into small vesicles, possibly triggered by a transient rise in $Ca^{2+}$ (Figure 13-4). Intermediate filament proteins of the nuclear lamina become hyperphosphorylated by a cyclin-Cdk complex and this triggers disassembly of the nuclear envelope (further discussed in Chapter 17).

The nucleus contains a high level of the small GTP-binding protein *Ran1* which is released into the cytoplasm when the nuclear envelope breaks down. Ran 1 has been shown to be essential for spindle formation, probably by triggering the phosphorylation of microtubule-related proteins. The latter may include MAPs whose binding to microtubules is regulated by protein kinases; stathmin, which regulates polymerization (Chapter 11); and katanin, the protein mentioned in Chapter 12 that is able to sever microtubules. Both stathmin and katanin become active at this early stage of mitosis and, together with other proteins, they cause microtubules to dramatically shorten and become highly dynamic. Microtubules in a mitotic cell have turnover rates typically less than one minute, in contrast to 5 minutes to hours during interphase.

## Microtubules assemble into a spindle

With the break-up of the nuclear membrane, the array of microtubules in the cytoplasm changes. In interphase, as we saw in Chapter 11, most cells contain a cytoplasmic array of microtubules radiating out from the single

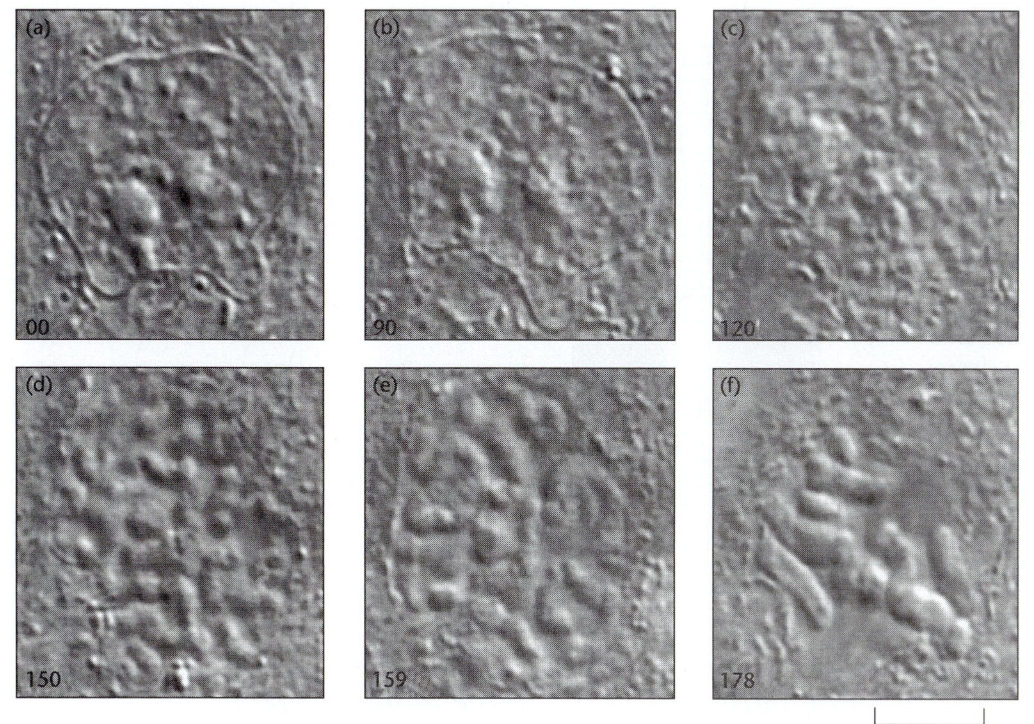


**Figure 13-4** Early stages of mitosis. Light microscope pictures of a mammalian cell in tissue culture as it proceeds from $G_2$, through prophase, to metaphase. In $G_2$ (a) the nucleus contains 1–3 nucleoli and numerous small chromatin granules. The latter appear to grow in size (b)–(d), eventually fusing into recognizable chromosomes. Breakdown of the nuclear envelope (e) occurs about 70 minutes after the first signs of chromosomal condensation, and in (f) a normal metaphase spindle is present. Elapsed time in minutes shown in lower left-hand corner. (From C.L Rieder and R.W. Cole, *J. Cell Biol.* 142: 1013–1022, 1998. © Rockefeller University Press.)

centrosome, together with a variable number of unattached microtubules. Toward the end of S-phase, and before nuclear membrane disintegration, the cell duplicates its centrosome to produce two daughter centrosomes, which initially remain together at one side of the nucleus. As prophase begins, the daughter centrosomes separate and move to opposite poles of the cell driven by centrosome-associated motor proteins that move along microtubules. Each centrosome serves to organize its own array of microtubules and the two sets of microtubules then interact to produce the mitotic spindle.

The rapidly growing and shrinking microtubules extend in all directions from the two centrosomes, exploring the interior of the cell. Some of these microtubules become stabilized against disassembly to form the highly organized mitotic spindle. Microtubules growing from opposite centrosomes bind to each other by means of linking proteins to produce the typical prolate ellipsoid shape from which the spindle derives its name (Figure 13-5). These interacting microtubules are called *polar microtubules* since they originate from the two poles of the spindle (the separated centrosomes).

The above description applies to a typical mammalian cell and the details can vary in other species. Plant cells and some embryonic animal cells do not have centrosomes and lack well-defined spindle poles. In these cells the local stabilization of microtubules by chromatin and the activity of microtubule-based motors are likely to be of paramount importance, as they are in the case of meiosis discussed later in the chapter.

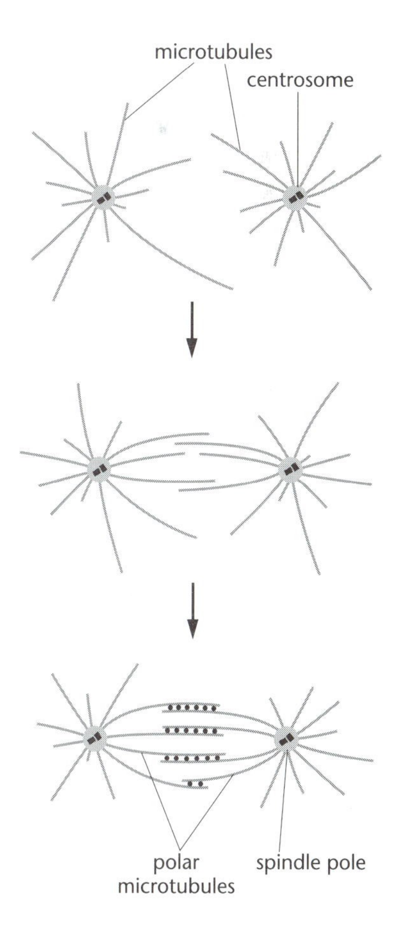


**Figure 13-5** How a bipolar spindle forms. Microtubules grow out in random direction from two centrosomes. When microtubules from opposite centrosomes meet, they can be stabilized and linked together by microtubule-associated proteins. Note that this highly simplified scheme omits chromosomes and the nuclear membrane, both of which are essential for the actual process.

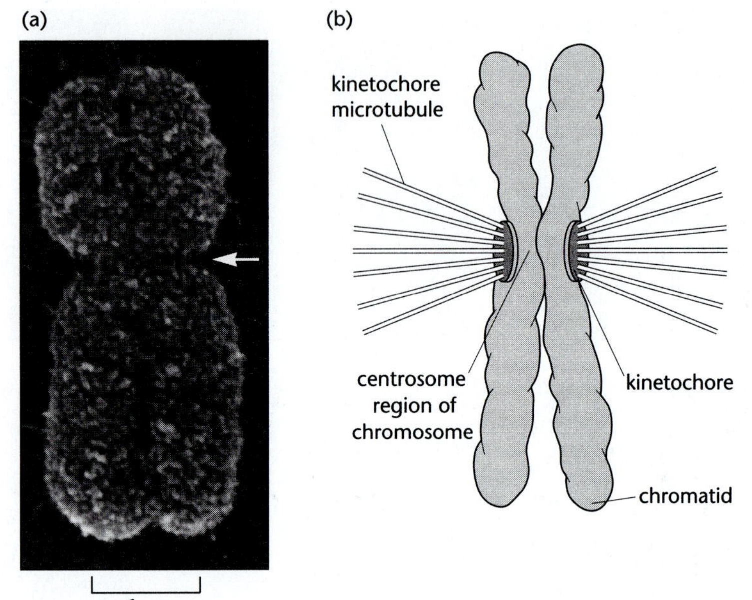


**Figure 13-6** Centromeres and kinetochores. (a) Scanning electron micrograph of a human chromosome at mitosis. Two sister chromosomes (called chromatids when they are attached to each other) are joined along their length. The constricted region (arrow) is the centromere where kinetochores assemble. (b) Schematic diagram of a mitotic chromosome showing its two chromatids attached to kinetochore microtubules via the kinetochores. Each kinetochore forms a protein plaque on the surface of the centromere. The plus ends of the microtubules bind to the kinetochores. (a, courtesy of Terry Allan.)

## *Chromosomes attach to the mitotic spindle*

During prophase, chromosomes condense into discrete visible structures. Each chromosome then consists of two identical bodies (chromatids) containing the two identical DNA molecules produced in a previous round of replication. As seen most clearly at metaphase (the stage of the cell cycle at which chromosomal analysis is usually carried out), stubby chromatids are joined together at a waistlike constriction known as a *centromere* (not to be confused with the centrosomes) (Figure 13-6a). In most higher eucaryotes, the centromere is also the location of a *kinetochore*, a specialized structure to which spindle microtubules attach (Figure 13-6b). Microtubules captured by the kinetochore, termed *kinetochore microtubules*, link the chromosome to the spindle pole (Figure 13-7).

Since kinetochores on sister chromatids face in opposite directions, they tend to attach to microtubules from opposite poles of the spindle, so that each replicated chromosome becomes attached to the two spindle poles. The number of microtubules per kinetochore varies a great deal between species. Human kinetochores are associated with 20 to 40 microtubules, whereas certain plant kinetochores bind well over 100 microtubules. At the other extreme, yeast kinetochores have only one microtubule (demonstrating that a single kinetochore microtubule is sufficient to move a chromosome during mitosis).

Kinetochores also vary in structure between species. In contrast to the well-defined disk of material seen in mammalian cells, some chromosomes in plant and lower animal cells have their entire poleward facing surface covered with microtubules during mitosis. The nematode *Ascaris* has multiple kinetochores on the same chromosomes in its germ line cells and also displays 'chromosome diminution' in cells other than germ line cells, during which chromosomes in somatic cells fragment and only the kinetochore-associated portions are retained. Examples such as this show that individual chromosomes can, if necessary, be handled in special ways.

Kinetochore assembly depends entirely on the presence of the centromeric DNA sequences. In the absence of these sequences, kinetochores will not assemble and chromosomes fail to segregate properly. The specific DNA sequences that determine the site of microtubule association in a kinetochore have been identified in yeast cells. Interestingly, each of the 17 chromosomes of this species of yeast has a slightly different centromeric sequence, raising the interesting possibility that each chromosome

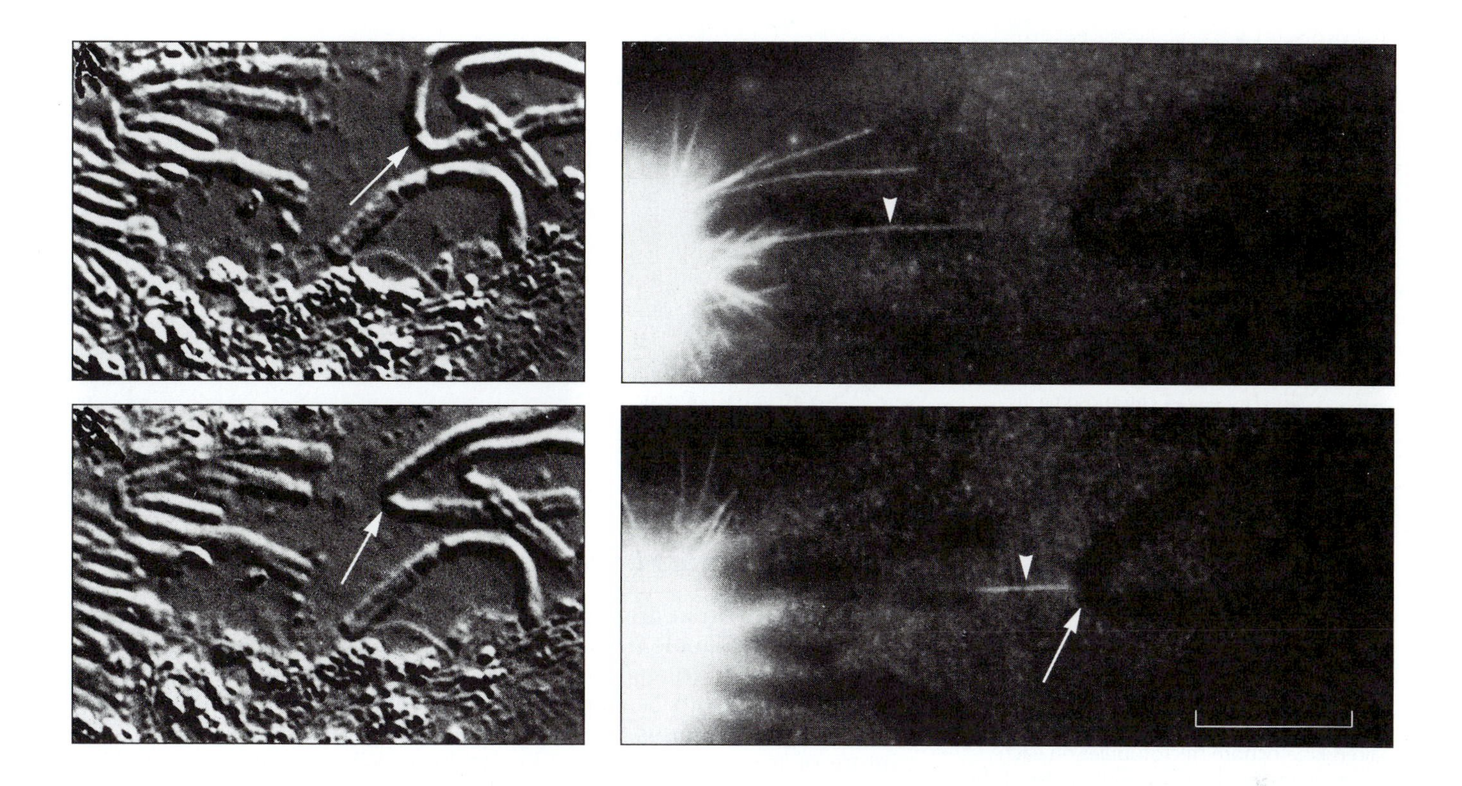

Figure 13-7 Attachment of microtubules to the kinetochore. Two micrographs on the left show condensed chromosomes in a newt cell undergoing mitosis. Arrows point to attachments between chromosomes and microtubules (invisible in this image). On the right a similar preparation shows attachment of a fluorescently labeled microtubule to a chromosome at two focal planes. The calibration bar is 5 μm. (Micrographs courtesy of Conly Rieder.)

might have a distinct signature that allows its interaction with microtubules to be individually determined. On the other hand, there are also regions of strong homology in the sequences, and centromeres can be inverted or swapped from chromosome to chromosome with no loss of function.

## *Chromosomes are both pushed and pulled by spindle microtubules*

We have just seen that spindle microtubules, gaining access to the nuclear compartment, attach to chromosomal kinetochores. The chromosomes now begin to move around, as if jerked first this way and then that. Eventually they align at the equator of the spindle, halfway between the two spindle poles, thereby forming the *metaphase plate* (Figure 13-8).

What are the forces that move chromosomes during mitosis? One obvious possibility is that they are moved by the continual growth and shrinkage of microtubules. A continual balanced addition and loss of tubulin subunits is needed to create the spindle in the first place and to maintain it during chromosomal movements. Even in an apparently static metaphase array of chromosomes, discussed below, tubulin subunits continue to treadmill within the spindle microtubules. Furthermore, a variety of experiments with purified components demonstrate that microtubule assembly and disassembly can generate both pushing and pulling forces, respectively. Microtubules attached to membrane fragments in egg extracts can push or pull them depending on whether conditions favor microtubule growth or shrinkage. Similar experiments have also been done with microtubules attached to isolated chromosomes.

A pushing force exerted by growing microtubules could be generated by a thermal ratchet type of mechanism mentioned in Chapter 4. Pulling could be driven by microtubule depolymerization, provided an appropriate form of attachment exists between the microtubules and the membrane or chromosomes to be moved. One protein that localizes to the microtubule plus ends and might contribute to their chromosomal attachment is called CLIP-170.

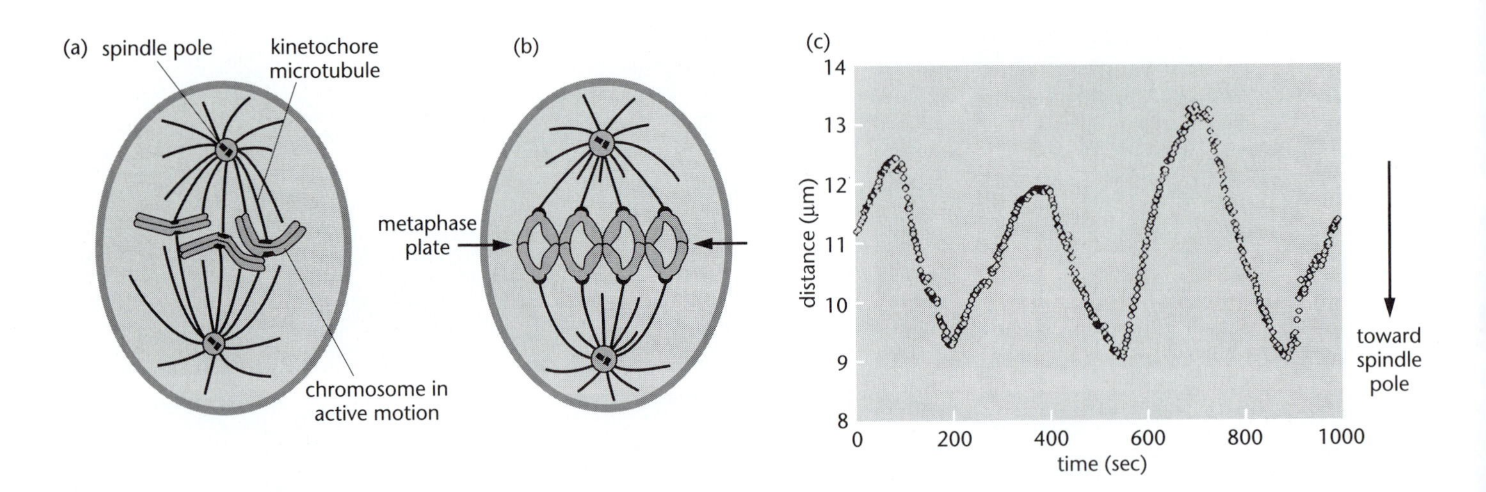


**Figure 13-8** Metaphase. (a) Breakdown of the nuclear envelope allows chromosomes to attach to spindle microtubules via their kinetochores and undergo active movement. (b) Eventually the chromosomes become aligned at the equator of the spindle, midway between the spindle poles. The paired kinetochore microtubules on each chromosome attach to opposite poles of the spindle. (c) Record of the position of a chromosome attached to kinetochore microtubules, showing movements toward and away from a pole and the abrupt switching between them.

## Molecular motors have multiple roles in mitosis

The second potential driving force for the movement of chromosomes is that from microtubule motors. Kinesins, cytoplasmic dyneins, and dynactin are found at kinetochores and other parts of the mitotic spindle and have been shown to be important for different stages of mitosis. Thus, during prophase, minus-end directed motors appear to contribute to the capture of microtubules by the kinetochore. When a randomly probing microtubule by chance passes close enough to a kinetochore, the kinetochore first binds to the side of the microtubule and then slides rapidly toward the spindle pole. This lateral attachment later converts to an end-on attachment in which the kinetochore seems to act like a sliding collar. A metaphase kinetochore maintains an association with tubulin subunits near the end of the microtubule while allowing loss of free tubulin molecules to occur from that end. (A possible model for the attachment of microtubules to a kinetochore is shown below, in Figure 13-11).

Molecular motors can also influence the assembly dynamics of microtubules. We mentioned in the previous chapter the kinesin Kin I that acts catalytically to dissociate tubulin. In a mitotic spindle, this enzyme could move to the plus ends of microtubules, including those at the kinetochore, and there catalyze a rapid depolymerization. A second kinesin, *chromokinesin*, is located not on the spindle microtubules but on the chromosomes themselves. This kinesin probably contributes to forces that push chromosomes away from the spindle poles, sometimes called a 'polar wind.' This force can be seen clearly if chromosomes in a mitotic cells are fragmented by laser microsurgery—the pieces of chromosome unattached to the spindle migrate steadily away from the spindle pole, as if being pushed.

## The kinetochore–microtubule link is stabilized by tension

The initial attachment of a chromosome usually takes place when it is close to a spindle pole, when microtubules attach only to one kinetochore. The other kinetochore remains free of microtubules until it is captured by microtubules growing from the other pole. Mistakes can occur during these early stages: two kinetochores might become attached to the same pole, or a single kinetochore might capture microtubules from both poles. However, these incorrect configurations, which would cause a failure of the chromosome to segregate properly if they persisted, are almost always rectified. The balanced arrangement in which sister kinetochores are linked to sister mitotic poles has the greatest stability.

Why the balanced arrangement should be the most stable is suggested by experiments that probe the mechanical attachment of chromosomes to the spindle.[1] By introducing a fine glass needle into a cell entering metaphase, one can force the two sets of kinetochore microtubules on a

[1] These experiments were all performed with cells undergoing meiosis—the nuclear divisions that produce germ cells—and it is an assumption that the results apply also to mitosis.

single chromosome to engage with the same spindle pole (Figure 13-9). Under normal conditions, such an arrangement is unstable; the error-correction mechanism mentioned above comes into play. However, if movements of an improperly oriented chromosome are resisted by a glass needle, then the chromosome is stabilized and the linkage between the kinetochore and its associated microtubules is made stronger. It seems that when a chromosome in a normal cell is attached to both poles, the *pulling* actions of the kinetochore microtubules on either side of the chromosome balance each other. Mechanical tension develops that can stabilize the interaction of the two kinetochores with their respective microtubules (and is transduced into changes in protein phosphorylation, as we see below).

The orientation of a kinetochore at the end of metaphase determines to which of the newly forming cells it will migrate. Consequently, the apparently aimless dance of chromosomes just described, like the shaking of a set of dice, will produce a random assortment of chromatids at each division. This is especially important in the nuclear divisions occurring during meiosis, since these generate the germ line cells involved in sexual reproduction. In human terms, the chance association of one meiotic kinetochore with a particular spindle pole determines which of our mother's or father's genes we inherit.

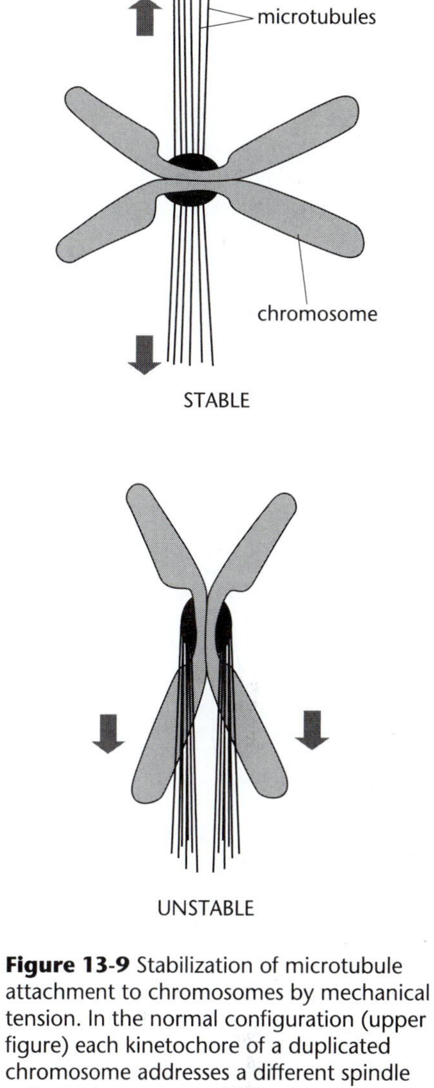


**Figure 13-9** Stabilization of microtubule attachment to chromosomes by mechanical tension. In the normal configuration (upper figure) each kinetochore of a duplicated chromosome addresses a different spindle pole. As the two sets of kinetochore microtubules pull on the chromosome, they develop tension, and this is believed to stabilize the linkage. However, the random nature of microtubule growth in the spindle can result in errors in which the two kinetochores address the *same* pole (lower figure). Such configurations appear to be unstable and do not persist. However, if the pulling action of the two sets of kinetochore microtubules is opposed by a microneedle introduced into the cell, then the 'unstable' arrangement is longer lasting and the paired chromosomes are carried, improperly, to the same spindle pole.

## *Activation of a protease signals commencement of anaphase*

Metaphase often lasts 30 minutes or more, the chromosomes maintaining a restless equilibrium at the metaphase plate like runners waiting for the gun. *Anaphase* then begins abruptly with the synchronous fission of centromeres and separation of pairs of replicated chromosomes. It is as though the links holding the replicated chromosomes together are suddenly cut. Once they are free to move, the daughter chromosomes are pulled toward their respective poles.

The delay at the metaphase plate is one of the checkpoints referred to at the beginning of the chapter. It is the cell's way of ensuring that all chromosomes have become suitably attached to each of the two poles before proceeding—a form of quality control. Even a single errant chromosome that has failed to make a connection—and which, in consequence, could produce abnormal daughter cells—is enough to delay the onset of anaphase. But how can a cell tell that every one of its multiple chromosomes (8 in a fly; 46 in a human; 104 in a carp) is successfully attached? The answer seems to depend on the ability of a kinetochore to generate a 'wait' signal so long as it is not correctly attached to the spindle.

Certain proteins in the kinetochore are phosphorylated at the start of metaphase but become dephosphorylated after that chromosome has attached to the two spindle poles (perhaps, as just mentioned, due to the development of mechanical tension). This change in phosphorylation is believed to be monitored by one or more rapidly diffusing proteins, called *m*itotic *a*rrest *d*eficient (MAD) because mutant cells with defects in these proteins no longer obey the checkpoint. MAD proteins associate transiently with the kinetochore and it has been suggested that they might move rapidly from one kinetochore to the next like a bee visiting flowers, thereby monitoring their state of phosphorylation. So long as even a single kinetochore is unattached, and therefore still phosphorylated, the MAD proteins will carry a 'wait' signal. Only when all the chromosomes have become suitably attached and under tension will the inhibitory message carried by MAD be turned off.

Anaphase itself is initiated when a large complex of proteins termed the anaphase-promoting complex, or *APC*, becomes active. APC contains enzymes able to ubiquitinate specific proteins in the chromosome and hence target them for proteolytic degradation by a proteasome. One protein targeted by APC is the mitotic cyclin discussed above, and the

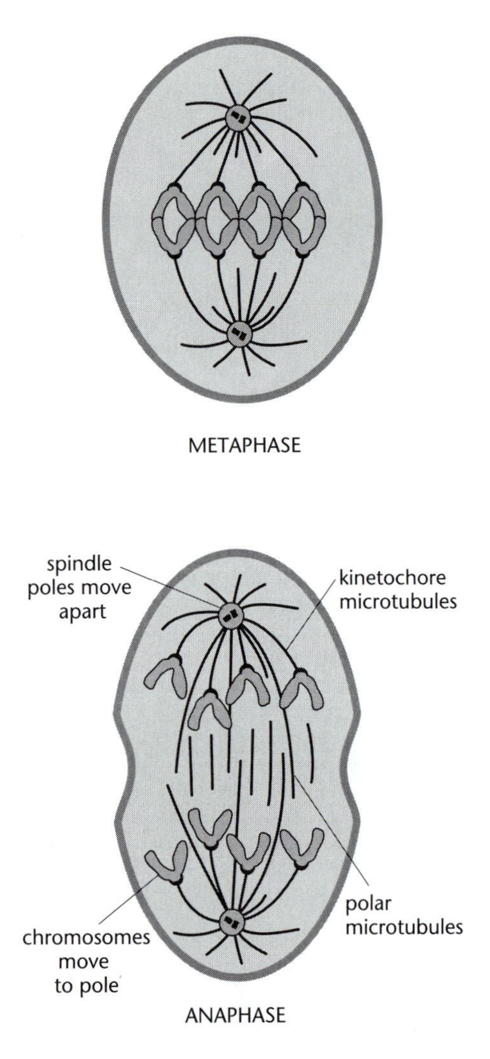


**Figure 13-10** Metaphase to anaphase. Triggered by a cytoplasmic signal, the paired kinetochores on each chromosome separate, allowing the two daughter chromosomes to be pulled toward their respective poles, typically at about 1 μm per minute. Two types of movement can be distinguished during anaphase, the movement of chromosomes to the poles, due to the shortening of kinetochore microtubules (anaphase A), and the separation of poles by the elongation of polar microtubules (anaphase B).

degradation of this protein turns off the kinase activity of the associated Cdk. Another target may be one or more proteins linking the paired chromosomes together. Proteolysis of these linking proteins could free the chromosomes from one another, allowing them to separate under the action of the spindle microtubules. The operation of the APC complex, what its substrates are, how it is regulated, where it is located in the cell, are all important questions for the future.

## *Anaphase consists of two distinct mechanisms*

Anaphase typically lasts only a few minutes. During this time, the chromosomes move at a slow but steady rate of about 1 μm/min toward their respective poles, where they cluster to form the nucleus of a new cell. Their movement is the result of two independent processes within the spindle (Figure 13-10). The first is the poleward movements of chromatids accompanied by the shortening of the kinetochore microtubules, usually referred to as *anaphase A*. The second is the separation of the two poles accompanied by the elongation of the polar microtubules, known as *anaphase B*.

Anaphase A and anaphase B operate by different mechanisms. They are sensitive to different types of drugs: a low concentration of chloral hydrate, for example, prevents the lengthening of polar microtubules (anaphase B), but has no effect on the poleward motion of the chromosomes (anaphase A). Moreover, in different organisms, the relative contribution of the two events to the final separation of the chromosomes varies considerably. In mammalian cells, spindle elongation begins shortly after the chromosomes have begun their voyage to the poles, so that the two processes are mostly coincident. In other cells, such as yeasts, the chromosomes have reached their final destination before anaphase B begins.

In mammalian cells, the spindle elongates 1.5–2-fold during anaphase. In the plant *Tradescantia*, spindle length hardly changes at all, and greater than 90% of chromatid separation is accounted for by anaphase A. The opposite extreme is found in some protozoa where the spindle elongates to such an extent that the poles become separated by 15 times their original distance in the metaphase spindle.

## *Motor proteins in the kinetochore pull a chromosome to a spindle pole*

Movement of chromosomes from the equator of the cell to the poles in anaphase A is usually accompanied by a shortening of kinetochore microtubules. This depolymerization may contribute to the chromosome-to-pole movements, which are inhibited if microtubule depolymerization is blocked by the addition of taxol. Conversely, the movements are accelerated by the addition of very small amounts of nocodazole, which cause the microtubules to depolymerize. One major site of loss of tubulin subunits, interestingly, is the kinetochore itself. If tubulin molecules, labeled with biotin, are injected into metaphase cells, they initially add to kinetochore microtubules at the point of attachment to the kinetochore, as previously described. Subsequently, however, the labeled molecules are lost from the same region as anaphase commences.

Evidence for the involvement of motor proteins has also been obtained from an experiment in which the spindle of demembranated spermatocytes was cut with a glass needle between the chromosomes and one pole and the cut regions were swept away. The chromosomes in such a cut spindle continue to move poleward at normal rates, ceasing only when they come within a few micrometers of the cut edge. Again, this is consistent with the idea that a motor associated with the kinetochore enables it to move along the spindle microtubules.

We saw above that in order to account for the erratic movements of chromosomes at metaphase, the kinetochore attachment must be of a special kind. It must maintain a firm mechanical link between the spindle and the chromosome while at the same time allowing addition and loss of tubulin molecules to the microtubule to occur. Motor proteins are known to operate at the kinetochore, but how their action can be coordinated with the growth or shrinkage of kinetochore microtubules is unclear. One model that has been suggested is that a kinetochore is like a sliding collar, in which motor proteins maintain contact with tubulin molecules at or near the end of the microtubule and walk along them while still allowing addition or loss of tubulin subunits to occur (Figure 13-11). A similar mechanism at the spindle pole could account for subunit loss at that end of the microtubules.

## *The mitotic poles are pushed apart by microtubule sliding*

The second anaphase movement, anaphase B, separates the two poles. As the two poles move apart, the polar microtubules between them lengthen, apparently by the assembly of tubulin molecules onto their pole-distal, or plus ends. The extent of spindle pole separation at anaphase and the extent of overlap of polar microtubules in the midzone of the spindle both vary from one species to another.

Painstaking reconstruction of complete spindles from hundreds of serial thin sections examined in the electron microscope indicates that, at the start of anaphase, polar microtubules from each half spindle overlap in the central region near the spindle equator. In some cells, such as diatoms, the packing of microtubules originating from the opposite poles is strikingly orderly and precise. Similar reconstructions made toward the end of anaphase show that the region of overlap has grown much shorter as though sliding had occurred between the two sets of microtubules. In addition there is, in most cells, an increase in length of the polar microtubules, which presumably have grown by polymerization at their free plus ends. Several marking experiments have confirmed the picture of polymerization and sliding in association with spindle elongation.

The driving force for anaphase B is thought to be provided by two sets of motor proteins operating on the polar microtubules. One set acts on the long polar microtubules that form the spindle itself: these motor proteins slide the polar microtubules from opposite poles past one another at the equator of the spindle, pushing the poles apart. The other force seems, at least in some species, to operate on the microtubules that extend from the spindle poles but point away from the chromosomes toward the cell cortex. These motor proteins are thought to be associated with the cell

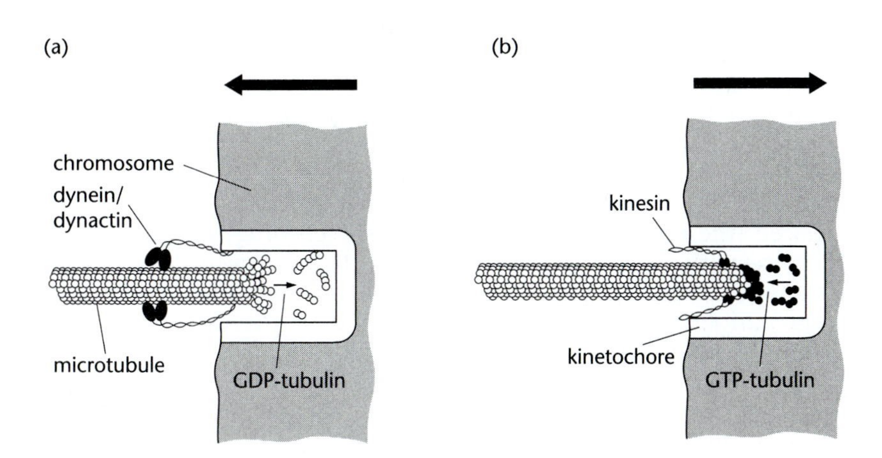


**Figure 13-11** Possible model for the attachment of microtubules to a kinetochore. (a) Chromosomes move *toward* the spindle pole by the action of dynein and dynactin while, at the same time, microtubules shorten at the kinetochore by the loss of GDP tubulin. (b) Chromosomes move *away* from the pole by the action of kinesin motors and by the addition of GTP-tubulin subunits to the microtubule. Note that the same kinetochore can alternate between the two directions of movement, and so must contain both types of molecular motor.

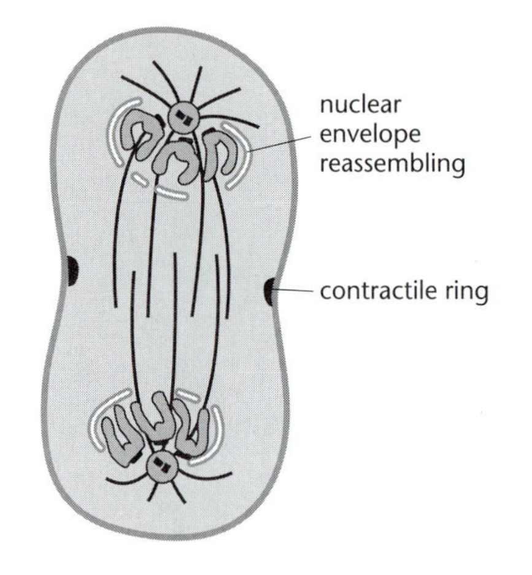


**Figure 13-12** Telophase. As anaphase is completed, the two sets of chromosomes arrive at the poles of the spindle. A new nuclear envelope reassembles around each set, completing the formation of two new nuclei and marking the end of mitosis. The division of the cytoplasm begins with the assembly of the contractile ring.

cortex and to pull each pole toward the cell periphery. Consistent with this idea, anaphase speeds up in some cells when the interpolar microtubules of late anaphase are cut, for example by laser microirradiation.

## *Nuclei reform during telophase*

In the final stage of mitosis, known as *telophase* (from the Greek word *teleos*, meaning complete), chromosomes that have been drawn to their separate poles congregate into two new clusters (Figure 13-12). The chromosomes become less condensed and kinetochore microtubules are no longer visible. The interpolar microtubules left behind in the middle of the spindle become progressively constricted by the advancing contractile ring. A nuclear envelope forms.

In many cells a residue of the spindle remains long after other mitotic events have been completed. Known as a *midbody*, this consists of two bundles of interdigitating microtubules linked at a central dense plaque. As the two daughter cells move apart, blebs of cytoplasm stream from the midbody into one or other of the two cells. The final bundle is extremely stable and the microtubules it contains are not readily depolymerized by drug treatment or exposure to high pressures: it seems likely that this region is either shed from the cell or dissolved by proteolysis.

At the start of mitosis, as we saw, the outer and inner nuclear membranes break down into closed membrane fragments, while the thin network of filaments beneath the inner nuclear membrane (the nuclear lamina) and the nuclear pores disperse. The reverse process defines the start of telophase. Fragments of nuclear membrane associate with the surface of individual chromosomes, partially enclosing each of them before fusing to re-form the complete nuclear envelope. The nuclear lamins, the intermediate filament protein subunits that were phosphorylated during prophase, are now dephosphorylated and reassociate to form the nuclear lamina (Chapter 17).

Once the nuclear envelope has re-formed, nuclear pores pump in nuclear proteins and the nucleus expands. The condensed chromosomes decondense into their interphase state and gene transcription resumes. A new nucleus has been created and mitosis is complete and all that remains is for the cell to complete its division into two.

## *The mitotic spindle dictates the plane of cytoplasmic cleavage*

Cell division entails more than the separation of chromosomes and the formation of new nuclei. It is also the time at which membranes, cytoskeleton, organelles, and soluble proteins, are segregated between the two daughter cells. This is achieved by *cytokinesis*, the process by which the cytoplasm is cleaved into two. Cytokinesis usually begins in anaphase but is not completed until after the two daughter nuclei have formed. Whereas mitosis involves a transient microtubule-based structure, the mitotic spindle, cytokinesis in animal cells involves a transient structure based on actin filaments—the contractile ring, described in Chapter 7.

The contractile ring invariably forms perpendicular to the long axis of the mitotic spindle, so that the two daughter cells receive identical and complete sets of chromosomes. If, at the first sign of furrowing, the spindle is displaced (the earliest experiments used cell centrifugation, but it can also be done with a glass needle), then the starting furrow disappears and a new one develops at a site corresponding to the new spindle location. How the mitotic spindle dictates the position of the cleavage furrow remains a mystery, but two favored models both invoke stimulation of the actin cortex by microtubules. In one model astral microtubules cause the neighboring cortex to relax; in another spindle microtubules near the cell equator cause the cortex to contract.[2]

[2] Evidence from nematode embryos suggest that a kinesin is needed to form the midbody and establish a cleavage furrow.

The usual situation in most dividing cells is that the mitotic spindle is located centrally in the cell. The cleavage furrow forms around the equator of the parent cell so that the two daughter cells are of equal size and contain similar molecules. During embryonic development, however, there are many instances where the spindle is positioned asymmetrically, so that the furrow creates two cells that differ in size. This happens in the early division of sea urchins and *Caenorhabditis elegans*, for example, as well as in many instances of meiotic divisions. One aster of the spindle becomes centered while the other becomes smaller and moves closer to the cortex. Present evidence suggests that in such cases there are molecular landmarks in the cortex to which aster microtubules attach and then pull on the pole (Chapter 20).

## *During the division of plant cells, new wall formation takes place within the spindle*

Unexpectedly, in view of the vital part played by mitosis, the same detailed pattern of movements is not found in all eucaryotic cells. Whereas a spindle composed largely of microtubules is always present, chromosomal segregation can take place without breakdown of the nuclear membrane, as in many species of algae and protozoa, in the slime molds, and in fungi such as yeast.

In cells of higher plants, nuclear membrane breakdown occurs, but the spindle poles are not focused as in mammalian cells. The plant mitotic apparatus is broad and diffuse, shaped more like a barrel than a spindle. The structure of the mitotic poles is also different in plants, since they lack the pair of centrioles, which are prominent features of the poles of most higher animal cells. The mode of formation of dispersed spindles is discussed below in the context of meiosis. Cytokinesis in higher plant cells is completely different from that in animal cells. Plant cells are enclosed in rigid cell walls, and divide from the inside rather than from the outside, by building a new wall between the two daughter cells.

Because of the rigidity of the cell wall, the plane of division is crucially important to the future shape of the entire plant. It is significant, therefore, that the plane of division is marked even before mitosis commences by a remarkable transitory assembly of microtubules, known as a *preprophase band*. The band forms beneath the plasma membrane in the precise position at which a new cell wall will be located, even when the divisions are asymmetrical, as in the organogenesis of structures such as stomata. Microtubules of the preprophase band disappear in the course of mitosis, leaving behind a localized accumulation of actin.

The new wall usually assembles in association with the residual polar spindle microtubules, which form an open cylindrical structure between the two daughter nuclei. Small vesicles derived from the Golgi apparatus and filled with cell wall precusors, move along the microtubules to the equatorial region where they fuse to form a disk-shaped vesicle called the *cell plate*. As the disk, which also contains actin, grows, microtubules assemble at its periphery at the expense of more centrally placed microtubules, which disassemble at the same time. Rearrangement of these microtubules allows the supply of wall precursors to continue to accumulate, so the cell plate can grow. Eventually the growing disk fuses with the plasma membrane and cellulose microfibrils are then laid down to complete the new cell wall (Figure 13-13).

## *Meiotic divisions in eggs proceed without a centrosome*

Mitosis is not the only event in which chromosomes are moved within the cell. Sexual reproduction also requires extensive rearrangements of the genetic material under the guidance of microtubules. During the process of *meiosis*, for example, sperm and eggs are produced by a special

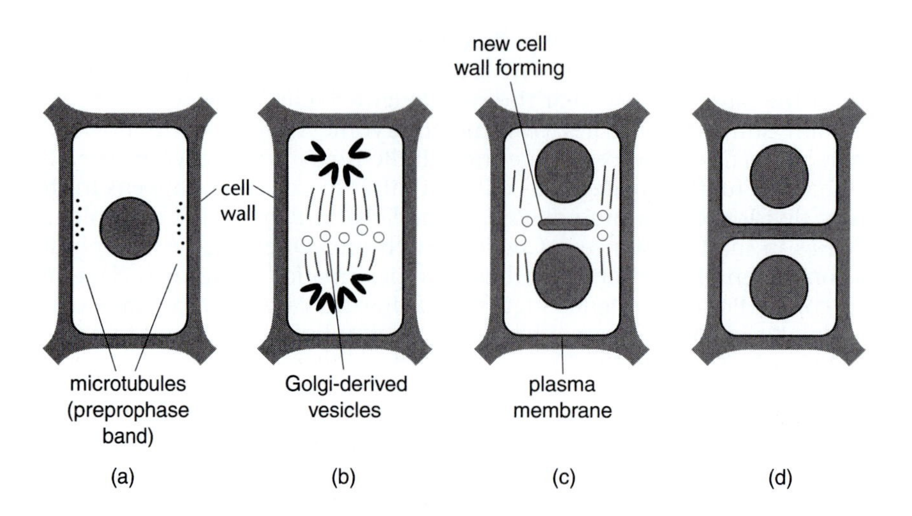


**Figure 13-13** Cytokinesis in a higher plant cell. (a) Before mitosis begins, a ring of microtubules around the waist of the cell—the preprophase band—marks the location of future division. (b) At the beginning of telophase, after the chromosomes have segregated, microtubules guide membrane vesicles to the center of the spindle. (c) The vesicles, derived from the Golgi apparatus and filled with cell wall material, fuse to form the new cell wall, which grows outward to reach the plasma membrane and original cell wall. (d) New and old membrane and cell wall fuse, completely separating the two daughter cells.

kind of cell division in which the number of chromosomes is precisely halved. The actual process is complicated and entails first the duplication of chromosomes, then the association of maternal and paternal chromosomes to form 'tetrads' (which allow the exchange of genetic material by chromosomal crossovers). Two successive divisions then produce the final haploid egg or sperm cell.

The two divisions of meiosis are accomplished by means of a microtubule-based spindle similar to that described above for mitosis. However, in the generation of egg cells the meiotic spindle forms without a centrosome, in this respect resembling mitosis in plant cells. The mechanism of assembly of this structure must be subtly different, since it cannot depend on the growth and shrinkage of microtubules nucleated from a central location.

During female meiosis, microtubules grow randomly around the chromosomes, apparently being nucleated in the cytoplasm. There is evidence that γ-tubulin, which in a mitotic cell is mainly located at the centrosome, is distributed throughout the cytoplasm of a meiotic cell. Furthermore, most microtubules form close to the chromosomes themselves, as though induced to form by some influence diffusing from the chromatin. One intriguing proposal is that protein phosphatases associated with chromatin create a protein dephosphorylation gradient in the nearby cytoplasm, and that this triggers the nucleation of microtubules close to the chromatin. Assembly into spindles could also occur spontaneously through the action of microtubule motors—a possibility we will return to in Chapter 19.

Of course, self-organization of microtubules into poles cannot on its own explain the generation of a bipolar spindle. Why are microtubules self-organized by motors into two poles and not three or four, for instance? In mitotic systems, as we saw above, bipolarity comes from the presence of two centrosomes that form the two poles. In female meiotic systems, bipolarity appears to come from interactions between antiparallel microtubules. Once motors have organized microtubules into arrays with uniform polarity, plus-end motors can then begin to cross-link antiparallel microtubules, eventually driving microtubules into two poles.

## *Fusion of egg and sperm nuclei is orchestrated by microtubules*

The sets of haploid chromosomes generated by meiosis, one set from each of two individuals, recombine during fertilization to produce a diploid cell. During this process, chromosomes are moved within the cytoplasm by means of transient arrays of microtubules resembling, but not identical to, the mitotic spindle.

Fertilization begins with the attachment of a sperm to the surface of the egg. Following entry of the sperm into the egg cytoplasm, the two 'pronuclei' have to move together in order to fuse. In sea urchin eggs, an array of radial microtubules, called the sperm aster, forms around centrioles imported by the sperm and (it seems) pushes the sperm nucleus toward the center of the egg. As microtubules from the sperm aster make contact with the egg pronucleus, it begins to move rapidly toward the sperm pronucleus.

As the two pronuclei come into contact, the two centrioles separate and form two centrosomes. The sperm aster dissolves and another temporary microtubule-containing array, the *interim apparatus*, forms. This too dissolves prior to nuclear breakdown and the first mitotic division of the zygote.

This complicated sequence of events illustrates the sophisticated control the cell extends over its microtubules (three distinct arrays forming and dissolving in sequence) and the value of these arrays for critical cellular processes. While the mechanisms of these events are not understood in detail, it seems probable that, as in mitosis, assembly–disassembly cycles of microtubules and microtubule-based motor proteins will be important.

## Bacteria can manipulate the mechanism of reproduction

We began this chapter with bacteria and it is appropriate to finish with them as well. Although their own mode of cell division appears simpler than that of plant or animal cells, certain bacterial species display an intimate knowledge of meiosis. The bacterium *Wolbachia* spends its life within the ovaries and testes of insects, being transmitted from mother insects to their offspring through the egg's cytoplasm. Because the sperm is almost empty of cytoplasm, male insects—although they can be infected—are unable to pass on their bacteria. So *Wolbachia* prefers females, and it has evolved the capacity to interfere with its hosts' sex lives and gender ratios. In wood lice, for example, these bacteria manage to transform infected males into functional females. In flies and mosquitoes, they prevent infected males from fertilizing uninfected eggs.

In some wasp species *Wolbachia* has eliminated males entirely by somehow disrupting the first cell division in the wasp's egg. That makes the egg diploid, which in most wasps causes the egg to develop as a female. In some species the infection condemns the insects to perpetual asexual multiplication. The wasps' asexual state may, however, be cured by treating them with antibiotics that kill the symbiotic bacteria.

How *Wolbachia* manages to perturb the complex machinery of meiosis to its own ends is a fascination question—one of many for future researchers to answer.

## Outstanding Questions

*What drives the irregular chromosomal movements within an interphase nucleus? Motor proteins, what are they, and against what framework do they move? How are the ends of microtubules attached to kinetochores? How are such attachments made and broken by the cell cycle clock? Proteolysis plays an essential part in the timing of mitosis—what if any part does proteolysis play in other cell movements? Are changes in* $Ca^{2+}$ *important for the timing or efficiency of cell division? If so, how do they act? How does the mitotic spindle signal to the cell cortex so as to position the contractile ring precisely in the midplane of the cell? What triggers APC activation, and hence anaphase? How is the sequence of events during anaphase ordered? What triggers the sudden separation of sister chromatids at anaphase? How is the midbody removed at the end of cell division? Mitosis can tolerate major changes in the expression level of many of its molecular components—what regulatory mechanisms make this possible?*

## *Essential Concepts*

- Bacteria divide by a simpler process than eucaryotic cells, but one of the proteins involved, FtsZ, is clearly related in structure to tubulin.
- The eucaryotic cell cycle consists of several different phases. These include S phase during which the nuclear DNA is replicated and M phase during which the nucleus divides (mitosis) and then the cytoplasm divides (cytokinesis).
- Events in the eucaryotic cell cycle are controlled by proteins called cyclins that rise and fall in amount at specific times. Cyclins activate protein kinases, called Cdk's, which then phosphorylate and hence regulate multiple other target proteins.
- The unidirectional nature of the cycle is due to the periodic activation of proteolytic machinery, which degrades cyclins and other components at specific points in the cycle.
- The onset of M phase is signaled by the breakup of the nuclear membrane and formation of a mitotic spindle made of microtubules, which segregates chromosomes to opposite poles of the cell. In higher animals, the spindle forms on two centrosomes, which have duplicated and moved to opposite sides of the nucleus.
- Upon dissolution of the nuclear envelope, microtubules of the spindle attach to chromosomes at discrete structures called kinetochores situated on either side of the chromosomal centromere.
- Two sets of kinetochore-associated microtubules extending from the mitotic poles maneuver chromosomes to a position midway between the two poles.
- The movement of chromosomes by the spindle is driven both by microtubule motor proteins and by microtubule polymerization and depolymerization.
- A control mechanism monitors the positioning of chromosomes on the midplane of the dividing cell, perhaps by a tension-dependent dephosphorylation, and generates a signal when all chromosomes have become suitably aligned.
- The signal activates a complex of proteins called APC that initiates anaphase—the separation of chromosomes—by triggering a series of proteolytic events.
- At anaphase, the tension in chromosomes is suddenly released as sister chromosomes detach from each other and are pulled to opposite poles. The two poles also move farther apart, possibly because of sliding forces developed between the parallel arrays of microtubules.
- During the final phase of mitosis, telophase, the nuclear envelope reforms around each group of separated chromosomes and the mitotic spindle disassembles.
- Similar arrays of microtubules to those found in mitosis are used during meiosis to segregate chromosomes into germ line cells, and during fertilization to facilitate the fusion of male and female pronuclei.

## Further Reading


Barton, N.R., Goldstein, L.S.B. Going mobile: microtubule motors and chromosome segregation. *Proc. Natl. Acad. Sci. USA* 93: 1735–1742, 1996.

Erickson, H.P. Atomic structures of tubulin and FtsZ. *Trends Cell Biol.* 8: 133–137, 1998.

Erickson, H.P. FtsZ, a prokaryotic homolog of tubulin? *Cell* 80: 367–370, 1995.

Gorbsky, G.J. Cell cycle checkpoints: arresting progress in mitosis. *Bioessays* 19: 193–197, 1998.

Hyams, J.S., Brinkley, B.R. Mitosis: Molecules and Mechanisms. London: Academic Press, 1989.

Hyman, A.A., Karsenti, E. Morphogenetic properties of microtubules and mitotic spindle assembly. *Cell* 84: 401–410, 1996.

Inoué, S., Salmon, E.D. Force generation by microtubule assembly/disassembly in mitosis and related movements. *J. Cell Biol.* 6: 1619–1640, 1995.

Karsenti, E. Severing microtubules in mitosis. *Curr. Biol.* 3: 208–210, 1993.

King, R.W., et al. How proteolysis drives the cell cycle. *Science* 274: 1652–1659, 1996.

Merry, N.E., et al. Cytoskeletal organization in the oocyte, zygote, and early cleaving embryo of the strip-faced dunnary (*Sminthopsis macroura*). *Mol. Reprod. Dev.* 41: 212–224, 1995.

Morgan, D.O. Regulation of the APC and the exit from mitosis. *Nat. Cell Biol.* 1: E47–E53, 1999.

Nanninga, N. Morphogenesis of *Escherichia coli*. *Microbiol. Mol. Biol. Rev.* 62: 110–129, 1998.

Nédélec, F.J., et al. Self-organization of microtubules and motors. *Nature* 389: 305–308, 1997.

Nicklas, R.B. How cells get the right chromosome. *Science* 275: 632–637, 1997.

Nicklas, R.B., et al. Kinetochore chemistry is sensitive to tension and may link mitotic forces to a cell cycle checkpoint. *J. Cell Biol.* 130: 929–939, 1995.

Nicklas, R.B., et al. Tension-sensitive kinetochore phosphorylation *in vitro*. *J. Cell Sci.* 111: 3189–3196, 1998.

Nogales, E., et al. Tubulin and FtsZ form a distinct family of GTPases. *Nat. Struct. Biol.* 5: 451–458, 1998.

Ohba, T., et al. Self-organization of microtubule asters induced in *Xenopus* egg extracts by GTP-bound Ran. *Science* 284: 1356–1358, 1999.

Powers, J., et al. A nematode kinesin required for cleavage furrow advancement. *Curr. Biol.* 8: 1133–1136, 1998.

Rieder, C.L., Salmon, E.D. The vertebrate cell kinetochore and its roles during mitosis. *Trends Cell Biol.* 8: 310–318, 1998.

Samuels, A.L., et al. Cytokinesis in tobacco BY-2 and root tip cells. A new model of cell plate formation in higher plants. *J. Cell Biol.* 130: 1345–1357, 1995.

Sharp, D.J., et al. Antagonistic microtubule-sliding motors position mitotic centrosomes in *Drosophila* early embryos. *Nat. Cell Biol.* 1: 51–54, 1999.

Skibbens, R.V., et al. Kinetochore motility after severing between sister centromeres using laser microsurgery: evidence that kinetochore directional instability and position is regulated by tension. *J. Cell Sci.* 108: 2537–2548, 1995.

Stouthamer, R., et al. Antibiotics cause parthenogenetic *Trichogramma* (Hymenoptera/Trichogrammatidae) to revert to sex. *Proc. Natl. Acad. Sci. USA* 87: 2424–2427, 1990.

Stouthamer, R., et al. Molecular identification of microorganisms associated with parthenogenesis. *Nature* 361: 55–68, 1993.

Strome, S. Determination of cleavage planes. *Cell* 72: 3–6, 1993.

Vernos, I., Karsenti, E. Motors involved in spindle assembly and chromosome segregation. *Curr. Opin. Cell Biol.* 8: 4–9, 1996.

Walczak, C.E., Mitchison, T.J. Kinesin-related proteins at mitotic spindle poles: function and regulation. *Cell* 85: 943–946, 1996.

Walczak, C.E., et al. A model for the proposed roles of different microtubule-based motor proteins in establishing spindle bipolarity. *Curr. Biol.* 8: 903–913, 1998.

Waters, J.C., et al. Mad2 binding by phosphorylated kinetochores limits error detection and checkpoint action in mitosis. *Curr. Biol.* 9: 649–652, 1999.

Wilde, A., Zheng, Y. Stimulation of microtubule aster formation and spindle assembly by the small GTPase Ran. *Science* 284: 1359–1362, 1999.

# CHAPTER 14

# Cilia

Packed into a cylinder about 200 nm in diameter—not much wider than a muscle myosin molecule is long—is an amazingly intricate molecular machine. The core of a eucaryotic cilium or flagellum is built from well over 200 specific types of proteins that work together to generate repeated and often complex propagating waves. Like a myofibril, this compact and regular bundle can be isolated free of other components of the cell and its action can be studied in detail. Moreover, cilia are also highly accessible to mutational analysis, notably in the biflagellate alga *Chlamydomonas*. We are still far from understanding how this marvel of miniaturization works in detail, but a combination of biochemistry, molecular genetics, and high-resolution electron microscopy has given us at least a broad picture of how a cilium works.[1]

[1] Although there are differences between cilia and flagella in length, distribution, waveform, and other respects, the molecular basis of their movement is closely similar. For many purposes the two names can be used interchangeably.

We will see in this chapter that the primary force-generating event in a cilium is the ATP-driven cyclic movements of ciliary dynein—the first microtubule-based motor protein to be discovered. Dynein movements cause microtubules in the core of a cilium to slide against each other, and this produces bends in the cilium. However, microtubule sliding by itself is not enough to generate complex three-dimensional waveforms, and the intrinsic motion of the dynein arms is coupled and corralled by an odd collection of protein side-arms, spokes, and cross-bridges. Exactly what each part of this baroque engine does is sometimes hard to discern, and mutational analysis often indicates that individual proteins can be deleted without causing major dysfunction. In this regard cilia, and the other motile microtubule bundles described later in this chapter, reflect the historical aspect of cytoskeletal structures. The assortment of protein molecules they contain is the result of an accumulation over many years of evolution and the function of many individual proteins has changed radically over the generations.

## Ciliary axonemes contain the machinery of wave propagation

We mentioned in Chapter 1 that eucaryotic flagella have the ability to propagate waves independently of the rest of the cell. They can be detached from individual sperm cells or from single-celled organisms such as *Chlamydomonas* by means of a focused laser beam or from suspensions of large numbers of cells by applying either shear forces or unusual solution conditions. Suspended in a buffer containing Mg-ATP, these isolated structures continue to propagate waves along their length, showing that they contain the machinery of motion.

A further step in the analysis is to remove the plasma membrane. If sea urchin sperm (which swim freely in sea water) are exposed to the nonionic detergent Triton X-100, the plasma membrane is stripped away. The sperm are now dead but the core of protein in their flagella remains intact. Remarkably, this core of protein, known as the *axoneme*, will once again generate swimming movements if ATP is added. Reactivated axonemes swim with a waveform and beat frequency indistinguishable from those of live controls (Figure 14-1).

Evidently the axoneme, like a myofibril, can generate movement independently of the rest of the cell. There is, however, an important difference. A myofibril contracts only if the $Ca^{2+}$ level is raised above about $5 \times 10^{-6}$ M and it then contracts only once, irreversibly. By contrast, an axoneme propagates waves continually whether $Ca^{2+}$ is present or not (although the rate of wave propagation will be altered, as we see below). Since any propagating wave must involve cyclic movements at particular regions of the axoneme, local control over these movements must reside in the axoneme itself. Unlike the case of muscle, the membrane of a cilium or flagellum is not needed to coordinate the movements.

## The ciliary axoneme is built from a 9+2 array of microtubules

One of the first biological discoveries to be made with an electron microscope was the unique pattern of microtubules in the ciliary axoneme. Known as a '9+2' pattern, this consists of a peripheral ring of nine *doublet microtubules*, each a pair of microtubules fused along their length, and two

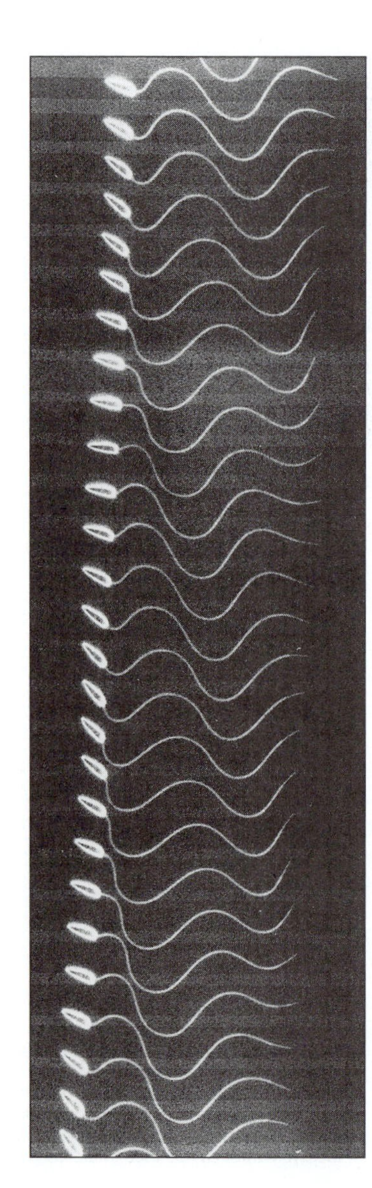

**Figure 14-1** The repetitive wavelike motion of a single flagellum. A tunicate sperm is seen in a series of images captured by stroboscopic illumination at 400 flashes per second. (Courtesy of Charles Brokaw.)

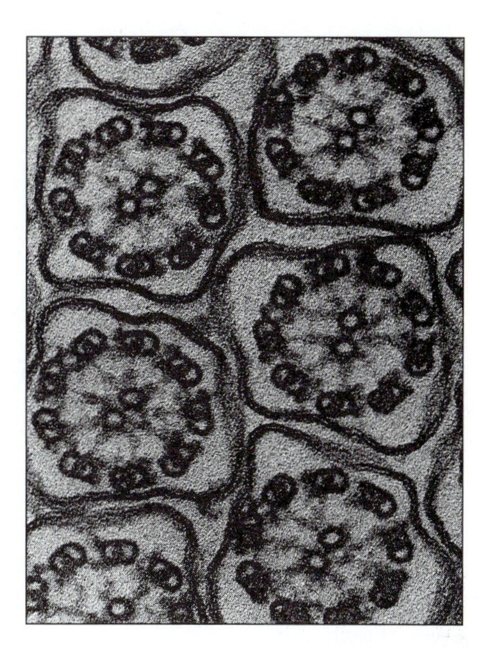

**Figure 14-2** The arrangement of microtubules in a cilium or flagellum. Cross-section through a field of close-packed cilia illustrating their distinctive 9+2 arrangement of microtubules. (Courtesy of Peter Satir.)

conventional (or singlet) microtubules in the center of the ring (Figure 14-2). Each doublet microtubule was later seen, with improved methods of staining and fixation, to comprise one complete microtubule (also known as an *A-tubule*) with 13 protofilaments as described earlier for cytoplasmic microtubules (see Figure 11-5) fused to one incomplete microtubule (or *B-tubule*) made of 10 or 11 protofilaments. Each of the two centrally located microtubules (the *central pair*) has the usual number of 13 protofilaments. All of the microtubules in the 9+2 bundle extend continuously for the length of the cilium, which is usually about 10 µm but can be as great as 200 µm in some cells. Sperm flagella can be even larger, reaching over a centimeter in some insects.

The distinctive 9+2 pattern is seen in sectioned material from widely different sources. Many sperm cells from organisms as diverse as lower plants and mammals show this pattern of microtubules in their flagella (although there are also many variants, as we see later in the chapter). Cilia from protozoa to man generally display the same unmistakable ultrastructural signature.

A related '9+0' pattern, in which the axoneme has nine double microtubules but lacks the central pair, is found in sensory detector organs. Olfactory receptors, for example, have a surface protrusion that is basically a nonmotile cilium, while rod cells in the vertebrate retina have a modified cilium between the cell body and the segment rich in photosensitive membrane. Hair cells of the inner ear often carry an individual nonmotile cilium on their surface in addition to the arrays of actin-containing stereocilia (Chapter 6).

The function of these cilialike structures appears not to be to generate movement but simply to maintain a surface protrusion. Presumably, once the basic program for making a cilium appeared during evolution it was strongly conserved thereafter, being activated whenever an extension of the cell surface was needed. We discuss the molecular basis of cilia assembly in the next chapter.

## *Ciliary microtubules are decorated with side-arms, spokes, and cross-links of protein*

Cross-sections of cilia reveal an elaborate infrastructure of protein structures that link the microtubules together and provide the machinery of movement (Figure 14-3). Each structure has a distinctive periodicity along the axoneme (Table 14-1).

*Dynein arms*, projecting like crab claws from each doublet microtubule, are capable of extending to the neighboring B-tubule and of providing the crucial force generation for ciliary bending. Two sets of dynein arms can be distinguished on each A-tubule—the outer dyneins and the inner dyneins. There is also a thinner connection between adjacent doublets in the form of a fine protein link that extends around the perimeter of the axoneme like a hoop around a barrel. Known as a *nexin link*, this acts like an elastic belt holding the bundle of microtubules together.[2]

The *central apparatus* at the hub of the axonemal cross-section contains two microtubules linked at intervals along their length. Two sets of arms, one pair on each microtubule, curve around in a protective fashion to form a sheath around the central pair. Radial spokes project inward from the nine outer doublets, each terminating in a globular head close to the central pair. Note that the axoneme as a whole is not precisely

[2] The proteins in this linker have not so far been isolated, so that the name 'nexin' sometimes applied to them is probably premature, especially since another protein also called nexin is an inhibitor of serine proteases and widely distributed in extracellular matrix.

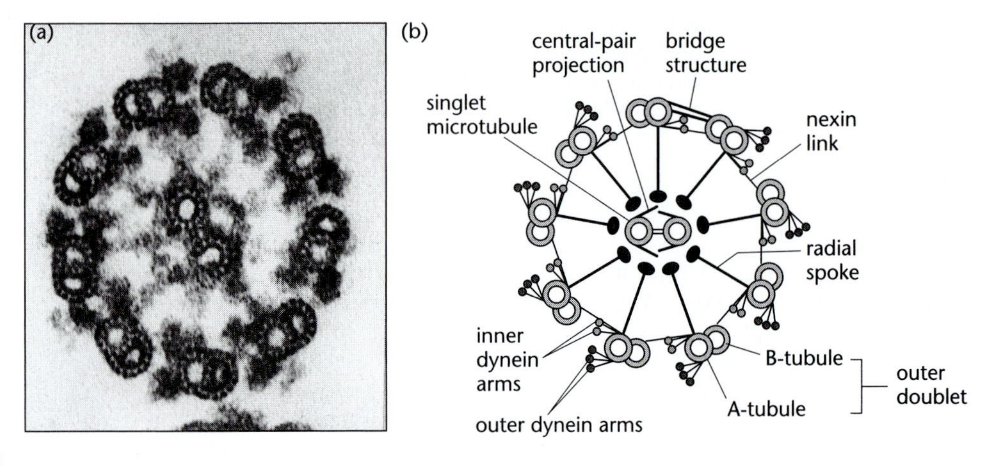


**Figure 14-3** Schematic section through a cilium. The nine outer microtubules (each a special paired structure) carry two rows of dynein arms. The heads of these dyneins appear in this view to reach toward the adjacent microtubules (they have a clockwise orientation indicating that we are looking from the cell out toward the tip of the cilium). In a living cilium, dynein heads periodically make contact with the adjacent microtubule and move along it, thereby producing the force for ciliary beating. Various other links and projections shown are proteins that hold the bundle of microtubules together and convert the sliding motion produced by dyneins into bending. (a, courtesy of Lewis Tilney.)

symmetric, but the circular ring of A- and B-tubules and their dynein side-arms does give an approximate rotational symmetry, thereby permitting electron microscope images to be enhanced by optical averaging techniques. Each section through an axoneme therefore has a unique handedness reflecting the polarity of the cilium itself. If a cilium or flagellum is viewed as if looking out from the cell, the dynein arms all point clockwise (as in Figure 14-3).

In addition to features that repeat along the length of the cilium or flagellum, there are also distinct structures at either end. The tip of a cilium is characterized by protein complexes that cap the plus ends of microtubules, resembling in this respect the proteins of the kinetochore discussed in the previous chapter. This region must also contain the machinery necessary to assemble the ciliary axoneme, a topic we will return to in Chapter 15.

## *Tektin forms one of the protofilaments in ciliary doublet microtubules*

Many of the axonemal structures just mentioned can be selectively extracted in buffers—dynein, for example, was originally isolated in the 1960s by extraction of cilia with concentrated salt solutions. However, there are also insoluble 'protofilaments' in an axoneme that cannot be dissolved in either high- or low-salt solutions. These have been shown to be made of a novel set of highly insoluble filamentous proteins, known as *tektins*, which have molecular weights in the region of 50 kDa and have

**Table 14-1 Major protein components of ciliary axonemes**

| Structure | Function | No. of proteins | Periodicity |
|---|---|---|---|
| microtubules | principal structural components of axoneme, the outer doublets provide tracks for motors causing the bending and for motors responsible for intraflagellar transport | 4–6 | 4 nm (tubulin monomer), 8 nm (tubulin dimer) |
| dynein arms | cause microtubule sliding | 25–40 | 24 nm average, inner arms are staggered by radial spokes |
| nexin links | hold adjacent microtubles in the proximal portion of an axoneme together | ? | 96 nm |
| radial spokes | relay signals from the central pair to the dynein arms, thereby controlling bending | 20 | 24 + 32 + 40 nm |
| central apparatus | rotates, perhaps through kinesins, and regulates dynein action via radial spokes | 20 | 16 nm |
| plus end capping | may control growth of axoneme | 12 | NA |

These are the major structures seen in the electron microscope: more than 200 different proteins are found in ciliary axonemes. Periodicities are from sea urchin sperm.

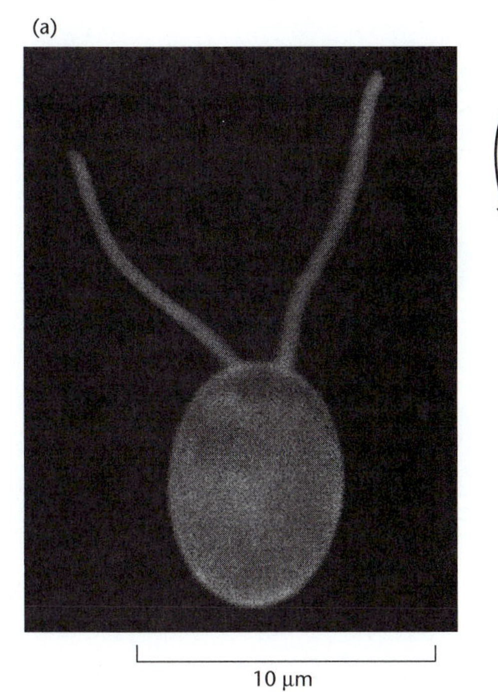


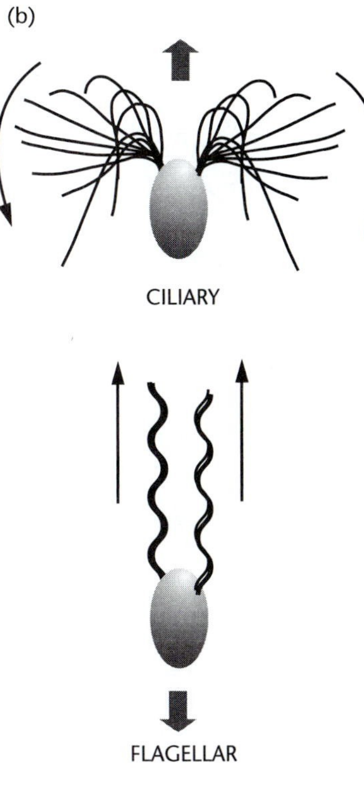


**Figure 14-4** *Chlamydomonas* flagella. (a) The green alga *Chlamydomonas* swims by means of paired flagella. (b) The flagella normally beat in a repetitive action reminiscent of a human breast-stroke. However, in an escape response, the two flagella reverse and propagate sinusoidal waves. (Micrograph courtesy of Robert A. Bloodgood, University of Virginia.)

some similarities to intermediate filaments. Tektins extend along the length of the ciliary microtubules and appear to be an integral part of the microtubule wall. In fact, one of the protofilaments close to the junction between B-tubule and A-tubule is probably made not of tubulin but of three slightly different tektins.

What are the functions of tektins? Their location in the axoneme and their similarity to intermediate filaments suggest that they might add mechanical strength to the axoneme. A framework of tektin filaments with nexin links between them could serve as a resilient structural skeleton. Alternatively, or perhaps in addition, tektins could act like a molecular ruler. Since these filaments are located close to the boundary between A- and B-tubules in a cilium, they are in a good position to influence the assembly of these structures. Sequences of tektins indicate that they have an extended coiled-coil structure with a central rod region of length around 48 nm (Chapter 17). Along each tektin rod, cysteine residues occur with a periodicity of approximately 8 nm, coincident with the axial repeat of tubulin dimers in microtubules. In principle, a protofilament made of tektin chains arranged in serial order could provide positional information for the periodic attachment of structures at multiple overall repeats, including those of the dynein arms and the radial spokes.[3]

## Genetic analysis reveals the biochemical complexity of flagella

Mutational analysis is an especially powerful way to examine the mechanism of cell movement. In an assay for cell motility, wild-type cells can crawl or swim away from their starting point, while those that remain must be either dead or paralyzed. If motility depends on a single well-defined structure such as a flagellum, then this can be examined directly for lesions due to the genetic defect.

A favorite object for such studies is the unicellular green alga *Chlamydomonas reinhardtii*, which has two flagella that propel it through the water (Figure 14-4). Many mutants of *Chlamydomonas* with impaired motility have been isolated, a convenient method being to screen for cells

[3] The intermediate chains of the dynein molecule are known to mediate its attachment and docking to flagellar microtubules, and may also contribute to the precise periodicity along the axoneme.

**Figure 14-5** Summary of the *Chlamydomonas* mating cycle, showing dikaryon formation.

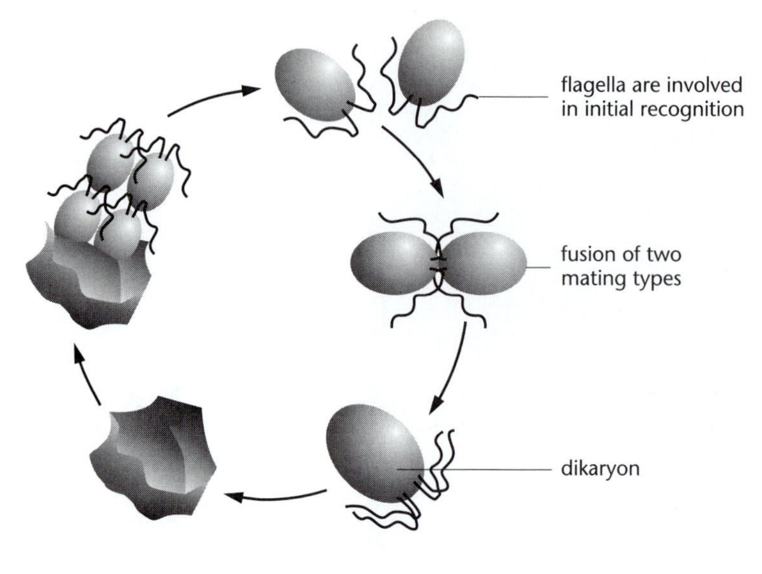


unable to move toward light. In some mutants the mechanism of flagellar assembly is defective and flagella are absent; in others flagella are present but are either nonmotile or slow-moving. Structural abnormalities can often be seen in electron micrographs: in one class of nonmotile mutant, for example, the only detectable change is the absence of dynein arms; in another the radial spokes are missing; a third lacks both the central microtubules and the inner sheath. In all three types of mutant the isolated membrane-free axonemes fail to beat in the presence of ATP.

Flagellar mutants provide essential information on the proteins that compose the axoneme. For example, a two-dimensional gel analysis of mutants lacking either the inner or the outer dynein arm shows that the two lack different sets of polypeptides and the inner and outer arms are therefore biochemically distinct. Another type of mutant, which lacks the entire radial spoke complex, has 12 missing polypeptides and a mutant that lacks just the radial spoke head is missing just 6 of the 12.[4]

In order to identify the function of a mutant protein with greater confidence one can use some of the unusual events that occur during the sexual cycle of *Chlamydomonas*. When *Chlamydomonas* gametes of opposite mating type are mixed, pairs of cells fuse to form a temporary dikaryon with four flagella (Figure 14-5). This cell will swim for several hours before resorbing the flagella and commencing zygote formation. If wild-type and paralyzed strains are crossed, the temporary dikaryon initially has two motile and two nonmotile flagella. In many such crosses, however, about 1 hour after cell fusion, the two paralyzed flagella begin to beat. Apparently access to the wild-type cytoplasm has provided a functional copy of their defective gene product. By performing the above mating with radioactively labeled wild-type cells and unlabeled mutant cells, it has been possible to identify the specific polypeptide made defective by the mutation—a procedure known as *dikaryon rescue*.

## *Ciliary bending depends on the sliding of adjacent microtubules*

Everyone now accepts that ciliary beating is driven by the sliding of adjacent sets of microtubules against each other. One of the earliest pieces of evidence came from an examination of the tip of a bending cilium in the electron microscope. If bending was produced by changes in length due to the compression or stretching of component microtubules then the distinctive 9+2 pattern should be preserved along the entire length of the

[4] Why should a mutation in a single gene cause the loss of 6 or 12 polypeptides from the cilium? The most likely answer is that the mutated protein is necessary for normal assembly of the remaining proteins—in its absence the other proteins are synthesized but rapidly degraded, since they are not incorporated into the axoneme.

**Figure 14-6** The movement of dynein causes a flagellum to bend. (a) In an intact flagellum, the microtubule doublets are tied to each other by flexible links of protein so that the repetitive action of their dynein arms produces bending. (b) If the outer doublet microtubules and their associated dynein molecules are freed from other components of a sperm flagellum and then exposed to ATP, the action of dynein arms causes the doublets to slide against each other, telescope fashion.

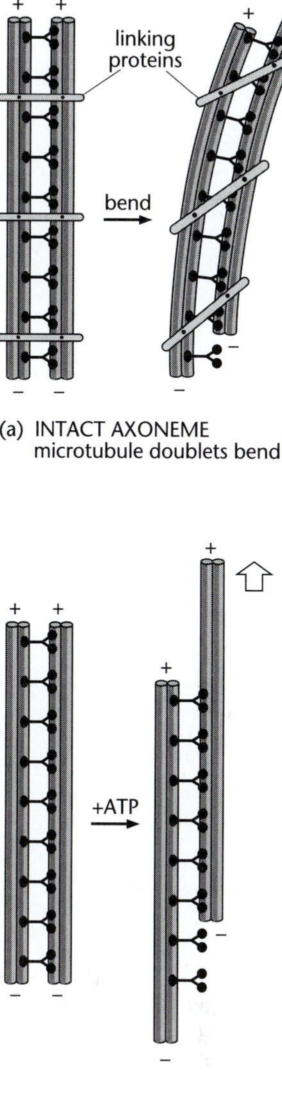


cilium. What was found, however, was that microtubules on the inside of the ciliary bend extended farther at the tip than microtubules on the outside of the bend, a clear indication that the microtubules remain the same length and slide in relation to each other.

Evidence was also obtained from longitudinal sections. Since the side-arms and other projections are arranged at regular intervals along the axoneme, they act as distance markers, like the lines on a measuring tape. Suitably aligned sections of a bending cilium show directly that slippage occurs.

A different approach to the mechanism of ciliary bending was obtained by examining isolated sperm flagella from which the outer membrane had been removed by detergent treatment. As already mentioned, if ATP is added to such extracted axonemes they propagate waves similar to those seen in intact cells. Gold particles attached to such axonemes show differential motion due to microtubule sliding. Moreover, if the structure of the axoneme is weakened by a brief exposure to a proteolytic enzyme such as trypsin, then ATP addition produces a different effect. As seen under dark-field illumination, the axonemes do not bend but elongate, telescoping out to reach up to nine times their original length. They elongate because adjacent microtubules, freed of their normal cross-links by trypsin digestion, slide over each other on addition of ATP (Figure 14-6).

## *Microtubule sliding is produced by dynein arms*

In a trypsin-treated axoneme, the radial spokes and interdoublet links have been destroyed by the protease. Only the dynein arms are left intact, so they must be responsible for the sliding and consequent disintegration.

The earliest evidence to show that dynein arms produce sliding was obtained by Gibbons and Rowe in 1965. Axonemes from sea urchin sperm tails were extracted in a low-ionic-strength buffer lacking divalent cations, which solubilizes dynein arms while other components such as the radial spokes and nexin cross-links are left intact. When tested in the usual way for motility, the extracted axonemes were no longer able to produce waves in the presence of ATP. However, if material extracted from the axoneme was added back to the depleted axonemes then the normal ultrastructural morphology was restored, together with the ability to generate movement. The extracted material later provided a source of ciliary dynein.

Mutational analysis provided similar evidence by a different route. Mutants of *Chlamydomonas* with defective radial spokes or central pairs are paralyzed, but outer doublet microtubules will slide when treated with protease and exposed to ATP. The same is not always true of mutants lacking inner dynein arms. In some cases these not only have paralyzed flagella but also fail to slide and elongate following trypsin treatment of their axonemes (see Figure 14-6).

The two sets of dynein arms, inner and outer, have different protein compositions and structures, and they display slightly different functions. Although the first *Chlamydomonas* mutants lacking inner or outer dynein arms to be examined were paralyzed, this is not always the case. Mutants lacking either inner or outer arms can beat their flagella, showing that either is sufficient to generate movement. Interestingly, however, the beat frequencies of such mutants, and the rates at which their trypsinized axonemes slide apart, are different from those of wild-type axonemes. Mutants of *Chlamydomonas* missing the outer dynein arms swim with reduced speed and their axonemes slide apart at approximately half of wild-type rates. Inner arm mutants swim with near normal beat frequencies and slide at near normal rates, but they have different flagellar waveforms.

## Dynein is a high-molecular-weight ATPase that binds to microtubules

The studies on cilia just described led to the eventual isolation and characterization of *ciliary dynein*, the first microtubule motor protein discovered. Ciliary dynein is a very large multisubunit protein with a head region containing a tubulin-activated ATPase activity and a tail portion that mediates ATP-insensitive binding to microtubules. In the electron microscope, isolated outer arms from *Tetrahymena* or *Chlamydomonas* appear like a fist holding three balloons, three globular heads attached to a common base. Other ciliary dyneins from sources such as sea urchin sperm are often double-headed, and some inner arm species are single-headed (Figure 14-7). Resolving their components by polyacrylamide gels reveals three (or two) heavy chains each greater than 500 kDa and multiple lower-molecular-weight components, some of which have been localized to the base of the unit. The heavy chains have been sequenced and contain four regions (known as P-loops) that are commonly associated with ATP binding and hydrolysis and several $\alpha$-helical coiled-coil domains.

As for other motor proteins, the head of the ciliary dynein molecule is the region that generates mechanical force, containing both the microtubule-binding site and the ATPase catalytic activity. Kinetic analysis of the movements of dynein suggests that it operates in a fundamentally different cycle to the myosin/actin cycle in muscle, resulting in the walking of dynein heads along adjacent microtubule doublets. Since there are two dynein arms at each position along an A-tubule, a line of five or so dynein heads, like a team of horses, will advance together along the B-tubule (see Figure 14-3). Axonemal dyneins are likely to work coordinately in the flagellum or cilium, spending little time actually applying force to the microtubule, reminiscent of the myosin–actin filament system. In contrast, the functions of cytoplasmic dyneins discussed in Chapter 12 are closer to those of conventional kinesins. It is reasonable to suppose that

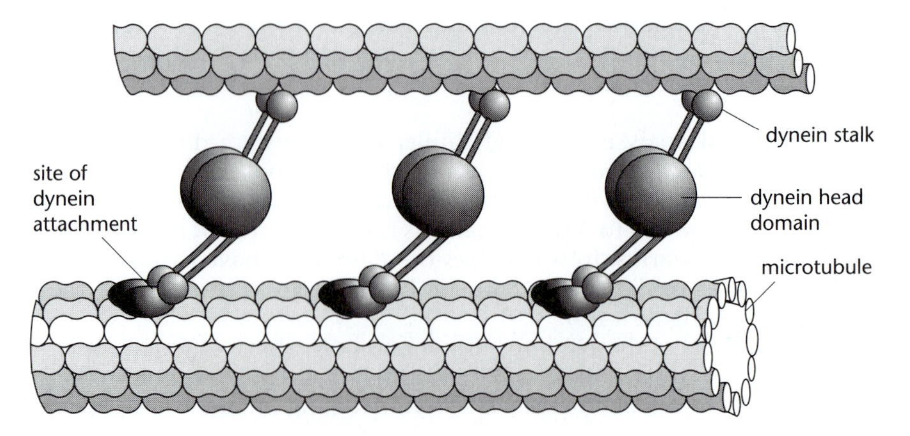


**Figure 14-7** Ciliary dynein. Ciliary dynein is a large protein assembly (nearly 2 million Da) composed of 9 to 12 polypeptide chains, the largest of which are two or three heavy chains each greater than 500 kDa. The heavy chains form the major portion of the globular head and stem domains, and many of the smaller chains are clustered around the base of the stem. The base of the molecule binds tightly to an A-tubule in an ATP-independent manner, whereas the large globular heads have an ATP-dependent binding for a B-tubule. As indicated here, the actual binding site is thought to be on a stalk about 10 nm long projecting from the large head domain. When the heads hydrolyze their bound ATP they move toward the minus end of the second microtubule, thereby producing a sliding force between the adjacent microtubule doublets in a cilium or flagellum. The two-headed form of ciliary dynein, formed from two heavy chains, is illustrated here.

the latter motor proteins will be processive, perhaps working in a hand-over-hand manner, since it would be inefficient for the organelle to dissociate from the microtubule repeatedly en route.

Curiously, the actual site of contact between the dynein head and the microtubule along which it moves may not be the large head domain itself but a short rigid stalk projecting from it. Both sequence analysis and high-resolution electron microscopy reveal that the tubulin binding site exists on a region of coiled-coil sequence terminating in a 6 nm globular region. Exactly how or why the dynein might use such an unusual form of attachment is not known, but it has been suggested that the head domain is so large (~3300 amino acids and 13–14 nm) that multiple heads cannot simultaneously make contact with a microtubule. The stalk might have evolved to alleviate this problem.

## Cilia contain a diversity of dynein molecules

In a now familiar pattern, molecular genetics reveals an astonishing diversity of dyneins, even in a single cilium. To begin with, most cilia have two completely separate sets of dyneins—the outer dynein arms and the inner dynein arms—and each set contains many dynein isoforms. Eight distinct inner dynein heavy chains can be resolved biochemically from *Chlamydomonas*, for example, each with its own complement of intermediate and light chains. Current evidence suggests that the inner arms are organized in precise groups that repeat in a 96 nm pattern along the axoneme, in exact register with the paired radial spokes.

Astonishingly, the location of the various isoforms along the length of an axoneme may also be nonuniform, so that particular types of dynein are more abundant in the proximal, or middle, or distal portions of the axoneme. Evidence is beginning to emerge for specific anchoring proteins that attach this or that dynein to a particular location in the axoneme. There are also hints that when the cell turns or otherwise changes its swimming behavior, individual dynein components may be modified selectively in one or other of the two *Chlamydomonas* flagella. How all these different types of dynein work together in an axoneme is a mystery at present.

## Slide-to-bend conversion depends on radial spokes

Although the basic movement of an axoneme is that of microtubules sliding, some degree of control is needed before a sinusoidal or more complex waveform can be produced. Given the way the dynein arms are arranged, if they all became active at the same instant the axoneme would simple twist into a coil. In order to form a planar bend, dynein arms on one side of the axoneme should be active while those on the other side should be inactive—preferably detaching to allow passive sliding in a direction opposite to the power stroke (Figure 14-8).

Before we discuss more complex mechanisms of axoneme regulation, it is worth pointing out that the passive mechanical properties of a cilium by themselves could contribute to its waveform. Even a single

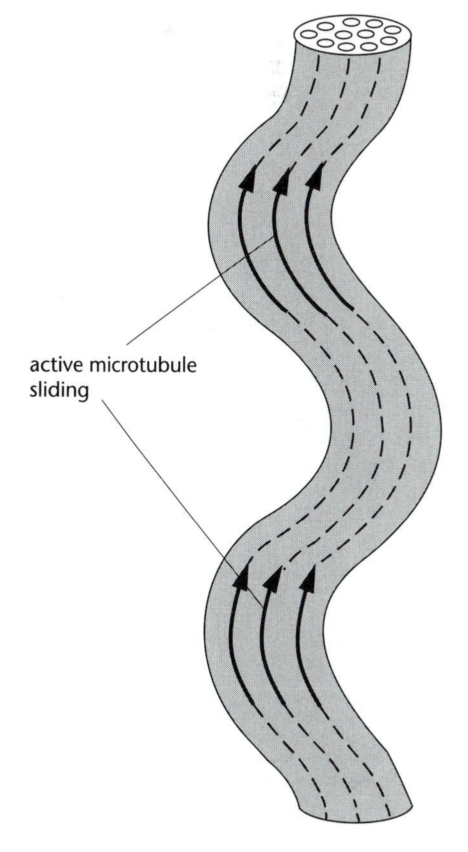


**Figure 14-8** Regions of active microtubule sliding during wave propagation. A planar, approximately sinusoidal, wave in the flagellum is depicted in plan view (as though lying flat on the page). Regions of active sliding are indicated by arrows. Only selected dyneins in the region of the arrows are active—those on the other side of the axoneme and elsewhere are not actively engaged with adjacent doublets and are carried passively by the movements of the structure.

microtubule moving on a surface through the action of bound dynein molecules is thrown into sinuous wavelike motion if its forward progression is blocked by a barrier. The waves in this case are due to two opposing forces, one due to dynein action the other to the flexural rigidity of the microtubule. In a similar manner, the transmission of mechanical events in an flagellar axoneme could cause dynein arms on one side of the axoneme to be detached while those on the other are fully active.

A typical ciliary beat pattern, with its idiosyncratic waveform, is more difficult to explain by passive mechanics. Here we need the equivalent of a distributor in an automobile engine to regulate the sequence of dynein arm activation. The controlling element must be part of the axonemal structure itself, and there are hints that it may be associated with the central apparatus. Observations of cilia on the surface of *Paramecium* show that the central pair of microtubules rotates as a unit by 360° in each beat cycle. Subsequent observations confirm that central pair rotation occurs whenever the ciliary beat is three-dimensional, that is, when the effective and recovery strokes occur in different planes. Axonemes *lacking* the central pair are typically either immotile or perform a simple circular swirling motion.

How does rotation of the central pair trigger dynein arms to fire in sequence? The obvious answer is that some controlling influence is relayed by the radial spokes. Evidence in favor of this interpretation has been obtained from mutants of *Chlamydomonas* lacking the head region of radial spokes. These mutants are nonmotile and axonemes prepared from the flagella are incapable of reactivation with ATP. But the axonemes will still slide following brief exposure to trypsin. Evidently, the dynein arms in such mutants are competent to produce movement, but are inhibited by a defect in the radial spoke head.

## *Ciliary beating is influenced by $Ca^{2+}$ and cAMP*

In general, as already remarked in Chapter 3, cell movements probably evolved in order to carry cells into more favorable locations. Thus we find that the frequency with which cilia and flagella beat responds to environmental factors, especially in free-living organisms. Many factors affect the pattern and frequency of ciliary beating, each of which will operate a distinctive cascade of signaling events within the cell. However, a common element in many cases so far studied is changes in cytoplasmic calcium ions. The responses of *Chlamydomonas* to light described in Chapter 3, for example, are triggered by light-induced changes in cytosolic $Ca^{2+}$ in its two flagella. The complexity of the response is illustrated by the fact that the two flagella show different responses. The flagellum located nearest to the eyespot (see Figure 3-12) is more active at low (nanomolar) calcium concentrations, whereas the other flagellum is more active at high (micromolar) calcium concentrations. Changes in calcium influx produced by incoming light lead to an increased amplitude of one of the two flagella and consequent reorientation of the cell toward the light.

In the avoidance response of *Paramecium* to a physical barrier there is a brief (1–2 second) reversal of its surface cilia (see Figure 3-7). This reversal is accompanied by a change in membrane potential, which can be detected by an intracellular recording electrode. If the posterior portion of the cell is stimulated by tapping with a glass rod, then the normal resting potential of about –30 mV becomes more negative and the cilia beat faster. This change is due to an efflux of $K^+$ ions from the plasma membrane, making the inside more negative. Conversely, if the anterior end of the cell is stimulated (as occurs when it swims into an obstacle) the membrane is depolarized, and a reversal of ciliary beat occurs. Depolarization under these conditions is due to an influx of $Ca^{2+}$ ions.

The role of intracellular calcium has been shown directly by reactivation of demembranated paramecia by ATP. Below 1 μM $Ca^{2+}$, reactivated cell models swim forward, but above this level the power stroke of the cilia is reversed and the cell models swim backward. In a living cell the intracellular level of calcium ions is maintained at a low level by the action of calcium ion pumps in the cell membrane. But stimulation of the membrane, for example when the cell encounters an obstacle, causes $Ca^{2+}$ ions to rush into the cell down their electrochemical gradient. The calcium concentration rises above the threshold for ciliary reversal and the cell backs away from the source of irritation. As soon as the stimulus is removed, the pumps take effect and forward swimming is resumed.[5]

Other second messengers also affect the frequency of cilium beating, especially cyclic nucleotides. In *Paramecium*, for example, cAMP is the signal to increase the frequency of the ciliary beat, and in permeabilized preparations cAMP is absolutely required to regenerate motility. Both $Ca^{2+}$ and cAMP work on cilia lacking their plasma membrane, so whatever their molecular targets they must all be physically bound to the microtubule framework itself.

## Inner dynein arms are regulated by phosphorylation

Summarizing at this point: we have seen that the ciliary axoneme bends through the concerted action of dynein arms, which causes neighboring microtubules to slide. Waveforms of defined shape are possible because the cell can regulate the activity of its dynein molecules and hence its microtubule sliding. Evidence from *Chlamydomonas* mutants and reactivated axonemal models points to the region of contact between radial spokes and dynein inner arms as the most likely site of control.

How is this control achieved? The most likely mechanism is by protein phosphorylation. Thus, mutant axonemes lacking radial spokes are normally paralyzed, but they can be made to move again by adding a protein kinase inhibitor. Restoration of inner arm dynein activity can also be achieved by adding a protein phosphatase (which should remove phosphate groups from proteins). Both a protein phosphatase and a cAMP-dependent protein kinase can be isolated from ciliary axonemes.

A contemporary model of regulation of the ciliary axoneme thus predicts that dynein arms are inhibited by phosphorylation. The site of phosphorylation is probably not the dynein head itself but one of the many polypeptide chains associated with it, the prime candidate being a 138 kDa intermediate chain. In the intact axoneme, radial spokes are predicted to inhibit the phosphorylation of dynein through a specific interaction between the spoke head and a regulatory protein in the dynein complex. Factors such as cAMP and $Ca^{2+}$ ions could act on ciliary beating through a similar mechanism, by modifying the action of endogenous protein kinases and phosphatases. What these kinases and phosphatases are, where they are located, and how their enzymatic activity controls the activity of dynein, are all topics for future research.

## The left–right asymmetry of the body depends on ciliary beating

Mutational defects in cilia and flagella are not restricted to lower organisms. A hereditary defect in humans known as *Kartagener's syndrome* is accompanied by, and perhaps caused by, defects in axonemal structure. Human patients with this condition have chronic bronchitis (because the ciliated epithelia of the airways fail to clear mucus and dust particles) and the males are sterile (due to nonmotile sperm). Cilia and flagella from affected individuals have structural defects, lacking dynein arms and sometimes the central sheath.

Kartagener's syndrome often includes the condition known as *situs inversus* in which the heart and other internal organs are on the wrong

[5] Some mutants of *Paramecium* are unable to move backward even under the most extreme provocation. In these mutants, called *pawn* mutants since they move like the pawns in chess, the defect seems to be in the membrane detection system rather than the axoneme itself.

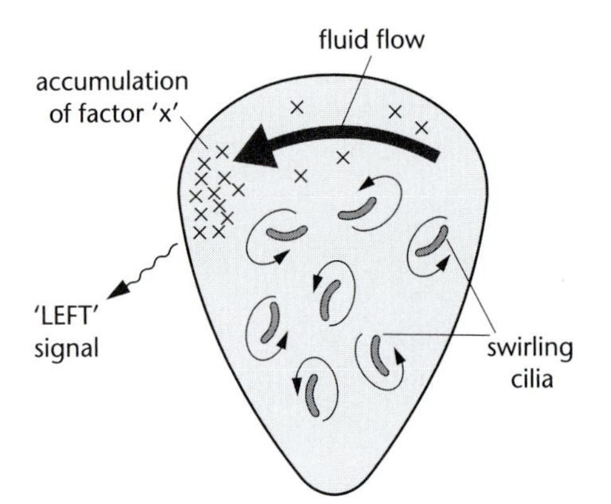


**Figure 14-9** Model for left–right determination. The diagram represents the node region of a mammalian gastrula. Cilia on epithelial cells lining the node rotate counterclockwise, with a continual swirling motion. An as yet unidentified molecule (factor 'x') is supposed to be carried in this flow and to accumulate at the left side of the node. There it interacts with receptors to trigger a cascade of reactions leading to the expression of 'left-specific' genes.

side of the body. This bizarre finding led to an early suggestion that the beating of a cilium early in embryonic development might establish the normal left–right asymmetry of the body. Recent findings add weight to this idea.

Left–right asymmetry in a mammalian embryo appears to be set up during gastrulation, in a transiently formed epithelial chamber known as the *node*. Cells in the node carry cilia on their surfaces that display an unusual type of movement in which they swirl in vortical fashion rather than beat. These cilia are also atypical because they have a 9+0 axonemal structure in which the central pair of microtubules is absent (in this and other features they resemble primary cilia, described in the following chapter). Interestingly, mouse mutants in which the left–right asymmetry of the body is disrupted have been found either to lack nodal cilia or to have nodal cilia that are immotile. One such mutant, prepared by a genetic knockout, lacks a specific kinesin KIF3 thought to be crucial for the assembly of the ciliary axoneme, as we describe in the next chapter.

These results lead to a possible mechanism for the establishment of left–right asymmetry in mammalian development. The idea is that the beating of nodal cilia at a crucial stage of development causes a flow of fluid that concentrates specific molecules at the left of the embryo (Figure 14-9). Binding of these specific molecules to receptors at the left side of the node then triggers expression of left-defining genes.

## Many naturally occurring variants of sperm flagella exist

Far more bizarre variations in axonemal structure exist. In fact, the simple 9+2 pattern is reliably found only in those eucaryotic flagella in which rapid motion through water is at a premium. Cells that swim in other environments have other design limitations and their axonemes are often extensively modified. A few examples were mentioned in Chapter 1, including the 'internal' flagellum of *Trypanosomes*, and the hairy flagella of *Chrysomonads*.

Sperm flagella, which bear the full brunt of natural selection, are remarkably varied in size and structure. Organisms such as sea urchins that utilize external fertilization, in which egg and sperm cells are released into the surrounding water, usually have the canonical 9+2 axoneme. By contrast, organisms that employ internal fertilization produce sperm that have to travel much shorter distances. The nature of their odyssey depends on the geography of the copulatory apparatus and many specialized forms of flagella have evolved.

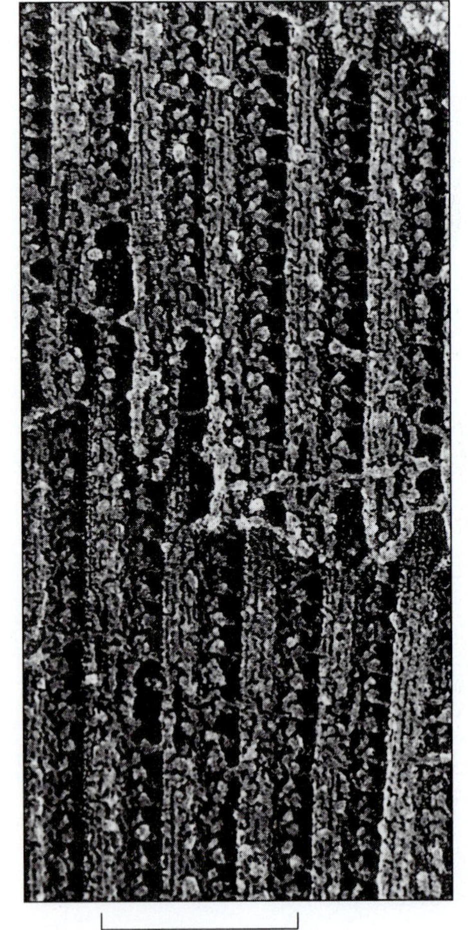


**Figure 14-10** Axoneme of a gall midge sperm. The axoneme was quick-frozen, fractured, and then examined by electron microscopy. Parallel microtubules in planar arrays are revealed by this treatment, each microtubule linked to its neighbor by series of repeating dynein arms. Dynein links consist of a slender stalk to which is attached a bilobed head. (Micrograph courtesy of P. Lupetti et al, *Cell Motil. Cytoskeleton* 39: 303–317, 1998.)

Swimming ability is often far less important for sperm performing internal fertilization, and their flagella are often sluggish and largely vestigial. To mention a few examples: nematodes and crustacea have aflagellate sperm; the sperm of eel and some fish have a 9+0 axoneme (that is, nine outer doublets with central pair absent); platyhelminths have biflagellate sperm; while various species of insects display anything from aflagellate (0) sperm, through nonmotile sperm with 9+0 axonemes, to axonemes containing huge bundles of several hundred microtubules (Figure 14-10).

Human sperm have a flagellum with a central 9+2 axoneme encircled by nine outer dense fibers. The latter are composed of a distinct family of proteins (known as the parergins) that have been suggested to be related to keratins in composition and to function as elastic stiffening agents. The anterior portion of the axoneme is surrounded by a series of crescent-shaped mitochondria arranged in a helical sheath. More distally the axoneme is enclosed by additional cytoskeletal structures including a fibrous sheath and two columnar bundles of filaments. The latter structures presumably restrain the bending of the flagellum and limit its waveform.

## *Large bundles of parallel microtubules bend through coordinated dynein action*

The sliding-microtubule principle is used in other structures than axonemes, as shown by the vigorous undulations of certain primitive protozoa. The flagellate *Saccinobacculus*, which inhabits the hind gut of termites and the wood-eating roach, contains within its cell body a ribbonlike bundle of microtubules that twists and undulates, enabling the cell to crawl over surfaces (Figure 14-11a). Microtubule bundles of this kind, known as *axostyles*, contain thousands of parallel microtubules arranged in a series of flat laminae, stacked on top of each other to form a relatively thick ribbon.

Freeze-etch pictures of the *Saccinobacculus* axostyle show that microtubules in individual lamina are associated through protein bridges

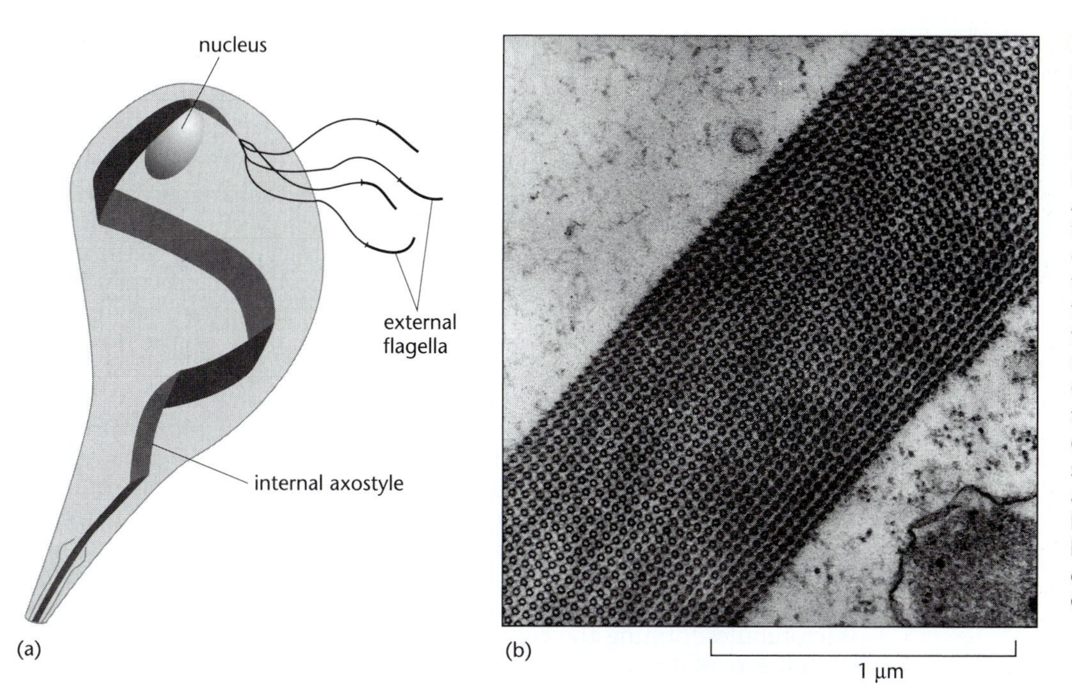


**Figure 14-11** Axostyle. (a) Diagram of the protozoan *Saccinobaculus* showing the large internal bundle of microtubules (axostyle) in its cytoplasm. Active bending and twisting of the axostyle causes the cell as a whole to wriggle its way through the dense matrix of a termite gut, which is its normal habitat. (b) Cross-section through the axostyle viewed by electron microscopy, showing the multiple arrays of microtubules, linked by protein cross-bridges. (Micrograph courtesy of A.V. Grimstone.)

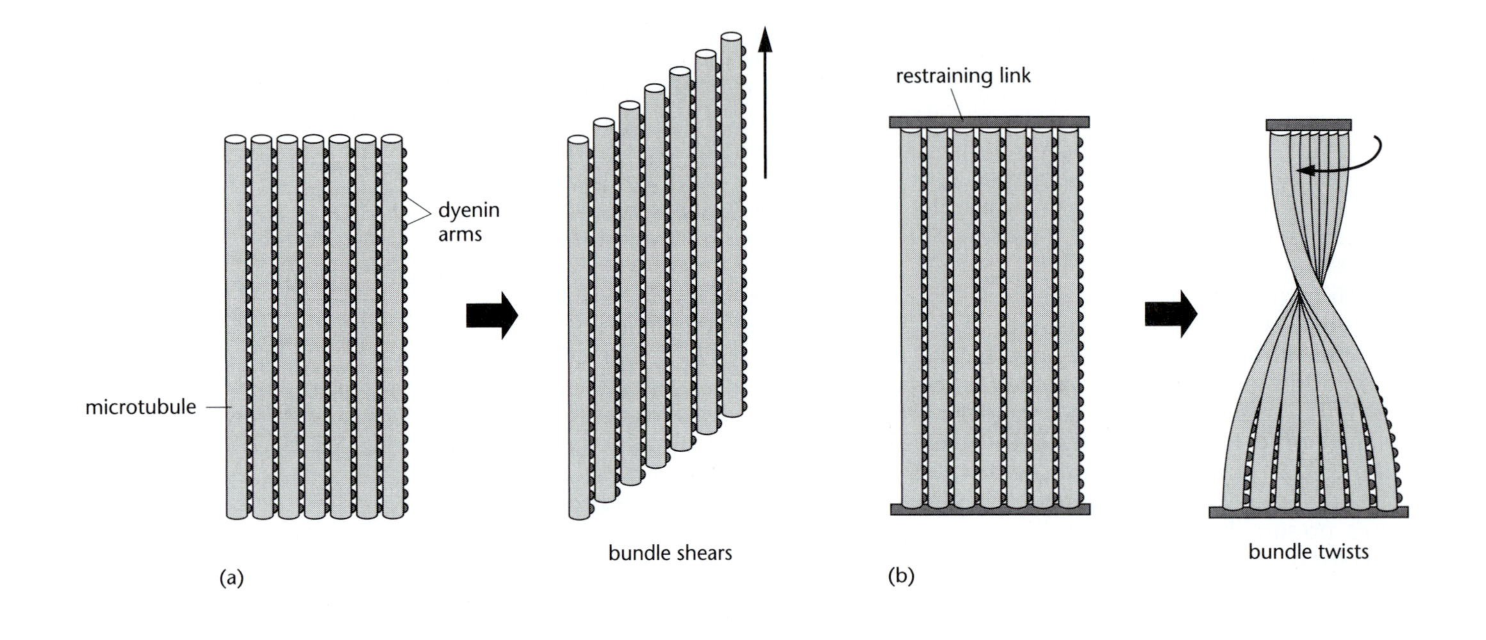


**Figure 14-12** Mechanism of twist generation in an axostyle. (a) The action of multiple dynein arms in a parallel array of microtubules will cause the array to shear. (b) If the motion of microtubules is constrained by protein links at either end of the bundle, shearing converts into a twisting motion.

resembling dynein arms (Figure 14-11b). They can be extracted under similar conditions to those for dynein and, like ciliary dynein, they undergo a change in tilt on removal of ATP (while the axostyle as a whole straightens out). By contrast, the links between microtubules of *adjacent* lamellae are more delicate and ill-defined. It has been proposed that the dynein arms within each lamella change their angle of attachment, causing the adjacent microtubules to slide as they do within the ciliary axoneme. However, since microtubules in the lamellae are anchored both at the top and the bottom of the axostyle, the sliding motion produces a helical twist rather than a bending (Figure 14-12).

Other axostyles generate other movements, including rotation. A bizarre flagellate from the termite gut rotates its head region several times a second, apparently by means of a centrally located axostyle connecting the head to the body! How the rotation occurs, and why, has not been satisfactorily answered.

## Protein bridges can specify the geometry of microtubule bundles

In contrast to the loosely defined arrangement of microtubules in an axon or dendrite, those in cilia and in many other microtubule-containing structures of protozoa have a very precise and often complex pattern in cross-section. The formation of such precise arrays often depends on a template of nucleating structures that specifies the geometry of the microtubules as they assemble. It is also important to have rigid cross-linking proteins that maintain the correct geometry throughout the length of the bundle and there are some instances in which these appear to be the major determinant of form.

In the well-known heliozoan *Actinophrys sol*, the body is roughly spherical with a large central nucleus and a cytoplasm filled with small vacuoles of various kinds. At the periphery of the nucleus are sites that initiate *axopodia*—stiff filaments composed of bundles of microtubules that radiate out as much as 100 μm in many different directions. Electron microscopy shows that the core of an axopodium consists of a curious 'jelly roll' arrangement of two sheets of microtubules, spirally interwound and joined by a series of cross-links (Figure 14-13). In other species, the microtubules forming the complex core of the tentacles originate from a central core of dense material and the nucleus is pushed to one side of the

**Figure 14-13** Cross-section of a feeding tentacle of *Actinosphaerium*. The microtubules in this electron micrograph form two spirally wound sheets, held together by a system of links. Short links join adjacent microtubules in the two spiral sheets; longer links join microtubules along 12 radii and in parallel arrays offset from these radii by an angle of 30°. Formation of the array appears to be through a self-assembly mechanism that depends on the specific angles between cross-links on microtubules (see Figure 14-14). (Micrograph courtesy of John Tucker.)

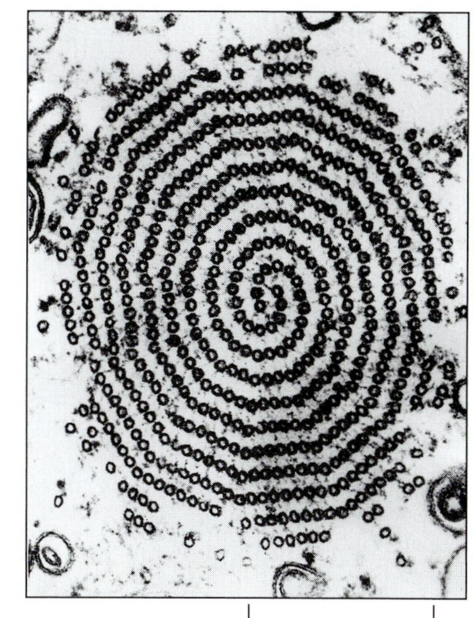


cell, and often is lobulated and penetrated by the bundles of microtubules passing through.

Axopodia are rigid structures but they retract if the microtubules in their core disassemble, as occurs during feeding. Contact with a small organism causes the tip of the axopodia to dilate and become cup-shaped. Singly or in groups they envelope the prey and draw it back into the cell body, while at the same time their axial bundles of microtubules disappear. Microtubule loss and axopodial retraction can also be induced under laboratory conditions by lowering the temperature or exposing the organism to an elevated concentration of $Ca^{2+}$. Disintegration is extremely rapid and under appropriate conditions can be reversed by a reassembly of microtubules that is almost as fast.

The response of heliozoans to cooling and rewarming provides an interesting illustration of the role of specific cross-links in determining microtubule patterns. Microtubules that have disintegrated due to the low temperature grow back upon warming from the central dense region. Initially they are only poorly organized: groups of microtubules are found in small constellations with spacings between microtubules of 7 nm and 30 nm (Figure 14-14). Gradually these small clusters coalesce into larger and larger arrays until the characteristic double-coil with 12-fold symmetry is produced. One of the most important factors in the genesis of this pattern is probably a continual selection for the most stable array, a process analogous to crystallization.

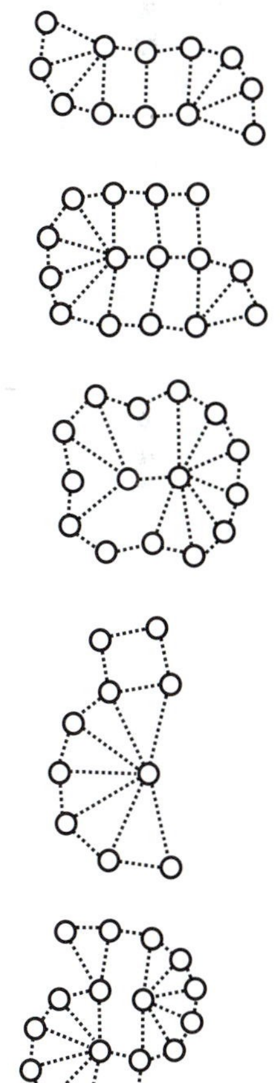

## Outstanding Questions

*What is the atomic-level structure of the dynein molecule? Which region of this molecule makes functional contact with a microtubule (is it the large globular head or the smaller stalk)? Why are there so many different subtypes of dynein in a cilium? How is the sequence of dynein movements in a cilium coordinated? Is it by physical, chemical, or electrical signals? Which kinases and phosphatases modulate beat frequency in response to $Ca^{2+}$ and cAMP? Where are they situated, what are their substrates, and how does phosphorylation affect movement? What drives the transport movements on the surface of cilium? Which factors determine the left–right asymmetry of the mammalian body plan? Are nodal cilia crucial for the establishment of this asymmetry? Would it be possible to create mutants with inverted symmetry? Why do so many cells produce primary cilia and what is their function? Can primary cilia move, and if so, why?*

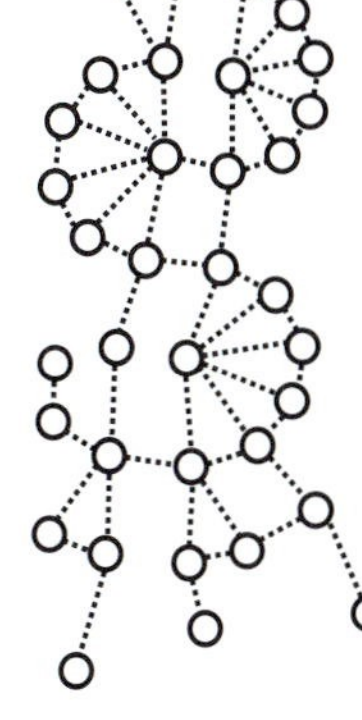

**Figure 14-14** Assembly intermediates of an axopodium. Clusters of microtubules (open circles) seen in cross-section of a feeding tentacle of *Actinosphaerium* after brief cold treatment. The broken lines show how cross-links of two lengths could generate the observed patterns. Random clusters such as these slowly reform the canonical pattern shown in the previous figure, perhaps by a selection of patterns that maximize the number of cross-links.

## *Essential Concepts*

- Cilia have a cylindrical core containing a ring of nine doublet microtubules. These cause the cilium to bend by sliding against each other.
- Side-arms of the motor protein ciliary dynein extend from each microtubule doublet and make contact with the adjacent doublets.
- Contact stimulates an ATPase in the dynein head and, as shown in cell-free motility assays, causes it to walk along the microtubule doublet toward its minus end.
- Over 200 accessory proteins are associated with the flagellar axoneme. These form a variety of axonemal structures such as the dynein arms, the radial spokes, and nexin links that decorate the microtubules.
- One of the protofilaments of the outer doublet microtubules is made not of tubulin but a fibrous protein, tektin, which is related in structure to intermediate filaments. Tektin might act as a molecular ruler to position other projections along the axoneme.
- A central apparatus comprising a central pair of microtubules with attached structures runs down the middle of the axoneme. A series of radial spokes project from the central apparatus toward the inner dynein arms.
- Both the frequency of the ciliary beat and, in some case, its waveform can change in response to environmental influences. Changes are mediated by an increase or decrease in the concentration of second messengers, especially calcium ions and cAMP.
- The often complex waveform of the ciliary beat results from a coordinated inhibitory control over dynein arms by the radial spokes. The spokes may modulate the activity of kinases and phosphatases that act on dynein.
- Axonemes vary widely in the sperm of different species and in modified cilia that act as sense organs in many animal species.
- A human deficiency, called Kartagener's syndrome, is caused by defects in ciliary axonemes. Patients with this condition suffer from chronic bronchitis and immotile sperm.
- Another function often associated with Kartagener's syndrome is an inversion of the left–right asymmetry of the body. Normal asymmetry has now been shown to arise from the asymmetric beating of cilia at an early stage of embryogenesis.
- Even more divergent types of axoneme occur in the motile axostyles found in protozoa, often composed of thousands of parallel microtubules. The assembly of these structures depends on nucleating structures and specific cross-links

## Further Reading


Afzelius, B.A., et al. Flagellar structure in normal human spermatozoa and in spermatozoa that lack dynein arms. *Tissue and Cell* 27: 241–247, 1995.

Brokaw, C.J. Computer simulation of flagellar movement. *Cell Motil. Cytoskeleton* 42: 134–148, 1999.

de Loubresse, N.G., et al. A contractile cytoskeletal network of *Paramecium*: the infraciliary lattice. *J. Cell Sci.* 351: 351–364, 1988.

Fawcett, D.W., Porter, K.R. A study of the fine structure of ciliated epithelia. *J. Morphol.* 94: 221–230, 1954.

Gee, M., Vallee, R. The role of the dynein stalk in cytoplasmic and flagellar motility. *Eur. Biophys. J.* 27: 466–473, 1998.

Gibbons, I.R., Rowe, A.J. Dynein: a protein with adenosine triphosphate activity from cilia. *Science* 149: 424, 1965.

Grimstone, A.V., Cleveland, L.R. The fine structure and function of the contractile axostyles of certain flagellates. *J. Cell Biol.* 24: 587–400, 1965.

Habermacher, G., Sale, W.S. Regulation of flagellar dynein by phosphorylation of a 138-kD inner arm dynein intermediate chain. *J. Cell Biol.* 136: 157–176, 1997.

King, S.J., Dutcher, S.K. Phosphoregulation of an inner dynein arm complex in *Chlamydomonas reinhardtii* is altered in phototactic mutant strains. *J. Cell Biol.* 136: 177–191, 1997.

Lohret, T.A., et al. A role for katanin-mediated axonemal severing during *Chlamydomonas* deflagellation. *Mol. Biol. Cell* 9: 1195–1207, 1998.

Luck, D.J.L. Genetic and biochemical dissection of the eukaryotic flagellum. *J. Cell Biol.* 98: 789–794, 1984.

Nojima, D., et al. At least one of the protofilaments in flagellar microtubules is not composed of tubulin. *Curr. Biol.* 5: 158–167, 1995.

Nonaka, S., et al. Randomization of left–right asymmetry due to loss of nodal cilia generating leftward flow of extraembryonic fluid in mice lacking KIF3B motor protein. *Cell* 95: 829–837, 1998.

Norrander, J.M., et al. Structural comparison of tektins and evidence for their determination of complex spacings in flagellar microtubules. *J. Mol. Biol.* 257: 385–397, 1996.

Porter, M.E. Axonemal dyneins: assembly, organization, and regulation. *Curr. Biol.* 8: 10–17, 1996.

Redeker, V., et al. Polyglycylation of tubulin: a posttranslational modification in axonemal microtubules. *Science* 266: 1688–1691, 1994.

Ringo, D.I. Flagellar motion and the fine structure of the flagellar apparatus in *Chlamydomonas reinhardtii*. *J. Cell Biol.* 33: 543–571, 1967.

Satir, P. Studies on cilia III. Further studies on the cilium tip and 'sliding filament' model of ciliary motility. *J. Cell Biol.* 39: 77, 1968.

Satir, P., et al. The control of ciliary beat frequency. *Trends Cell Biol.* 3: 409–412, 1993.

Smith, E.F., Lefebvre, P.A. The role of central apparatus components in flagellar motility and microtubule assembly. *Cell Motil. Cytoskeleton* 38: 1–8, 1997.

Summers, K.E., Gibbons, I.R. Adenosine triphosphate-induced sliding of tubules in trypsin-treated flagella of sea-urchin sperm. *Proc. Natl. Acad. Sci. USA* 68: 3092, 1971.

Tamm, S. $Ca^{2+}$ channels and signalling in cilia and flagella. *Trends Cell Biol.* 4: 305–311, 1994.

Tamm, S.L., Tamm, S. Rotary movements and fluid membranes in termite flagellates. *J. Cell Sci.* 20: 619–639, 1976.

Walczak, C.E., Nelson, D.L. Regulation of dynein-driven motility in cilia and flagella. *Cell Motil. Cytoskeleton* 27: 101–107, 1994.

Yang, P., Sale, W.S. The $M_r$ 240,000 intermediate chain of *Chlamydomonas* flagellar inner arm dynein is a WD-repeat protein implicated in dynein arm anchoring. *Mol. Biol. Cell* 9: 3335–3349, 1998.

# CHAPTER 15

# Centrioles and Basal Bodies

One of the most fascinating questions to ask about cilia is, how does the cell make them? We have already seen that cilia and flagella vary enormously in the number, position, and length in different cells. Thus, a sea urchin sperm has one long flagellum, whereas an epithelial cell in the human body might have its entire apical surface covered with a lawn of short cilia. Some protozoa have rings of cilia arrayed like the arms of a hydra around an oral cavity, whereas other protozoa walk over surfaces by means of leglike bunches of cilia. It is difficult enough to imagine how the more than 200 polypeptides in an axoneme can be brought together and assembled in the correct amounts and position to make the distinctive 9+2 structure of an axoneme. When we realize that the lengths of cilia and flagella and their positions on the cell surface are precisely regulated, the task seems even more challenging

An essential element in the growth of cilia and flagella is an enigmatic structure known as a *basal body*—a compact barrel of short microtubules that nucleates the assembly of the axoneme. Basal bodies are closely related to the centrioles found in the centrosome (or spindle pole) and under some circumstances the one can convert into the other. Evidently, in order to produce a set of cilia the cell must first create basal bodies in sufficient numbers and then position them beneath the plasma membrane in the correct location of the cell surface. Once in position, each basal body has to be supplied with tubulin, dyneins, radial spoke proteins, and so on, and it must then guide their assembly into a cilium. The protein building blocks have to be delivered to the growing point of the cilium and assembly must continue until the correct length for the particular cilium or flagellum is reached. The molecular bases of these various steps, so far as they are presently understood, form the subject of this chapter.

**Figure 15-1** Ciliated cell. Scanning electron micrograph of a fetal mouse esophagus showing an isolated cell that has developed cilia of uniform length on its surface. In other regions, shorter microvilli are present. (Courtesy of Raymond Calvert.)

## *Flagella and cilia grow rapidly to a characteristic length*

Despite their complex structure, cilia and flagella often form with impressive rapidity. In the fetal mouse esophagus, for example, each epithelial cell develops about 250 cilia during the first 12 hours after birth, with individual cilia growing in a few hours to a predetermined length (Figure 15-1). A sea urchin embryo becomes a free-swimming blastula in as little as 8 hours after fertilization, generating about a thousand cilia in 5 hours or less. Even more rapid ciliogenesis is associated with the replication of protozoa: division of *Paramecium* entails the rapid proliferation of fields of surface cilia; as *Chlamydomonas* enters mitosis, its two flagella are resorbed and subsequently restored rapidly in the two postmitotic daughters.

*De novo* formation of cilia or flagella is a part of the complex life cycles of many protozoa. *Naegleria gruberi*, some strains of which may cause fatal meningitis in man, grows normally as an amoeba, but it can develop flagella under specific conditions. For example, if a laboratory culture is treated with deionized water, the amoeboid cells round up, develop basal bodies and paired flagella, and swim into the surrounding medium. The entire process takes 12 hours, with flagellar elongation itself taking only about 30 minutes.

Many cells have the capacity to shed their cilia or flagella in response to environmental stimuli. In *Chlamydomonas*, for example, spontaneous deflagellation follows exposure to acidic solutions or those containing high concentrations of calcium ions. These cause the nine outer microtubule doublets to be severed at the transition zone between the axoneme and the basal body.[1]

Cilia and flagella are replaced rapidly by cells after this experimental removal, and in almost every case, initially rapid but decelerating kinetics have been observed. Thus, flagellar regeneration in *Chlamydomonas* begins without a significant pause at about 0.2 μm/min and gradually slows as the flagella approach their original length of 10–12 μm.

The final length of cilia and flagella is genetically determined, being characteristic of the cell and even the position on the cell. The two flagella of *Chlamydomonas* are normally about 10 μm long, but mutants have been isolated in which they are shorter than usual, or much longer (20–30 μm). However, the final length of a flagellum is not so rigidly defined that it cannot be influenced by external conditions, and a variety of experimental treatments—such as a brief exposure to trypsin, or to hypotonic conditions—can increase flagellar length. The mechanism by which cells control the length of their cilia and flagella does not seem to rely on the regulation of supply of tubulin and dynein. As discussed below, flagella will grow to a predetermined length even when these proteins are present in large excess in the cytoplasm.

## *A pool of assembly-competent axonemal proteins exists in the cytoplasm*

Rapid production of large numbers of cilia or flagella requires an abundant source of axonemal proteins. Newly made axonemal proteins first enter the cytoplasm in an unassembled state, becoming part of a pool of assembly-competent precursors. This pool, which is also fed by proteins formed by the disassembly of existing cilia or flagella, provides the immediate source of protein for axonemal formation.

In some cases, such as sea urchin eggs, the pool of ciliary precursors is very large, providing enough material for multiple rounds of ciliogenesis. In other cells, the pool size is smaller, and may become exhausted unless replenished by new synthesis. Thus, if *Chlamydomonas* cells are deflagellated in the presence of cycloheximide, a drug that inhibits most protein synthesis, they regenerate flagella of half the normal length. This experiment suggests that there is a pool of axonemal protein sufficient to make

[1] Severing of microtubules during deflagellation is probably caused by katanin—the enzyme mentioned in Chapter 13 that also mediates active severing of microtubules during mitosis.

(a)

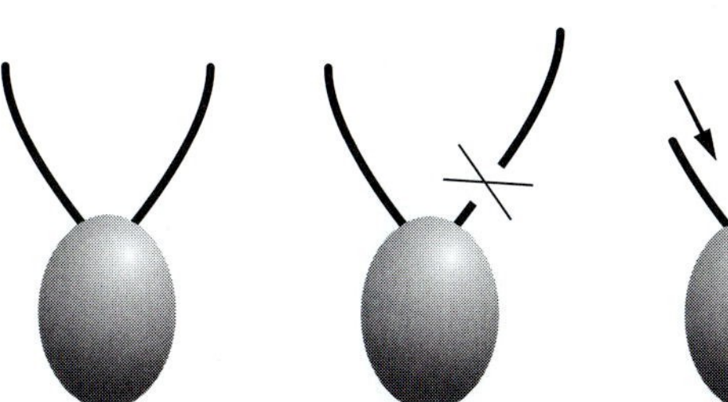

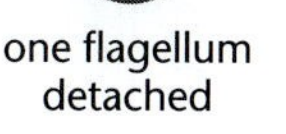

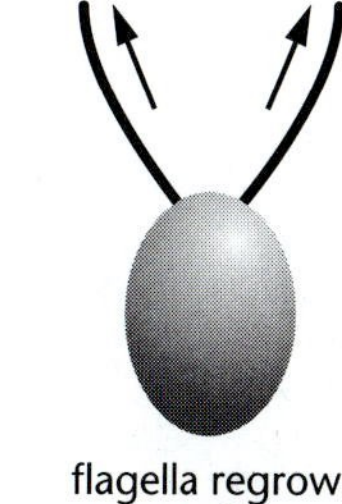

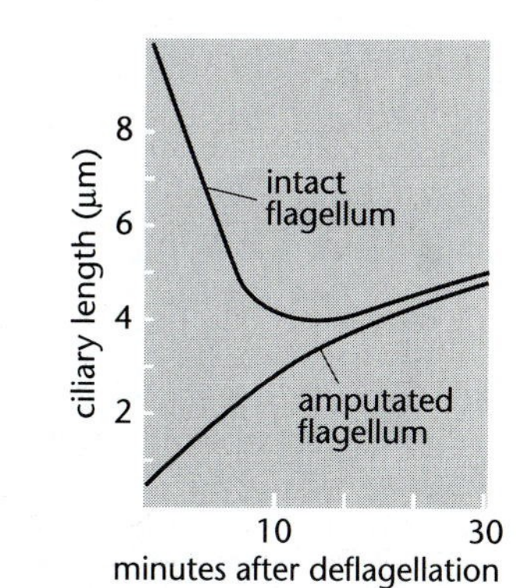


**Figure 15-2** Regulation of flagellar length in *Chlamydomonas.* (a) Detachment of one flagellum in the presence of cycloheximide (an inhibitor of protein synthesis) causes the surviving flagellum to shorten. When it reaches the length of the regenerating flagellum, then both grow in parallel. (b) Time course of flagellar regeneration.

half the normal length of flagella in the cytoplasm before deflagellation. Moreover, resorption of flagella—which can be induced by removing $Ca^{2+}$ from the medium or exposing cells to low concentration of specific drugs such as caffeine or IBMX (an inhibitor of phosphodiesterases)—enhances the rate of flagellar regeneration, presumably because it increases the pool of axonemal precursors in the cytoplasm.

Interplay between the assembled ciliary proteins and the cytoplasmic pool is revealed in ingenious experiments in which just one of the two flagella is removed by mechanical detachment in the presence of protein synthesis inhibitors (Figure 15-2a). In this situation, the remaining flagellum often decreases in length transiently until the newly formed flagellum reaches the same length. Thereupon the two flagella grow out in parallel—subunits of one axoneme evidently being used to produce a second axoneme (Figure 15-2b).

Amputation of flagella stimulates a rapid and coordinate synthesis of flagellar proteins. A particularly striking example occurs in *Chlamydomonas* gametic cells, a differentiated cell type produced after nitrogen starvation. Gametes have greatly reduced numbers of ribosomes and a low basal level of protein synthesis, which makes it possible to label and distinguish the synthesis of many flagellar proteins. Within 30 minutes of deflagellation the synthesis of α- and β-tubulins, dyneins, and some 100 other flagellar proteins is greatly enhanced. Synthesis rates for the major flagellar proteins remain maximal for 60–90 minutes then return to earlier levels by 120–240 minutes.

## *Cilia and flagella are nucleated by basal bodies*

Sections of cilia and flagella examined in an electron microscope show that the axoneme continues down into the cell and terminates in a basal body—a cylindrical structure about 500 nm long and 250 nm in diameter, made of nine groups of triplet microtubules and accessory structures (Figure 15-3). The nine triplets have a distinct polarity and, like the axoneme, a unique handedness (Figure 15-4).

Basal bodies are closely similar to the centrioles found in the centrosome of many cells, but they are associated with different structures. Most centrioles, as we saw in Chapter 11, are surrounded by dense amorphous material that sometimes reflects their ninefold symmetry. This pericentriolar material, which includes rings of γ-tubulin, is crucially important as the major site of microtubule nucleation in interphase and mitotic cells. Basal bodies lack this amorphous material but are instead attached to the axoneme and to contractile structures, described below, that anchor them to other parts of the cytoskeleton.

In the experiments just described, in which the flagella or cilia are removed by mechanical or drug treatment, it is clear that basal bodies,

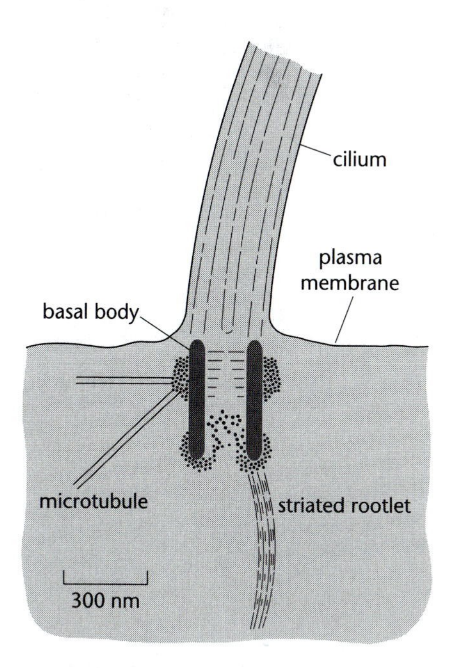


**Figure 15-3** Basal body. Schematic section through a cilium shows the position of its basal body and associated structures.

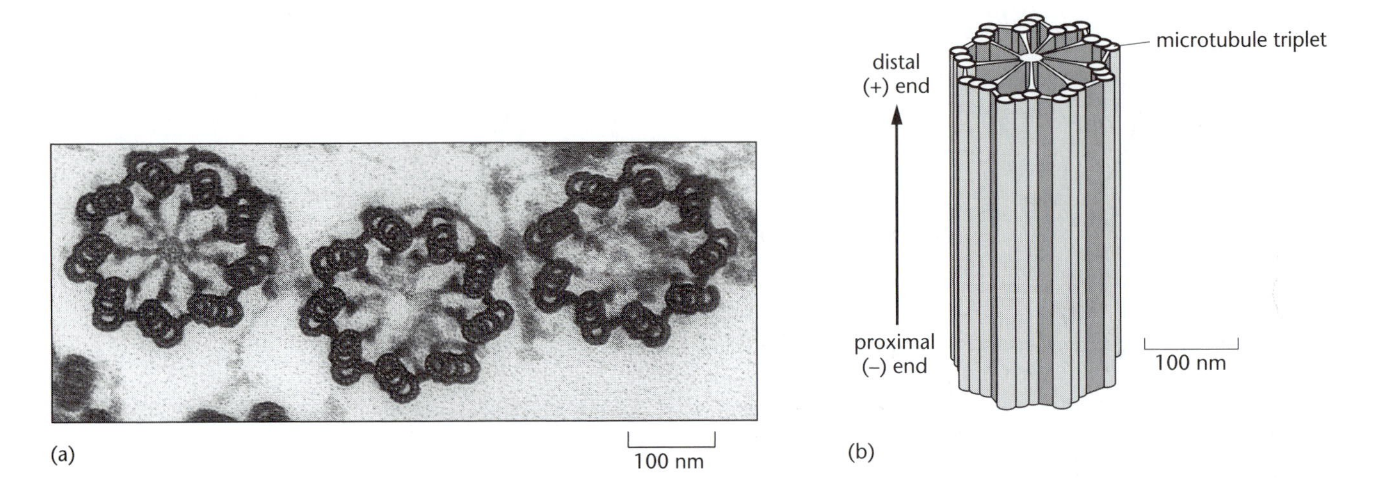


**Figure 15-4** Structure of a basal body. (a) Electron micrograph of a cross-section through three basal bodies in the cortex of a protozoan. Each basal body forms the lower portion of a ciliary axoneme. A centriole has a closely similar structure. (b) Schematic drawing of a centriole or basal body. It is composed of nine sets of triplet microtubules, each triplet consisting of one complete microtubule fused to two incomplete microtubules. Other proteins form links that hold the cylindrical array together. Note that the structure as a whole has a distinct polarity and handedness due to the asymmetric positioning of the nine triplets. The *distal* end of the centriole, corresponding to the plus end of its triplet microtubules, is the end that grows a cilium. The *proximal* end, corresponding to the microtubule minus ends, faces into the cytoplasm and is often anchored by striated rootlets, as discussed later in the chapter. (Micrograph courtesy of D.T. Woodrum and R.W. Linck.)

remaining in position specify the location and orientation of newly formed cilia and flagella on the cell surface. The twin flagella of *Chlamydomonas*, for example, emerge from the anterior end of the cell at a precisely defined angle of close to 90°, the two flagellar basal bodies being held rigidly at the appropriate angle by striated rootlets (described below). In the cortex of ciliates such as *Paramecium*, or close to the luminal surface of ciliated epithelial cells, huge fields of regularly arranged basal bodies are maintained rigidly in position by links made of cytoskeletal proteins (Figure 15-5).

Removal of cilia by mechanical or drug treatment leaves the basal bodies beneath the membrane. They may then be isolated by disrupting the cell, followed by differential centrifugation, and the isolated basal bodies will then nucleate the growth of a bundle of microtubules using tubulin dimers from mammalian brain. As in the case of fragments of axonemes, regrowth under these conditions produces only singlet microtubules, indicating that additional factors are necessary to nucleate the detailed structure of a native axoneme.

## Axonemal parts preassemble near the basal body

The assembly and maintenance of the ciliary axoneme presents a eucaryotic cell with a logistical problem of some magnitude. As already stated, the organelle is made of more than 200 distinct polypeptides which, after being made in the cytoplasm, must find their way to the flagellum. Furthermore, radioactive labeling experiments (some performed in the late 1960s) reveal that most newly synthesized proteins in a regenerating *Chlamydomonas* axoneme are added to its distal tip. So they not only have to find their way to the basal body but also must move up the axoneme to its tip before being incorporated. Indeed, cilia are now realized to utilize an assembly line of considerable sophistication with parts of the axoneme being prefabricated near the base of the cilium and then carried to the end of the growing structure before being put into position.

Assembly begins with the synthesis of the messenger RNA molecules (mRNAs) encoding axonemal proteins and their translation into protein molecules. Synthesis of mRNA (transcription) takes place in the cell nucleus but there is evidence that manufacture of the axonemal proteins (translation) may occur close to the basal body. In the case of muscle development described in Chapter 10, and in other examples to be mentioned in Chapter 19, mRNA molecules have been found to move through the cytoplasm to specific locations. Similarly, mRNA molecules encoding ciliary proteins have been found (in *Naegleria* and elsewhere) to move close to basal bodies before being translated.

Once proteins are made close to the basal body they assemble into prefabricated parts of axonemal structures, such as radial spokes or inner

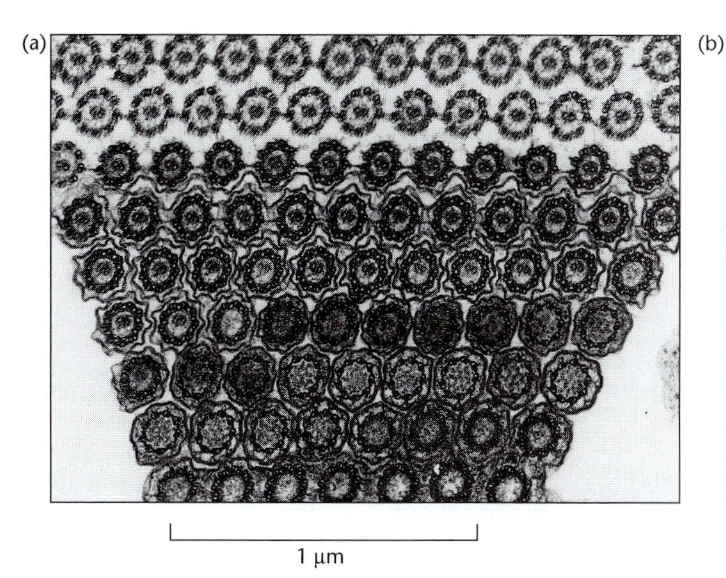


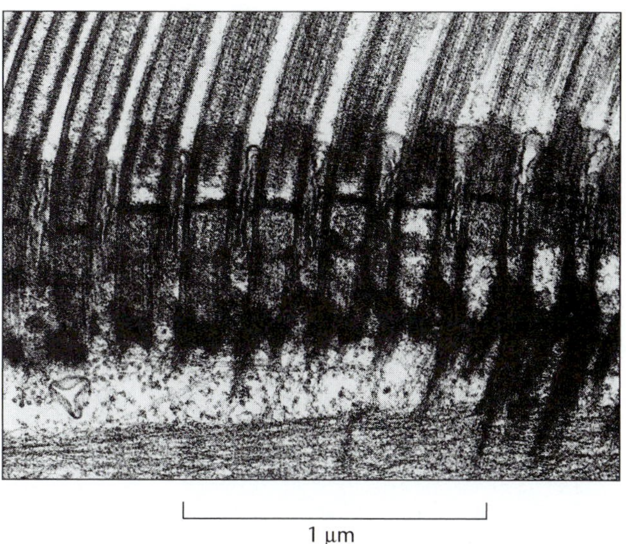


**Figure 15-5** Array of basal bodies. (a) Field of cilia forming the macrocilia of the ctenophore *Beroë* shown in a grazing section that moves from the nine doublets plus central pair of the axoneme (*top*) to the nine triplets of the basal body (*bottom*).
(b) Vertical section through the cilia showing their basal bodies and ciliary rootlets. (Courtesy of Sidney Tamm.)

dynein arms. These complex structures are then delivered to the correct location in the growing cilium or flagellum by a motor-driven process termed *intraflagellar transport* (*IFT*).

## *Motor proteins move rafts of protein to the tip of the cilium*

Close scrutiny of living cilia by interference optics reveals a continual traffic of small structures sandwiched between the plasma membrane and the outer microtubule doublets of the axoneme. These structures, which move up and down the cilium at speeds of 2–4 μm/sec, consist of a varying number of electron-dense particles ('IFT particles') assembled into linear arrays called 'rafts.' Although the detailed composition of rafts is still under investigation, they appear to include preassembled parts of the axoneme together with 15 or so polypeptides that are not axonemal precursors. The rafts also appear to be associated with motor proteins one of which, the product of the *fla10* gene, seems to be especially important for ciliary transport. The *fla10* gene product is a heteromeric kinesin (termed kinesin II, or KIF3) with two distinct motor domains linked to a third domain (see Table 12-1). IFT particles collect together into rafts of varying lengths, which then move up microtubule doublets to the tip of the cilium.

When they arrive, the rafts presumably disperse and their parts are incorporated into the axoneme. Small particles containing unused proteins then return by the same route but in the opposite direction, being in this case moved by a cytoplasmic dynein (Figure 15-6). The complete cycle continues without ceasing, even in cilia and flagella that have grown to a stable length. Indeed, because of this transport, the contents of the axoneme are being continually turned over—with fresh molecules adding to the tip and then moving down to the base. Inactivation of the specific kinesin II or cytoplasmic dynein involved in this process therefore has important consequences even for nongrowing cilia. Blocking kinesin both prevents particle movement and causes flagellar shortening, whereas blocking the cytoplasmic dynein leads to cilia engorged by an accumulation of transport particles.

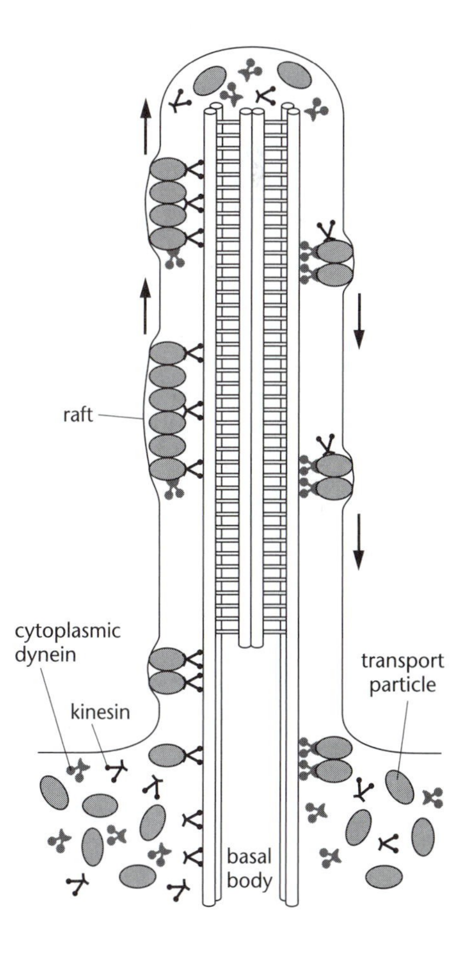


**Figure 15-6** Active transport in the assembly of a cilium. Protein structures needed for the axoneme are preassembled into particles at the base of the cilium. They then move by means of an associated kinesin to the tip of the structure, where they are incorporated into the growing axoneme. Unused components, together with products of axonemal turnover, return by the same path, in this case carried by a cytoplasmic dynein.

**Figure 15-7** Primary cilium. Side view of a primary cilium about 10 μm long on the surface of a kidney cell in tissue culture. The conspicuous swelling at the tip is typical of primary cilia on these cells, but its significance is unknown. (Courtesy of Conly Rieder.)

The details of this fascinating process are important for an understanding of how cilia grow to specific lengths. Clearly, the assembly of transport particles and their movements up and down the growing axoneme are prime targets for agents that regulate axonemal assembly, such as calcium ions, G proteins, and protein kinases. Moreover, the mechanisms used to control ciliary growth might give us clues to the assembly of other complex structures in a cell. It is now clear, for example, that the targeted knockout of the heteromeric kinesin in mouse retina causes slow degeneration of the rod outer segments, which are based on a modified cilium.

## *Many cells produce an apparently nonfunctional primary cilium*

In sea urchin embryos, ciliary growth seems to occur in three distinct stages. First to form are short cilia, about 7 μm long and lacking the central pair of microtubules. These immotile cilia then mature and grow into 18 μm long, typical 9+2 beating cilia found over most of the surface of the embryo. At a specific region of the cell, however, a tuft of cilia elongates further to about 30 μm. Injection of embryos with antibodies to kinesin suppresses all but the short, immotile cilia, suggesting that kinesins might be needed to supply intermediates for the elongation of outer doublets and for beating.

Interestingly, short cilia lacking the central pair (9+0) occur naturally in many cells. These *primary cilia* grow spontaneously from one of the two centrioles of many vertebrate cells during interphase. They have been detected in cells of most tissues of the vertebrate body, including brain, liver, lungs, spleen, smooth muscle, and testis.[2] They are also frequently encountered in cultured cell lines, the length being a characteristic of the strain (Figure 15-7). Primary cilia are resorbed during mitosis, usually during prophase, and reappear early in $G_1$. Their function is unknown, although in some situations they could perform a sensory function, such as monitoring fluid flow. Note that although primary cilia are usually assumed to be nonmotile, some 9+0 cilia can indeed move, as we saw in the previous chapter (see Figure 14-9).

The development of primary cilia occurs either deep-seated within the cell, from the region of its centrosome, or at the cell periphery, one of the two centrioles having migrated to a position close to the plasma membrane. Contact with membrane appears to be obligatory for the development of the axonemal shaft and, in the case of a site deep in the cell, the membrane usually comes from the Golgi apparatus. In light of experiments mentioned above, it is tempting to suppose that the differences between primary cilia and normal cilia might arise from the absence, or inactivation, of certain kinesins involved in intraflagellar transport.

## *Contractile bundles control the position and orientation of flagella*

As mentioned above, basal bodies are usually anchored to other cytoskeletal structures in the cell. One of the most widespread and interesting attachments of this type are *striated rootlets*—bundles of 6 nm filaments that show calcium-sensitive contractile or elastic behavior (Figure 15-8). These filaments are composed in part of a 20 kDa protein *centrin* (also known as caltractin), a calcium-binding protein related to calmodulin, and troponin C (Chapter 9). Centrin undergoes a large change in conformation when it binds $Ca^{2+}$, so that bundles of centrin filaments contract when intracellular levels of $Ca^{2+}$ rise above a certain level.

[2] Interestingly, the two centrioles in a pair differ in their ability to form a primary cilium. As discussed later in the chapter, they can also be distinguished morphologically and differ in lineage, since one was formed as the daughter of the other.

In *Chlamydomonas* and related green algae, centrin forms calcium-sensitive fibers that connect basal bodies to one another and to the nucleus. Fibers made of centrin are also located in the transitional region between the flagellar base and the basal body.

The bridge of centrin fibers between the two basal bodies in *Chlamydomonas* appears to control the orientation of the two flagella. Mutants with defective connecting fibers, for example, are unable to coordinate flagellar beating. In wild-type cells an increase in $Ca^{2+}$ (caused by collision with an object, or exposure to intense light) causes the bridge to shorten and the angle between the two basal bodies to decrease. This throws the two flagella forward from their customary 'breast stroke' position, briefly producing parallel flagellar-type waves that drive the cell backward. A similar change in orientation can be produced in permeabilized cell models by increasing the level of $Ca^{2+}$ above $10^{-7}$ M. An even more dramatic reversal is seen in other unicellular organisms (Figure 15-9).

Centrin in other locations may function during the spontaneous deflagellation shown by *Chlamydomonas*. The stellate fibers sever the ciliary axonemes by displacing them laterally and the striated rootlets pull the nucleus toward the basal body following deflagellation. The latter movement might facilitate delivery of mRNA molecules to the base of the flagellum where they can be translated 'on site'—surely the most efficient strategy for rapid regeneration.

Proteins closely related to centrin are found in other structures in protozoa, such as the contractile stalks of *Vorticella* (Chapter 4) and a lattice of protein filaments that hold basal bodies in the cortex of *Paramecium*. Numerous centrin isoforms are also found in vertebrates, including humans. In some cells, striated bundles of centrin are found attached to basal bodies, presumably helping to maintain their correct orientation (see Figure 15-8). Centrin can also be detected by means of antibodies in the lumen of centrioles. It has been suggested that in the latter location, centrin could be involved in the calcium-induced release of microtubules from the centrosome or in the replication of centrioles, but no-one really knows.

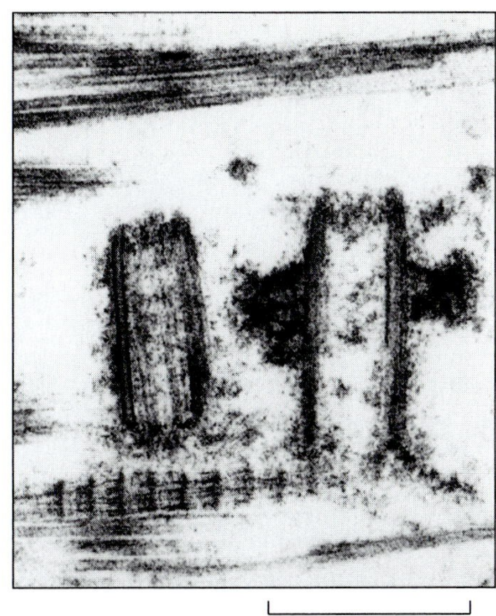


**Figure 15-8** Striated rootlets. Section through the base of a microtubule array in a supporting cell of the guinea-pig ear. Two centrioles are seen, one of which is anchored by a striated bundle of centrin filaments. (Micrograph courtesy of John Tucker.)

## *Basal bodies can form centrioles, and vice versa*

We see that centrioles and basal bodies are associated with two distinct types of microtubule array—the sunburst of microtubules that radiates into the cytoplasm in interphase or mitotic cells, and the tightly structured cylindrical array of microtubules in the ciliary axoneme. In both cases, the organelle appears to act as a MTOC (microtubule-organizing center) but its precise function differs in the two cases.

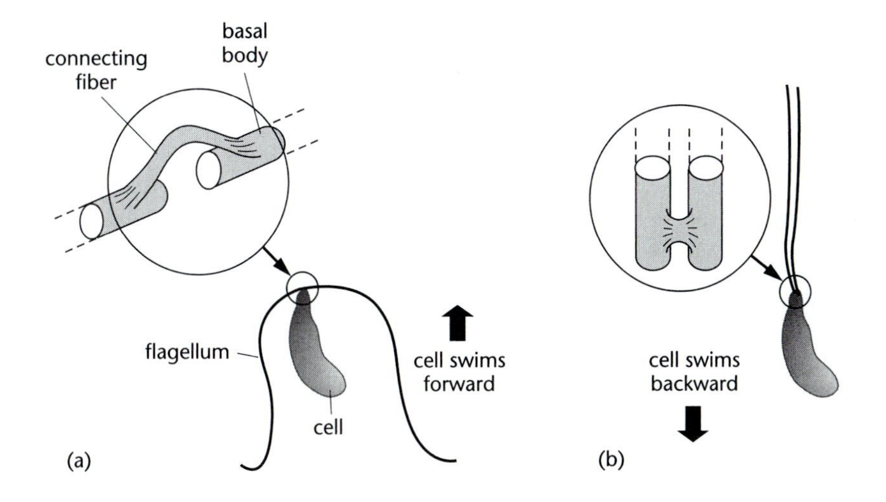


**Figure 15-9** Reorientation of basal bodies in the green alga *Spermatozopsis similis*. (a) During forward swimming the basal bodies are oriented antiparallel at an angle of 180° to each other. (b) During backward swimming the basal bodies are in a parallel orientation. The reorientation is caused by a $Ca^{2+}$-induced contraction of the band of filaments connecting the two basal bodies. A similar change, although less extreme occurs in *Chlamydomonas*.

In the centrosome of interphase cells and the pole of mitotic cells, centrioles do not nucleate microtubules directly. Although microtubules radiate from the general area of the centriole, they actually terminate in the rings of γ-tubulin embedded in a loose matrix of protein surrounding the centriole. A major component of this matrix is a filamentous protein called pericentrin (unrelated to the calcium-binding protein centrin mentioned above). Centrioles do seem to stabilize the form of the centrosome, and to maintain its compact form. But they cannot be essential for the formation of mitotic or cytoplasmic arrays of microtubules, since plant cells and cells in the early stages of embryogenesis of some amphibians lack centrioles

Basal bodies, by contrast, are indispensable. They provide nuclei for the assembly of microtubules in ciliary axonemes and act as templates for its distinctive ninefold symmetry. Here the complete centriolar morphology appears to be essential; if it is absent no cilia are formed.

These observations argue that the principal function for which centrioles evolved was to make cilia. Their association with centrosomes and mitotic poles might have arisen subsequently as a convenient way of maintaining progenitors of basal bodies through successive cell divisions. In support of this thesis, individual centrioles frequently convert into basal bodies in the normal course of cellular differentiation, as in the formation of cilia for use in sensory structures. Other examples include the growth of the sperm tail axoneme during spermatogenesis, which in many species arises from one of the two centrioles that is retained during male meiosis, and growth of the two anterior flagella of *Chlamydomonas* from the pair of centrioles formed during a preceding cell division.[3] It is relevant to note also that cells of higher plants, which lack centrioles in their centrosomes, are also notable for the absence of ciliated structures.

## *New centrioles usually form close to mature centrioles*

Cilia come from basal bodies and basal bodies come from centrioles. So where do centrioles themselves come from? How are they formed during the cell cycle? And how are large numbers of basal bodies generated in cells with fields of cilia on their surfaces?

In most animal cells, the pair of centrioles is closely associated with a *centrosome*—an irregular matrix containing the filamentous protein pericentrin together with rings of γ-tubulin, which act as nucleation sites for microtubules (see Chapter 11). The centrosome replicates independently of the nucleus and appears to have a controlling influence on the division of the cell as a whole. If the nucleus is physically removed from a sea urchin egg, or if nuclear DNA replication is blocked by the inhibitor aphidicolin, cycles of centrosome doubling and division proceed almost normally. In the eggs of *Drosophila* treated with aphidicolin, proliferating centrosomes dissociate from the blocked nuclei and move through the cytoplasm to coordinate cleavage. The result is cells with centrosomes but no nuclei.

In most animal cells, centrosomal duplication is accompanied by the separation of the pair of centrioles and the growth of new centrioles beside the old ones. New centrioles are usually at right angles to the old ones, and lie near their basal part. The primordial centriole, called the procentriole, first appears as a condensation of dense material about the same diameter as a centriole but devoid of microtubules. Soon after its first appearance, a cartwheel structure with ninefold symmetry can be seen within the dense material. The new centriole then elongates by accretion of material at its free end and polymerization of tubulin to form first singlet microtubules, then doublets and triplets to build the centriole's walls. Some of the tubulin molecules that assemble into these triplet microtubules are of δ-tubulin, an isoform limited to centrioles and basal bodies. The newly formed centriole separates from the parent centriole

[3] The opposite conversion can also take place: if a preparation of basal bodies is injected into a sea urchin egg it will generate multiple mitotic spindles. But this is not a common occurrence in the living cell.

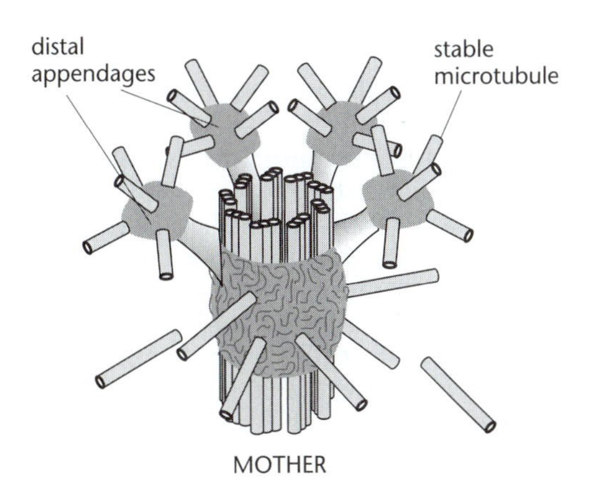


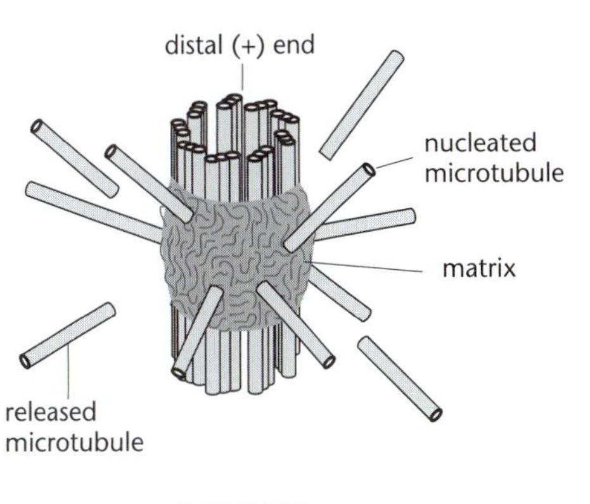


**Figure 15-10** Mother and daughter centrioles. Both types of centrioles have the ability to nucleate microtubules from an associated matrix of γ-tubulin and pericentrin. However, only the mother centriole can establish a stable radial array of microtubules, such as that in interphase cells and the mitotic spindle. In this case, the microtubules are anchored to a series of 'appendages,' about nine in number, that project from the distal end of the mother centriole. The appendages (only four are shown here, for clarity) contain proteins such as ninein that stabilize the minus ends of microtubules released from the matrix.

but maintains its perpendicular orientation to it. Then the two members of the original centriole pair separate and move to opposite poles of the dividing cell, each with its newly formed daughter centriole.

## *Mother and daughter centrioles differ in structure and function*

The two centrioles of a pair are not identical. They differ in detailed anatomy and in their capacity both to nucleate cytoplasmic microtubules and to seed a primary cilium. During mitosis in animal cells, one centriole of each pair—the 'mother' centriole—is often oriented perpendicular to the spindle axis and provides a focus for the converging microtubules of the mitotic spindle. In the embryonic division of some organisms, moreover, the two mother centrioles at opposite poles of the spindle are always oriented at right angles to one another, this orthogonal relationship anticipating the direction of centrosome separation at the next division.

If the centrioles of an animal cell in tissue culture are followd in a microscope, for example using specific fluorescent probes, then mother and daughter centrioles display different movements. On completion of mitosis, the mother centriole moves directly to a location close to the forming nucleus and remains in a relatively static position. The daughter centriole, however, can move away from the mother, wandering erratically, as though on an elastic tether, as far as the plasma membrane or the midbody. Eventually, as the cell enters $G_1$ phase, movements of the daughter become less marked and it returns to a position close to its mother.

Both mother and daughter centrioles are associated with a matrix of γ-tubulin and pericentrin and both are capable of nucleating microtubule assembly from this matrix. However, the microtubules produced in this manner are easily lost from the centriole, perhaps being released by the action of the microtubule-severing protein katanin. A *stable* array of microtubules, such as that seen in a typical interphase animal cell or in a mitotic spindle pole, forms only from the mother centriole, the microtubules in this case being anchored to a set of nine or so appendages projecting from its distal end (Figure 15-10). The present idea is that microtubules nucleated at the centrosomal matrix are released and then captured by proteins associated with the appendages on the mother centriole. Microtubules might also be captured by other structures in the cytoplasm or on the plasma membrane, especially if they have been released by a peripatetic daughter centriole that has wandered away from the cell center.

## *Parent/daughter differences also exist in basal bodies*

The distal end of a centriole corresponds to the plus ends of the triplet microtubules forming its cylindrical wall, and it is this end that is potentially able to grow into a cilium. In cells that nucleate a primary cilium, for example, the cilium grows from a mother centriole that has docked onto the plasma membrane by its distal end.

Parent/daughter differences also exist in basal bodies and are used by *Chlamydomonas* to generate asymmetries in its cytoskeleton. Although largely bilaterally symmetric, the two sides of *Chlamydomonas reinhardtii* can be distinguished by the location of their eyespot, which is always on the same side of the cell as the daughter basal body. At mitosis, the basal body pairs segregate in a precise orientation that maintains the asymmetry of the cell and results in mitotic poles that have an invariant handedness.

Using the eyespot as a marker, it is possible to distinguish the two flagella of a *Chlamydomonas* cell and show that they too are nonidentical. They differ physiologically in their response to $Ca^{2+}$ ions, and they also have a separate genetic specification. Mutants have been found that fail to assemble one of the two flagella—specifically the one attached to the daughter basal body. Ultimately, it is likely that many aspects of cellular asymmetry will be traced to the fact that each basal body or centriole exists in only one enantiomeric (mirror image) form.

## *Centrioles can appear* de novo

In most animal cells, as we have seen, new centrioles form close to preexisting centrioles. Moreover, it has been found that if the centrosome is removed from a cultured tissue cell by microsurgery, then the cell appears unable to produce new centrioles (although it can produce a microtubule-organizing center close to the nucleus, and an astral array of microtubules).

But there are situations in which centrioles arise *de novo*. During the differentiation of *Naegleria*, mentioned above, production of flagella is preceded by the appearance of two basal bodies where previously there were none. Eggs of frogs and sea urchins lack centrioles but can be induced to form them by brief exposure to hypotonic solutions or to deuterium oxide. Similarly, the first four or five cleavages during early mouse development occur in the absence of centrioles: the eggs do not possess centrioles and the sperm basal body is lost or inactivated at fertilization. Centrioles reappear during later development.

Spermatogenesis in some plants is possibly the most striking demonstration of the appearance of centrioles *de novo*. Many kinds of algae, mosses, and ferns possess multiflagellated sperm whose axonemes include a 9+2 arrangement of microtubules that grow from basal bodies closely similar to those found in animal cells. The centrioles, which are absent elsewhere in the life cycle, appear at the poles of the mitotic spindles that precede spermatogenesis and they are passed to the young spermatid.

In some species that form multiflagellated sperm, the basal bodies appear on an unusual electron-dense spherical structure known as a *blepharoplast*. In the fern *Marsiliea*, for example, the blepharoplast is an electron-dense spherical structure about 1 μm in diameter, which is penetrated by numerous channels. The number of channels, about 100–150 is the same as the number of flagella in the mature spermatozoids and their diameter approximates that of a centriolar cartwheel. During metaphase and anaphase of the last mitosis before sperm manufacture, the internal channels are lost and the surface of the blepharoplast becomes covered with procentrioles. These later develop into mature centrioles, which then form basal bodies for the flagella axonemes.

Similar sequences of changes are observed in other multiflagellated plant sperms, including those of the Ginkgo tree, which form approximately 1000 flagella and those of the cycads with over 10,000.

## *Could centrioles have had a symbiotic origin?*

The apparently independent replication of centrosomes and centrioles led cell biologists in the past to suggest that these structures might carry some form of templating information vital for their replication—information that is not provided by the genetic material in the nucleus. This postulate is still debated, but present evidence argues against a physical association of a specific DNA with basal bodies. Although centrioles isolated from *Chlamydomonas* do in fact carry a small amount of DNA, this seems likely to be a contamination with nuclear DNA (perhaps due to the centrin links between centrioles and the nucleus) rather than a unique sequence.[4]

Another possibility, harder to test, is that centrioles might have originated in a separate organism. Could this provide a mechanism for the continuity of centrosomes and centrioles and perhaps even suggest a symbiotic origin of centrioles? Recall that mitochondria and chloroplasts, which actually carry separate strands of DNA encoding some of their structural proteins, are widely accepted to have arisen by a symbiotic association. Is it possible that centrioles, in a similar way, arose from an association between a primitive eucaryote and a motile bacterium? An association of this kind can actually be found in several present-day species of flagellated protozoa inhabiting the termite gut. The protozoan *Myxotricha paradoxa*, for example, swims by means of spirochetes (Chapter 16) that attach themselves in an ordered array to the surface of the protozoan and beat in synchrony.

Unfortunately, there is little hard evidence for this theory. Centrioles, as just mentioned, are unlikely to carry significant amounts of DNA. The construction of a bacterial flagellum is completely different from that of a eucaryotic flagellum (as we will see in the next chapter), and the two structural proteins, flagellin and tubulin, fail to show any significant similarity in sequence or structure. It is also difficult to imagine an evolutionary path by which a bacterial flagellum might turn into a microtubule. On the whole, it seems more likely that, if basal bodies did have a symbiotic origin, then that symbiont was a eucaryotic cell, or at least a cell whose flagella contained microtubules.

## *Outstanding Questions*

*What is the detailed composition and architecture of centrioles and basal bodies? Of which proteins are they made, and what are their individual functions? How did centrioles appear in evolution? Why do they have ninefold symmetry and why are they made of microtubule triplets? What happens to a sperm centriole when it enters the egg? How is a daughter centriole made? What mechanisms position it at right angles to the parent? What is the significance of the movements of daughter centrioles: where do they go, and how are they driven? How does a daughter centriole mature into a mother centriole? How do the two differ in their capacity to nucleate microtubules? How is centrin used in vertebrate cells, and what is its function in a centriolar lumen? How are precise arrays of basal bodies produced? What molecular events trigger assembly of an axoneme? How does the cell package prefabricated structures for transport into a growing cilium? Which motors move IFT rafts within an axoneme? What tells structures when to leave the raft and to add onto the axoneme? How does a cell specify the length of its cilia and flagella?*

[4] The association of basal bodies with specific RNA molecules is altogether more likely, and could arise from the targeting of mRNA molecules.

## Essential Concepts

- The length and number of flagella and cilia are inherited characteristics of a particular cell type. Flagella regrow to their original length following experimental removal, often in the absence of protein synthesis.
- The site and orientation of cilia and flagella and their characteristic ninefold symmetry are determined by basal bodies. These are morphologically similar to centrioles, with a cylindrical array of nine triplet microtubules.
- Most axonemal proteins assemble into large complexes at the base of a cilium and are then transported by kinesin molecules to the tip of the cilium, where they form dynein arms, radial spokes, or other structures.
- Unused complexes are carried back to the base of the cilium by cytoplasmic dynein. The continual transport of subunits up and down the cilium permits not only growth but also turnover of axonemal components.
- Basal bodies are associated with a number of cytoskeletal structures including various types of striated rootlet. These contain centrin, a protein that undergoes a calcium-induced change in conformation and serves to control the orientation of flagella.
- Each centriole has a unique polarity and when a daughter forms it is oriented at right angles to the mother. The pair of centrioles is typically embedded in a centrosome.
- Mother and daughter centrioles are associated with different structures and perform different functions in the cell. Whereas both are capable of nucleating microtubules, only a mother centriole appears able to maintain a stable radial array of microtubules.
- Basal bodies in general arise from centrioles, which themselves usually form in close association with a parent centriole. However, many clear-cut examples exist in which centrioles appear in a cell that previously had none.
- Basal bodies and centrioles do not contain appreciable amounts of coding DNA. There is very little evidence to support the idea that these organelles arose during evolution from procaryotic symbionts.

## Further Reading


Bobinnec, Y., et al. Centriole disassembly *in vivo* and its effect on centrosome structure and function in vertebrate cells. *J. Cell Biol.* 143: 1575–1589, 1998.

Cole, D.G., et al. *Chlamydomonas* kinesin-II-dependent intraflagellar transport (IFT): IFT particles contain proteins required for ciliary assembly in *Caenorhabditis elegans* sensory neurons. *J. Cell Biol.* 141: 993–1008, 1998.

Dictenberg, J.B., et al. Pericentrin and γ-tubulin form a protein complex and are organized into a novel lattice at the centrosome. *J. Cell Biol.* 141: 163–174, 1998.

Fowkes, M.E., Mitchell, D.R. The role of presassembled cytoplasmic complexes in assembly of flagellar dynein sunbunits. *Mol. Biol. Cell* 9: 2337–2347, 1998.

Han, J.W., et al. mRNAs for microtubule proteins are specifically colocalized during the sequential formation of basal body, flagella, and cytoskeletal microtubules in the differentiation of *Naegleria gruberi*. *J. Cell Biol.* 137: 871–879, 1997.

Hayashi, M., et al. Real-time observation of $Ca^{2+}$-induced basal body reorientation in *Chlamydomonas*. *Cell Motil. Cytoskeleton* 41: 49–56, 1998.

Hepler, P.K. The blepharoplast of *Marsilea*: its *de novo* formation and spindle association. *J. Cell Sci.* 21: 361–390, 1976.

Johnson, K.A., Rosenbaum, J.L. Polarity of flagellar assembly in *Chlamydomonas*. *J. Cell Biol.* 119: 1605–1611, 1992.

Jones, J.C.R., Tucker, J.B. Microtubule-organizing clusters and assembly of the double-spiral microtubule pattern in certain heliozoan axonemes. *J. Cell Sci.* 50: 259–280, 1981.

Klotz, C., et al. Genetic evidence for a role of centrin-associated proteins in the organization and dynamics of the infraciliary lattice in *Paramecium*. *Cell Motil. Cytoskeleton* 38: 172–186, 1997.

Lefebvre, P.A., Rosenbaum, J.L. Regulation of the synthesis and assembly of ciliary and flagellar proteins during regeneration. *Annu. Rev. Cell Biol.* 2: 517–546, 1986.

Levy, Y.Y., et al. Centrin is synthesized and assembled into basal bodies during *Naegleria* differentiation. *Cell Motil. Cytoskeleton* 40: 249–260, 1998.

Maniotis, A., Schliwa, M. Microsurgical removal of centrosomes blocks cell reproduction and centriole generation in BSC-1 cells. *Cell* 67: 495–504, 1991.

Middendorf, S., et al. Identification of a new mammalian centrin gene, more closely related to *Saccharomyces cerevisiae* CDC31 gene. *Proc. Natl. Acad. Sci. USA* 94: 9141–9146, 1997.

Morris, R.L., Scholey, J.M. Heterotrimeric kinesin-II is required for the assembly of motile 9+2 ciliary axonemes on sea urchin embryos. *J. Cell Biol.* 138: 1009–1022, 1997.

Paintrand, M., et al. Centrosome organization and centriole architecture: their sensitivity to divalent cations. *J. Struct. Biol.* 108: 107–128, 1992.

Pazour, G.J., et al. A dynein light chain is essential for the retrograde particle movment of intraflagellar transport (IFT). *J. Cell Biol.* 141: 979–992, 1998.

Piel, M., et al. The respective contributions of the mother and daughter centrioles in centrosome activity and behavior in vertebrate cells. *J. Cell Biol.* 148: 317–329, 2000.

Ramanis, Z., Luck, D.J. Loci affecting flagellar assembly and function map to an unusual linkage group in *Chlamydomonas reinhartii*. *Proc. Natl. Acad. Sci. USA* 83: 423–426, 1986.

Rosenbaum, J.L., Child, F.M. Flagellar regeneration in protozoan flagellates. *J. Cell Biol.* 34: 345–350, 1967.

Rosenbaum, J.L., et al. Intraflagellar transport: the eyes have it. *J. Cell Biol.* 144: 385–388, 1999.

Roth, K.E., et al. Flexible-substratum technique for viewing cells from the side: some *in vivo* properties of primary (9+0) cilia in cultured kidney epithelia. *J. Cell Sci.* 89: 457–466, 1988.

Salisbury, J.L. Centrin, centrosomes, and mitotic spindle poles. *Curr. Opin. Cell Biol.* 7: 39–45, 1995.

Tuxhorn, J., et al. Regulation of flagellar length in *Chlamydomonas*. *Cell Motil. Cytoskeleton* 40: 133–136, 1998.

Vinella, D., et al. GTPase enters the ring. *Curr. Biol.* 3: 65–66, 1993.

Vorobjev, I.A., Chentsov, Y.S. Centrioles in the cell cycle. *J. Cell Biol.* 98: 938–949, 1982.

CHAPTER 16

# Bacterial Movements

Despite their small size and minimalist construction, most bacteria are nevertheless able to swim. Evidently, the cost of making and operating flagella accords cells a survival advantage over nonmotile species. This survival advantage arises because swimming can be directed by environmental influences. Although too small to sense a gradient along the length of the cell, and unable to swim long distances because of buffeting by Brownian motion, bacteria can bias their random swimming toward more favorable locations. What this favorable environment might be depends on the bacterium and where it lives. Some inhabit other organisms, as pathogens or symbionts; others live in soil or open water, or form part of complex biofilms. Each environment has its own exigencies that have led to the evolution of a particular species-dependent form of motility.

In broad terms, however, there is a general plan, which is shared by most motile bacteria and typified by the case of *Escherichia coli* and related bacteria. As described in this chapter, *E. coli* swim by means of rigid helical flagella, each driven by a minute rotary motor at its base. The flagellar motors are marvels of miniaturization, having a diameter less than 50 nm and able to spin at more than 300 revolutions per second. An equally diminutive sensory control system, a biochemical circuit made of proteins, records the salient features of the local chemical environment over the preceding few seconds and uses this information to provide sophisticated control over the motor. As a result, the bacterium changes its direction of movement in an informed manner, moving toward sources of food and away from potentially harmful environments.

All of the genes and proteins involved in chemotaxis of *E. coli* have been identified and isolated, making this the most completely understood

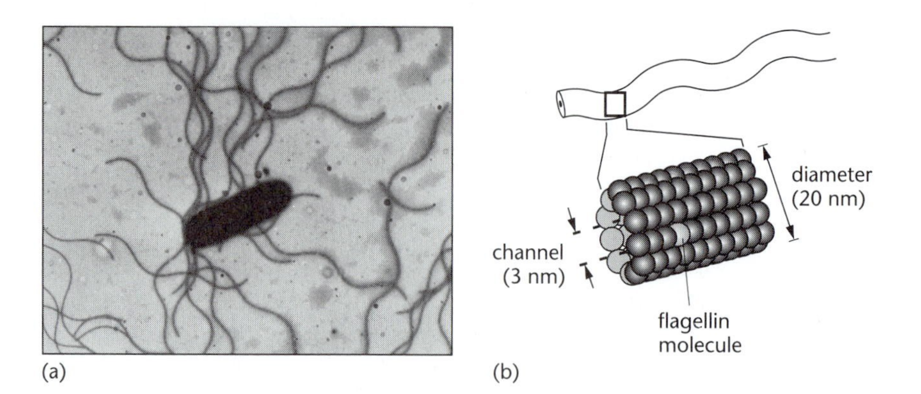


**Figure 16-1** Bacterial flagella. (a) Negatively stained image of a bacterium surrounded by flagella (some having detached from the cell). Enteric bacteria, such as *E. coli* and *Salmonella* have 6–10 helical flagella attached to their surface. (b) Schematic of a flagellum, showing the arrangement of its flagellin subunits. (a, courtesy of Shin-ichi Aizawa.)

form of motile cell behavior. We are also well on the way to understanding the related but often more complex systems used by other chemotactic bacteria, which we will examine in the latter part of the chapter. Some of these use motors and flagella similar to those in *E. coli* but in a different manner—as in helical spirochetes or swarming *Proteus*. Other forms of bacterial movement are radically different and still uncharted in molecular terms, such as gliding over surfaces or even, mysteriously, the capacity to swim rapidly through water without flagella or any other visible means of propulsion.

## Bacterial flagella are hollow cylinders made of flagellin

Bacterial flagella—the passive propellers of bacterial swimming—are smoothly curving cylinders of protein about 20 nm in diameter and of indefinite length, typically around 5–10 µm (Figure 16-1a). Their cylindrical wall is (in most species) built from a single protein. In *E. coli* close-packed flagellin molecules are built into 11 linear protofilaments arrayed around an apparently hollow central lumen (Figure 16-1b). No plasma membrane surrounds the flagellum, which is therefore in direct contact with the surrounding fluid.

Bacterial flagellins from species of *Salmonella* and *Escherichia* typically have molecular weights of 50–60 kDa. They will spontaneously assemble into polymer, producing flagella with the same cylindrical structure and long-pitch helical waveform as those produced by the bacterium. The waveform of a flagellum is a genetically determined characteristic of a particular species of bacterium, expressed through the self-assembly properties of its flagellin. Many mutants with defective flagellin have been isolated, including those in which the flagella are straight (and therefore nonfunctional) or have a tight right-handed ‘curly’ or ‘coiled’ waveform (also of limited efficiency).

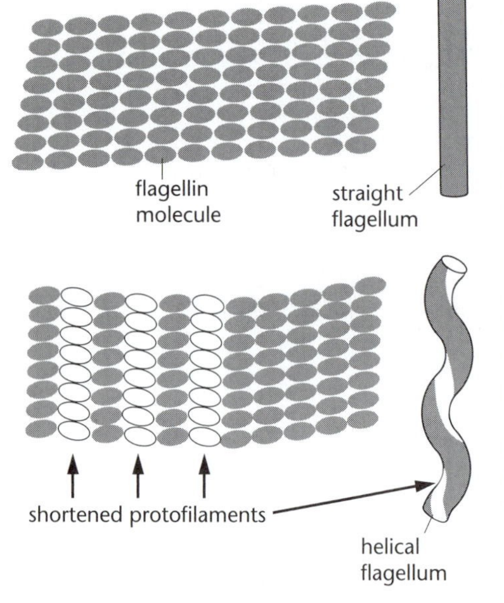


**Figure 16-2** Bistable flagellin molecules. The patterns on the left show the arrangement of flagellin molecules in cylindrical flagella, opened out to show the 11 protofilaments in columns. The resulting flagella are shown on the right. Flagellin molecules are colored gray or white to show their conformation. When all the flagellin molecules have the same shape, as in the top pattern, the resulting flagellum must be a linear cylinder. However, if one or more of the protofilaments becomes shorter in length (lower pattern), due to its subunits adopting a second conformation, the flagellum becomes distorted into a helix. (Based on Calladine, 1975.)

It can be argued, for geometrical reasons, that bacterial flagellin must exist in more than one conformation. This is because the regular packing of identical subunits can produce a linear cylinder (like a microtubule) but not a hollow tube that curves in a helical fashion. One widely accepted model proposes that each protofilament in a flagellum adopts, uniformly along its entire length, one of two conformations, with the actual waveform depending upon how many protofilaments have one conformation and how many the other (Figure 16-2). This model can explain why bacterial flagella flip from one waveform to another as a result of the torsional stress of flagellar rotation (especially during motor reversal, as discussed below).

## Flagellin molecules travel through the hollow core of the flagellum

Bacteria that have lost their flagella by breakage grow new ones, flagellin molecules for this purpose being synthesized in the cell and assembled at the distal tip of the flagellum. Thus, if bacteria are pulse-labeled with radioactive amino acids, radioactive proteins appear first at the distal end of the flagella. The situation is formally similar to that in growing eucaryotic cilia and flagella where, as we saw in the previous chapter, assembly also occurs at the tip. However, when we consider that the bacterial flagellum is nothing but a naked tube of protein, we realize that its growth is unlikely to depend on molecular motors transporting material from the cell to the flagellar tip.

How then do flagellin molecules travel from the cell body to the tip of a flagellum 10 μm away? Surprisingly, the answer is that they move, probably by diffusion, down the hollow central canal of the flagellum. The hole down the middle of an *E. coli* flagellum is only about 3 nm in diameter, which is too small to admit a folded flagellin molecule but could accommodate an unfolded molecule. Specific proteins associated with the flagellar apparatus are thought to act like an export agent that send flagellin molecules down the flagellum, and one of their tasks may be to unfold flagellin molecules at the start of their journey. Assembly at the other, distal, end of the growing flagellum depends on a protein called HAP2 (product of the *fliD* gene) that catalyzes both the controlled refolding of flagellin molecules and their assembly into the flagellar wall. The longer a flagellum gets, the slower it grows, as expected from a diffusion-limited process.[1]

Transport of flagellin molecules in the flagellar lumen raises many intriguing questions. What are the physical conditions within a hollow tube of protein of these dimensions? Will water molecules inside a flagellum be more highly organized—more 'ice-like'—than that in bulk solution? By what mechanism can protein molecules be driven down a small tube in which there is no room for motor proteins? If flagellin molecules can move down the middle of flagella, then why should not microtubules, which are similar in form to bacterial flagella and have a larger hole down the middle, also carry internal traffic?

The base of the flagellum is connected to its motor via a short region known as the flagellar hook (Figure 16-3). This structure is both flexible and strong, enabling the flagellum to change its orientation on the cell surface while still transmitting rotational force (torque) to the flagellum. This is why it is possible to attach mutant bacteria, lacking flagella, to a glass coverslip by an antibody to hook protein—a technique that, as we mentioned in Chapter 1, provided the first experimental evidence for the rotation of the flagellar motor.[2]

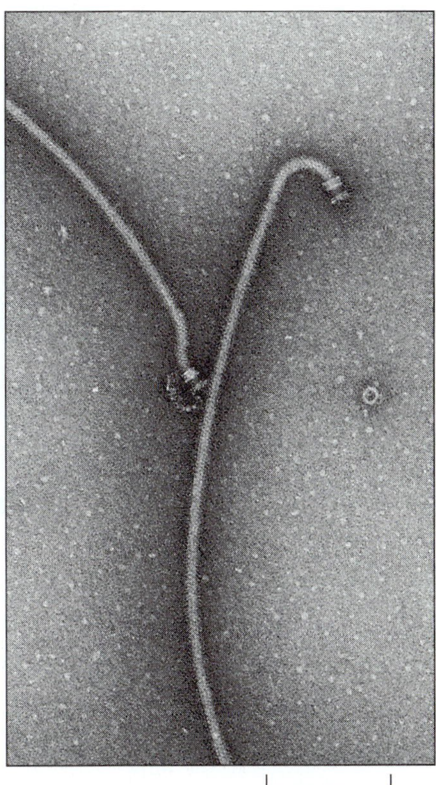


**Figure 16-3** Flagellum plus hook. Negatively stained image of a flagellum from *Salmonella* showing its terminal hook region. At the end of the hook a single flagellar motor can be seen. (Courtesy of Shin-ichi Aizawa.)

## *The flagellar motor is driven by a flux of protons*

Each of the six or so motors on the surface of an *E. coli* cell is a roughly cylindrical structure about 50 nm in diameter (Table 16-1). A flagellar motor can rotate at more than 100 revolutions per second (typically 300 rps for *E. coli* in an unloaded condition) and can reverse its direction

**Table 16-1 Flagellar motors**

| | |
|---|---|
| size | diameter ~30 nm, height ~50 nm |
| number of genes | about 40 genes are needed to make the motor plus its flagellum |
| number of proteins | about 20 different proteins make a motor |
| number of torque generators | 8 |
| number of protons per revolution | ~1200 |
| speed of rotation | 300 Hz (18,000 rpm) for *E. coli*, some bacteria spin at more than 100,000 rpm |
| efficiency* | less than 5% at low load, but 50–100% at high load |

* *Efficiency* is the ratio of useful work done by the motor to the energy supplied to it by the proton gradient.

1 Interestingly, mutant bacteria in which HAP2 is missing fail to make flagella and secrete flagellin molecules into the surrounding medium. The flagellin molecules presumably 'leak' through the unstoppered holes in each flagellar complex.

2 Amazingly, it seems that flagella continue to grow while their motor is rotating. We have to envisage flagellin molecules moving from the cytoplasm through the motor and hook region, through the lumen of the flagellum and then adding to the tip of the flagellum while, at the same time, the entire structure is rotating at 300 times per second!

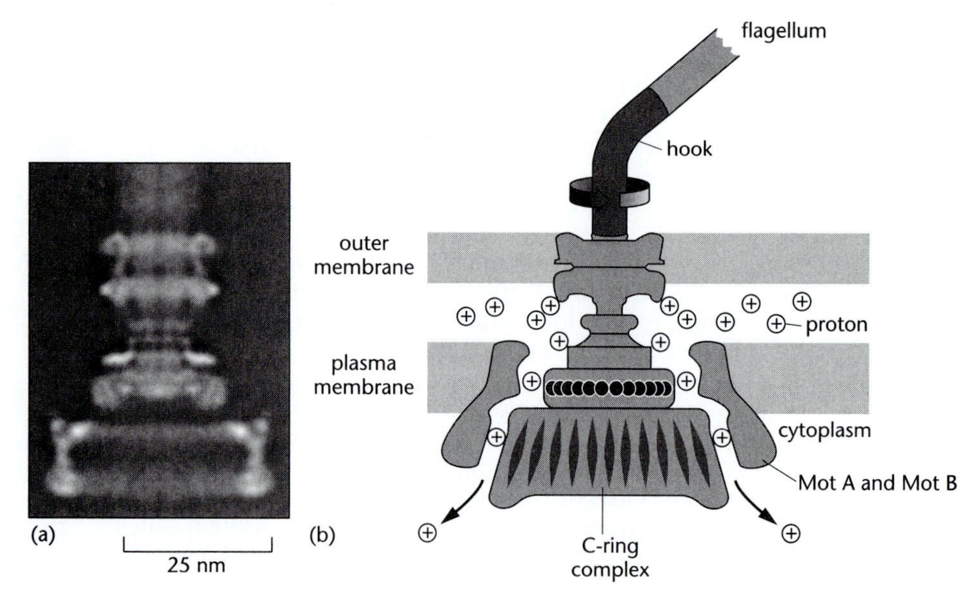


**Figure 16-4** Flagellar motor. (a) The flagellar motor from *Salmonella typhimurium*, image obtained using a low-dose electron microscope and image-averaging methods.
(b) Schematic diagram of the motor. (Image courtesy of David DeRosier.)

of rotation in less than a tenth of a second. The structure of the motor is, of course, far too small to see in the light microscope, but it can be examined by electron microscopy, either in sections of whole bacteria or in isolated preparations of flagella plus their associated motors (see Figure 16-3).

A cylindrical rod forms the central part, or 'drive shaft' of the flagellar motor, passing through two hollow rings of protein embedded in the outer membrane and terminating in a circular ring or bush of protein (Figure 16-4). The hollow rings of protein probably work like bearings and are not essential for force production (they are absent in some species of bacteria). The lower part of the central rod is associated with a circular complex of proteins, the C-ring complex ('C' for cytoplasm). This protrudes into the cytoplasm and contains elements responsible for both force production and also for controlling the direction of rotation of the motor. The outer face of the C ring is believed to interact with a studlike ring of proteins, called MotA and MotB, embedded in the plasma membrane, to contribute directly to torque production.

In contrast to motor proteins such as myosin or kinesin, the flagellar motor is not driven by the hydrolysis of ATP. In fact it is driven by an even more fundamental source of energy—a gradient of protons across the plasma membrane (in some species of bacteria sodium ions are used instead). The proton gradient is generated by the controlled oxidation of food molecules and is the primary source of ATP in both bacteria and mitochondria. During the oxidation process, protons are pumped across the plasma membrane, out of the bacterium. The resulting difference in concentration of $H^+$ ions across the membrane becomes a source of energy that can be harnessed by allowing the protons to return down their concentration gradient through a specific transmembrane complex called *ATP synthase*. The function of this complex is to couple the movement of protons back into the cell to the formation of ATP from ADP and inorganic phosphate.[3]

In similar fashion, the motor at the base of each bacterial flagellum channels protons from the outside of the cell to the inside and uses them, rather as water is used in a turbine, to drive the rotation of the flagellum. The location of the proton channels is believed to be the ring of MotA and MotB surrounding the basal portion of the C ring.

[3] Curiously, the cytoplasmic portion of the ATP synthase also rotates (shown experimentally by attaching actin filaments to individually isolated ATP synthase complexes and observing their rotation in a light microscope). ATP synthase is a smaller and simpler structure than the flagellar motor, and not surprisingly rotates at a slower rate —why it should move at all is debated.

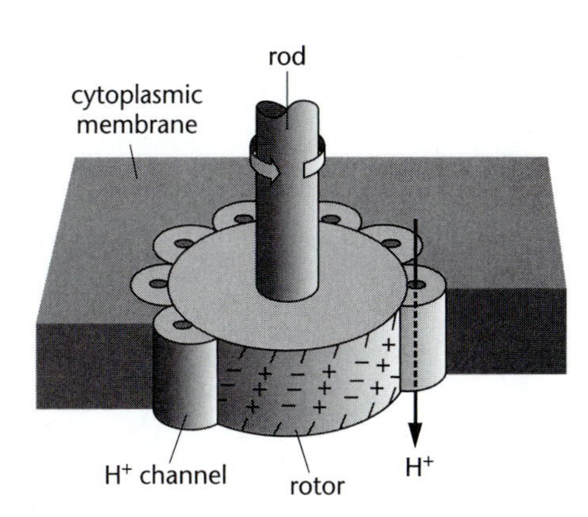


**Figure 16-5** Electrostatic model for the flagellar motor. Theoretical scheme showing how alternating lines of charges on the rotor could be used to turn the flagellar motor. A series of channels around the perimeter of the rotor allow protons to enter the cell. As positive charges move into the cell, they attract negative charges on the rotor. Because the lines of charge are tilted with respect to the channels, this attraction causes the rotor to rotate in its default, counterclockwise direction. Motor switching, in this model, could be effected by changing the tilt of charges on the rotor (not shown).

## *How does the flagellar motor work?*

We have seen that the flagellar motor is driven by protons moving across the bacterial membrane. But precisely how this movement of charge is coupled to the rotation of the flagellum is not known. It is particularly difficult to envisage how a motor can change its direction of rotation from clockwise to counterclockwise, since protons move in the same direction across the membrane in both states.

Many theoretical models have been proposed for this fundamental process. In some, rotation and switching depend on ratchetlike constraints produced by individual protons as they move through the system. Switching may then either be produced by steric (conformational) changes or may be electrostatic in nature. Models of this type are often said to be *tightly coupled*, meaning that there is a one-to-one relationship between the passage of each proton and the rotation of the motor.

Another class of models exists in which the torque-generating mechanism is more like a turbine than a ratchet. These models are less rigid in design and are said to be *loosely coupled*. For example, suppose the rotor has alternating lines of positive and negative charges on its surface tilted with respect to the rotor axis. If now protons stream into the cell, passing though channels adjacent to the rotor, charge interactions will cause the rotor to turn (Figure 16-5). But the rotation will depend on the overall flux of charge, not on the movement of individual protons.

Switching in a loosely coupled model could be caused by changing the angle of tilt of the channels or of the lines of charge on the rotor, but it could also be electrostatic in nature. Theoretical analysis shows that if each channel contains a number of negatively charged groups with which protons associate transiently as they pass into the cell, then the overall effect on rotation will depend on how many sites are occupied—which in turn will be affected by the acidity, or p$K$ of the charged groups.

If protons are only weakly bound to the channel sites (low p$K$) then the effective movement is that of positive charges moving into the cell and the rotor will turn in the default, counterclockwise (CCW), direction. But if protons are tightly bound to the fixed sites (reflecting a high p$K$) then as they hop from site to site into the cell they will in effect create a movement in the opposite direction of 'proton holes,' so the rotor will change its direction of rotation (Figure 16-6). Control over the numbers of protons in the channels could, in this hypothetical model, be exerted by biochemical interactions with the proteins forming the proton channel.

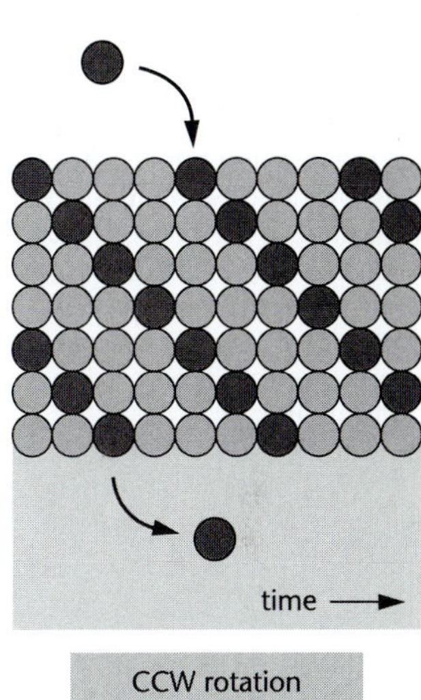


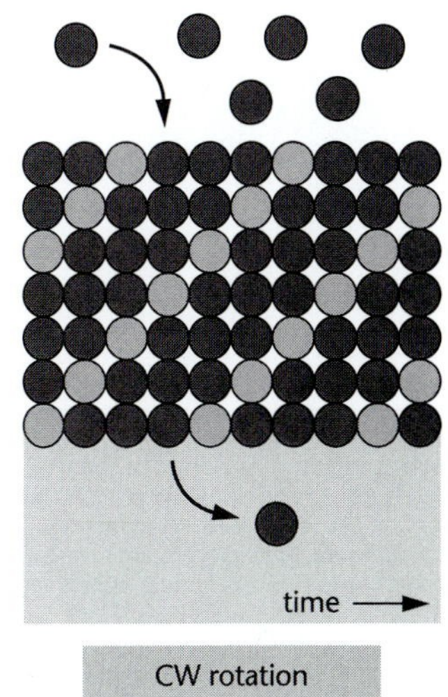


**Figure 16-6** Electrostatic switching. In this theoretical model, a motor changes its direction of rotation by changing the acidity, or p$K$, of groups in the proton channel. Each column of circles represents a series of proton binding sites in a channel. A site is dark if occupied by a proton, light if not. Time is shown from left to right, and net movement of protons is downward (into the cell). At low p$K$ (upper figure) binding sites are sparsely occupied, so the flux of protons creates a movement of positive charge into the cell. In the lower figure, sites are densely occupied (high p$K$) and the movement of protons creates an effective movement of negative charges in the opposite direction (see Berry, 1993).

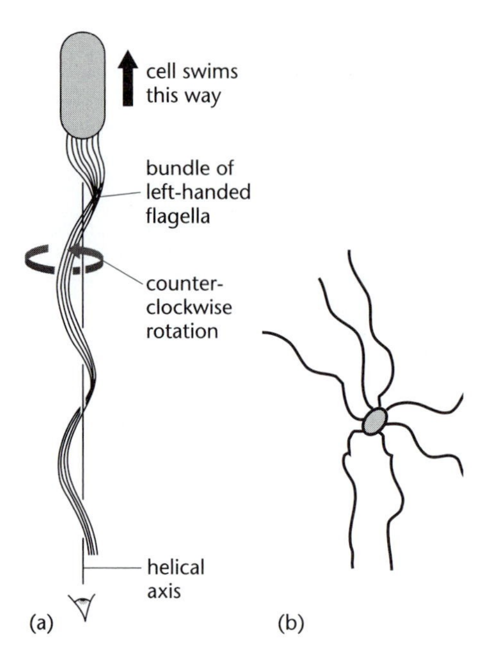


**Figure 16-7** Geometry of flagellar rotation. (a) Left-handed flagella rotating counterclockwise (looking from the outside of the cell toward the flagellar base) push the cell forward. Rotation in this direction also causes the flagella to collect into a synchronously moving bundle that generates a smoothly swimming motion. (b) When left-handed flagella are rotated clockwise they pull on the cell; the flagella fly apart and also undergo localized changes to a right-handed waveform. These transient reversals produce the disordered motion known as tumbling.

## *Transient reversals of the motor steer the bacterium*

Since the bacterial flagellum is a helical structure, rotation in one direction has a different hydrodynamic consequence to rotation in the other. In both *E. coli* and *Salmonella*, flagella are left-handed, so that counterclockwise rotation of the motor (viewed as though looking down the flagellum into the cell) generates a force that pushes the cell forward. Rotation in this direction therefore generates smooth swimming (Figure 16-7). Clockwise rotation, by contrast, disrupts the smooth progression of the bacterium through the water. Instead of driving the cell with a coordinated parallel bundle, the flagella fly apart, attempting to pull the cell in many directions at once. They also undergo abrupt transitions in waveform in which the long-pitch left-handed helix changes to a tight right-handed helix. This change in waveform starts at the motor and propagates outward, but under normal conditions it reverses before it reaches the tip. During these transitions the flagellar bundle is dispersed and the cell body moves chaotically—that is, it tumbles.[4]

It may be recalled that as bacteria swim they alternate between periods of smooth swimming, known as 'runs' and short bursts of chaotic changes in direction termed 'tumbles' (Chapter 3). These two forms of movement are produced by flagellar rotation in the two possible directions—counterclockwise rotation produces runs and clockwise rotation produces tumbles. Switching between the two directions of rotation can be seen in bacteria that have been tethered to a surface through a flagellar hook. Bacteria attached in this manner rotate continuously through the operation of their motor, turning in directions that correspond to episodes of running or tumbling. Thus a wild-type bacterium, in the absence of chemoattractant or repellent, turns mainly counterclockwise but with brief episodes of clockwise rotation. Mutants that tumble continuously when they swim turn exclusively clockwise when they are tethered to a coverslip; mutants that run continuously turn exclusively counterclockwise.

## *The direction of flagellar rotation is influenced by chemoattractants*

We have emphasized that the *function* of bacterial swimming (the reason this elaborate apparatus appeared during evolution) is to carry the cell to more favorable locations, especially sources of food. Cells of *E. coli* can recognize about 30 distinct substances in their environment, which fall into five categories as shown by competition experiments. These substances are recognized by four types of transmembrane receptor molecules, designated Tar, Tsr, Trg, and Tap (Table 16-2). Different receptors operate independently, so that bacteria will swim toward a distant source of serine, for example, even in the presence of a constant high concentration of ribose or aspartate. On the other hand, a distant source of serine will not be detected if the surroundings contain a high concentration of alanine, since this competes for binding to the same receptor (Tsr).[5]

**Table 16-2 Some attractants and repellents for *Escherichia coli***

| Receptor | Attractant | Repellent |
|---|---|---|
| Tsr | serine, cysteine, alanine | indole, $H^+$ |
| Tar | aspartate, maltose, glutamate, $H^+$ | $Ni^{2+}$, $Co^{2+}$ |
| Tap | dipeptides | $H^+$ |
| Trg | galactose, ribose, $H^+$ | |

In addition to the four receptors listed here, a fifth receptor, Aer, related in sequence to the others but lacking the transmembrane domain, is responsible for detecting oxygen.

[4] In some motor mutants that rotate their flagella in a clockwise direction all the time, the flagella adopt completely right-handed helices, which then form bundles and cause run swimming behavior. In such bacteria chemotactic responses are inverted, so that attractants cause tumbles and repellents cause runs.

[5] *E. coli* also responds chemotactically to glucose, its favorite food, in this case using the normal glucose uptake pathway to trigger downstream signals.

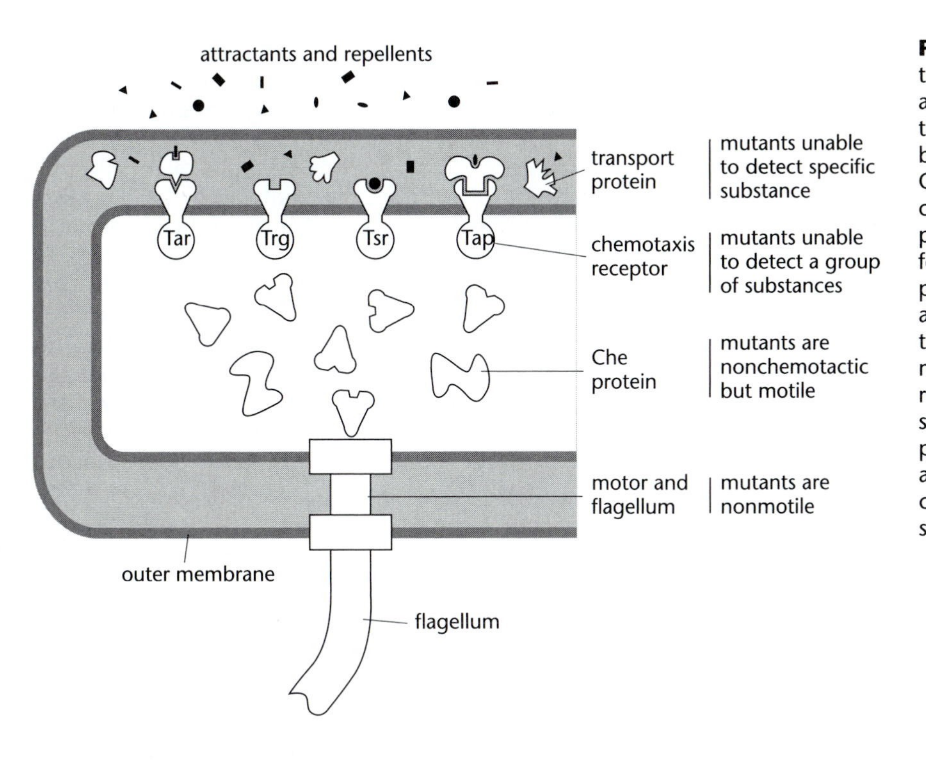


**Figure 16-8** Principal stages in signal transduction during bacterial chemotaxis. An attractant or repellent molecule diffuses into the periplasmic space of the bacterium and binds to a specific transport protein. Occupation of this binding site causes a conformational change in the transport protein, enabling it to bind in turn to one of four types of chemotaxis receptor in the plasma membrane. (Note, not all repellents and attractants act through an intermediate transport protein, some bind directly to the membrane receptors). Occupation of the receptor binding site activates an internal signal that is relayed by the cascade of Che proteins to the flagellar motor. In the case of an attractant, the motor rotates in a counterclockwise direction and the cell swims smoothly.

The link between flagellar rotation and oriented swimming is made plain when tethered bacteria are exposed to chemoattractants. In the unperturbed state, before chemoattractant is added, tethered cells undergo sporadic reversals in the direction of rotation, corresponding to the intermittent tumbles shown during normal swimming. A sudden increase in the concentration of a chemoattractant suppresses rotation in the tumbling direction, whereas a decrease in chemoattractant greatly increases the frequency of tumbles. When tumbles are suppressed, a cell perseveres in its current direction of motion. Repellents such as butyric acid and nickel ions work in a similar way to attractants except that they cause responses that are inverted. Thus, a rise in the concentration of $Ni^{2+}$ produces an avoidance response (increased tumbling rate) and a fall in $Ni^{2+}$ produces an attractant (run) response.

## *Four categories of genes control chemotaxis*

Because of their short generation time and haploid genome, bacteria are ideal organisms for genetic analysis. This is especially the case for functions affecting motility, which, as already described in connection with eucaryotic flagella (Chapter 14), offer powerful selection methods to the investigator. Mutations have been obtained in essentially all of the genes responsible for the chemotactic response in *E. coli* (the genetic map has been saturated). These can be grouped into four principal categories, corresponding to the four stages in sensory transduction (Figure 16-8).

1. Mutants in the first category are unable to detect specific substances. For example, the cell might ignore a gradient of the sugar maltose but respond normally to all other sugars and amino acids. Behavioral perturbations of this kind typically arise from a defect in a protein that binds to the chemotactic substance in question. In the case of maltose, for example, it is known that a specific protein located in the periplasmic space binds to maltose. The chemotactic machinery uses the complex of maltose with its binding protein as an indication that maltose is present. A defect in the maltose-binding protein would thus make the bacterium selectively blind to this sugar alone.

2. Mutations in the second category affect chemotaxis to *all* substances in a given class, as defined in competition experiments. For example, a mutant bacterium might fail to recognize small peptides while still recognizing and moving toward sugars and amino acids. Mutations in this category affect one of five types of membrane-bound chemotaxis receptor proteins, designated Tar, Tsr, Trg, Tap, and Aer (see Table 16-2). The first four are transmembrane proteins with binding sites on their outward-facing, extracellular domain for one or more specific substances and a cytoplasmic domain that interacts with signaling proteins in the interior of the cell. Binding of the cognate ligand sends a message (as a propagated change in conformation) from the outside of the cell (or, to be more precise, from the periplasmic space) to the bacterial cytoplasm. The fifth receptor, Aer, monitors intracellular energy levels rather than the external environment, and is used to detect and move toward oxygen.
3. A third type of chemotaxis mutant fails to send the correct signals from membrane receptors to the flagellar motor. Bacteria in this category are able to swim normally but show an aberrant response to attractants and repellents—they might swim smoothly all the time, for example, or tumble incessantly. Other more subtle mutations affect the adaptive response to chemotactic stimuli (described below). A bacterium with such a mutation might respond correctly to a sudden rise in concentration of a sugar or an amino acid but then persist in its mode of swimming instead of adjusting back to baseline behavior, as it should. Six genes in this category (designated *che* genes, for 'chemotactic') have been identified in *E. coli* and their function is discussed below.
4. The final category is nonmotile mutants that fail to show any chemotaxis because they are unable to swim effectively. These include bacteria that have lost their flagella; bacteria whose flagella are so distorted in structure that they cannot produce effective swimming movements; and bacteria with defects in the flagellar motor that paralyze it, slow it down, or prevent it from reversing direction.

## *Chemotactic signals are relayed by a chain of coupled protein phosphorylations*

How are changes in the concentration of attractants or repellents relayed to the flagellar motors? The answer lies in the chain of biochemical signals carried by Che proteins depicted in Figure 16-9. The central part of this signaling cascade is CheA ('A' in the diagram), a protein kinase that uses ATP to phosphorylate one of its histidines. The phosphorylated protein

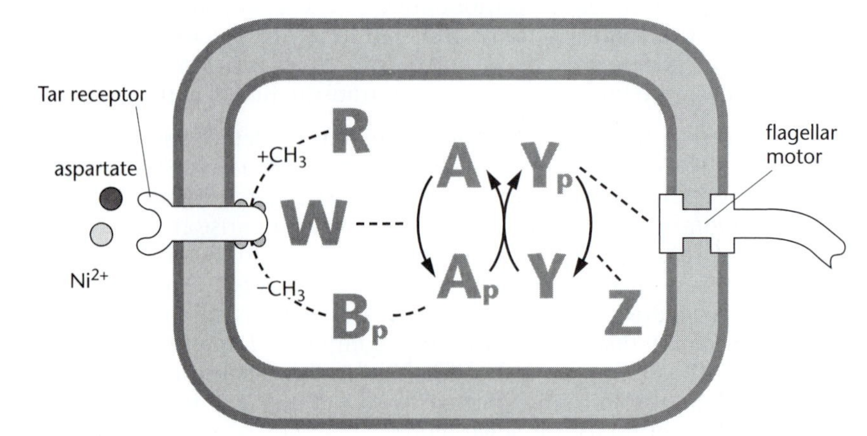


**Figure 16-9** Chemotactic pathway in *E. coli.* The letters refer to Che proteins (CheA, CheY, and so on). Binding of an attractant to a chemotactic receptor (in this case aspartate to Tar) inhibits CheA autophosphorylation. This in turn reduces the cytoplasmic concentration of CheYp leading to the default CCW flagellar rotation and run swimming behavior.

(Ap) is unstable and readily transfers its phosphoryl group to a second protein, so that in the cell CheA undergoes a continual, rapid cycle of phosphorylation and dephosphorylation. Although it seems wasteful, futile cycles of this kind are a universal feature of signal pathways in all living cells. They allow the concentrations of signal molecules (such as phosphorylated proteins) to change rapidly—the speed of response being directly related to the rapidity with which the cycle spins.

CheA transfers its phosphoryl group to a small, highly soluble protein CheY ('Y' in Figure 16-9), which then diffuses through the cytoplasm to a flagellar motor. Binding to the inside face of the flagellar motor, phosphorylated CheY (Yp) causes it to turn in a clockwise direction thereby producing a chaotic tumble in the bacterium.[6] Mutants lacking CheY protein swim continuously without tumbling, as do 'gutted' mutants that lack all of the genes involved in chemotactic signaling (the default rotation of the flagellar motor itself is counterclockwise). The action of CheYp is terminated by dephosphorylation in a reaction promoted by CheZ, ensuring that the reversal of flagellar rotation is brief.

The rate of CheA autophosphorylation, and hence the concentration of CheYp is stimulated by the *unoccupied* chemotactic receptors, working in conjunction with the linker protein CheW ('W'). The complete system maintains Ap, and hence Yp, at precisely that level at which the bacterium switches from runs to tumbles about every second, as already described.

Substances outside the bacterium influence this cascade of reactions by binding to the receptors and altering the rate of CheA phosphorylation. Attractants, such as serine or aspartate, reduce the rate of phosphorylation, causing Yp to fall and the bacterium to swim in a smooth undeviating fashion (since it has encountered a favorable environment). Conversely, repellents increase the rate of CheA phosphorylation, increase Yp, and hence cause a tumble, so that the bacterium tries to move away from the stimulus.

## *Adaptation of the chemotactic response depends on protein methylation*

We saw in Chapter 3 that adaptation is an essential part of chemotaxis, as it is for all sensory processes. *E. coli* can respond chemotactically to an impressively wide range of concentrations of attractants (from less than 10 nM to over 1 mM in the case of aspartate). It does so by continually normalizing its swimming to the new ambient concentration, so that further increases can be detected. Experimentally, adaptation can be observed by exposing cells to a fixed concentration of attractants: initially they suppress their tumbling and swim smoothly, but over the course of a minute or so (depending on the strength of the stimulus) the cell slowly starts to tumble again. Eventually the cell returns to its original pattern of alternating runs and tumbles, although we know that it must be changed internally in some subtle way. In other words, the cell has a biochemical record, or 'memory' of its previous surroundings.

Adaptation has been shown to depend on the methylation of the chemotactic receptors. Methylation increases the ability of the receptor to stimulate CheA phosphorylation—in other words, it counteracts the effect of attractant binding. An enzyme, CheR, transfers a methyl group to a free carboxyl group on a glutamic acid residue of the receptor protein. Up to five methyl groups can be transferred to a single receptor, the extent of methylation increasing at higher concentrations of attractant (where each receptor spends a larger proportion of its time with ligand bound) (Figure 16-10). When the attractant is removed, the receptor is demethylated by a second enzyme, CheB. In reality, both enzymes, CheR and CheB, work continually, simply modulating their rates of catalysis during a chemotactic response so that the cell reaches a new equilibrium at which the receptors have a new steady-state level of methylation.

[6] For simplicity we consider here only a single flagellum, although in a reality there are half a dozen or more on each cell. How the different flagella on a cell coordinate their action is not known for certain, although hydrodynamic factors may be important.

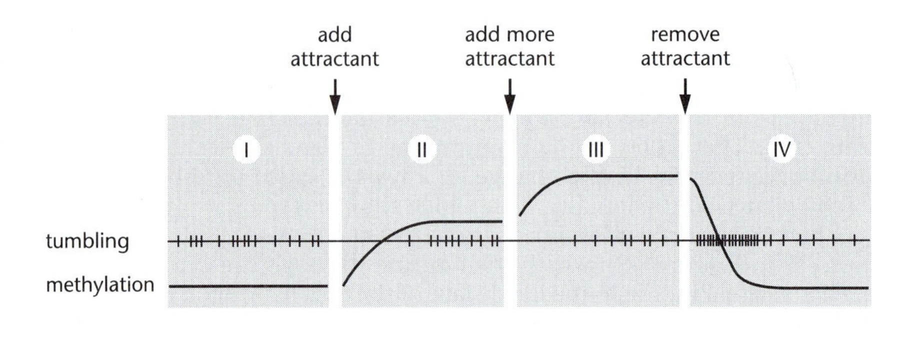


**Figure 16-10** Correlation of receptor methylation with behavioral response. In the absence of chemoattractant (I) the bacterium tumbles at a basal rate (each tumble is here represented as a vertical bar). Addition of an attractant (II) suppresses tumbles initially, but as methylation of the chemotaxis receptor rises, tumbling returns to its original base rate. Further addition of the same attractant (III) generates another suppression of tumbling and another increase in methylation (to a maximum of five methyl groups per receptor. (IV) Once the cell has adapted, removal of the attractant causes a transient tumbling response.

## Conformational changes are central to chemotaxis

Adaptation in bacterial chemotaxis has the distinctive feature that it is *exact*—that is, the swimming pattern of a bacterium returns almost precisely to its original pattern of runs and tumbles after being exposed to a whole variety of different stimuli over a wide range of concentrations. But the mechanism is puzzling, since it seems to require the bacterium to have some reference point by which it tells when the correct balance has been restored. One intriguing idea is that the methylation reactions might be sensitive to the *conformation* of the membrane receptor. Like most membrane receptors, the chemotactic receptors flip randomly between an active conformation in which they catalyze CheA phosphorylation, and an inactive conformation. If methylation by CheR occurs only on receptors in their inactive conformation, and if CheB demethylates only active receptors, then the system should automatically return to its previous state following any perturbation.

More generally, changes in protein conformation underpin the entire signaling process. Individual chemotactic receptors actually detect multiple environmental signals—the receptor Tar, for example, is used by bacteria to detect not only aspartate, but also maltose, nickel ions, hydrogen ions, and even changes in temperature. These diverse signals probably take effect through the same conformational transition, simply shifting the balance of probability closer to or farther from the active state. Conformational changes also underlie the downstream changes of the chemotaxis pathway. The active conformation of the receptor stabilizes the catalytically active form of CheA; phosphorylation of CheY stabilizes a conformation that interacts with the switch complex of the motor; the switch complex undergoes a conformational change to specify either clockwise or counterclockwise rotation of the motor.

## Cooperative interactions between receptors could increase their sensitivity

Conformational changes may have yet another function in the chemotactic response—that of increasing sensitivity. The chemotactic receptors in *E. coli* are not scattered randomly over the surface of the cell but rather are localized in a cluster, usually at one pole. The reason for this clustering is unknown, but the theoretical possibility has been proposed that it might increase the responsiveness of the cell. As already mentioned, bacterial chemotaxis has a very low threshold for attractants such as aspartate (cells can detect less than 10 nM aspartate) together with a very wide range of operation of at least four orders of magnitude in concentration. This combination of parameters is difficult to explain by conventional mechanisms, and computer simulations based on experimental binding and reaction rates usually predict thresholds at least ten times greater than that seen in living bacteria.

The theoretical possibility has been put forward, that receptors clustered on the bacterial surface could be sufficiently close to interact with

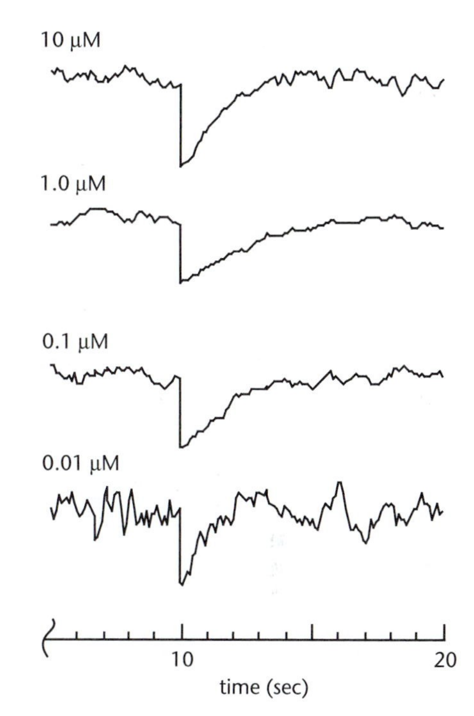


**Figure 16-11** Effects of receptor–receptor coupling. Theoretical result showing the behavior of an array of 2500 chemotaxis receptors, in which the conformations of neighboring receptors are coupled. Each trace shows the number of receptors in an active conformation in the presence of the indicated concentration of attractant. Because of exact adaptation, each trace begins with the same average number of active receptors. At 10 seconds, the concentration of attractant is doubled, causing a transient suppression of receptor activity followed by a return to baseline due to adaptation. Conformational coupling allows the set of receptors to respond to attractant over at least four orders of magnitude (From Duke and Bray, 1999.)

each other. That is, conformations might spread from one receptor to its neighbor, so that a single ligand binding event influences adjacent receptors and thereby increase the size of the downstream signal. Simulations based on this idea, show that if coupling between adjacent receptors has a critical value, then the array of receptors will have the desired combination of low threshold and wide range of responsiveness (Figure 16-11).

## *Many variants of the chemotaxis system exist*

Chemotaxis of *E. coli* is arguably the best understood form of cell behavior. All of the intracellular proteins involved in this chain of responses, from receptor to flagellum, have been purified and all have been sequenced at the DNA level. Structural information is available for most of the proteins, and the enzymatic reactions they catalyze have been analyzed kinetically. More than 60 mutants lacking identified proteins, singly or in combination, have been isolated and their chemotactic responses documented. Computer simulations of the molecular events in bacterial chemotaxis can now reproduce the behavior of wild-type and mutant bacteria with impressive accuracy.

Despite this wealth of detailed information, there is still much to learn. The basic mechanism of the motor is not known, nor is the structure of the receptor complex or how it operates. We have only the sketchiest notion of how the different molecules mediating the chemotactic response are spatially positioned in the cell, and how they operate in the face of the violent movements due to thermal energy.

When we look at motile species other than the enteric bacteria *E. coli* and *Salmonella*, we commonly find the same basic mechanism of chemotaxis but with subtle variations. The number of flagella and their position on the cell varies widely, as does the mode of their rotation. The light-sensitive bacterium *Rhodobacter sphaeroides*, for example, has tufts of flagella at each end of the cell, which rotate in opposite directions. The flagellar motors in this organism are closely related to those described above, but the intermittent tumbling needed for chemotaxis is achieved not by reversing the rotation as in *E. coli* but by stopping.

Components of the signal transduction pathway also vary widely. Bacterial chemotaxis receptors are commonly homologous to those of *E. coli* and are similiarly regulated by methylation. But the number of types can be very large (10 in *Bacillus subtilis*; 12 in *R. sphaeroides*) reflecting particular preferences for nutrition and habitat. In the few species that have been studied in detail, the central signaling pathway of CheW, CheA, and CheY seems to be in place, but to this have been added other inputs and different combinations of signaling proteins. For example *Helicobacter pylori*, the causative agent of stomach ulcers, has three CheW-CheY fusions but no CheB or CheR. The protein CheZ, which as we saw above facilitates removal of phosphate groups from CheYp, has so far been found only in *E. coli* and its relatives. It will be fascinating to determine the sequences of these various related species and trace back the evolutionary origins of their chemotactic machinery.

## *Bacteria communicate*

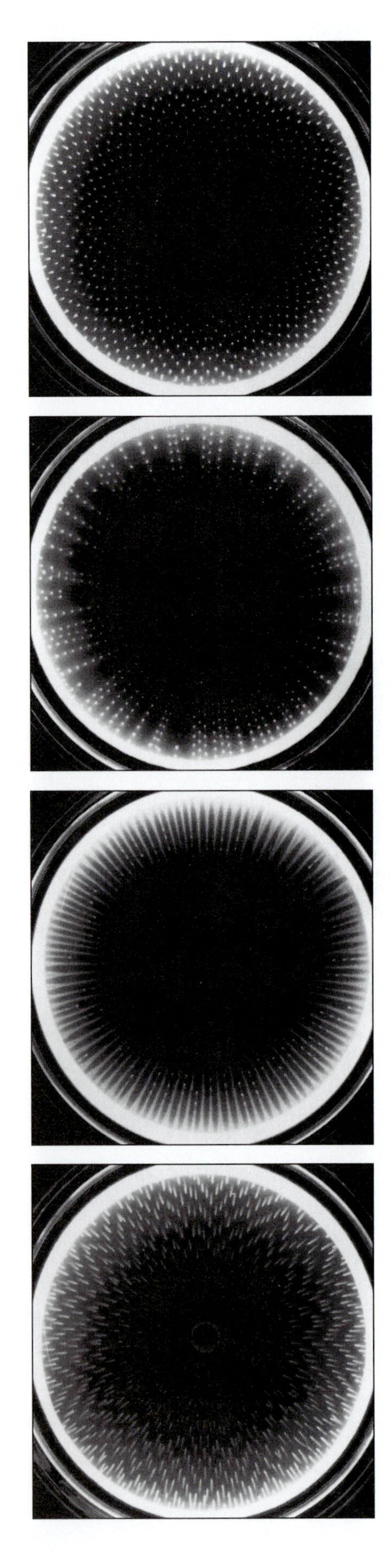

An intriguing phenomenon that depends on chemotactic behavior is the formation of large-scale patterns in cultures of bacteria. Wild-type *E. coli* cells, inoculated into the middle of a semi-solid agar with a single nutrient, move out as a traveling ring of cells as they consume the nutrient. However, if high levels of the metabolite succinate are used as the nutrient, the cells secrete aspartate (which serves as an attractant) and the stable ring breaks up into spots of bacterial density. The precise pattern depends upon succinate concentration but can show remarkably regular and intricate radial lines, concentric circles, and spirals (Figure 16–12).

Although these patterns have no known biological function, many dense aggregates or colonies are found under natural conditions. The ubiquitous soil bacteria myxobacteria, for example can collect into multicellular structures called *fruiting bodies* in which thousands of spores, resistant to heat, desiccation, and long-term starvation, are packaged and released (see Figure 2-13). Pathogenic bacteria also frequently collect together for the purpose of invading living tissue, as in the plague bacillus *Yersinia pestis*, which accumulates in lymph nodes, and *Pseudomonas* bacteria in burn wound infections. Bacteria in oceans and freshwater lakes are typically located close to surfaces or in 'stable consortia' of more than one photosynthetic species. Species of the sulfur bacterium *Thiovulum* form colonies like veils over the bottom of marine harbors. Remarkably, these bacteria anchor themselves to the sediment by long threads of mucus and use their flagella not for swimming but to drive currents of oxygen-rich water through the colony. Thin biofilms consisting of bacteria enmeshed in polysaccharide are of great environmental and economic importance. Biofilms on plant roots are thought to protect plants from infection by pathogens and may be involved in nonsymbiotic nitrogen fixation. On the other hand, unwanted biofilms plague the oil and shipping industries and are a health hazard in surgical wards.

In general, the production of a stable colony or aggregate of bacteria depends on two factors—cell motility and the ability of one cell to communicate with its neighbors. Motility is needed to establish the colonies in their correct location and to maintain their density and any internal organization. Communication between bacteria guides individual cells to the colonies and coordinates their activities, as in the formation of myxobacteria fruiting bodies under conditions of starvation. The signals may be diffusible molecules, released by one cell and taken up by another, or surface-bound molecules that promote the adherence of cells within a colony.

Remarkably, secreted molecules passing between bacteria can also give them a sense of the size of the population to which they belong (a 'smell of the crowd')—this ability being known as *quorum sensing*. The luminescent marine bacterium *Vibrio fischerii*, for example, secretes an autoinducer that triggers light production. When the bacteria are freely distributed in the ocean, the autoinducer is very dilute and no light is produced. However, if the bacteria are harbored by a species of squid in its light-producing organ, then the level of autoinducer rises and the bacteria

**Figure 16-12** Patterns formed by *E. coli.* Bacteria were inoculated in the center of a dish of semi-solid agar containing intermediates of the citric acid cycle such as succinate. Symmetrical patterns made of spots and stripes of bacteria formed by the local aggregation of bacteria, probably under the influence of secreted attractants. (From E.O. Budrene and H.C. Berg, *Nature* 349: 630–633, 1991. © Macmillan Magazines Ltd.)

(a)

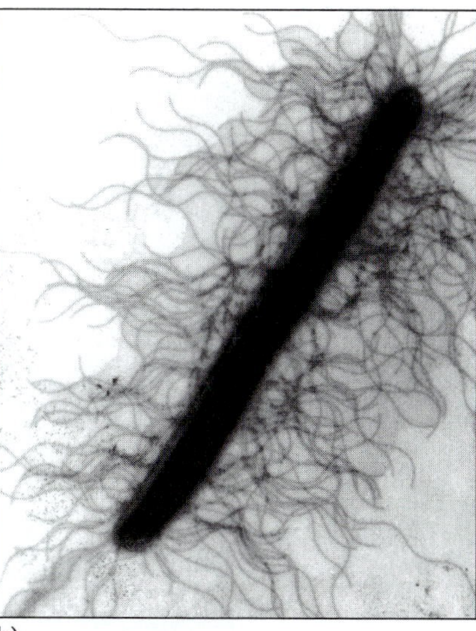

(b)

**Figure 16-13** Proteus swarming. (a) A small drop of *Proteus mirabilis* inoculated onto a semisolid agar plate produces an ever-enlarging bull's-eye colony pattern, shown here after 24 hours. The pattern is due to successive waves of swarmer cells moving out from the center. (b) Single cell of *Proteus mirabilis* in its swarming mode. The cell is highly elongated (about 45 μm in length) and carries numerous flagella on its surface. (Micrographs courtesy of Robert Belas.)

produce light. Both organisms draw benefit from this symbiosis, the squid uses light production as camouflage (to simulate moonlight to potential prey beneath), whereas the bacteria achieve a protected and nutritionally rich niche.

## Bacteria use flagella to swarm over surfaces

Bacteria obviously derive great benefit from being able to perform chemotaxis in a nutritionally depleted environment, such as a public water supply. But why should *E. coli* swim in the intestine of a mammal where it is bathed in nutrients? The answer is that it doesn't! Bacteria in such environments do not swim because they have shut off the synthesis of the entire apparatus of flagella, motors, and signaling proteins. The proteins responsible for these functions are controlled by a small set of genes that are under the control of catabolite repression—the system that turns off the synthesis of selected proteins when a preferred metabolite (in this case glucose) is present in high concentrations.

Under different circumstances, some bacteria can control their flagella in another direction, by enormously *increasing* their numbers. Bacteria such as *Proteus mirabilis*, *Vibrio*, and even *E. coli* under some conditions, produce flagella in large numbers when they are cultured on a suitable surface. Simple round colonies of *Proteus* on soft agar, for example, develop streams of cells that swarm radially in waves over the surface (Figure 16-13a). Eventually these swarms of cells revert to new rounded colonies at distant locations. Underpinning these changes in colony morphology, individual cells change from simple rod-shapes about 2 μm in length and bearing a few flagella into highly elongated cells 20 μm or more in length with more than 1000 flagella (Figure 16-13b). Swarming is produced by the motor-driven rotation of these flagella, but precisely how is not understood.

## Spiral bacteria migrate by screwing

Another kind of cell movement is shown by species of bacteria with flagella located at their poles. This includes species of *Halobacteria*, which have tufts of flagella projecting from each end of the cell, and various forms of spirochetes in which the flagella originate from the poles and

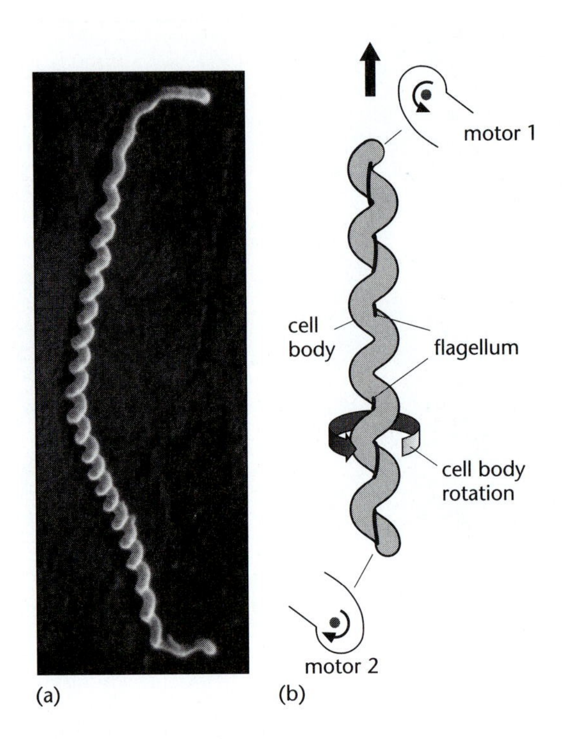


**Figure 16-14** Spirochete swimming. (a) Scanning electron micrograph of the spirochete *Leptospira* showing its right-handed helical form. (b) Simplified diagram showing the geometrical relationships between the cell body and the two flagella (here combined into a single rod). Rotation of the cell body in the direction shown by the circular arrow, carries the spiral-shaped cell upward. In order to maintain this motion, motors at either end of the cell must turn in opposite directions. Insets show the *intrinsic* rotation of the two motors viewed from outside the cell. (a, courtesy of O. Carleton et al, *J. Bacteriol.* 137: 1413–1416, 1979. © American Society for Microbiology.)

fold back along the cell, being trapped beneath an external membrane. Although in both cases the flagella and the motors by which they are driven are closely similar to those in *E. coli*, the hydrodynamics of their motion are different.

Spirochetes in particular move in a bizarre way. The cells are exceedingly long and thin and their cell wall is often cast into a helix. In species of *Leptospira* (the causative agent of one type of jaundice and other human diseases) the cells may be 20 μm long and 0.15 μm in diameter (Figure 16-14a). Other species of spirochete show a variety of complex cell shapes and swimming patterns, but generally rotate as they move, the rotation of their flagella causing them to screw through gel-like media.

*Leptospira* has one flagellum at each end of the cell. Each flagellum, a few micrometers long, is attached to a motor and folds back toward the center of the cell along the helical axis of the cell body. These flagella are thought to be more rigid than the cell body, so that the cell body conforms to their shape. In geometrical terms, the cell body may be thought of as a flexible cylinder (like a length of rubber tubing) wound in a right-handed helix around the two flagella (a centrally located rod). Torque produced by the gyration of flagella at the two cell ends is balanced by the cell body rolling in the opposite direction. Thus, if the motor at the anterior end rotates in a counterclockwise direction (seen from outside the cell), it will drive the cell body in the opposite direction (Figure 16-14b). The motor at the other end of the cell must turn in the opposite direction (again viewed from outside the cell) so that the two motors do not cancel each other out. How this coordination of the two motors is achieved is not known.

## *Cyanobacteria glide by secreting mucilage*

Many species of bacteria have the capacity to move over surfaces without using flagella or cilia. This form of movement, known as *gliding*, is shown by myxobacteria as they swarm and feed (Chapter 2), by species of cytophaga, and by the cyanobacteria, a large group of photosynthetic bacteria found in many freshwater and marine environments. The speed and other characteristics of gliding vary widely between different species, and may have more than one mechanism.

In filamentous cyanobacteria, chains of cells glide with a slow uniform motion in a direction parallel to the cell's long axis, occasionally interrupted by reversals. During gliding, cells advance without any obvious deformation of their surface but with the production of slime, a complex polysaccharide material left behind as a mucilaginous trail. The possibility thus arises that the movement of the cell could actually be driven by the extrusion of slime, which binds to and pushes against the surface.

Support for this idea comes from a detailed analysis of the organelles responsible for slime secretion. Using an electron microscope, sections of filamentous cyanobacteria show pore complexes that extend from the cytoplasm to the outside. The pores in species such as *Anabena* are arranged in a ring around the cross wall, or septum, between adjoining cells. The pores are 14–16 nm in diameter and radiate outward from the cytoplasm at an angle of 30–40° to the plane of each septum.

Experiments in which slime produced by the cells is marked by particles reveal that it is produced from the pores precisely in the correct location to generate the observed movement. However, not all pores are used all of the time, and some measure of control seems to be necessary to ensure that the secreted material pushes in the correct direction. If a chain of *Anabena* cells glides sideways, only those pores on the trailing side of the filament secrete slime while those facing the direction of travel are inactive. How the action of particular pores is controlled is unknown.

## *Disease-causing bacteria and viruses harness the actin cytoskeleton to enter cells*

Gliding is just one of the remarkable variety of motile and other mechanisms bacteria employ to enter and infect plant and animal cells. Eucaryotic cells are full of food molecules of all kinds and, over the course of evolution, bacteria have become expert at gaining access to this rich source of nutriment. Their propagation often depends on specific interactions with the host cytoskeleton, as is the case of *Wolbachia* that modify the host mitotic spindle (Chapter 13).

Since the actin cortex plus attached plasma membrane presents the primary barrier to most cells, it is an important target for many pathogenic bacteria. *Listeria*, for example, spread from one cell to the next by triggering filopodia on the surface of the first cell, which are then engulfed together with their associated bacterium by a neighboring cell (Chapter 5). Similarly, *Salmonella typhimurium*, which causes food poisoning, and *Yersinia pestis*, the causative agent of bubonic plague, induce the cells lining the intestine to engulf them and take them up into their cytoplasm.

The uptake of bacteria and viruses into cells is not precisely the same as the phagocytosis performed by macrophages, since bacteria taken up by epithelial cells are able to avoid the usual destruction in lysosomes. How bacteria achieve this is not completely understood, but some appear able to mimic the action of natural growth factors on receptors on the cell surface. The virus that causes Lassa fever and the bacterium that causes leprosy, for instance, both enter cells by interacting with dystroglycan, a protein that links the actin cytoskeleton to the extracellular matrix in many cells (and is itself implicated in several forms of muscular dystrophy). *Salmonella* injects a protein into cells that activates the small G proteins Rac and Cdc42 (described in Chapter 6) and hence initiates cytoskeletal rearrangements and ruffling. The ruffles, which are stabilized by a second injected protein, then carry the bacteria passively into the cell enclosed in a membrane sheath (Figure 16-15).

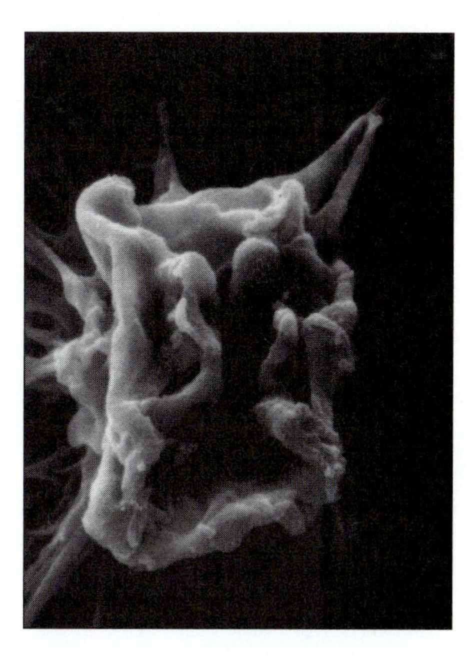

**Figure 16-15** Cell surface ruffling in response to bacterial infection. A single bacterial cell of the species *Shigella flexneri* is seen on the surface of a human cell in culture. The bacterium has induced membrane ruffles and filopodia on the surface of the host cell prior to engulfment. (Courtesy of Phillippe Sansonetti and Guy Tran Van Nhieu, *Trends Microbiol.* 4(6): cover photo, June 1996.)

## *Synechococcus swims without flagella*

Finally a word should be said about the remarkable ability of some species of cyanobacteria to swim without any visible means of propulsion. In 1985, the curious observation was made that five isolates of the cyanobacterium *Synechococcus* from the Sargasso Sea and one from the South Atlantic were capable of rapid swimming. The motile cells were coccoid to rod-shaped, 1–2 μm in length, and were seen to move at speeds of to 25 μm/sec along straight or looping paths, rotating around their long axis as they moved in the manner of flagellated bacteria. Strikingly, the cells appeared to lack flagella or any other organelle that might be associated with motility. Examination of the cells in the electron microscope and shearing experiments (which should remove flagella if present) all failed to indicate the presence of surface flagella.

How these organisms swim is still a mystery although there are some clues. Individual *Synechococcus* cells that become attached to a glass surface rotate about the point of their attachment, rather like tethered *E. coli* cells. This suggests that something on the surface spins in the manner of a motor, and there is some evidence that it might be driven by a flux of protons. An abundant cell-surface protein has been isolated that binds $Ca^{2+}$ and when this is deleted, the cell loses the ability to swim, although not to rotate.

Why *Synechococcus* should want to swim is another puzzle. Rapid though it is, a speed of 25 μm/sec will carry a cell only about 2 meters in 24 hours (even if it could swim in a straight line)—an inconsequential distance with respect to light in the open ocean. Moreover, because of their small size, individual cells of *Synechococcus* behave like colloidal particles in sea water, their movements being dominated by the physical mixing processes in the water. One suggestion is that the motility of these bacteria enables them to respond chemotactically to small particulate aggregates of nutrient-rich material that cycle continuously in warm and temperate oceans.

## *Outstanding Questions*

*Does* E. coli *have an internal structural framework made of protein? If so, which proteins form filaments, and how are they organized? Do bacteria have motor proteins? How are flagellin molecules guided into the lumen of a flagellum and how do they travel to its tip? What is the process by which they become incorporated into the wall of the flagellum? How does the movement of protons across the plasma membrane drive the flagellar motor? What conformational changes occur in a flagellum as it reverses direction? How are the multiple flagella on a bacterial surface coordinated? What controls the clustering of chemotactic receptors and what function does this have for chemotaxis? How does CheY control motor switching? How do the chemotactic signal pathways of other bacteria work? What is the mechanism of bacterial swarming? How do neighboring cells in a swarm become entrained so that they move together? What drives gliding of myxobacteria and cytophaga? How does* Synechococcus *swim?*

## *Essential Concepts*

- Bacteria such as *Escherichia coli* swim by means of flagella made of single helical tubes of the protein flagellin.
- Each flagellum is attached through a flexible hook to a motor embedded in the bacterial membrane which is driven by a flux of hydrogen ions across the membrane.
- Because each flagellum has a helical handedness (usually left-handed), rotation in only one direction allows smooth swimming. Transient reversal of the motor produces chaotic changes in direction known as tumbling.
- Alternation between swimming and tumbling is influenced by the presence of chemoattractants and repellents in the medium and forms the basis of the chemotactic response.
- Analysis of nonchemotactic mutants of *E. coli* shows that attractants activate one of five kinds of chemotactic receptor in the plasma membrane, either directly or as a complex with a transport protein in the periplasmic space.
- Each receptor molecule is part of a complex of proteins on the inner face of the plasma membrane that generates a diffusible protein, CheYp, at levels depending on the ligand-bound state of the receptor.
- CheYp diffuses to the flagellar motors, where it influences the direction of flagellar rotation and hence the swimming behavior of the bacterium.
- The same cascade of signals also triggers the slower adaptation of the response through the reversible methylation of the chemotactic receptors. Methylation alters the rate at which CheYp is produced and hence modulates swimming performance.
- We probably understand the chemotaxis of *E. coli* in greater detail than any other form of cell movement, but many unsolved mysteries about its detailed performance remain.
- Many commonly occurring species of bacteria use a chemotactic mechanism that is basically similar to that of *E. coli*, but with numerous subtle modifications.
- Both in the laboratory and in natural environments, bacteria often accumulate into large colonies, such as biofilms. Their social behavior depends both on motility and on the ability of bacteria to communicate with each other.
- Some species of bacteria use flagella in other ways. *Proteus* uses large numbers of flagella like oars to swarm over surfaces; the flagella of spirochetes are enclosed in a membrane, enabling these cells to screw their way through viscous media.
- Species of cyanobacteria have been discovered that swim rapidly through sea water without flagella or other obvious surface appendage. The mechanism (and function) of this form of motility is currently a mystery.

## *Further Reading*


Adler, J. Chemoreceptors in bacteria. *Science* 166: 1588, 1969.

Aizawa, S.I. Flagellar assembly in *Salmonella typhimurium*. *Mol. Microbiol.* 19: 1–5, 1996.

Alam, M., Oesterhelt, D. Morphology, function and isolation of halobacterial flagella. *J. Mol. Biol.* 176: 459–475, 1984.

Armitage, J.P. Bacterial tactic responses. *Adv. Microb. Physiol.* 41: 229–289, 1999.

Barkai, N., Leibler, S. Robustness in simple biochemical networks. *Nature* 387: 913–917, 1997.

Berg, H.C. Random Walks in Biology. Princeton, NJ: Princeton University Press, 1993.

Berry, R.M. Torque and switching in the bacterial flagellar motor. *Biophys. J.* 64: 961–973, 1993.

Brahamsha, B. An abundant cell-surface polypeptide is required for swimming by the nonflagellated marine cyanobacterium *Synechococcus*. *Proc. Natl. Acad. Sci. USA* 93: 6504–6509, 1996.

Calladine, C.R. Construction of bacterial flagella. *Nature* 255: 121–124, 1975.

Cluzel, P., et al. An ultrasensitive bacterial motor revealed by monitoring signaling proteins in single cells. *Science* 287: 1652–1657, 2000.

Cossart, P., (Ed). Cellular Microbiology. Washington: American Society for Microbiology, p. 326, 2000.

Cudmore, S., et al. Actin-based motility of vaccinia virus. *Nature* 378: 636–639, 1995.

DeRosier, D.J. The turn of the screw: the bacterial flagellar motor. *Cell* 93: 17–20, 1998.

Dramsi, S., Cossart, P. Intracellular pathogens and the actin cytoskeleton. *Annu. Rev. Cell Biol.* 14: 137–166, 1998.

Duke, T.A.J., Bray, D. Heightened sensitivity of a lattice of membrane receptors. *Proc. Natl. Acad. Sci. USA* 96: 10104–10108, 1999.

Falke, J.J., et al. The two-component signaling pathway of bacterial chemotaxis: a molecular view of signal transduction by receptors, kinases, and adaptation enzymes. *Annu. Rev. Cell Dev. Biol.* 13: 457–512, 1997.

Fenchel, T., Glud, R.N. Veil architecture in a sulphide-oxidizing bacterium enhances countercurrent flux. *Nature* 394: 367–369, 1998.

Francis, N., et al. Isolation, characterization and structure of bacterial flagellar motors containing the switch complex. *J. Mol. Biol.* 235: 1261–1270, 1994.

Goldstein, S.F., et al. Structural analysis of the *Leptospiracae* and *Borrelia burgdorferi* by high-voltage electron microscopy. *J. Bacteriol.* 178: 6539–6545, 1996.

Hoiczyk, E., Baumeister, W. The junctional pore complex, a prokaryotic secretion organelle, is the molecular motor underlying gliding motility in cyanobacteria. *Curr. Biol.* 8: 1161–1168, 1998.

Jones, C.J., Aizawa, S.I. The bacterial flagellum and flagellar motor: structure, assembly and function. *Adv. Microb. Physiol.* 32: 110–163, 1991.

Khan, S., et al. Architectural features of the *Salmonella typhimurium* flagellar motor switch revealed by disrupted C-rings. *J. Struct. Biol.* 122: 311–319, 1998.

Losick, R., Kaiser, D. Why and how bacteria communicate. *Sci. Am.* 276: 68–73, 1997.

Morgan, D.G., et al. Structure of bacterial flagellar filaments at 11 Angstrom resolution. *J. Mol. Biol.* 249: 88–110, 1995.

Piperno, G., Mead, K. Transport of a novel complex in the cytoplasmic matrix of *Chlamydomonas* flagella. *Proc. Natl. Acad. Sci. USA* 94: 4457–4462, 1997.

Rebbapragada, A., et al. The Aer protein and the serine chemoreceptor Tar independently sense intracellular energy levels and transduce oxygen, redox and energy signals for *Escherichia coli* behavior. *Proc. Natl. Acad. Sci. USA* 94: 10541–10546, 1997.

Schuster, S.C., Khan, S. The bacterial flagellar motor. *Annu. Rev. Biomol. Struct.* 23: 509–539, 1994.

Shi, W., and Zusman, D.R. The two motility systems of *Myxococcus xanthus* show different selective advantages on various surfaces. *Proc. Natl. Acad. Sci. USA* 90: 3378–3382, 1993.

Shimkets, L.J. Social and developmental biology of the myxobacteria. *Microbiol. Rev.* 54: 473–501, 1990.

Takeuchi, A. Electron microscope studies of experimental salmonella infection. I Penetration into the intestinal epththelium by *Salmonella typhimurium*. *Am. J. Pathol.* 50: 109–136, 1967.

Thomas, D.R., et al. Rotational symmetry of the C ring and a mechanism for the flagellar rotary motor. *Proc. Natl. Acad. Sci. USA* 96: 10134–10139, 1999.

Waterbury, J.B., et al. A cyanobacterium capable of swimming motility. *Science* 230: 74–76, 1985.

Zhao, D., et al. Role of the *S. typhimurium* actin-binding protein SipA in bacterial internalization. *Science* 283: 2092–2095, 1999.

# PART THREE

## Integration of Cell Movements

CHAPTER 17

# Intermediate Filaments

Most cells in animal tissues contain a network of insoluble filaments that preserves their shape even after actin filaments and microtubules have been extracted. This network is composed of *intermediate filaments*, so named because their diameter of about 10 nm is intermediate between those of thick and thin filaments of striated muscle, the tissue in which they were first discovered—they are also 'intermediate' in diameter between actin filaments (7 nm) and microtubules (25 nm). Although intermediate filaments are not directly responsible for generating movements, they have important effects on the shape and mechanical properties of many cells and can constrain and modify their movements.

Extending from the nuclear envelope to junctions in the cell membrane, intermediate filaments constitute a tough, interconnected network that gives the cell mechanical integrity. In specialized skin cells, intermediate filaments of keratin accumulate in the cytoplasm, becoming extensively cross-linked into an insoluble matrix. When these keratin-containing cells die, they remain as complex multicellular structures tightly adherent to each other. The intermediate filaments in this way form the outermost layer of skin, hair, nails, horn, and feathers, providing the animal with a barrier against heat and water loss as well as supplying it with camouflage, armament, and ornamentation.

We will see in this chapter that intermediate filaments are much more varied in sequence and structure than either actin filaments or microtubules. They exist in many different forms, more than 50 different types of intermediate filament proteins being found in humans, for example. Although their primary role is mechanical, this does not mean that they are simple. In fact they produce fibers of highly sophisticated construction: uniform and conserved in their basic design but highly variable in detailed chemistry and mechanical properties, enabling them to match the requirements of each particular cell.

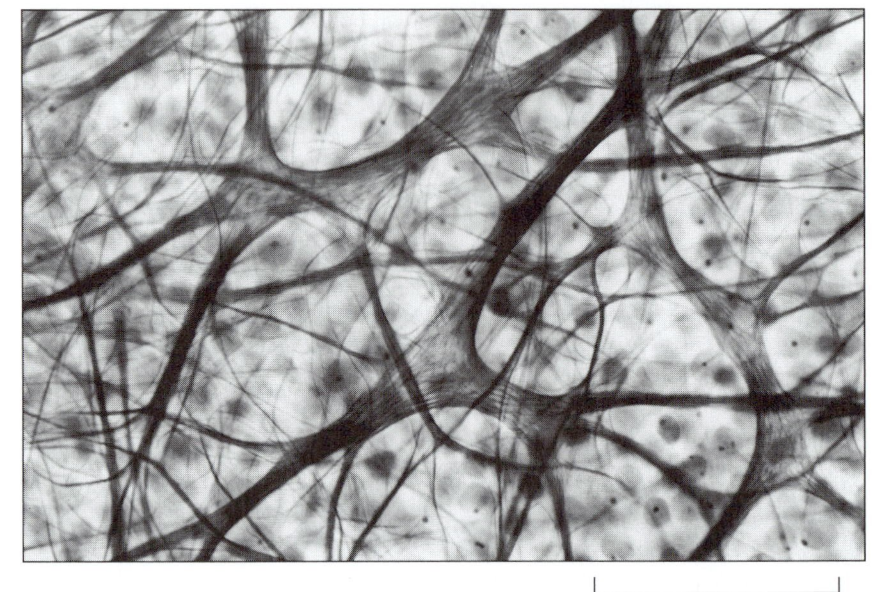


**Figure 17-1** Neurofilaments in a rabbit retina. Staining of a section of rabbit retina with reduced silver picks out neurofilaments running through the cell bodies and dendrites of the retinal nerve cells. (Courtesy of Leo Peichl and Brian Boycott.)

## *Intermediate filaments are tough and insoluble*

In 1843, the histologist Robert Remak discovered long, thin structures called neurofibrils in living nerve cells. Later improvements in staining techniques, especially using the reductive precipitation of metallic silver onto fixed tissue, picked these elements out selectively and with dramatic clarity, revealing them to be distributed throughout the cytoplasm of most nerve cells (Figure 17-1). The identification of neurofibrils as a distinct form of cytoskeletal filaments came only much later, since it required the use of the electron microscope. Intermediate filaments were first identified on morphological grounds in developing muscle cells in the late 1960s and seen to be the components of neurofibrils shortly after that.[1]

Because of their stability, intermediate filaments can often be obtained as a residue following extraction of other cell components. In a typical procedure, cells are lysed to remove soluble components, then DNA and RNA are removed by enzymatic digestion. Actin filaments, myosin filaments, and microtubules are extracted with concentrated salt solutions, leaving a tangled mass of filaments, with individual diameters about 10 nm (Figure 17-2).

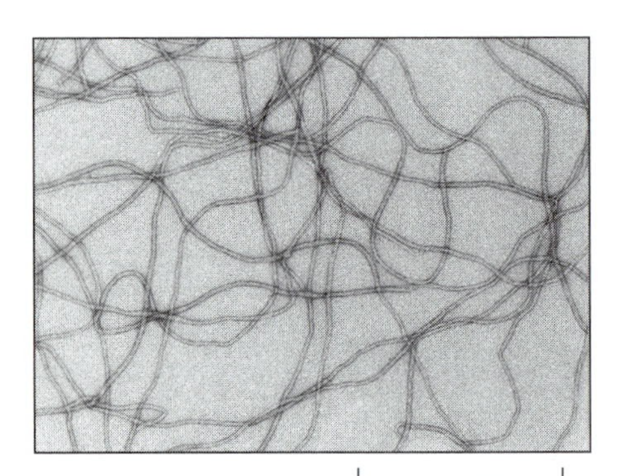


**Figure 17-2** Intermediate filaments. Tangled mass of filaments seen in the electron microscope following negative staining. (Courtesy of Roy Quinlan.)

As indicated by this procedure, intermediate filaments are insoluble in most physiological buffers, however they can be dissolved in solutions of a denaturing agent such as urea and then fractionated by chromatography to yield pure proteins. A gentler way to disperse the filaments is to dialyze them against dilute solutions containing phosphate ions (for example, 1 mM sodium phosphate at pH 7.5), leading to the fraying of the filaments, which unravel into progressively thinner strands. Eventually the filaments are converted into short fibrous elements that then break up into rod-shaped dimeric molecules, about 50 nm long and with globular regions at both ends.

## *The protein subunits of intermediate filament proteins are coiled-coil dimers*

Intermediate filaments have an important place in the history of protein structure determination. The earliest x-ray diffraction patterns of protein were obtained in the 1930s from wool fibers, which are made of keratin. These patterns indicated a regular arrangement of amino acids in the polypeptide chain, later modeled as a regular helix of amino acids and

[1] Thick bundles of keratin filaments, known as tonofibrils, were also seen by nineteenth-century histologists in sections of skin. The intense staining of these massive bundles had a spiky appearance and the layer in which they occurred was called the *stratum spinosum*.

called an α-helix. This same folding motif was subsequently found to be widespread in most proteins, although few have such a high content as keratins and other members of the intermediate filament family.

Largely because of their insolubility, little was known about the detailed structure of intermediate filaments until their amino acid sequences were determined. This revealed firstly that the protein subunits of different types of intermediate filaments vary widely in size and sequence details; and secondly, that despite this variation they are nevertheless built on a common plan. All have a highly conserved central region of about 310 amino acids flanked by sequences at the N- and C-terminal ends that are variable in size and sequence (but characteristic of particular intermediate filament types). The basic subunit of all intermediate filaments is a coiled-coil dimer of two polypeptide chains in register (Figure 17-3).

The amino acid sequence of the central, rod domain region contains multiple repeats of a sequence of the type *abcdefg* (heptad repeats), where *a* and *d* are usually nonpolar amino acids, such as valine, isoleucine, or alanine. As already discussed for myosin, this pattern of amino acids enables two α-helices to wind together into a coiled-coil. The dimensions of the central rod domain of intermediate filament proteins, about 46 nm, are exactly what one would predict from the sequence data.[2]

Analysis of the amino acid sequence of the rodlike regions of different intermediate filament proteins shows that they can be further subdivided into four distinct α-helical coiled-coil segments with short nonhelical linker regions between them (see Figure 17-3). In addition to the heptad repeat, most sequences also show a series of zones of alternating positive and negative charge about 9.5 residues apart, giving these intermediate filaments highly charged surfaces. The alternating bands of charge in the rod sections of adjacent molecules may contribute to the binding interactions that help to hold the assembled intermediate filament together.

Superimposed on the highly conserved features of intermediate filament proteins there are systematic differences in sequence that define five distinct classes (Table 17-1). Particular cell types usually express two or more classes of intermediate filament protein, each producing its own filament. The keratins of epithelial cells are unusual in that they are formed

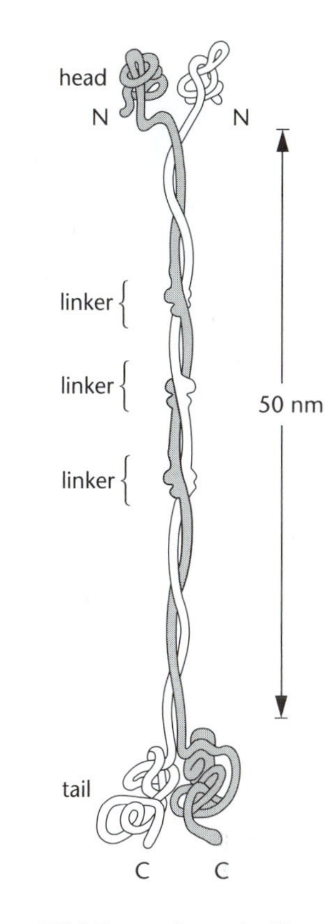


**Figure 17-3** Form of a typical intermediate filament protein. The central rodlike coiled-coil domain is interrupted by short linker regions. Nonhelical domains at either end are highly variable in size and contain both conserved and highly variable subregions.

**Table 17-1 Intermediate filament proteins**

| Class | Protein (MW) | Location |
|---|---|---|
| type I<br>type II | keratins (40–67 kDa) | Abundant in epithelial cells and their derivatives (such as hair and nails). Filaments consist of heteropolymers of type I and type II keratins. Different pairs are expressed in different tissues. |
| type III | vimentin (54 kDa) | Vimentin is widely expressed in mesenchymal cell types and in various transformed cell lines. |
| | desmin (53–54 kDa) | Desmin is found in smooth muscle and at the Z discs of striated muscle. |
| | GFAP (50 kDa) | GFAP is expressed in glial cells and astrocytes. |
| | peripherin (57 kDa) | Peripherin is found in neurons in the peripheral nervous system. |
| type IV | neurofilaments NF-L (62 kDa); NF-M (102 kDa); and NF-H (110 kDa) | NF-L, NF-M, and NF-H are coexpressed and form heteropolymers in axons, dendrites, and cell bodies of adult neurons. |
| | α-internexin (66–70 kDa) | α-Internexin forms homopolymers in developing neurons. |
| type V | lamin A (60–70 kDa)<br>lamin B (63–68 kDa) | Lamins form a fibrous meshwork, the nuclear lamina, on the inner surface of the nuclear membrane in higher eucaryotes. Multiple lamin isotypes are produced by differential RNA splicing. |

[2] The length of helical rod of intermediate filament proteins can be calculated as follows: the length of the 'coiled-coil' sequence is about 310 amino acids. The rise per residue of an α-helix is 0.15 nm. The length of each α-helical domain is therefore $310 \times 0.15$ nm = 47 nm, with a slight reduction in length due to the helical twist.

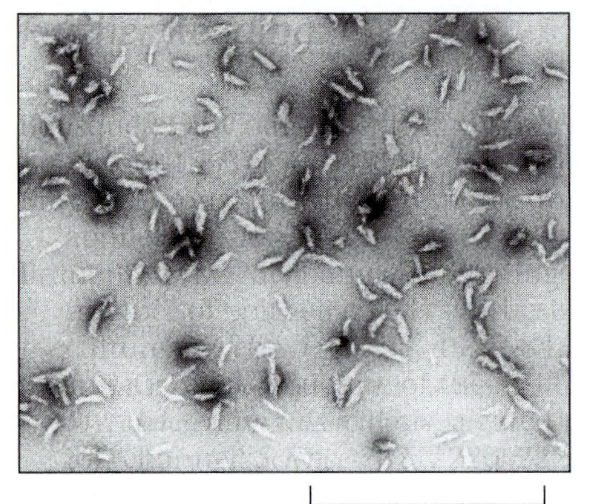


**Figure 17-4** Small aggregates of intermediate filament proteins. Discrete, unit-length, filaments of *Xenopus* vimentin visualized by negative staining. These formed after 2 seconds of incubation at room temperature. (Courtesy of Harald Herrmann.)

by the obligatory copolymerization of two types of protein (type I plus type II keratins). Almost all eucaryotic cells have lamins in their nucleus and usually at least one other type of intermediate filament in their cytoplasm. Some cells make two different types of cytoplasmic intermediate filaments: for example, some epithelial cells have distinct arrays of vimentin and keratin intermediate filaments. Mixed filaments containing two or more proteins of the same class are often found, such as vimentin and desmin, or the neurofilament triplet. The characteristic intermediate filament complement of a cell is maintained in tumors derived from that cell, providing a useful method of typing tumors that supplements conventional histological techniques. The various classes of intermediate filaments are further discussed below.

## *Intermediate filament assembly involves longitudinal annealing*

Homodimers of vimentin, desmin, or keratin are soluble in dilute phosphate-containing solutions. In solutions of physiological ionic strength they spontaneously assemble into long filaments that resemble intermediate filaments in size and rigidity. This process, which has been studied using a 'stopped-flow' approach adapted for the electron microscope, shows that after the first few seconds, short filaments (*unit filaments*) about 70 nm long and about 16 nm wide are formed in large numbers (Figure 17-4). These unit-length filaments rapidly anneal longitudinally to produce segmented intermediate filaments about 300 nm in length. Over the next several minutes, the segmented intermediate filaments elongate further and start to compact, yielding uniform, smooth-looking intermediate filaments with a fairly uniform width.

During the assembly process, pairs of dimers initially associate into stable tetramers, with individual dimers lying side by side but antiparallel (Figure 17-5). Higher-order interactions—which are not completely defined and may vary according to the type of intermediate filament protein and assembly conditions—then allow the tetramers to assemble into the unit-length filaments. In most cases there seem to be 8 tetramers (32 polypeptide chains, or 16 coiled-coils) in the cross-section of each unit filament. Longitudinal annealing of the unit filaments then proceeds by the interdigitation of free ends, followed by compaction of the segmented immature fibrils. There is evidence that the intramolecular arrangement of an intermediate filament is not rigidly fixed but can vary with conditions. It may change with mechanical strain, for example, a phenomenon called 'strain hardening.'

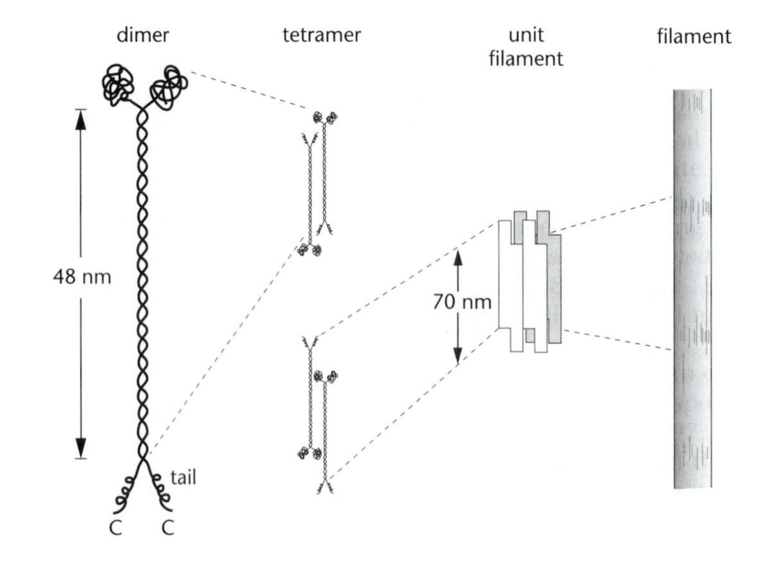


**Figure 17-5** Stages in assembly of intermediate filaments. Coiled-coil dimers associate together into an antiparallel tetramer, which then aggregate into the unit length filaments shown in Figure 17-4. In the following step, unit filaments are thought to associate (anneal) end-to-end to produce the intermediate filament.

Genetic engineering experiments in which intermediate filament genes are truncated or point-mutated indicate that sequences close to either end of the central domain are crucial for intermediate filament assembly. Even conservative substitutions in these locations (that is, replacements of one amino acid by another of similar size and chemistry) can interfere with filament formation. Synthetic oligopeptides containing these critical sequences also have a markedly disruptive effect on intermediate filaments in the cell if injected into the cell in sufficient quantity.

Since intermediate filaments are insoluble in buffers whose ionic conditions resemble those of living cytoplasm there must be mechanisms to prevent their subunits associating in an uncontrolled fashion as they are synthesized. The principal method by which this is achieved is through phosphorylation. Most intermediate filament proteins have sites in their N-terminal domain that, when phosphorylated by a protein kinase, reduce the ability of the protein to assemble into filaments. In some cases phosphorylation will even reverse the formation of filaments. Vimentin filaments phosphorylated under controlled conditions by cAMP-dependent kinase disperse even in buffers of physiological ionic strength, whereas removal of the phosphate groups allows the filaments to reform. Another example is that of the cyclic dissociation and assembly of the nuclear lamina, which depends on the phosphorylation and dephosphorylation of lamins as discussed below.

## *Head and tail domains of the protein subunits enhance filament assembly*

At either end of the conserved α-helical central region of an intermediate filament protein molecule are its variable, nonhelical head and tail domains. These range in size from the short C-terminal domain of the smallest keratin to the large and multiply phosphorylated C-terminal domain of mammalian neurofilaments. Within these terminal regions, some regions are highly variable whereas others are conserved.

Since all intermediate filament proteins assemble to form morphologically similar 10 nm structures, it seems doubtful, from sequence alone, that the head and tail domains can be essential for polymerization as such. However, they are able to modulate the process. Thus, although amino-terminal truncated desmin and vimentin can each copolymerize into intermediate filaments when mixed with wild-type proteins, they cannot form proper intermediate filaments on their own. Similarly, neither headless neurofilament proteins nor headless lamins assemble normally. Loss of the tail has more subtle and hard-to-detect effects, but may affect bundling, since filaments formed from tailless keratins have a tendency to unravel under conditions in which wild-type filaments do not. In addition to these affects on the intermediate filament itself, it is likely that the head and tail regions affect its interaction with other components of the cell, interactions that will often depend on the specific context of a particular cell.

It is interesting to compare the design of intermediate filament proteins with that of actin filaments and microtubules. Actin and tubulin have been highly conserved in amino acid sequence during evolution and, although they perform many varied functions in cells, changes in the properties of actin filaments or microtubules are conferred by a plethora of actin-binding and microtubule-associated proteins. Intermediate filaments, by contrast, may be thought of as having constant and variable parts within the same molecule. Their 'polymerizing region' is mostly (but not entirely) confined to one region of the molecule, leaving the remainder free to vary with evolution. In a sense, the variable terminal domains of intermediate filament proteins serve the same function as, say MAP2 or filamin—modulating the function of an otherwise unchanging structural element.

## *Intermediate filaments are not polar*

It is important to note that intermediate filaments do not have a structural polarity. They are built from subunits (tetramers) that are themselves symmetrical, being the product of an antiparallel association of dimers (see Figure 17-5). There is consequently no way that one could distinguish, on structural grounds, one end of an intermediate filament from another. Nor could a second molecule, alighting on an intermediate filament, identify one direction from another solely from the local features of the filament.

The consequences of this seemingly innocuous feature are far-reaching, since it implies that motor proteins of the kind associated with actin filaments and microtubules, or even the kinds of processive enzymes associated with DNA, such as DNA polymerase, will not work on intermediate filaments.[3] These motors depend on the polarity of their filament track to point their motile machinery in the correct direction. In practice, as mentioned previously for myosin, they do not even bind to a filament that is presented with the incorrect orientation.

In theory, one could imagine proteins sliding along intermediate filaments by thermal motion (some proteins do move in this way along DNA). The resulting two-dimensional Brownian motion would be highly inefficient and lack directionality, but could perhaps be biased by cytoplasmic structures. Another notional scheme would be to use conventional motors 'launched' from one position of the cell (say a nuclear pore) along an intermediate filament. So long as such a protein did not let go of its filament (it was highly processive) it could preserve its direction of movement. It must be said, however, that there is no evidence for either possibility. So far as we know, intermediate filaments do not have motors, and are not used as a basis for transport in living cells.

## *Assembly of nuclear lamins is controlled by phosphorylation*

The defining characteristic of any eucaryotic cell, whether animal or plant, is its nucleus, the organelle containing the cell's DNA and associated proteins. Eucaryotic nuclei are enclosed by a thin envelope made from a double-walled membrane continuous with the smooth endoplasmic reticulum, and an underlying supporting *nuclear lamina* composed of a fibrous protein network (Figure 17-6). The nuclear lamina can be isolated

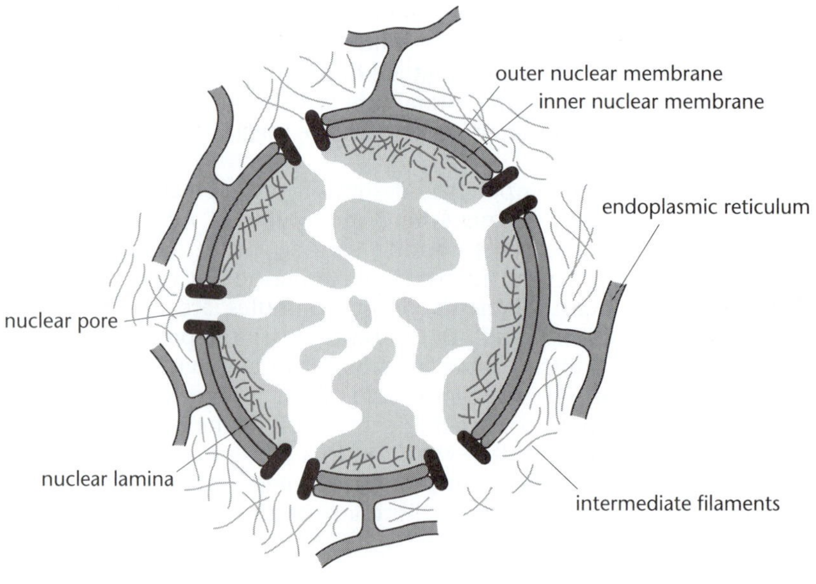

**Figure 17-6** The interphase nucleus. Diagrammatic view of a cross-section through a typical cell nucleus. The nuclear envelope consists of two membranes, the outer one being continuous with the endoplasmic reticulum. Two networks of cytoskeletal filaments provide mechanical support for the nuclear envelope—cytoplasmic intermediate filaments such as vimentin and a fibrous network of lamin intermediate filaments lining the internal face of the nuclear membrane. How the two are connected is presently unclear.

outer nuclear membrane
inner nuclear membrane
endoplasmic reticulum
nuclear pore
nuclear lamina
intermediate filaments

[3] Another filament lacking polarity is the MSP filament of nematode sperm (Chapter 8). These also appear to lack motor proteins, although the same reservations apply as to intermediate filaments.

as a membrane-free sheet containing specialized proteins that form pores in the nuclear membrane. In regions away from the nuclear pores, the sheet consists of a two-dimensional lattice of proteins filaments 8–10 nm in diameter made of type V intermediate filaments, or *lamins* (Figure 17-7).

Lamins have been cloned from organisms ranging from marine invertebrates and insects to mammals and there are hints that they occur also in plants. Mammalian cells contain four lamin genes, which generate seven or more distinct isoforms through alternative RNA splicing. Although the sequences of lamins place them firmly in the family of intermediate filaments, they do differ from proteins such as vimentin or neurofilament proteins in having a 42-residue insert in their central α-helical rod domain. Interestingly, a similar insert is also found in the cytoplasmic intermediate filaments of invertebrates, such as worms and snails, suggesting that lamins might represent an evolutionarily ancient form of the protein.

We saw in Chapter 13 that a major event defining the entry of a cell into mitosis is the breakdown of its nuclear envelope. Lamins are important players in this event, being dispersed by a transient phosphorylation of several serine residues. The 'soluble' mitotic form of the nuclear lamina consists of tetramers of phosphorylated lamins analogous to the assembly intermediates described above. When the cell has segregated its chromosomes, the lamins are dephosphorylated and an intact nuclear envelope reforms around the now separated chromosomes. Whether lamin phosphorylation actually triggers nuclear membrane breakdown or is a downstream consequence of another switch is not yet certain.

Lamins have an undisputed function in supporting the nuclear envelope, providing it with stability and mechanical strength, and may also provide sites to which cytoplasmic intermediate filaments are anchored. However, their functions are not only mechanical, and there are indications that lamins also participate in the genetic processes of the nucleus. An intact lamina is required for DNA replication and specific interactions have been reported between replication factors and lamins. Whatever the original function of nuclear lamins may have been, their intimate juxtaposition with chromatin over billions of years would have opened the door to their being used for other (even genetic) purposes.

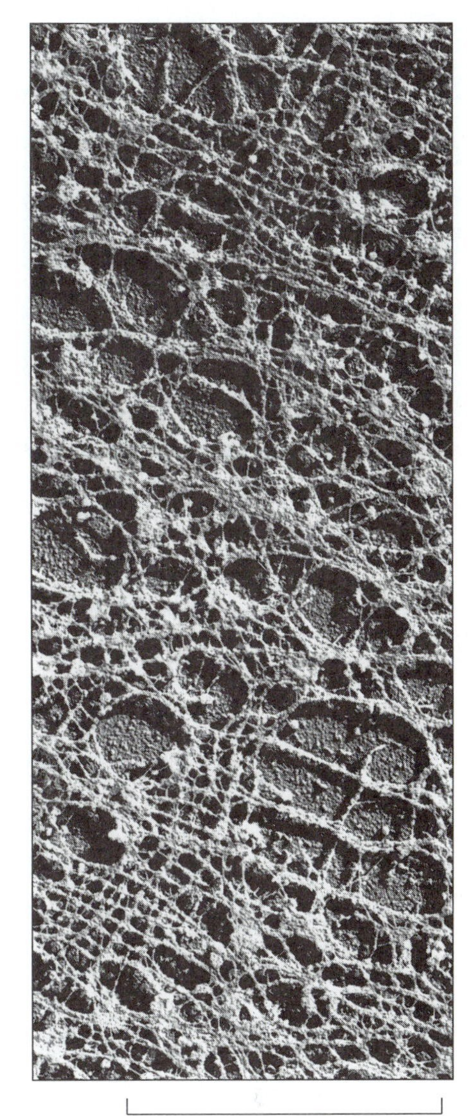


**Figure 17-7** Nuclear lamina network. Isolated nuclear membranes from frog oocytes were extracted with detergent, freeze-dried, shadowed with metal, and examined by electron microscopy. In some regions nearly orthogonal meshworks of lamin filaments can be seen. (Courtesy of Ueli Aebi.)

## *Keratin filaments form a meshwork in epithelial cells*

Epithelial cells contain two related families of *keratins* (the proteins of skin and hair), sometimes known as type I, or acidic keratins, and type II, or basic keratins.[4] Keratins are by far the most variable type of intermediate filament protein, with at least 20 distinct genes for those in human epithelia and another dozen genes for keratins specific to hair and nails.

A single epithelial cell can make many different keratins, all of which copolymerize into a single keratin filament system. The simplest epithelia, found in early embryos and in some adult tissues such as the liver, contain only a single type I and a single type II keratin. Epithelia in other locations, such as the tongue, bladder, and sweat glands, contain six or more keratins, the particular blend depending on the cell's location in the organ. The diversity is most pronounced in skin, where distinct sets of keratins are expressed by the cells in the different layers of the epidermis. There are also keratins characteristic of actively proliferating epithelial cells. This heterogeneity of keratins is clinically useful: in the diagnosis of epithelial cancers (carcinomas), the particular set of keratins expressed can be used to determine the epithelial tissue in which the tumor originated and thus help to decide the type of treatment that is likely to be most effective.

A primary function of keratins is to provide a tough outer layer of the animal body. During the terminal differentiation of skin cells, most organelles and cytoplasmic proteins are lost, whereas the keratins are

[4] Some workers prefer the term *cytokeratin* pointing out that 'keratin' in common usage refers to the extracellular material of hair or horn, which includes both intermediate filaments and a variety of matrix proteins.

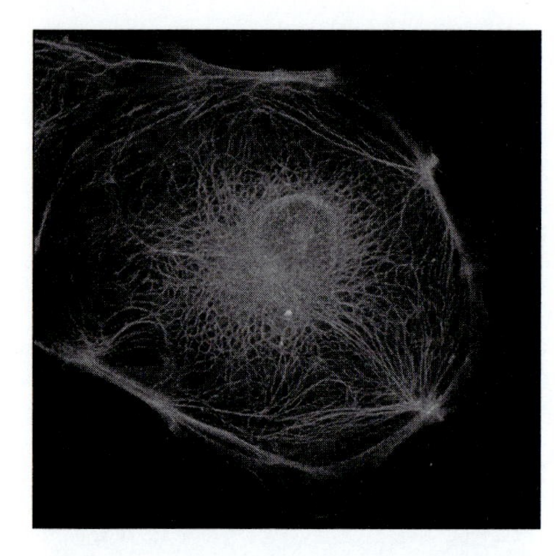

**Figure 17-8** Keratin filaments. An epithelial cell in tissue culture has been stained with antibodies to keratin. A lacelike network of filaments is seen extending throughout the cell, and attached to specific sites on the plasma membrane. (Courtesy of Mary Osborn, MPI Goettingen.)

progressively packaged into larger bundles and cross-linked by covalent bonds into some of the most stable protein–protein interactions known in nature. Association of keratins with tissue-specific linker proteins generates a fabulous variety of materials. Spiders webs, silkworm silks, rhinoceros horns, and human hair are all made from the same basic ingredient but each has a distinctive flexibility, tensile strength, and water permeability.

Living epithelial cells, as we have mentioned, also contain keratin filaments, although they are less bundled and less cross-linked than extracellular keratins. Filaments of keratin appear to be anchored to the cytoplasmic surface of the nucleus and to extend to the vicinity of the cell membrane, usually where the cell experiences mechanical stress (Figure 17-8). Specific isoforms of keratin are often found at specific locations of a cell, suggesting that fine-tuning of mechanical responses is important. A case in point is seen in the mammalian inner ear, where supporting cells provide a supracellular framework for the hair cells (which detect incoming sound, see Chapter 6). Pillar-shaped supporting cells contain bundles of microtubules attached via keratin filaments to actin filaments in the cell cortex. The precise composition and location of the keratin filaments differs in each of three morphologically distinct types of supporting cell, suggesting that each has subtly different micromechanical requirements.

## *Keratin filaments are anchored in desmosomes*

Keratin filaments in the cytoplasm of an epithelial cell are often anchored to specific structures in the plasma membrane, principally cell junctions known as *desmosomes*. These are buttonlike contacts of great mechanical strength made of transmembrane linker glycoproteins and a dense plaque of intracellular attachment proteins. The transmembrane linkers contain cadherins—a family of self-binding, $Ca^{2+}$-requiring membrane proteins already encountered in the context of actin-rich adherens junctions (Chapter 6). Attachment to cadherins depends on proteins such as desmoplakins, which anchor loops of keratin filaments and thereby establish a continuous mechanical link from cell to cell throughout the tissue. Indeed, the entire epithelial layer can be seen as a basketwork of keratin filaments running from cell to cell (albeit bridged by desmosome junctions) that serves, among other things, to hold the nuclei in central locations within the cells (Figure 17-9).

A second type of cell junction, known as a *hemidesmosome*, is found at the base of multilayered epithelia and anchors intermediate filaments to the extracellular matrix. Hemidesmosomes have a similar appearance in the electron microscope to desmosomes, although their protein composition is quite distinct. Thus, in place of cadherins they have integrins—the transmembrane proteins rich in focal adhesions (Chapter 6). And in place of desmoplakins, they use BPAG1 and plectin to link intermediate filaments to the plaque of membrane proteins.

## *Vimentin filaments extend throughout the cytoplasm*

Type III intermediate filament proteins include *vimentin*, *desmin*, *glial fibrillary acidic protein* (*GFAP*), and *peripherin*, each of which has a characteristic distribution in certain cell types (see Table 17-1). Unlike keratins, type III intermediate filament proteins can self-assemble into homopolymers. They can also form mixed polymers with other type III proteins or (under artificial conditions) with neurofilament proteins. Type III proteins will not, however, coassemble with keratins, so that cells expressing the two kinds of proteins always form two distinct intermediate filament networks in the cytoplasm.

Of the proteins in this family, vimentin is the most abundant and ubiquitous. In fact, it is commonly the first intermediate filament to be

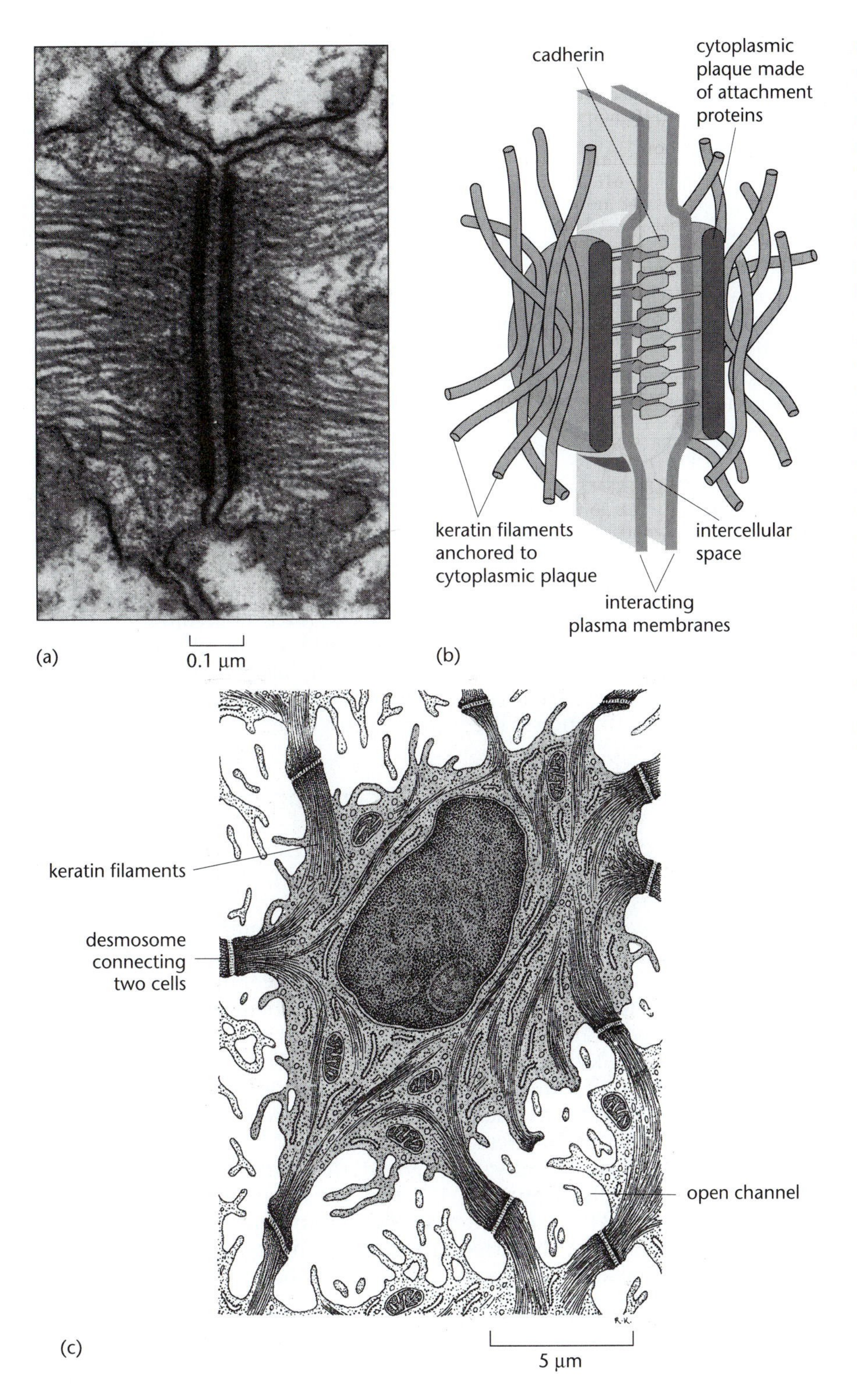


**Figure 17-9** Desmosomes. (a) An electron micrograph of a desmosome joining two cells in the epidermis of a newt, showing the attachment of keratin filaments. (b) Schematic drawing of a desmosome. On the cytoplasmic surface of each interacting plasma membrane is a dense plaque composed of a mixture of intracellular attachment proteins, to whose inner face keratin filaments are anchored. Proteins of the cadherin family are anchored in the outer face of each plaque and span the membrane to bind the two cells together. (c) Drawing from an electron micrograph of a section of human epidermis, showing the bundles of keratin filaments that traverse the cytoplasm of one of the deep-lying cells and are inserted at the desmosome junctions that bind this cell to its neighbors. Between adjacent cells in this deep layer of the epidermis there are also open channels that allow nutrients to diffuse freely through the metabolically active tissue (a, from D.E. Kelly, *J. Cell Biol.* 28: 51–59, 1966; b, adapted from B. Alberts et al, *Essential Cell Biology*. New York: Garland Publishing, 1998; c, from R.V. Krstić, Ultrastructure of the Mammalian Cell: An Atlas. Berlin: Springer, 1979.)

expressed in cells (apart from the nuclear lamins) during embryonic development, being progressively replaced by other cell type-specific proteins as cell differentiation proceeds. Conversely, when cells are grown in tissue culture and start to lose their differentiated characteristics they often revert to the expression of vimentin. Vimentin filaments ramify throughout the cytoplasm of cultured fibroblasts, for example, even though in this case they are not firmly anchored at the periphery by desmosomal junctions.

Vimentin filaments tend to associate with microtubules in the cell and if the latter are disrupted by means of drugs, then the network of

**Figure 17-10** Desmin filaments in immature muscle. Skeletal muscle cells undergoing differentiation in tissue culture show numerous longitudinally oriented, meandering desmin filaments in their cytoplasm. (From T. Schultheis et al, *J. Cell Biol.* 114: 953–966, 1991. © Rockefeller University Press.)

vimentin filaments collapses into a ball close to the nucleus. The intimate association between these two systems is further underscored by the finding that small complexes containing vimentin can move out to the periphery of the cell along cytoplasmic microtubules. These precursor complexes seem to be driven by a kinesin, but their molecular composition and the process by which they assemble into intact filaments are not known.

The nature of the association between vimentin filaments and the nucleus is also poorly understood. How are vimentin filaments on one side of the nuclear envelope attached to the nuclear lamins on the other side? Are they attached directly or linked through other proteins? Is there, in either case, a continuous chain of protein passing, for example through the nuclear pores? Or might there be a novel type of junction analogous to desmosomes linking the two sides of the nuclear membrane? Questions of a similar nature apply to the other end of vimentin filaments as well, since fibroblasts in tissue culture generally lack well-formed cell-to-cell junctions embedded in their plasma membrane. Could linker proteins such as plectin anchor them to the actin cortex?

## *Desmin filaments integrate muscle contraction*

Desmin is the characteristic intermediate filament of muscle cells. In vertebrate skeletal muscle it is found (together with vimentin and cross-linking proteins) at the periphery of each Z-disc. Adjacent Z-discs in neighboring myofibrils are thereby tied together laterally and can be isolated as a sheet following homogenization of glycerinated muscle. The linkage extends out to the plasma membrane (sarcolemma) where desmin filaments insert into specialized membrane attachment regions known as *costameres*. By holding the sarcomeres of adjacent myofibrils together in this fashion and linking them to the plasma membrane, desmin filaments allow contractions of individual myofibrils to be coordinated. Microvariations in timing, extent, and duration of contraction of individual myofibrils will be 'ironed out' by these mechanical links.

Desmin filaments ramify throughout the cytoplasm of developing skeletal muscle cells (Figure 17-10) and are even more abundant in smooth-muscle cells. In the latter, as we saw in Chapter 10, bundles of actin and myosin are dispersed throughout the cytoplasm, extending between dense plaques in the cytoplasm or associated with the plasma membrane. Attached to the dense plaques are large numbers of desmin filaments, which are in a suitable location to tie together the separate actions of the individual contractile elements. This tension-bearing function is illustrated by experiments in which the smooth-muscle cell is stretched mechanically. Desmin filaments in a stretched smooth-muscle cell come to lie in a centrally located position, implying that they establish a tough series of continuous threads that resist tension running throughout the cytoplasm (Figure 17-11).

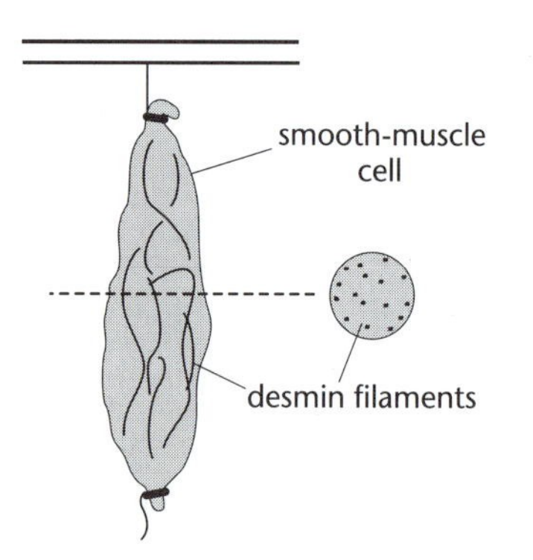


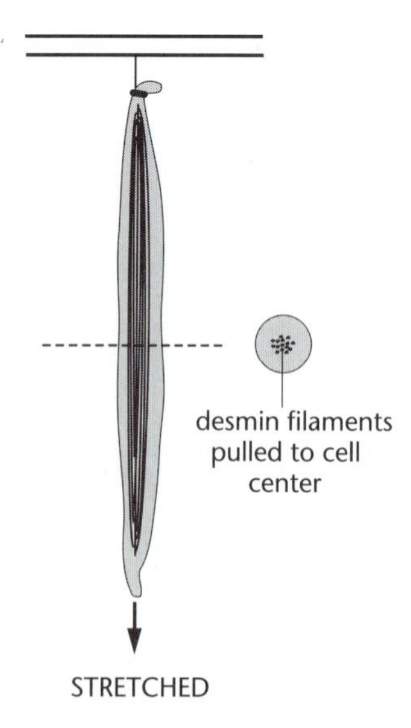


**Figure 17-11** Stretch-induced redistribution of desmin filaments. A single smooth-muscle cell was suspended vertically and stretched by means of a weight attached to its lower end. The desmin filaments move to a central position in the cytoplasm, suggesting that they are the principal tension-bearing structures.

## *Neurofilaments give tensile strength to long axons and determine their diameters*

Intermediate filaments are vitally important structural elements for axons in the peripheral nervous system. Mammalian axons often exceed 1 meter in length, whereas their diameter is usually less than 10 μm, an axial ratio of 1:100,000. These thin threads of cytoplasm could easily break as the animal moves (think of the sciatic nerve of a pole-vaulter). Neurofilaments provide the main protection against rupture, supplying a continuous internal core of fibrous protein that limits the extension of the nerve. Measurements of the tensile strength of axoplasm of giant axons from a marine worm show that neurofilaments are as strong on a weight-to-weight basis as filaments of wool (as indeed they should be, since they are made of closely related proteins).

Many of the classical staining methods developed in the late nineteenth century by Bodian, Cajal, and others are now realized to be specific for neurofilaments. These methods typically result in the deposition of metallic silver or gold onto the bundled neurofilaments, allowing them to be visualized against a clear background (see Figure 17-1). The basis for this selective staining is still not understood, although it probably depends on a selective affinity of the neurofilament protein itself for the metal atoms.

Neurofilaments (NF) are made from intermediate filaments of type IV (see Table 17-1). In humans, three categories are expressed, NF-L (light), NF-M (medium), and NF-H (heavy), with predicted sizes of 62, 102, and 110 kDa. The differences in size reflect major differences in the lengths of their glutamate- and lysine-rich carboxy domains. Neurofilaments are obligate heteropolymers, built from all three types of NF protein. NF-L appears to form the filament backbone, while NF-M and NF-H seem to associate more peripherally and form extensive interfilament cross-links.[5]

Electron microscopy of axons shows that neurofilaments have brush-like side-arms projecting from their filamentous backbone (Figure 17-12). The side-arms are an integral part of the neurofilament structure, formed from the carboxy-terminal domain of neurofilament peptides. The two larger (NF-M and NF-H) subunits have domains that can extend up to 30 nm away from the backbone of the filament. These domains also contain many sites for phosphorylation in a repetitive sequence motif of lysine-serine-proline present in about 20–30 copies in NF-H and about 10 in NF-M.

The degree of phosphorylation of NF-M and NF-H changes with development and location in the neuron, the highest level being found in the neurofilaments of mature axons. This and other evidence has led to the idea that these side-arms perform a space-filling function, expanding apart because of repulsion between their phosphorylated side-chains. In this view neurofilaments play a major part in determining the diameter of the axon (and hence, indirectly, the velocity of conduction of action potentials, which increases with axon diameter). Whether the awesome combinatorial possibilities inherent in the multiple phosphorylation sites of neurofilaments are fully exploited is an open question.

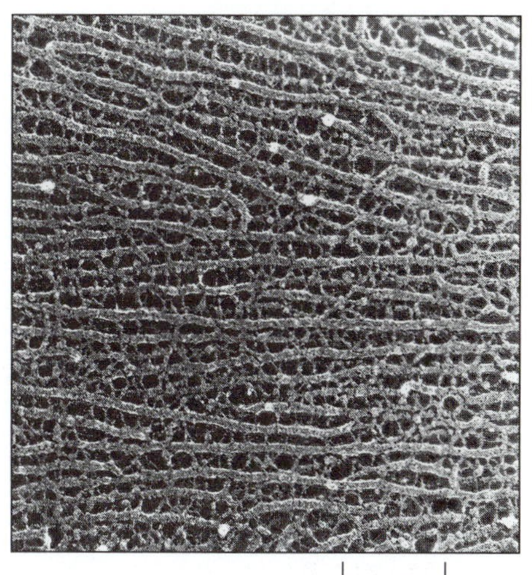


**Figure 17-12** Neurofilaments. An electron micrograph of a rat axon that has been extracted with detergent reveals the high density of neurofilaments in its cytoplasm and their extensive interconnection by fine protein cross-links. (Courtesy of Nobutaka Hirokawa, University of Tokyo.)

## *Linker proteins connect intermediate filaments to the rest of the cell*

In the 1960s, it was proposed, on the basis of high-resolution electron microscopy, that the cytoplasm of animal cells contains a highly interconnected three-dimensional network of filaments. Innumerable fine connections were seen between the three principal types of cytoskeletal filaments and other structures in the cell, such as the membrane and nucleus. Since the molecular identity of these links was unknown at the time, they were often dismissed as artifacts of fixation. However, more

[5] Other intermediate filament types are expressed in neurons at specific locations and times of development. For example, α-internexin, a type IV protein, appears in cells that are committed to neuronal differentiation but then decreases in amount as neurofilaments are made. Peripherin is a class III intermediate filament rich in peripheral axons.

recent studies of proteins that associate with intermediate filaments now offer a variety of possible candidates for these links. Indeed, an entire family of linking proteins, known as the *plakins* because of their relation to the desmoplakins mentioned above, has now been discovered. These bind together not only intermediate filaments, but also microtubules and actin filaments, the nuclear lamina and the cytoplasmic plaques of desmosomes. Plakins are now believed to act as versatile cross-linking agents that hold together different parts of the cytoskeleton into a single, integrated structure.

The first known and best-characterized, member of the plakin family is *plectin*, a large (greater than 500 kDa) dumbbell-shaped protein consisting of a 190 nm rod section formed from a double-stranded coiled-coil with globes about 9 nm diameter at either end. The plectin molecule has binding regions for vimentin, lamin B, and MAPs, and for other plectin molecules (Figure 17-13). It occurs widely in tissues of higher vertebrates, often appearing as cell-specific varieties generated by alternative RNA splicing. In some cell types plectin associates mainly with intermediate filaments, whereas in others it is localized to peripheral regions, especially at desmosomal and hemidesmosomal junctions, and even has a function in controlling actin filaments.

Another desmoplakin relative is BPAG1, originally identified as an autoantigen in patients suffering from bullous pemphigoid, a disease affecting the skin and mucous membranes. Immunoelectron microscopy showed that BPAG1 is localized mainly on the cytoplasmic plaque domain of hemidesmosomes at the site of keratin intermediate filament anchorage.

Interestingly, both plectin and BPAG1 exist in different isoforms generated by differential RNA splicing. As for troponins (Chapter 10), alternative splicing is a means to generate a diverse set of related proteins that is able to perform subtly different functions in different cells. In the case of BPAG1, for example, different forms are found in epithelia and nervous tissue. The neuronal form of BPAG is distinguished by having actin-binding domains in addition to other binding regions. The targeted disruption of this particular isoform results in severe degeneration in the nervous system, perhaps because of the loss of crucial links between neurofilaments and actin.

Other intermediate filament-binding proteins are known, such as the *filaggrins*, a family of histidine-rich proteins of mammalian epidermis. Filaggrins are specifically expressed in terminally differentiating mammalian epidermis as highly phosphorylated precursor proteins, the negative charges of the phosphate groups presumably counteracting the positively charged histidine residues. These precursor proteins are later processed by proteolysis to yield functional filaggrins, which then form highly stable cross-links between keratin filaments.

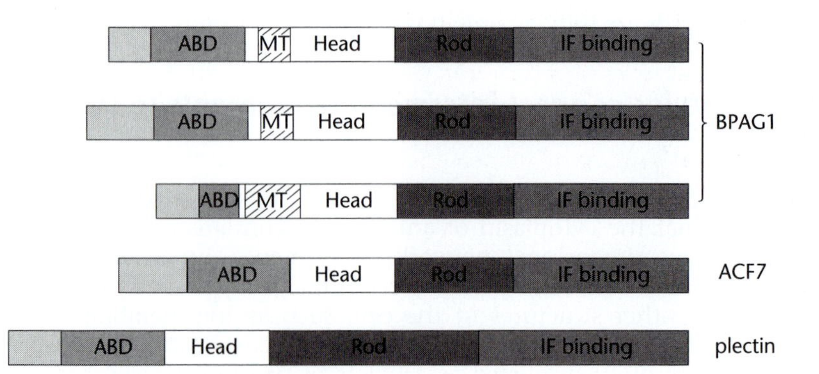


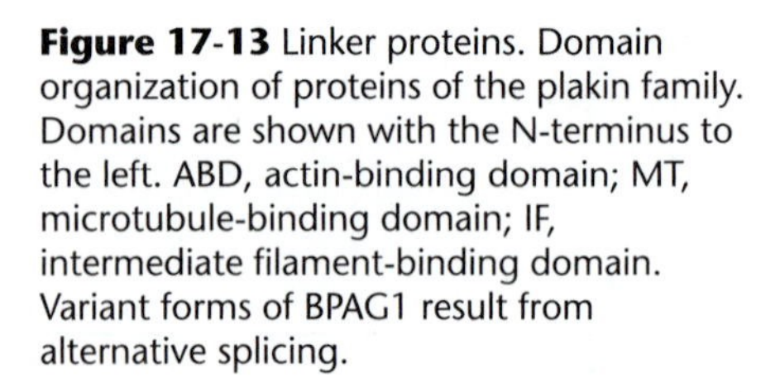

**Figure 17-13** Linker proteins. Domain organization of proteins of the plakin family. Domains are shown with the N-terminus to the left. ABD, actin-binding domain; MT, microtubule-binding domain; IF, intermediate filament-binding domain. Variant forms of BPAG1 result from alternative splicing.

## *Keratins are implicated in many genetic diseases*

We have seen that keratins provide higher animals with their tough outer coats, with specializations for thermal regulation, armament, camouflage, and sexual display. Any bird, rodent, or human that lacked all keratins would be in a sorry condition and unlikely to survive. Loss of *individual* keratins, however, might be expected to be less serious, given the large repertoire of keratins in the genome (over 30 genes in humans), and specific nonlethal deficiencies can indeed be found.

Two keratins expressed in the innermost layer of the epidermis—K5 and K14—have been implicated in a range of human skin disorders. Mutations in these genes engineered into mice show that their normal function is to impart mechanical integrity to epidermal cells. When the structural framework for these cells is disorganized, the cells become fragile and prone to rupture, leading to extensive blistering. The symptoms in mice are strikingly similar to those shown by humans with a human skin disorder known as *epidermolysis bullosa simplex* and patients with this disorder have now been found to have point mutations in either K5 or K14 keratins. Other disorders caused by keratin mutations include skin blistering on the protective thick skin of hands and feet and stress-induced degeneration of hair and nails, the lining of the mouth and esophagus, and the corneal covering of the eye.

## *What is the function of cytoplasmic intermediate filaments?*

In contrast to keratins, the function of cytoplasmic intermediate filament proteins such as vimentin and desmin has been a vexing issue for many years. They are abundant and widespread, yet there are reports that cells and even entire organisms can survive, apparently in good health, without them. In the early stages of mammalian development, moreover, embryonic cells have relatively few intermediate filaments in their cytoplasm and intermediate filaments have yet to be found in *Drosophila*, plants, or fungi.[6]

It now seems likely that cytoplasmic intermediate filaments do have distinct functions, although they can be quite difficult to detect. Knockout mice lacking vimentin develop and reproduce without major deficiencies but show changes in wound healing and in resistance to hypertension. Mice lacking desmin develop functional, contracting muscle but have a propensity to die of heart failure. Genetically engineered animals that lack neurofilaments survive and have functional nervous systems, but they display uncontrollable quivering and have breeding and behavioral defects.

It is worth recalling, in this context, the wide gulf that exists between the functional testing of cells and organisms under laboratory conditions and the struggle for survival that most organisms undergo under natural conditions. A deficiency that reduces the efficiency of wound healing or that makes an animal slower or more prone to heart attack might be very difficult to detect in the laboratory, requiring subtle tests and a large statistical base. But animals with these same deficiencies would be at a pronounced disadvantage in the wild and their capacity for survival would be significantly impaired. From the standpoint of evolution, a mutation that reduces survival by even a few percent may be expected to be lost from a population in time.

[6] Plants possess several proteins similar in sequence to intermediate filament proteins, but no actual filaments have so far been found.

## Outstanding Questions

*How do subunits pack together to form an intermediate filament? How variable is their internal structure? Does mechanical stress change the packing of subunits, and if so how? Do any movements take place along intermediate filaments? If so, what drives them? How are cytoplasmic intermediate filaments such as vimentin linked to the nucleus? Do they bind directly or indirectly to nuclear lamin (and if the latter, what are the linker proteins)? Are they linked through nuclear pores? Do keratin filaments and neurofilaments ever depolymerize? If so when, and how? Does lamin dispersal trigger nuclear envelope break-up? How are lamins involved in chromosomal positioning and replication? What is the significance of the movement of vimentin-containing particles along microtubules? Which proteins link intermediate filaments with actin filaments and microtubules? How are they positioned? Are they sensitive to filament polarity? Is their function purely mechanical? Where are neurofilaments assembled in a nerve cell? Are they transported in an unassembled form (and if so, what is it)? Why do neurofilaments have a distinct tripartite structure? What is the function of the hyperphosphorylation of neurofilament chains?*

## Essential Concepts

- Intermediate filaments are tough, insoluble polymers about 10 nm in diameter found in the cells of most animal tissues, made from fibrous intermediate filament proteins.
- There are many tissue-specific forms of intermediate filaments. Keratin filaments supply the cutaneous covering of the animal. Nuclear lamins provide a vital structural base for chromatin in the nucleus. Vimentin, desmin, and neurofilaments tailor the shapes and mechanical properties of particular cells.
- The family of intermediate filament proteins has more than 50 distinct members, each with a central rod domain, of conserved amino acid sequence, flanked at both ends by non-$\alpha$-helical domains. Dimeric subunits associate with one another in large overlapping arrays to form the intermediate filament.
- In contrast to the subunits of actin filaments and microtubules, those of intermediate filaments are highly variable in size and amino acid sequence. Most of the variation occurs in the terminal domains, which project from the surface of the intermediate filament and contribute tissue-specific functions.
- Assembly of intermediate filaments is regulated by phosphorylation of terminal domains. In some cases, such as lamins and vimentin, phosphorylation can lead to the disassembly of filaments once they have formed.
- Proteins associated with intermediate filaments such as desmoplakin, filaggrin, plectin, and BPAG1 form links to other parts of the cytoskeleton, thereby creating an integrated system throughout the cytoplasm.
- The primary function of intermediate filaments is to structure cytoplasm and to resist stresses externally applied to the cell. Mutations that weaken this structural framework increase the risk of cell rupture and cause a variety of human disorders.

## Further Reading

Bershadsky, A.D., et al. The state of actin assembly regulates actin and vinculin expression by a feedback loop. *J. Cell Sci.* 108: 1183–1193, 1996.

Bousquet, O., Coulombe, P.A. Missing links found? *Curr. Biol.* 6: 1563–1566, 1996.

Colucel-Guyon, E., et al. Mice lacking vimentin develop and reproduce without an obvious phenotype. *Cell* 79: 679–694, 1994.

Danowski, B.A. Costameres are sites of force transmission to the substratum in adult rat cardiomyocytes. *J. Cell Biol.* 118: 1411–1420, 1992.

Foisner, R., Wiche, G. Intermediate filament-associated proteins. *Curr. Opin. Cell Biol.* 3: 75–81, 1991.

Fuchs, E. Keratin and the skin. *Annu. Rev. Cell Dev. Biol.* 11: 123–153, 1995.

Fuchs, E., Cleveland, D.W. A structural scaffolding of intermediate filaments in health and disease. *Science* 279: 514–519, 1998.

Fuchs, E., Weber, K. Intermediate filaments: structure, dynamics, function and disease. *Annu. Rev. Biochem.* 63: 145–182, 1994.

Goldman, R.D., et al. The function of intermediate filaments in cell shape and cytoskeletal integrity. *J. Cell Biol.* 134: 971–981, 1996.

Herrmann, H., Aebi, U. Intermediate filament assembly: fibrillogenesis is driven by decisive dimer–dimer interactions. *Curr. Opin. Struct. Biol.* 8: 177–185, 1998.

Hirokawa, N., et al. Cytoskeletal architecture and immunocytochemical localization of microtubule-associated proteins in regions of axons associated with rapid axonal transport. *J. Cell Biol.* 101: 227–239, 1985.

Houseweart, M.K., Cleveland, D.W. Cytoskeletal linkers: new MAPs for old destinations. *Curr. Biol.* 9: R864–R866, 1999.

Lörke, S., et al. Expression of neurofilament proteins by horizontal cells in the rabbit retina varies with retinal location. *J. Neurocytol.* 24: 283–300, 1995.

Mogensen, M.M., et al. Keratin filament deployment and cytoskeletal networking in a sensory epithelium that vibrates during hearing. *Cell Motil. Cytoskeleton* 41: 138–153, 1998.

Moir, R.D., et al. Disruption of nuclear lamin organization blocks the elongation phase of DNA replication. *J. Cell Biol.* 149: 1179–1192, 2000.

Prahlad, V., et al. Rapid movements of vimentin on microtubule tracks: kinesin-dependent assembly of intermediate filament networks. *J. Cell Biol.* 143: 159–170, 1998.

Wang, L., et al. Rapid movement of axonal neurofilaments interrupted by prolonged pauses. *Nat. Cell Biol.* 2: 137–141, 2000.

Wiche, G. Role of plectin in cytoskeletal organization and dynamics. *J. Cell Sci.* 111: 2477–2486, 1998.

Yang, Y., et al. An essential cytoskeletal linker protein connecting actin microfilaments to intermediate filaments. *Cell* 865: 655–665, 1996.

# CHAPTER 18

# Cell Mechanics

If living cells were like miniature mechanical toys, built from a framework of metal wires and driven by tiny electrical motors, they would be much easier to study. Their movements would then be a question of *engineering*—the application of well-established physical principles to materials of known mechanical properties. We would be able to say precisely what forces the cell exerted and experienced, what shapes it could take, and in which direction it would travel. Cell motility would become an exact science.

The reality, of course, is very different. Cells are soft, wet structures enclosed by membranes and filled with an aqueous slurry of large and small molecules. Their composition and internal construction change continually in response to subtle chemical stimuli, making them unpromising objects for an engineer. Despite this we know that, in the last analysis, cells are subject to the same physical laws as man-made objects. In principle we *should* be able to treat them as a problem in mechanical engineering—one capable of exact solution.

In this chapter we describe some tentative steps that have been taken in this direction, examining the mechanics of protein filaments, both as individual elements and as cross-linked bundles, and the ability of polymerizing filaments to generate force (a process that must be considered in statistical terms). Molecular motors are considered from a mechanical standpoint, individually and as part of a framework of protein filaments. Moving to a more detailed and complex level, we then consider the properties of the meshwork of interconnected protein filaments in the cytoplasm. This three-dimensional network behaves in some ways as a gel and in others as a viscoelastic fluid, and is strongly influenced by the water balance inside cells. Finally, we look at the membranes of the cell, especially the plasma membrane, and try to assess their contribution to the mechanical properties of the entire system.

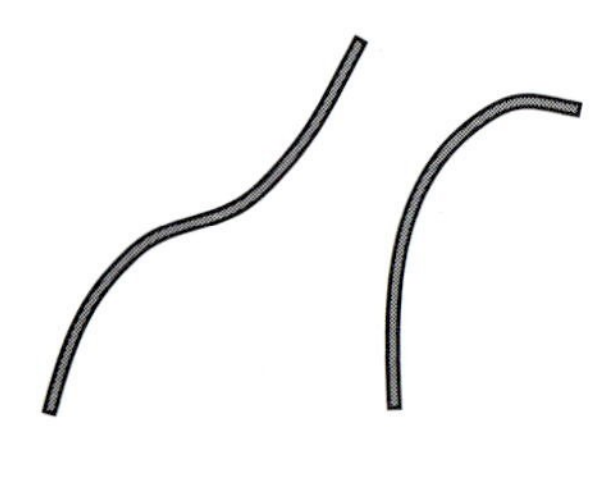


**Figure 18-1** Filament stiffness. Characteristic shapes of protein filaments suspended in water at room temperature. The three types of protein filament differ in persistence length $L_p$ (see Table 18-1). The bending shown here would be expected of a filament about 10 μm in length.

## *Cytoskeletal filaments are flexible rods*

A single protein filament suspended in water writhes and bends under the influence of thermal forces, a form of Brownian motion (Chapter 1). How much it bends, and hence its instantaneous shape, depends on the temperature and the particular type of filament. Thus an actin filament 10 μm long bends more than a microtubule of the same length but less than a single coiled-coil (Figure 18-1). These differences in shape can be measured by a parameter known as the *persistence length* $L_p$, related to the stiffness of the filament. Persistence length is the length of filament needed before the bending caused by thermal energy completely randomizes the orientation of the two ends. More precisely, $L_p$ is given by the equation

$$\langle \bar{u}(s) \times \bar{u}(0) \rangle = \exp\left(\frac{-s}{L_p}\right)$$

where $\bar{u}(s)$ and $\bar{u}(0)$ are unit vectors drawn tangent to the polymer contour at either end, $\langle\ \rangle$ denotes a time average over thermal fluctuations, and $s$ is the physical length of filament measured from one end, $s = 0$.

It is also possible, although difficult, to micromanipulate single protein filaments and measure their mechanical properties directly. Actin filaments, for example, can be held by small beads attached at either end, and then stretched or bent by means of a laser trap.[1]

All measurements made so far reveal major differences between the three principal kinds of cytoskeletal filaments. Actin filaments are flexible structures, poorly able to resist either bending or tensile forces (they break when exposed to the equivalent of 10 myosin heads). Microtubules are also poor at resisting tension, but much stiffer than actin and therefore the elements of choice where bending or buckling forces have to be resisted. Intermediate filaments are more flexible even than actin filaments but have great tensile strength.

## *Protein filaments have a similar elasticity to that of hard plastics*

One of the beauties of mechanics is that different descriptions of a structure have a quantitative relationship to one another. Thus, the persistence length, displayed in the shape of a flexing filament, can be calculated from its cross-sectional profile and the flexibility of the material of which it is made. The relevant equation is

$$L_p = EI/kT$$

where $I$ is the second moment of the cross-section (proportional to the area squared), $E$ is Young's modulus (a measure of the elasticity of the protein itself) and $kT$ is the energy of thermal motion, already introduced in Chapter 1. Comparing different types of filaments, we can calculate $L_p$ by observing their behavior in isolation (Table 18-1). $I$ can be deduced from the geometry of the filament (for example, by knowing how the protofilaments pack) and this gives us the modulus $E$. The value obtained for cytoskeletal filaments is around $1\text{–}3 \times 10^9$ Pa,[2] showing that the proteins of which these are made are less elastic than rubber ($E = 0.001 \times 10^9$ Pa), but more flexible than bone ($E = 20 \times 10^9$ Pa) or steel ($E = 200 \times 10^9$ Pa).

In fact, the elasticity of actin, tubulin, flagellin, and coiled-coil proteins is similar to that of hard plastics such as nylon or polycarbonate. The similarity also extends to the strength of these materials, which for both cytoskeletal proteins and hard plastics allows a maximum strain in the range of 1–10% length extension. Interestingly, we can use this figure to estimate roughly how much a cytoskeletal filament of a specified thickness can bend. The minimum radius of curvature of a single microtubule calculated in this way is roughly 1 μm, which in fact corresponds to the radius of the ring of microtubules in blood platelets, described in the

[1] A long filament flexing under the influence of thermal motion also resists being pulled straight, since this will decrease its entropy. This effect can contribute to the elasticity of a filament, as we saw previously in the case of the muscle protein titin (Chapter 9).

following chapter. Microtubules in living fibroblasts, flexing under the influence of cytoplasmic movements, have been observed to curve spontaneously to radii less than 1 μm, although breakages occur. Microfilaments according to the above calculation should be capable of very tight bends, with radii as small as 0.25 μm.

**Table 18-1 Persistence length of protein filaments**

| Filament type | Persistence length ($L_p$) |
|---|---|
| actin filament | 18 μm |
| microtubule | 6000 μm |
| coiled-coil | 0.1 μm |
| intermediate filament | 2 μm |
| bacterial flagellum | 1000 μm |

## Bundles of filaments make rigid stiffeners for the cell

The material strength provided by protein filaments can be greatly increased by bundling, such as in the assemblies of parallel microtubules in an axostyle described in Chapter 14, or the arrays of actin filaments inside a stereocilium described in Chapter 6. But how much stronger is this arrangement in quantitative terms?

We can imagine two extreme situations. In one, a bundle might be built from filaments so loosely associated that they slip against each other without resistance when the bundle bends. In this case the mechanical properties of the bundle should be given simply by multiplying the stiffness of each filament (the value of $EI$ in the above equation) by the number of filaments. At the other extreme, we could imagine the links between filaments to be of such rigidity and strength that they effectively extend the material of which the filament is made, making the bundle as strong as a single thick filament. In mechanical terms this will increase the second moment, $I$, which depends on the square of the cross-sectional area. We therefore expect the stiffness of such a bundle to increase as the *square* of the number of filaments it includes.

A case in point is the mechanical strength of intermediate filaments. Most kinds of intermediate filaments comprise bundles of around 16 coiled-coils in parallel (see Figure 17-5). Since we know that the persistence length of an individual coiled-coil is around 0.1 μm (see Table 18-1) we expect that under the first scheme (slippage) $L_p$ for the filament should be $16 \times 0.1$, or about 2 μm. By contrast, under the second scheme (highly cross-linked) it should be $16^2 \times 0.1$, or about 25 μm. Experimentally, neurofilaments have an $L_p$ about 1 μm and vimentin filaments about 3 μm, so by this criterion the coiled-coils in these structures are not highly cross-linked.

It is also possible to use motor proteins to link one cytoskeletal filament to another. We saw previously that bundles of actin filaments are held together by myosins, and bundles of microtubules by dyneins. The flexural rigidity of such bundles will be influenced by not only the number of cross-links but also their state of attachment. A sperm tail, for example, changes its flexural rigidity depending on experimental conditions, being far more rigid under conditions of rigor, when the dynein cross-bridges are locked in place, than under conditions of relaxation, caused by addition of ATP. In the language of mechanics, cross-linking by dynein increases the second moment of the cross-section of the sperm tail.

## Bundle formation is a complex physicochemical process

How filament bundles form in the first place is a complex issue in which thermodynamics and mechanics have a part. Even the simple mixing of filaments with linker molecules can produce a range of different final states. Depending on conditions, filaments might not associate at all, link into simple dimers, cluster into small or large bundles, or associate to produce a three-dimensional isotropic gel.

Some of the factors governing the type and extent of bundles of actin filaments have been explored experimentally. It may be recalled from Chapter 6 that many actin-binding proteins have the capacity to link filaments together. Proteins such as fascin and fimbrin are particularly well suited for making bundles because they are short and compact and

[2] The intensity of force distributed over an area is known as *pressure*, or *stress*, and measured as force per unit area. The basic unit of pressure or stress is the Pascal (Pa) equal to one Newton per square meter ($N/m^2$), or 10 dynes per square centimeter ($dyne/cm^2$). Youngs modulus, $E$, is the stress applied to the cross-section of a filament divided by the fractional change of length produced.

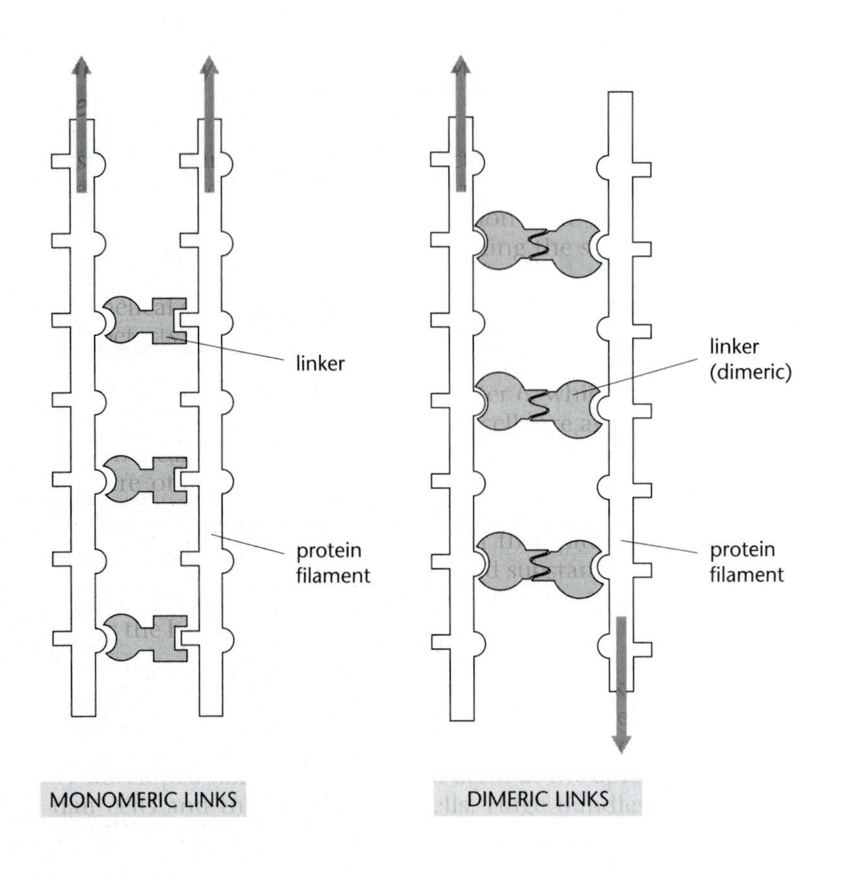


**Figure 18-2** Monomeric and dimeric linkages. Monomeric linker proteins have different binding sites at either end of the molecule, and can hold neighboring filaments with a parallel orientation. Dimeric linker proteins have identical binding sites and for reasons of symmetry tend to link filaments in an antiparallel orientation. Alternatively, if the linker molecule is long and flexible (as in the case of filamin) the net result will be a disordered three-dimensional network.

because they form *monomeric* cross-links between adjacent actin filaments (Figure 18-2). How many bundles are produced in a given mixture, and what their size becomes, also depend on such factors as the ratio of linker protein to actin filaments and the length of the filaments, which should be neither too long nor too short.

Kinetics are also important. A mixture of proteins might rapidly produce a tangled network of filaments because it forms cross-links so quickly that the filaments have no time to align into bundles. The network might be thermodynamically most stable as highly-ordered bundles, but in reality it may never achieve this state, especially if the cross-links are very tight and the filaments are very long.

We can expect kinetics to be crucial in a living cell, which has to produce bundles at the right place and the right time. So it is not surprising that cells employ a battery of control mechanisms, analogous to those used in the construction of cilia and other large structures, to direct the process. The bundle of actin filaments forming the core of an intestinal microvillus, for example, includes at least four distinct types of actin cross-linking protein (villin, fimbrin, spectrin, and α-actinin) each with a subtly different location and structural role (see Table 6-4). Despite its rigid appearance, this structure is highly dynamic and individual proteins turn over with half lives from 30 minutes to several hours.

## *Filament bundles can store strain energy*

The behavior of proteins filaments, both as individual structures and in large assemblies, is greatly influenced by conformational changes in their subunits. In an isolated filament, conformational changes can alter the packing lattice leading to an altered three-dimensional form, as we saw in the helical transitions of bacterial flagella mentioned in Chapter 16. Both actin and tubulin exist in two distinct conformations, stabilized by

**Figure 18-3** Movement driven by polymerization. Results of an experiment in which beads coated with ActA (a protein that triggers actin polymerization) were immersed in cytoplasmic extracts rich in actin. (a) Beads initially become entrapped in a symmetrical mesh of actin filaments and show only restricted random movements. (b) Sporadically, and at random, beads escape from the mesh of actin and are propelled through the extract by the coordinated growth of actin filaments on one side of the bead.

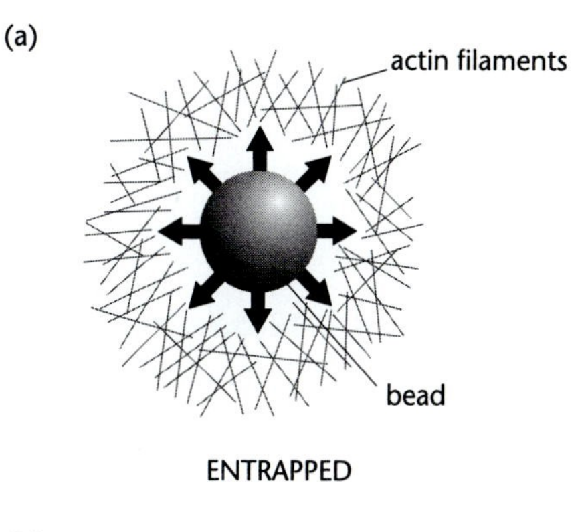


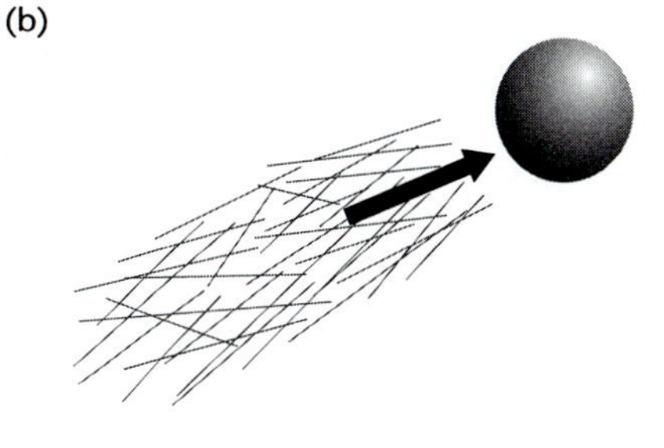


nucleotides, and their polymers show phenomena that depend on that fact. Probably the most important for cells is the dynamic instability of growth in microtubules, which has far-reaching consequences for the organization of the cytoplasm (Chapter 11).

When many filaments are packed together in a crystalline lattice, then conformational changes in their subunits can become coupled and spread through the structure. The large-scale extensions or contractions that result are among the most rapid and forceful encountered in living cells. Recall, for example, the springlike contraction of the centrin-based *Vorticella* axoneme described in Chapter 4 and the rapid extension of the bundle of cross-linked actin filaments in *Limulus* sperm described in Chapter 6. These two movements are triggered by large changes in calcium ion concentration, leading to an almost explosive coordinated change over the entire structure. Given the highly dynamic nature of individual protein molecules in a living cell, it is likely that concerted conformational changes of a more reversible kind will be of widespread occurrence. They may play an essential part in muscle contraction, for example, or in the bending of ciliary axonemes.

## *Systems of polymerizing filaments can exert a directed force*

Mention of kinetics reminds us that cytoskeletal filaments are highly dynamic structures that continually grow and shrink in the cell. We have already seen that this assembly and disassembly can produce large-scale movements, as in the movement of a chromosome or a bacterium through cytoplasm, or the extension of a filopodium. In broad terms, such movements are believed to be driven by a thermal ratchet mechanism, in which thermally driven movements create transient openings into which subunits can diffuse (see Figure 4-9). But this telegraphic description leaves unanswered many questions of molecular details and mechanics.

Witness a remarkable experiment in which micrometer-sized beads coated with a polymerization-inducing factor are suspended in a cytoplasmic extract containing fluorescently labeled actin (Figure 18-3). Each bead develops a symmetrical cloud of fluorescent actin filaments, which became progressively cross-linked by factors in the cytoplasm. As the cloud grows and the actin filaments press more tightly, the Brownian movements of the bead became increasingly restricted and short-range. But every now and then, at random, individual beads abruptly quit their shell of actin and migrate rapidly through the extract—like a rabbit breaking free of a snare. As the ballistic beads move, they produce a trail of cross-linked actin filaments, closely mimicking in this respect the behavior of *Listeria* bacteria moving within cells they infect (Chapter 6).

Experimental investigation of this remarkable behavior, supported by computer models, provides a possible explanation for the rocketlike behavior of the beads. The basic idea is that small displacements of the bead serve to coordinate the assembly and disassembly of actin filaments

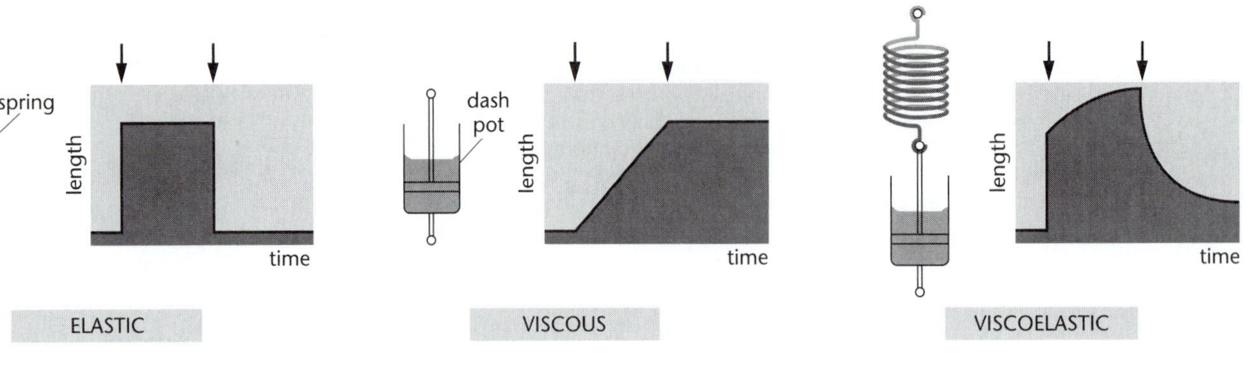


**Figure 18-4** Elastic, viscous, and viscoelastic materials. The mechanical properties of each material are illustrated by analogy to springs (elastic materials) and dashpots (viscous materials). Changes in length are shown as a function of time following the application of a fixed stress (such as the attachment of a weight) for the period between the two arrows. The molecules in *elastic* materials are linked together with large numbers of strong cross-links and deformations are shared among the many bonds, all of which change their length by a small amount. ('Rigid' materials differ from 'rubbery' materials in the numbers of cross-links between molecules and the extent to which the cross-links can be stretched.) The energy of deformation is stored in the molecules and bonds and removal of stress from an elastic material allows it to return to its original form. *Viscous* fluids, and plastic solids, are composed of molecules that are not firmly attached to each other. Stress applied to such materials therefore produces a flow, which can be measured by a viscosity. (Viscous materials are often represented as a dashpot in which a piston moves through a deformable fluid such as grease.) The flow dissipates the energy of deformation and, after removal of the deforming force, such materials show no tendency to recover their original shape. *Viscoelastic* materials, such as the cytoplasm, show both viscous flow and elasticity, so their behavior depends on time as well as the applied force. If we apply a sudden force to a viscoelastic material and measure its deformation we will find that it goes through a sequence of changes. Immediately after the load has been applied the stress is taken up by bonds between the polymeric molecules, as in an elastic material. With time, however, a proportion of the intermolecular bonds become free and parts of the molecules rotate and flow relative to each other. The deformation slowly increases at a rate determined by the viscosity of the material. This behavior is viscoelastic rather than plastic since removal of the stress is followed by a slow return toward the original dimensions as the molecules resume their more compact and more highly cross-linked states. Eventually the viscous flow ends as the deformation reaches a new plateau.

pressing onto its surface. A small push to the bead caused by the growth of one actin filament will move the bead and make more space for the addition of a subunit to a next-door filament—that is, neighboring filaments tend to grow together. Conversely, and for the same reason, monomer addition to actin filaments on *opposite* sides of a bead work against each other, the one tending (very slightly) to inhibit the other.

The success of the above strategy will depend on the size of the fluctuations at the filament ends, and it may be important that the filaments undergo dynamic instability. This will ensure that periods of growth and of shrinkage are relatively large and also, perhaps, make these durations responsive to the mechanical pressure between the actin filaments and the beads. Computer simulations indicate that if rates of dissociation fall with the development of compressive tension, they will automatically reach an appropriate value to drive the bead out of its shell.

## Networks of cytoskeletal filaments have viscoelastic properties

Even in the absence of cross-linking proteins, solutions of cytoskeletal filaments have complicated mechanical properties. At high dilution, the filaments writhe and diffuse independently in the manner already described. But as their concentration rises, filaments start to interfere with each other's movements. Because any one filament is more likely to encounter a second filament moving sideways than moving along its length, filaments become, in effect, constrained to lie in a 'virtual tube.' A single fluorescent actin filament in a concentrated solution can be seen to move predominantly lengthwise with a snakelike motion called *reptation*. At even higher concentrations, a new phenomenon appears in which the filaments associate side by side into rafts or islands of parallel protein filaments. Solutions of this kind are described as being 'nematic' or 'liquid crystal.'

Because of the interference between long, tangled filaments, a solution of protein filaments resists shearing forces. Thus we find that such a solution typically has a high viscosity, but that it may also have some elasticity, or ability to recover its shape following a sudden displacement. Recovery occurs because lateral interactions between adjacent molecules initially resist applied forces and then slowly yield.

Concentrated solutions of cytoskeletal filaments are said to be *viscoelastic*, having properties of both viscous solutions and elastic solids, the most important feature of viscoelastic behavior being the time-dependence of its deformations (Figure 18-4). Following application of a mechanical force, a viscoelastic substance shows an initial resistance to deformation, similar to that of an elastic solid, followed by a slow flow similar to that of a viscous fluid. Unlike a purely viscous material, however, a viscoelastic substance returns slowly toward its original form when the force is removed.

As one might expect, viscoelasticity is strongly enhanced by cross-linking proteins. Even one or two molecules of filamin per actin filament are enough to convert a viscous solution to a semisolid gel. As we saw in Chapter 6, filamin has two identical binding sites for actin, at either end of a long flexible molecule, and so is ideally suited to produce isotropic three-dimensional gels. Gels made from actin and filamin have the property (shown also by living cytoplasm) that their viscosity depends on the rate of deformation. When deformed rapidly (in seconds) these mixtures are 40 times more rigid than actin filaments alone. But when deformed slowly (over minutes or hours) they are the same as if filamin was not there. These time-dependent mechanical properties can be explained if protein links between actin filaments can rearrange and, given time, accommodate to an applied stress.

Bioengineers are beginning to analyze networks of polymers both experimentally and by computer simulations. One approach, for example, is to consider each strand of a network as a simple spring, subject to thermal fluctuations in length, and then to suppose these to be linked into a geometrically regular network. Calculations then permit one to predict how the ensemble will behave over time and how it will respond to external forces. Although they are still rudimentary, such simulations should make it possible ultimately to predict the material properties of actual networks of cytoskeletal polymers. Perhaps they will even give us clues to the mechanical properties of living cytoplasm—a much more complex issue?

## Cytoplasm is viscoelastic

The mechanical properties of living cytoplasm present major difficulties to an investigator because of the minute size of cells. However, there are cells, such as the slime mold *Physarum*, that can provide sufficient cytoplasm for conventional rheological measurements. Another approach is to implant small ferromagnetic particles into the cytoplasm of large cells and measure their movement in response to an externally applied magnetic field. In 1950, Crick and Hughes used this technique to demonstrate that cytoplasm is viscoelastic and has non-Newtonian properties, obtaining estimates of the modulus of elasticity of approximately 10 Pa (100 dyne/$cm^2$). More recently, magnetic spheres implanted in giant squid axon cytoplasm were subjected to magnetic fields applied both parallel to and at right angles to the long axis of the axon (Figure 18-5). Responses of the spheres indicated that the axonal cytoplasm is a complex viscoelastic fluid and also revealed a natural anisotropy, with a greater elastic modulus in the direction parallel to the long axis of the axon.

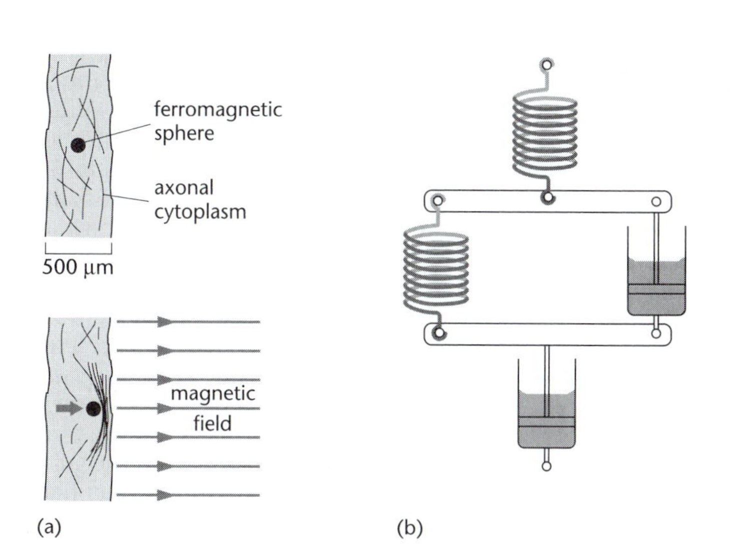


**Figure 18-5** Cytoplasmic viscoelasticity. (a) Cytoplasmic mechanics measured by the magnetic sphere method. The ferromagnetic sphere is implanted into the cytoplasm of a large cell (in this case a giant axon) and a magnetic force is applied. Movement of the sphere is followed through a microscope and, knowing the force applied by the magnet, the properties of the cytoplasm can be deduced. (b) Mechanical model for the cytoplasm. This system of springs and dashpots would give a similar mechanical response to that seen in the magnetic sphere experiments.

It is clear that the cytoplasm does not have the same composition in different cells. Squid axoplasm is dominated by neurofilaments and its mechanical properties therefore depend largely on the number and types of linkages between neurofilaments. The cytoplasm of *Physarum*, by contrast, is rich in actin filaments with few if any microtubules or intermediate filaments. Vertebrate axons and fish chromatophores are different again, with high densities of microtubules.

Cytoplasm also varies from one part of a cell to another. The most peripheral regions of cytoplasm, just under the plasma membrane, usually has a dense mesh of actin filaments (the cell cortex, see Chapter 6). This region excludes most membranous organelles and, when present in large amounts (as in the cortex of giant amoebae), presents a glassy transparent appearance in the microscope, referred to as a hyaline layer. By contrast, in more central locations of the cell, the cytoplasm of a typical animal cell is rich in membranous organelles, microtubules, and intermediate filaments. Most organelle transport also takes place in this inner compartment.

## Motor proteins make the cytoskeleton contract

A unique feature of the protein networks in living cells that distinguishes them from mixtures of synthetic polymers is that they can generate their own force. Proteins such as myosin and dynein are an integral part of the cytoskeleton and have the capacity to convert the energy of ATP hydrolysis into directional movement. Although some motor proteins carry membrane vesicles or other small organelles along protein filaments and therefore have only minor effects on the mechanics of the whole system, others physically distort the entire framework. Myosin can move actin filaments against each other, and dynein (or in some cases kinesin) can cause microtubules to slide. Depending on the orientation of the filaments and the way they are linked together, motor proteins can produce contractions, bending, or in some circumstances active extension.[3]

The most familiar and best-studied force is produced by actin and myosin in striated muscle. As we saw in Chapter 9, whole muscles, muscle cells (fibers), or even individual myofibrils can be dissected from a muscle and mounted in such a way that their contraction can be measured. Because of the highly regular arrangement of myosin and actin molecules in the myofibril, the detailed time-course and magnitude of tension transients can now be interpreted quantitatively in terms of the repeated action of myosin heads along actin filaments. 'Engineering' models of the myosin cross-bridge, using such parameters as the step size of the myosin head, the force generated per step, and the linear elasticity in the myosin neck region, correctly predict the performance of muscle fibers in a wide range of tests. In the future it may even be possible to extend the explanation to the atomic structures of actin and myosin.

Contractile structures in smooth muscle, and even more in nonmuscle cells, are less structured and correspondingly harder to handle. Stress fibers have been dissected out from fibroblasts and shown to be contractile and are responsible for the buckling of flexible substrata (see Figure 18-8). Fine microneedles have been inserted into large cells undergoing cleavage and used to estimate the force generated by the contractile ring. But in general the loose and random construction of these actin–myosin bundles makes both experiment and interpretation very difficult.

The ultimate, disordered contractile structure is a three-dimensional gel of filaments containing randomly positioned motor proteins (Figure 18-6). Networks of this kind can be made in the laboratory by mixing together actin filaments and myosin, and have highly nonlinear, complex behavior. Imagine, for example, such a gel in the cytoplasm, anchored to a small number of fixed points, such as cell junctions. As the gel contracts due to the action of its myosin molecules, filaments between fixed points

[3] Bundles of microtubules telescope apart during the last stages of mitosis (anaphase A). The elongation of cells such as retinal cells in fish may also depend on a sliding extension.

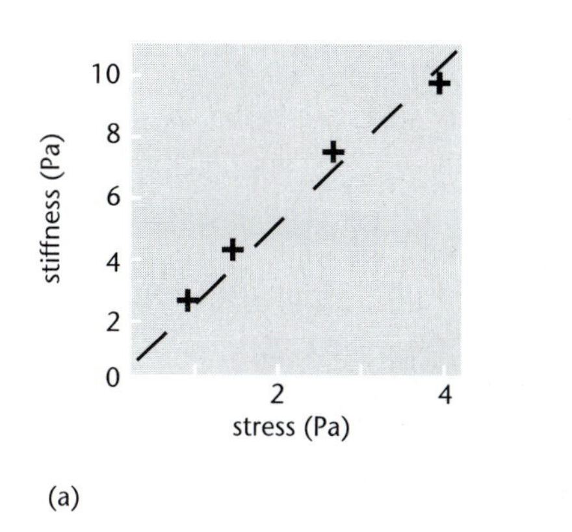


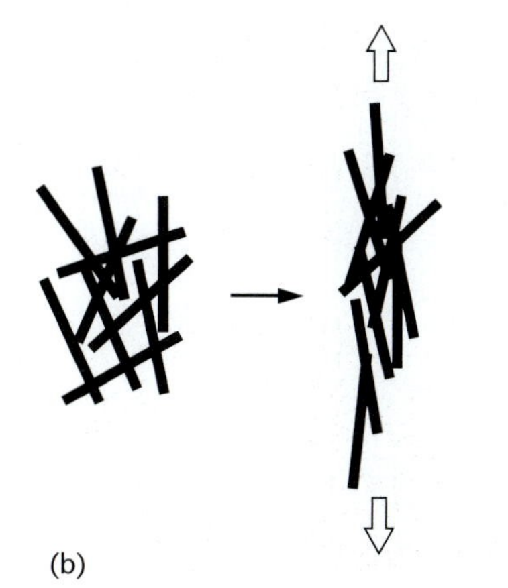


**Figure 18-6** Nonlinear performance of cytoplasm. (a) A magnetic bead attached to a cytoskeletal network experiences an elastic resistance (stiffness) that grows larger with the force applied. (b) Schematic views of a system of rigid filaments linked by flexible connectors (the latter being omitted for clarity) Application of tension to the assembly leads to the alignment of the elastic and compressive elements, leading to an increase in elastic resistance. Motor proteins, if present, would probably also work more effectively after stretching.

will become more highly aligned into bundles. Bundling will increase the efficiency of contraction, since more myosin molecules will work in the same direction. We might expect the force developed to increase with time—as indeed appears to be the case.

## *Cytoskeletal filaments form a tensegrity structure*

If animal cells contain bundles or networks of actin filaments and myosin that develop contractile force, what opposes this force? How does a cell achieve mechanical equilibrium and sustain a stable size and shape?

Nerve cells in culture illustrate a possible answer. The long axonal processes of newly growing axons (neurites) are supported mechanically by bundles of longitudinally aligned microtubules. Assembly of these bundles is a prerequisite of axonal elongation, and if the cell is treated with microtubule-disrupting agents such as nocadazole, the axon retracts rapidly. Retraction can, however, be blocked if the cell is simultaneously treated with agents that disrupt the actin cortex (such as cytochalasin). This suggests that in the normal axon the microtubule core and the actin cortex act against each other, one in compression and the other in tension.

An additional complication is introduced by the fact that a nerve cell in culture adheres to the culture substratum, mainly at its tip. This adhesion is necessary for axonal growth to occur, and if it is disrupted, the axon again retracts. Thus, the contractile force due to the cell cortex is balanced partly by the compression of cytoskeletal elements, especially microtubules, and partly by the tension developed in the cell-to-substratum adhesions (Figure 18-7).

In this respect, animal cells conform to the notion of tensional integrity, or *tensegrity*—an architectural term coined by R. Buckminster Fuller to describe three-dimensional structures (seen in tents, domed stadiums, and the work of some sculptors) in which a continuous network under tension is supported by compressive struts. This architectural analogy has been applied to the shape and structure of cells in tissue culture, and shown to account in a semiquantitative manner for some of their responses to perturbations, such as manipulations with fine needles. For example, it is known experimentally that when microtubules are depolymerized by drug treatment in a fibroblast, the network of actin filaments in the cortex develops a stronger contractile force and forms more focal adhesions.

These observations are *consistent* with a framework based on tensegrity, but other interpretations are possible. Some mechanical responses of the

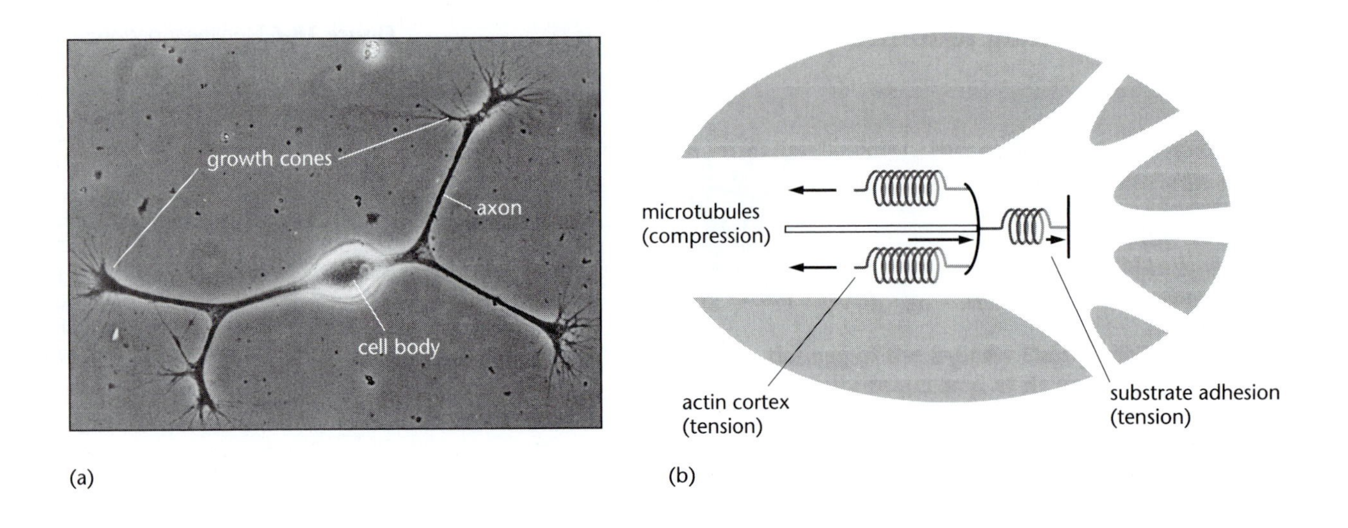


**Figure 18-7** Mechanical balance in a nerve cell. (a) Light micrograph of a single nerve cell in tissue culture. Short axons tipped by growth cones are growing out from the cell body. (b) Principal contributions to the mechanical balance of each axon.

cytoskeleton seem to be highly localized and not distributed across the whole cell. It is also clear that chemical signals play an essential part in integrating changes in different regions, as we see below.

## *Myosin II makes a major contribution to cortical tension*

The earliest measurements of the mechanical properties of living cells, some of them dating to the 1930s, focused on the deformation of very large cells. Sea urchin eggs, which are over 100 μm in diameter, spherical, and easily obtained in large numbers, were favorite objects of study. Individual eggs were mounted under a microscope, held in various kinds of jigs and clamps, and then depressed by probes such as strips of metal foil or flexible needles. The forces applied were estimated from the deflection of the foil or needle, and the deformations of the cell surface measured in the microscope.

Since then, techniques have become increasingly precise and rapid. Glass needles with tip diameters in the micrometer range can be maneuvered onto or into cells and their bending deflections measured with nanometer accuracy. Vibrating electrodes can be used to probe the surface elasticity of cells, changes in the amplitude and frequency of the tip being a sensitive indicator of the mechanical properties of the material it contacts. The tips of glass probes can be coated with specific antibodies or other molecules so that they adhere selectively to particular molecules such as integrins on the cell surface. Yet another approach is to grow cells on a thin elastic skin on the surface of silicone rubber. Cell contractions produce wrinkles in the underlying sheet, providing a graphic portrayal of the distribution of contractile forces (Figure 18-8).

One inescapable conclusion from these disparate measurements is that the surfaces of animal cells are under a steady level of tension. Moreover, given that the cortex is full of actin filaments, an obvious candidate for the source of this tension is myosin. Indeed, mutants of *Dictyostelium* lacking myosin II have a much reduced cortical tension, whereas a cell with a mutant myosin light chain kinase that works all the time has a

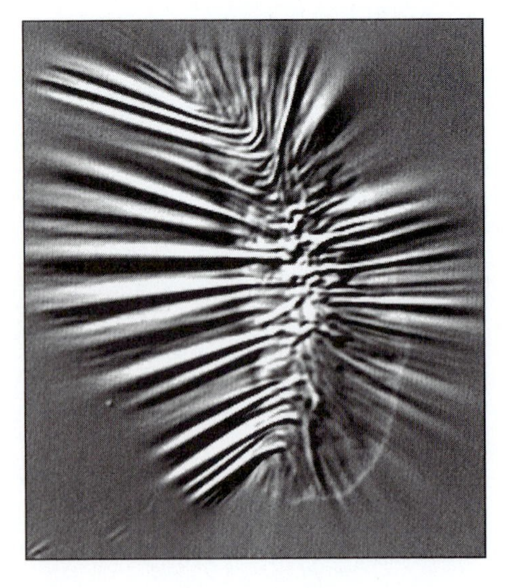

**Figure 18-8** Wrinkles in rubber caused by a cell. A goldfish keratocyte (skin cell) is moving from left to right on a transparent silicone rubber substratum. Traction forces generated by the cell compress the rubber into a series of wrinkles. The cell was viewed by Normarski differential interference contrast microscopy. The field of view is 60 μm. (Courtesy of Kevin Burton.)

surface that is stiffer than usual (due to a more active myosin II). Contractions of the cell also correlate with cell physiology in that the cortical tension of a dividing cell rises to a maximum immediately prior to cleavage and then falls. Chemotactic cells such as neutrophils become stiffer immediately after they sense an attractant, and then relax as they set off along the new trail.

**Table 18-2 Composition of cytoplasm**

| WATER | Percentage of the total (wet) weight of cells |
|---|---|
| bacteria | 70% |
| chicken heart fibroblasts | 77% |
| rat kidney cells | 74% |
| HeLa cells | 70% |
| axon | 87% |
| **PROTEIN** | **Percentage by weight (1% = 10 mg/ml)** |
| bacteria | 20–32% |
| muscle cells | 23% |
| red blood cells | 35% |
| mammalian tissue cells | 17–26% |
| squid axoplasm | 5% |

## *What is the structure of cytoplasm?*

The robust mechanical links existing between different parts of the interior of a cell show that the cytoskeleton is a continuous network. But what are the molecules that link actin filaments, microtubules, and intermediate filaments together, and what is their relationship to organelles, membranes, and other proteins? What, in short, is the structure of living cytoplasm?

The appearance of the cytoplasm in the electron microscope—the principal source of our mental images—depends critically on the technique of preparation. In conventional techniques the specimen is fixed with harsh chemicals, embedded in a plastic resin, and then stained by the deposition of heavy metal atoms. The stain has access only to those parts of the specimen exposed on the surface, and tends to accentuate membranes at the expense of both filaments and soluble proteins. Ironically, this technique became widely adopted over the years precisely *because* it gives images that are relatively clear and easy to interpret. However, if cells are extracted with nonionic detergent and then dried before examination in the microscope, structures such as mitochondria appear embedded in a dense polymorphic mesh, sometimes termed a microtrabecular lattice. Cytoskeletal filaments such as microtubules are shown encrusted with protein molecules and side-arms, and with thin strands linking them to membrane vesicles, mitochondria, and other filaments.

It is important to realize that even this second approach gives a simplified picture of the interior of a living cell. The protein concentration of a typical cytoplasm is around 20% by weight, with even higher values existing in special cells such as erythrocytes (Table 18-2). Attempts to come closer to the real situation, with molecules present in their native numbers and shapes, for example by rapid freezing, show a confused jumble of filaments, clusters, and single proteins (Figure 18-9). Moreover, we know that even this complex image cannot convey the rapid movement of molecules due to thermal energy. Every molecule performs a continuous manic dance as it is continually jostled from one location in the cell to another. A small protein will visit most locations within a bacterial cell in a tenth of a second, while even large structures like filaments and membranes shake and flex, occasionally breaking off pieces (by dissociation) and accruing new ones (by binding).

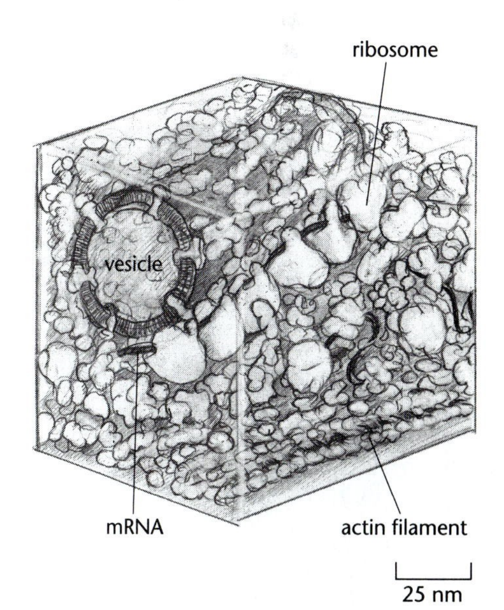


**Figure 18-9** Crowded cytoplasm. Artist's impression of a small volume in the interior of a eucaryotic cell.

## *Many 'soluble' proteins are associated with the cytoskeleton*

The concentration of protein in the cytoplasm (20–30% by weight) is much higher than that used in most biochemical experiments. Binding studies between actin and another protein, for example, are typically conducted at about 1 mg/ml protein (0.1%) and enzymatic activities are conventionally measured in solutions that are as dilute as possible. By contrast, the concentration of protein in the cytoplasm is closer to that at which proteins crystallize. One also has to remember that many other small and large molecules in the cell—metabolites, nucleic acids, and so on—in a sense compete with proteins for a limited supply of water molecules within the cytoplasm.

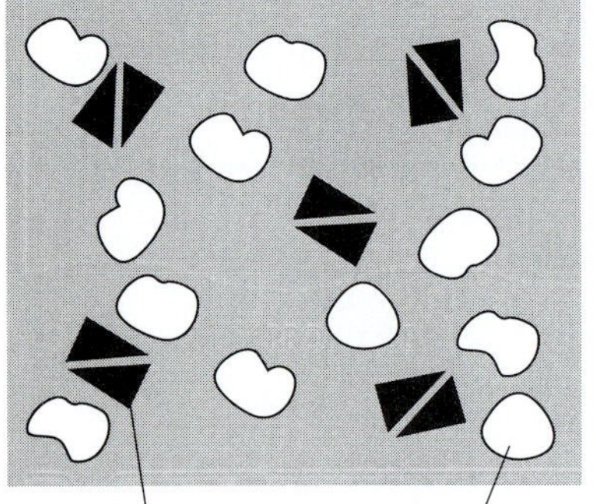


**Figure 18-10** Macromolecular crowding. The protein myoglobin normally exists in aqueous solution as freely diffusing monomers. Addition of an unrelated globular protein such as ribonuclease or lysozyme, however, causes myoglobin to associate into dimers due to competition for available water molecules. Similar effects are likely to promote extensive formation of multiprotein complexes inside living cells.

What are the consequences of such a high concentration of protein? Theoretically, it can be shown that as the number of water molecules in a protein solution decreases, protein molecules (1) tend to adopt a more compact conformation and (2) tend to bind to each other wherever possible. A striking illustration of this crowding effect is that myoglobin, which normally exists in dilute solution as free monomers, can be driven to form dimers by the addition of large amounts of an unrelated protein such as ribonuclease (Figure 18-10). Similarly, the crowded conditions in a cell should encourage even weak protein associations.

One of the earliest indications of this fact came from an experiment performed in 1968 on *Euglena*, a unicellular green alga. Suspensions of *Euglena* cells were centrifuged for 1 hr at 100,000 × *g* to stratify their contents; the cells were then frozen, sectioned, and stained histologically for eight enzymes (Figure 18-11). Enzymatic activities were mainly confined to the band rich in mitochondria and lysosomes, others were in the nucleus and ribosomal bands. Remarkably, *not one* of the enzymes appeared in the clear aqueous layer, even though this should have contained any freely soluble proteins present in the cytoplasm. On the other hand, if the cells were disrupted before centrifugation, then all of the enzymes showed at least some activity in the supernatant fluid.

There are other indications of a link between soluble proteins and the cytoskeleton. In skeletal muscle, for example, the enzymes of glycolysis (the pathway by which sugars are degraded anaerobically) are largely confined to the region of the thin filaments, suggesting an association with actin. Furthermore, even though the purified enzymes are freely soluble and show full enzymatic activity in dilute aqueous solution, they fractionate with myofibrils if a muscle is disrupted. Fruit flies carrying mutations that render their glycolytic enzymes unable to localize to their muscle sarcomeres are unable to fly, even though the purified enzymes are fully catalytic.

It is worth mentioning at this point that the nucleus might also have a framework of proteins that helps to organize the chromosomes. Indeed, one of the major nonchromosomal proteins in the nucleus is actin, estimated to be present in concentrations of around 5 mg/ml in the nuclei of amphibian oocytes. Most actin in an interphase nucleus seems to be either unpolymerized or in oligomeric form, but it can form filaments if the cell is stressed. What actin is doing in a nucleus is unclear, but it might be involved in some way in RNA transcription. Thus, antibodies to actin, or to actin-fragmenting proteins, injected into a nucleus cause a specific loss of the ability to make pre-messenger RNA without affecting the synthesis of ribosomal RNA. Recall also that actin plays an intimate part in the final stages of protein synthesis—in the translation of mRNA into protein occurring on ribosomes (Chapter 5).

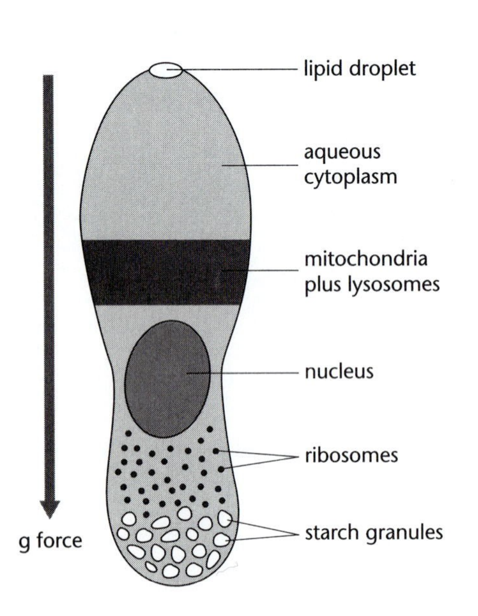


**Figure 18-11** Insoluble enzymes. In a classic experiment, *Euglena* cells were centrifuged at high speed, so that their contents become stratified. Histological staining for eight enzymes showed that they sedimented with formed organelles such as ribosomes or mitochondria. Not one of the enzymes was detected in the layer of aqueous cytosol, despite the fact that the same enzymes were released into an aqueous solution following disruption of the cell.

## *Cytoplasm has a low microviscosity and a high macroviscosity*

If so many enzymes are linked to membrane surfaces or the cytoskeleton, are there any truly diffusive proteins in the cell? The answer is clearly Yes. First of all, many fluorescent proteins injected into cells, or expressed by genetic engineering, spread rapidly, as if by diffusion. Their average rate of movement can be measured by bleaching a small region of cytoplasm and then observing the rate of recovery of fluorescence as fluorescent molecules diffuse back into the bleached region. When the rate of diffusion of fluorescently labeled bovine serum albumin is measured in this way, it gives a translational diffusion constant of about 10 $\mu m^2$/sec (Table 18-3). While this is about seven times smaller than the diffusion constant of this same protein in aqueous buffer, the difference can readily be explained by the high protein concentration of protein in the cell—we do not need to suppose that the BSA binds to structures inside the cell. Even actin and tubulin when injected in this manner diffuse initially through the cytoplasm, although they become progressively incorporated into static structures.

In contrast to the behavior of small nonbinding proteins, organelles and large particles in the cytoplasm move only in association with specific transport systems, if at all. It may be recalled that organelles in a living cell do not show Brownian movement, the visible manifestation of diffusion. The difference in behavior between organelles and small molecules is presumably due to the fact that the cytoplasm is an extensively cross-linked protein gel with an effective mesh size of 20–30 nm. Molecules that are small compared to the water-filled channels in the gel can move quickly through it by diffusion (cytoplasm has a low microviscosity). Organelles that are large compared to the cytoplasmic mesh, by contrast, must force apart links holding the structure of the gel together in order to move (cytoplasm has a high macroviscosity).

## *Proteins compete for water in the cytoplasm*

An important property of protein crystals is that water is present within them in two phases: *bound water*, which is associated with protein surfaces, and *bulk water*, which fills the remaining space. Bound water is not immobile like ice and its individual $H_2O$ molecules exchange rapidly with those in the surrounding medium. But it is less mobile than bulk water and has different solvent properties, being a worse solvent for most sugars and amino acids, for example. Since cells contain as much protein on a weight basis as many crystals (and other macromolecular constituents besides), it makes sense that they too will contain a significant proportion of bound water.

**Table 18-3 Diffusion in the cytoplasm**

| Molecule | Molecular weight | Diffusion coefficient $\mu m^2$/sec | |
|---|---|---|---|
| | | in water | in the cell |
| sucrose | 360 | 500 | 200 |
| green fluorescent protein | 27,000 | 90 | 10 (bacteria)<br>25 (eucaryotic) |
| albumin | 68,000 | 70 | 9 (eucaryotic) |
| tubulin | 100,000 | 60 | 7 (eucaryotic) |

See also Table 1-1. Protein measurements were made by fluorescent bleaching experiments; estimates of sucrose diffusion were obtained by injection followed by radioautography.

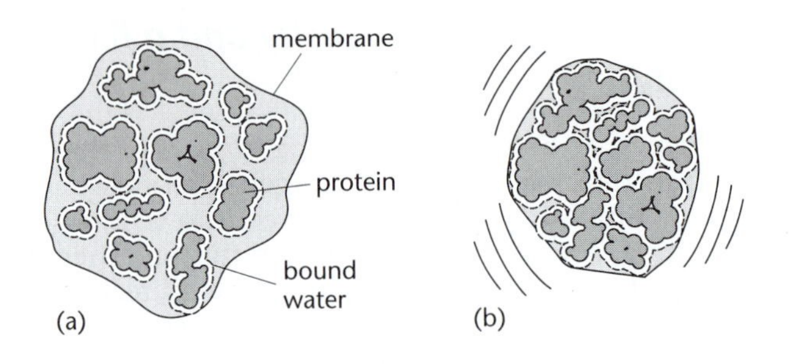


**Figure 18-12** Bound water. (a) Proteins and other macromolecules in a cell carry a loosely associated 'shell' of water that has slightly different properties from normal 'free' water. (b) Cells exposed to concentrated (hyperosmotic) solutions lose their free water due to osmosis, but the bound water is retained.

Indeed, experiments reveal that uncharged, soluble, molecules within mitochondria and bacteria distribute between two phases of water (Figure 18-12). One phase (equivalent to bulk water) is osmotically active and can change its volume over a large range, depending on the concentration of ions and small molecules in the medium. The other phase, corresponding to bound water, has a more or less constant volume and is retained, even when the organelle or cell is placed in strongly hyperosmotic conditions that tend to remove its water. In bacteria the amount of bound water corresponds to about 0.5 µl/mg dry weight of cell macromolecules (principally protein and RNA), whereas the amount of unbound water is some five times greater.

Two phases of water have also been detected in 'brine shrimps,' the dried gastrulae of the primitive crustacean *Artemia*. Brine shrimps, about 0.2 mm in diameter, are composed of about 4000 embryonic cells surrounded by a complex chitin shell. They occur naturally in highly saline lakes and can undergo severe dehydration and rehydration repeatedly without damage. Fully hydrated cysts contain about 58% water by weight but they can be desiccated to an amazing 2% water without irreversible damage. If such desiccated cysts are carefully rehydrated under controlled conditions, their metabolism recovers in two discrete stages. The first stage marks the onset of a limited number of metabolic reactions, whereas the second marks recovery of full metabolic activity and eventual embryonic development. As the water content rises above about 27%, the rates of metabolism and development increase but there is no further quantitative change.

Even mammalian cells are surprisingly insensitive to changes in the amount of water they contain and can survive the osmotically produced loss of almost a third of their total water without irreversible damage. Perhaps, as with *Artemia* gastrulae, it is not until the bound water is depleted that fundamental changes take place?

## *Hydrostatic pressure may drive cell protrusions in plants and lower eucaryotes*

The high concentration of ions, metabolites, and proteins in a cell means that if it is suspended in water, water molecules will try to enter by osmosis. For small molecules and ions, the *osmotic pressure* (the hydrostatic pressure required to stop this net flow of water) can be calculated from the *van't Hoff equation*

$$\pi = RT\,(C_1 + C_2 + C_3 + \cdots\cdots + C_n)$$

where $\pi$ is the osmotic pressure in atmospheres, $R$ is the gas constant, $T$ is the absolute temperature and $C_1$, ...., $C_n$ are the molar concentrations of the ions and small molecules. For concentrated solutions of proteins such as the cytoplasm, the osmotic pressure is much higher than indicated by this simple formula, partly because of the very large size of the molecules.[4] If a cell is to survive in a dilute aqueous environment, it must develop and manage a positive internal hydrostatic pressure that serves to bail out the excess water molecules entering due to osmosis.

[4] The van't Hoff equation applies to dilute solutions in which solute and solvent molecules are similar in size and chemical properties, criteria that do not apply to the cytoplasm. The relatively enormous size of protein molecules, in particular, means that they strongly retard the tendency of water molecules to escape from the cell, so their osmotic effect, molecule for molecule, is much larger.

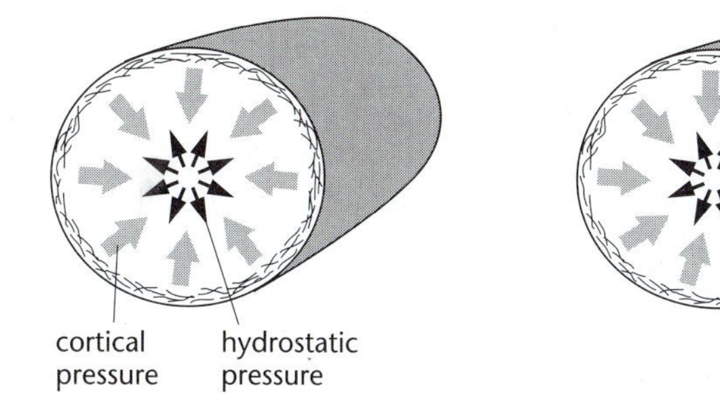


**Figure 18-13** Osmotic pressure and cortical contraction. In principle, reduction of the cortical tension in one region of an animal cell should allow the cell membrane to be pushed outward by hydrostatic pressure. Whether this mechanism actually operates inside vertebrate cells is still not established.

Bacteria, plants cells, and many lower eucaryotes have very high internal pressures—sometimes as high as 10 atmospheres ($10^6$ Pa)—necessitating the presence of a rigid cell wall. The plasma membrane, pressed hard against this wall by the osmotically generated hydrostatic pressure (sometimes called 'turgor' pressure), will readily expand at any local region in which the wall softens. It is widely assumed that osmotic pressure provides a driving force for the expansion of plant cells during growth, for example, or the elongation of fungal hyphae. We have already discussed (in Chapter 8) the evidence that suggests that the cytoplasmic streaming in cells such as giant amoebae and acellular slime molds is driven by hydrostatic pressure. Although these organisms do not have rigid cell walls, they do have thick and 'muscular' outer membranes well endowed with a thick actin cortex.

Whether vertebrate cells have osmotically generated pressures is less certain, since they normally exist in a milieu in which the osmotic pressure outside is very close to that of the cytoplasm. Presumably this is why animal cells live without an external wall but have instead a contractile cortex fastened to the inner face of the plasma membrane. It is possible that for some cells the contraction of this cortex will develop a small hydrostatic pressure that will be balanced by osmotic forces (Figure 18-13). Any expansion due to osmotic forces will therefore be opposed by the expansion of this cortex and at equilibrium there will be a balance of the two.

There is a second, more subtle, osmotic force that does not depend on the generation of turgor pressure. A tightly cross-linked meshwork of protein filaments immersed in water swells like a sponge due to water ingress. Tension will develop in the strands of the meshwork so that, if these strands are cut, they will tend to move apart. It is therefore possible that the gel of actin filaments forming the cell cortex could be made to expand through local liquefaction (solation), for example through a $Ca^{2+}$-stimulated action of gelsolin (Figure 18-14).

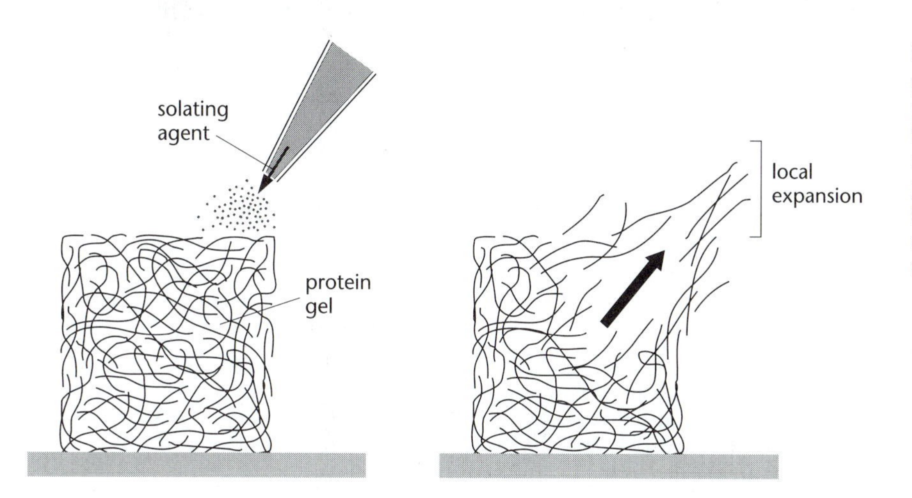


**Figure 18-14** Localized swelling of a protein gel by osmotic forces. Individual strands of a protein network are driven apart by the tendency of water to enter the gel. A locally applied solating agent will cut links in the protein gel, causing it to expand locally in that region. Note that this osmotic effect does not depend on a semi-permeable lipid membrane.

## *How do membranes contribute to the mechanical properties of the cell?*

Throughout this book we have maintained the view that the cytoskeleton is the primary source of mechanical strength inside cells. On their outside cells may develop tough cuticles, cell walls, and extracellular matrices, but inside (the argument goes) everything is sloppy and fluid, apart from the framework of protein filaments. But this is obviously too simple. Even in the absence of a cytoskeleton, cells will have *some* shape due to their content of other proteins and macromolecules, and to the presence of membranes.

The plasma membrane is especially interesting because it is so important for all aspects of cell physiology. All cell membranes are based on a thin sheet of lipid and protein molecules. The lipid bilayer serves primarily as a permeability barrier, whereas protein molecules fulfill most other functions, serving as receptors, enzymes, transport proteins, and so on. In a typical animal cell, protein accounts for about 50% of the area of plasma membrane, whereas in specialized membranes (such as the internal membranes of mitochondria or in retinal rod outer segments) protein can account for as much as 70% of the total area. Usually, the actin cortex is closely associated with the plasma membrane through proteins such as spectrin and ankyrin, and this dominates its mechanical properties. But it is possible to estimate the mechanical properties of the plasma membrane itself (that is, the lipid bilayer together with its intimately related, integral proteins) and these properties must be factored into to the performance of the cell as a whole.

A promising experimental approach is to maneuver submicrometer particles by means of optical laser tweezers close to the surface of cells growing in culture. The beads can be coated with specific antibodies so that they attach to specific molecules in the membrane. When the beads are set free, their movements can be followed by light microscopy. Lipid-attached particles skate randomly over the cell surface, indicating that the lipid molecules they are attached to undergo random diffusive movements. Protein-attached beads usually do the same initially, but some of them eventually become stationary, probably because they are anchored to the cytoskeleton (see Figure 8-10).

When beads attached to the plasma membrane are plucked away from the cell using optical tweezers, they pull out a thin tubular strand, or tether. The force required to pull out this tether and hold it in position is related to the tension in the plane of the membrane and changes in a predictable way with the osmotic swelling or shrinkage of the cell. Interestingly, changes in membrane tension seem to correlate with the capacity of cells to secrete substances or take them up by endocytosis. For example, when an animal cell begins to divide, the tension measured in the membrane increases, while at the same time the uptake of substances from the outside by endocytosis falls dramatically. Since endocytosis requires vesicles to be removed from the plasma membrane, it necessarily involves an decrease in surface area. It makes sense, therefore, that an increase in tension in the membrane would inhibit endocytosis, since it will resist area decrease. Conversely, the same increase in membrane tension should enhance membrane fusion and thereby facilitate secretion (Figure 18-15).

It is not yet possible to disentangle various contributions to the measured tension in the cell surface. The membrane tethers described above,

(a)
plasma membrane
vesicle

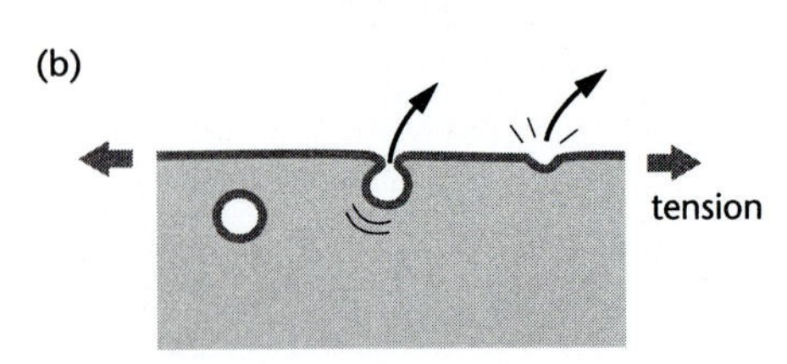


**Figure 18-15** How membrane surface tension affects uptake and secretion. (a) If the plasma membrane has a low surface tension, then vesicles can readily bud inward and move into the cytoplasm. (b) A high surface tension will favor changes that tend to increase the surface area—such as the fusion of vesicles *from* the cytoplasm.

**Table 18-4 Responses of cells to mechanical stimuli**

- Sheets of epithelial cells increase their rate of mitosis when stretched.
- Cells in connective tissue reorient, migrate, and divide more when the tissue is stretched.
- Endothelial sheets align with a flow of liquid over their surface and change the composition and arrangement of their actin cortex.
- Focal adhesions in cultured fibroblasts grow in the direction of externally applied tension.
- Muscle cells in culture make more actin and myosin when stretched repetitively. Isolated myofibrils generate tension-induced contractions when stretched.
- Nerve axons grow when stretched; growth cones can be guided by tension.
- Mechanical tension shapes major anatomical features of the developing brain.
- Spindle microtubules form firmer attachments to kinetochores when pulled.
- Surface protrusions are suppressed by increased tension in the cortical layer.
- All cells have receptors in their plasma membrane that generate signals when the membrane is stretched. Hair cells generate membrane potentials in response to minute deflections of their stereocilia.

for example, pull back with a force related to the in-plane tension but also reflect adhesion between the membrane and the underlying cytoskeleton. Evidently there is an interplay between the cytoskeleton and the membrane in mechanical terms. It could have a major impact on the properties of the cell, both in terms of the uptake and secretion of material and also, as we now see, in the regulation of ion transport into the cell.

## *Cells respond to sound, touch, gravity, and pressure*

The mechanical stress to which all living cells are exposed all of the time has immediate and far-reaching consequences for both physiology and biochemistry. The opposite is also true, and we saw in previous chapters that even subtle changes in the concentrations of signaling molecules can radically alter the shape and movement of a cell.

In plants, the forces of wind and water, or the expansion due to tissue growth, are resisted by tough cell walls. By contrast, most animal tissues readily deform under even small external mechanical forces, with consequent changes in the cytoskeleton. Even at the earliest stage of embryonic development, as a fertilized egg divides, daughter cells jostle for position, both pushing and being pushed as they increase in volume. As the embryo develops, programmed migrations and changes in shape stretch and swell the cells of which it is composed. Throughout the adult life of an animal, cells are continually stretched and compressed as a consequence of muscle contraction.

It is not surprising, therefore, that many cells have evolved specific mechanisms to detect and respond to mechanical forces (Table 18-4). Vascular endothelial cells, for example, derived from the lining of an artery and growing as a sheet of cells in a tissue culture chamber, display a marked sensitivity to the flow of liquid over their surfaces. If culture fluid is pumped over the surfaces of such cells, it causes them to arrest their division and develop spindle shapes aligned in the direction of the liquid flow (Figure 18-16). Cells treated in this fashion also synthesize more actin and myosin and develop a thicker membrane cortex. As another example, motor nerve axons in a mammalian embryo reach their synaptic targets

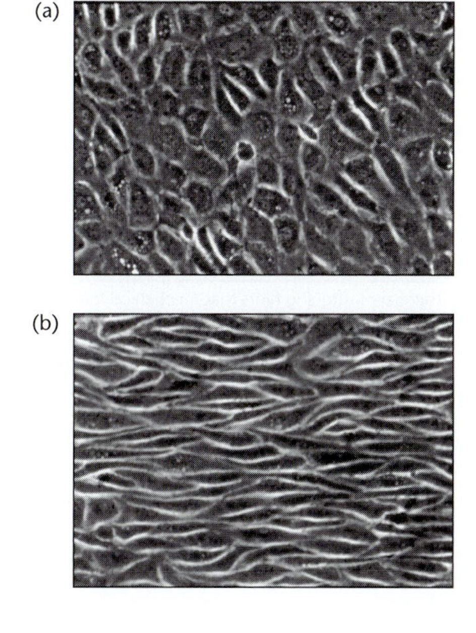


**Figure 18-16** Cells aligned by fluid flow. (a) Culture of endothelial cells obtained from bovine aorta. The cells form a confluent sheet but their orientations are random. (b) A sister culture exposed to a steady lamina shear stress of 20 dyne/cm$^2$ for 24 hours (a similar stress to that caused by normal blood flow). The cells have aligned in the direction of the fluid flow. (Courtesy of A.M. Malek and S. Izumo, 1996.)

when the cells are only a few millimeters long and are then physically stretched as the animal develops to its mature dimensions, tension in this case inducing axonal growth.

The sensing of physical forces in a cell's environment is primarily mediated by a specialized class of membrane proteins known as *mechanosensitive channels*. These have evolved the ability to transduce mechanical strain into an electrochemical response, enabling cells to respond to stimuli such as sound, touch, gravity, and pressure. Mechanosensitive channels are positioned in the membrane but attached to the cytoskeleton, the stresses being transmitted via specific molecular paths, much like chemical signals. One example, mentioned earlier, is the stretch-activated channels in hair cell stereocilia (Chapter 6). Because they are directly linked to adjacent stereocilia on the hair cell surface, these channels are responsive to amazingly small deflections. The response of endothelial cells of blood vessels to shear stress, mentioned above, is also thought to be due to the activation of mechanosensitive channels, eventually leading to the activation of protein kinases such as FAK—the kinase associated with focal adhesions—and the triggering of downstream events such as production of nitric oxide (NO).

Another way in which externally applied mechanical forces can change the cell is through a direct action on the cytoskeleton. It may be recalled, for example, that the link between a microtubule and the kinetochore in a mitotic spindle is stabilized by tension (Chapter 13), perhaps through a tension-induced structural change in the kinetochore/microtubule complex. Assemblies of actin and myosin undergo transient contractions when tweaked by a short pull, and in the laboratory this can produce oscillatory behavior in isolated myofibrils and strands of *Physarum* cytoplasm. Skeletal muscle cells, as we saw in Chapter 10, become thicker and stronger due to exercise, with synthesis of more actin and myosin. Another familiar example from vertebrate animals is the thickening and strengthening of bone by mechanical stress—which is due to a mechanically triggered proliferation of osteoblasts.[5]

## Outstanding Questions

*Can we explain the material properties of viscous solutions and gels composed of cytoskeletal filaments and associated proteins? Is it possible to simulate their behavior and predict their responses to applied stress on a computer? How is the polymerization of actin filaments harnessed so as to drive the plasma membrane outward, or to move an organelle through the cell? How important is dynamic instability to this coordination? How does a cell achieve mechanical equilibrium and sustain a stable size and shape? Does it contain elements permanently under tension or compression (as suggested by tensegrity models)? What is the structure of living cytoplasm, and what is the relationship between its overall mechanical properties and those of its component parts? Does osmotic pressure drive wall expansion in fungi? If so, how? What determines the biconcave disc shape of a mammalian red blood cell? How do endothelial cells sense the direction of fluid flow over their surfaces? In what form are actin, tubulin, and intermediate filament proteins transported within nerve axons? Where do they form polymers? What is the structure of the nuclear matrix? Why is actin present in the nucleus and what does it do? What is the origin of tension in the plasma membrane? What contribution do membrane proteins and the cytoskeleton make?*

[5] There are intriguing hints that mechanical forces might influence protein synthesis directly. Contact of cells with beads coated in extracellular matrix apparently leads to the local recruitment of ribosomes and mRNA coding for focal adhesion proteins.

## *Essential Concepts*

- Cell movements are driven by physical forces and governed by the mechanical properties of cell components, principally its cytoskeleton.
- Actin filaments, intermediate filaments, and microtubules behave as flexible rods in solution, with an elasticity similar to that of hard plastic. Microtubules are the most rigid of the three, whereas intermediate filaments have the greatest tensile strength.
- The use of protein filaments as stiffeners in the cell is made possible by their longitudinal association into bundles, typically held together by protein cross-links.
- The growth of protein filaments by polymerization can be harnessed to generate a force.
- Protein filaments form extensive three-dimensional networks in aqueous solution, with complex, viscoelastic properties. Cytoplasm is also viscoelastic, although its properties vary with the cell type and the region of the cell.
- Filament bundles and networks in the cell have the unique feature that they can develop force, due to the presence of molecular motors. Contraction in animal cells is resisted by a framework of compressive elements, such as microtubules.
- A variety of experimental approaches have been used to measure the mechanical properties of whole cells. These reveal not only the contractile nature of the cell surface but also a network of mechanical links throughout the cytoplasm.
- Cytoplasm has an extremely complex structure characterized by a very high concentration of macromolecules and a three-dimensional network of protein molecules and membranes that restricts the diffusion of vesicles and macromolecular complexes, while permitting molecules smaller than 20 nm to diffuse freely.
- The properties of cytoplasm reflect the high concentration of proteins it contains. A significant fraction of the water inside cells (~ 20%) is closely associated with the surfaces of proteins and has different properties from bulk water.
- For plants, fungi, and bacteria, hydrostatic pressure developed within the cytoplasm due to osmotic movements is a major driving force for cell expansion. Animal cells develop smaller internal pressures, but these could be sufficient to drive membrane expansion.
- Cell membranes also maintain a surface tension that, although smaller than the forces developed by the cytoskeleton, has a controlling influence on cell processes such as endocytosis and exocytosis.
- Living cells are continually exposed to mechanical forces and have evolved specific mechanisms to detect and accommodate to these forces. Mechanosensitive ion channels, for example, respond to stress in the cell surface and, by their affect on the cytoskeleton, trigger changes in metabolism and cell shape.

## Further Reading


Bousquet, O., Coulombe, P.A. Missing links found? *Curr. Biol.* 6: 1563–1566, 1996.

Bray, D. Axonal growth in response to experimentally applied tension. *Dev. Biol.* 102: 379–389, 1984.

Brunette, D.M. Mechanical stretching increases the number of epithelial cells synthesizing DNA in culture. *J. Cell Sci.* 69: 35–45, 1984.

Chang, G., et al. Structure of the MscL homolog from *Mycobacterium tuberculosis*: a gated mechanosensitive ion channel. *Science* 282: 2220–2225, 1998.

Cohen, W.D., et al. Elliptical versus circular erythrocyte marginal bands: isolation, shape conversion, and mechanical properties. *Cell Motil. Cytoskeleton* 40: 238–248, 1998.

Cole, K.S. Surface forces of the *Arbacia* egg. *J. Cell Comp. Physiol.* 1: 1–9, 1932.

Cooke, R., et al. A model of the release of myosin heads from actin in rapidly contracting muscle fibers. *Biophys. J.* 66: 778–788, 1994.

Crick, F.H.C., Hughes, A.F.W. The physical properties of cytoplasm: a study by means of the magnetic particle method. *Exp. Cell Res.* 1: 37–80, 1950.

Dai, J., Sheetz, M.P. Membrane tether formation from blebbing cells. *Biophys. J.* 77: 3363–3370, 1999.

Dennerll, T.J., et al. Tension and compression in the cytoskeleton of PC-12 neurites. II Quantitative measurements. *J. Cell Biol.* 107: 665–674, 1988.

Elowitz, M.B., et al. Protein mobility in the cytoplasm of *Escherichia coli*. *J. Bacteriol.* 181: 197–203, 1999.

Elson, E.I. Cellular mechanics as an indicator of cytoskeletal structure and function. *Annu. Rev. Biophys. Biophys. Chem.* 17: 397–430, 1988.

Furukawa, R., Fechheimer, M. The structure, function, and assembly of actin filament bundles. *Int. Rev. Cytol.* 175: 29–90, 1997.

Gavin, R.H. Microtubule–microfilament synergy in the cytoskeleton. *Int. Rev. Cytol.* 173: 207–242, 1997.

Gilbert, D.S. Axoplasm architecture and physical properties as seen in the *Myxicola* giant axon. *J. Physiol.* 253: 257–301, 1975.

Girard, P.R., Nerem, R.M. Shear-stress modulates endothelial-cell morphology and F-actin organization through the regulation of focal adhesion-associated proteins. *J. Cell. Physiol.* 163: 179–193, 1995.

Gittes, F., et al. Flexural rigidity of microtubules and actin filaments measured from thermal fluctuations in shape. *J. Cell Biol.* 120: 923–934, 1993.

Heidemann, S.R., et al. Direct observations of the mechanical behaviors of the cytoskeleton in living fibroblasts. *J. Cell Biol.* 145: 109–122, 1999.

Janmey, P.A. The cytoskeleton and cell signaling: component localization and mechanical coupling. *Physiol. Rev.* 78: 763–781, 1998.

Janmey, P.A., et al. The mechanical properties of actin gels. Elastic modulus and filament motions. *J. Biol. Chem.* 269: 32503–32513, 1994.

Käs, J., et al. F-actin, a model polymer for semiflexible chains in dilute, semidilute, and liquid crystalline solutions. *Biophys. J.* 70: 609–625, 1996.

Liu, G., et al. pH, EF-1α and the cytoskeleton. *Trends Cell Biol.* 6: 168–171, 1996.

MacKintosh, F.C., Janmey, P.A. Actin gels. *Curr. Opin. Solid State Mater. Sci.* 2: 350–357, 1997.

Mahadevan, L., Matsudaira, P. Motility powered by supramolecular springs and ratchets. *Science* 288: 95–99, 2000.

Malek, A.M., Izumo, S. Mechanism of endothelial cell shape change and cytoskeletal remodeling in response to fluid shear stress. *J. Cell Sci.* 109: 713–726, 1996.

Maniotis, A.J., et al. Demonstration of mechanical connections between integrins, cytoskeletal filaments, and nucleoplasm that stabilize nuclear structure. *Proc. Natl. Acad. Sci. USA* 94: 849–854, 1997.

McNeil, P.L. Cellular and molecular adaptations to injurious mechanical stress. *Trends Cell Biol.* 3: 302–307, 1993.

Money, N.P., Harold, F.M. Extension growth of the water mold *Achlya*: interplay of turgor and wall strength. *Proc. Natl. Acad. Sci. USA* 89: 4245–4249, 1992.

Oakely, C., and Brunette, D.M. The sequence of alignment of microtubules, focal contacts and actin filaments in fibroblasts spreading on smooth and grooved titanium substrata. *J. Cell Sci.* 106: 343–354, 1993.

Penman, S. Rethinking cell structure. *Proc. Natl. Acad. Sci. USA* 92: 5251–5257, 1995.

Pestic-Dragovich, C.S., et al. Regulation of cytoskeletal mechanics and cell growth by myosin light chain phosphorylation. *Am. J. Physiol.* 275: 349–356, 1998.

Raucher, D., Sheetz, M.P. Membrane expansion increases the endocytosis rate during mitosis. *J. Cell Biol.* 144: 497–506, 1999.

Rodionov, V.I., et al. Functional coordination of microtubule-based and actin-based motility in melanophores. *Curr. Biol.* 8: 165–168, 1998.

Salmon, E.D., et al. Diffusion coefficient of fluorescently-labeled tubulin in the cytoplasm of embryonic cells of a sea urchin: video image analysis of fluorescence redistribution after photobleaching. *J. Cell Biol.* 99: 2157–2165, 1984.

Sato, M., et al. Dependence of the mechanical properties of actin/α-actinin gels on deformation rate. *Nature* 352: 828–830, 1987.

Sato, M., et al. Rheological properties of living cytoplasm: endoplasm of *Physarum plasmodium*. *J. Cell Biol.* 97: 1089–1097, 1983.

Singhvi, R., et al. Engineering cell shape and function. *Science* 264: 696–698, 1994.

Takahashi, M., et al. Mechanotransduction in endothelial cells: temporal signaling events in response to shear stress. *J. Vasc. Res.* 34: 212–219, 1997.

Takeuchi, S. Wound healing in the cornea of the chick embryo. IV Promotion of the migratory activity of isolated corneal epithelium in culture by the application of tension. *Dev. Biol.* 70: 232–240, 1979.

Thoumine, O., et al. Elongation of confluent endothelial cells in culture—the importance of fields of force in the associated alteration of their cytoskeletal interactions. *Exp. Cell Res.* 219: 427–441, 1995.

Van Essen, D.C. A tension-based theory of morphogenesis and compact wiring in the central nervous system. *Nature* 385: 313–318, 1997.

van Oudenaarden, A., Theriot, J.A. Cooperative symmetry-breaking by actin polymerization in a model for cell motility. *Nat. Cell Biol.* 1: 493-499, 1999.

Vandenburgh, H., Kaufman, S. *In vitro* model for stretch-induced hypertrophy of skeletal muscle. *Science* 203: 265–268, 1979.

Venier, P., et al. Analysis of microtubule rigidity using hydrodynamic flow and thermal fluctuations. *J. Biol. Chem.* 269: 13353–13360, 1994.

Wang, N., et al. Mechanotransduction across the cell surface and through the cytoskeleton. *Science* 260: 1124–1127, 1993.

White, G.E., Fujiwara, K. Expression and intracellular distribution of stress fibers in aortic endothelium. *J. Cell Biol.* 103: 63–70, 1986.

Wojtas, K., et al. Flight muscle function in *Drosophila* requires colocalization of glycolytic enzymes. *Mol. Biol. Cell* 8: 1665–1675, 1997.

Wolpert, L. The mechanical properties of the membrane of the sea urchgin egg during cleavage. *Exp. Cell Res.* 41: 385–396, 1966.

Yanai, M., et al. Intracellular pressure is a motive force for cell motion in *Amoeba proteus*. *Cell Motil. Cytoskeleton* 33: 22–29, 1995.

Yang, Y., et al. An essential cytoskeletal linker protein connecting actin microfilaments to intermediate filaments. *Cell* 865: 655–665, 1996.

# CHAPTER 19

# Cell Shape

At first sight it seems paradoxical to discuss cell shape in a book on movements. But a moment's thought shows that in order to create an asymmetric form, movements must take place. Consider, for example, a highly branched nerve cell with many dendrites and an axon terminating in specifically located synapses (Figure 19-1). The shape of such a cell will depend on spatiotemporal changes in the length and position of microtubules as dendrites and axon extend from the cell body, and the actin-based amoeboidlike crawling movements of the growth cone that guide them to their correct locations. Nor does movement cease with the initial formation of a synapse, since many local refinements of size and detailed branching occur throughout life. Each dendrite and each synapse of a mature nerve cell has a distinct, possibly unique, complement of proteins. These must have at some time been transported from where they were made, often traveling long distances from the nerve cell body. Diffusion is slow and inefficient over anything but the shortest distances, so shape determination requires specific mechanisms of transport, typically driven by molecular motors.[1]

In this chapter we will examine some of the mechanisms that cells use to control their external shape. We begin with cells encased in rigid cell walls, such as plants and fungi, which change their shape by controlling synthesis and breakdown of their wall in selected regions. We then move to animal cells, and especially cells in the mammalian body, which have highly deformable surfaces and adopt a much richer and more varied repertoire of shapes. Despite differences in design, we will see that both plant and animal cells use their actin cortex to control local expansion of the cell surface. Typically, a cortical site, identified by some external or internal cue, serves as an assembly point for a cytoskeletal complex, which then directs the development of form and structure (*morphogenesis*). All eucaryotic cells also employ arrays of microtubules, positioned by microtubule-organizing centers, to give an overall polarity to the cell and control local differentiation of its cortex. Arrays of microtubules together with associated motors and organizing centers are able, in effect, to create a 'molecular morphogenetic field' inside a cell, in some ways analogous to the larger pancellular fields that control the development of embryonic form.

[1] The correlation between internal transport and degree of specialization is a general one that applies to human civilizations as well as to cells (see Adam Smith in *The Wealth of Nations*).

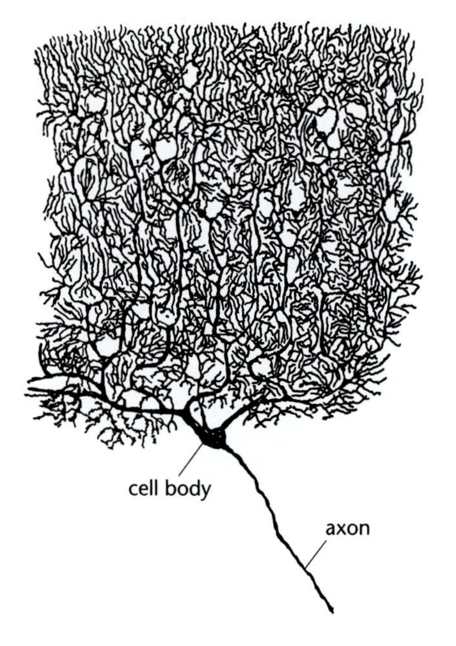


**Figure 19-1** Elaborate shape of a human nerve cell. Tracing of a silver-stained Purkinje neuron from the human cerebellum showing its elaborate, flattened dendritic tree and the initial part of its axon. Collateral branches and terminals of the axon are omitted. The shapes of nerve cells are generally characteristic of their location in the nervous system and in some cases can be individually identified, especially in invertebrate nervous systems. (Based on a drawing by S. Ramon y Cajal. Histologie du Système Nerveux de l'Homme et des Vertébrés. Paris: Maloine, 1909–1911.)

## Largest and smallest cells

The smallest cells are probably *mycoplasmas*—wall-less bacteria that lead a parasitic existence in close association with animal or plant cells. Some have a diameter of only 0.3 μm and can squeeze through filters with pore diameter 0.22 μm. *Mycoplasma genitalium* contains enough DNA to direct the synthesis of 479 genes, which may be the minimum that a cell needs to survive. Protozoan parasites can also be extremely small. *Theileria,* which infects cattle, is only 0.75 μm in diameter—less than 1/20 the volume of a small yeast cell—when it invades a lymphocyte.

What you consider to be the largest cell depends on your definition, since many very large syncytial structures contain multiple cell nuclei within a single plasma membrane. Most of the cells of a living tree, for example, are connected by thin strands of cytoplasm (called plasmodesmata), so that one could argue that they are, in effect, one giant cell.[2] In an adult human body, skeletal muscle cells are single multinucleate structures well over a centimeter in length, and motor neurons can be up to a meter in length. In birds, amphibians, and other animals, the egg is the largest cell.

## Bacteria, fungi, and plant cells have rigid cell walls

The shape of bacteria such as *Escherichia coli* is dominated by their rigid cell wall, which in turn reflects their very high internal pressure. It may be no accident that these cells resemble pressure vessels with cylindrical walls and hemispherical ends. Indeed, the shape of *E. coli* can be modeled using equations similar to those developed for cylindrical soap bubbles. But although the overall shape is symmetrical, the distribution of molecules in the cell is not. Thus, in *E. coli*, molecules of DNA are located in a distinct region of the cytoplasm and there is a polar accumulation of chemotactic proteins; as the cell enters division, a ring of cell division proteins such as FtsZ forms around its midline. Other bacteria display even more conspicuous regional differences, for example by forming a spore, a stalk, or a flagellum at one end of the cell.

Yeasts, which are unicellular fungi, are ideal for genetic studies of cell shape. They have a small genome (less than 1/100th that of a mammal), can exist in haploid form, and reproduce almost as rapidly as bacteria. Two yeast species have been especially important in studies of cell shape, the budding yeast *Saccharomyces cerevisiae* used by brewers and bakers, and the fission yeast *Schizosaccharomyces pombe*, whose second name comes from the African beer it is used to produce. Both species have rigid cell walls composed largely of the polysaccharide α-glucan and regulate their size and shape by modulating enzymes that degrade and synthesize the wall.

Plant cells also have thick walls, in this case made of cellulose. The primary cell wall contains several layers of *cellulose microfibrils*, which are bundles of cellulose molecules tightly linked together by hydrogen bonds. Orthogonally arranged microfibrils resist tension in the cell wall and a second type of polysaccharide, pectin, which is distributed throughout the wall, is better able to resist compression. The size and shape of a plant cell is influenced by its cytoskeleton, by osmotic pressure (turgor), and by mechanical stress. The latter two physical forces regulate enzymes responsible for the synthesis and breakdown of polysaccharides in the cell wall.

## Plant morphogenesis depends on oriented cell division and expansion

Plant cells, imprisoned within their cell walls, cannot crawl about and cannot move to different locations as the plant grows. But they can divide and they can swell, stretch, and bend. The morphogenesis of a developing

[2] A single colony of the fungus *Armillaria bulbosa* from a northern Michigan hardwood forest has a diameter in excess of 500 m. Since the hyphae of this species are usually syncytial, this could be the largest known cell.

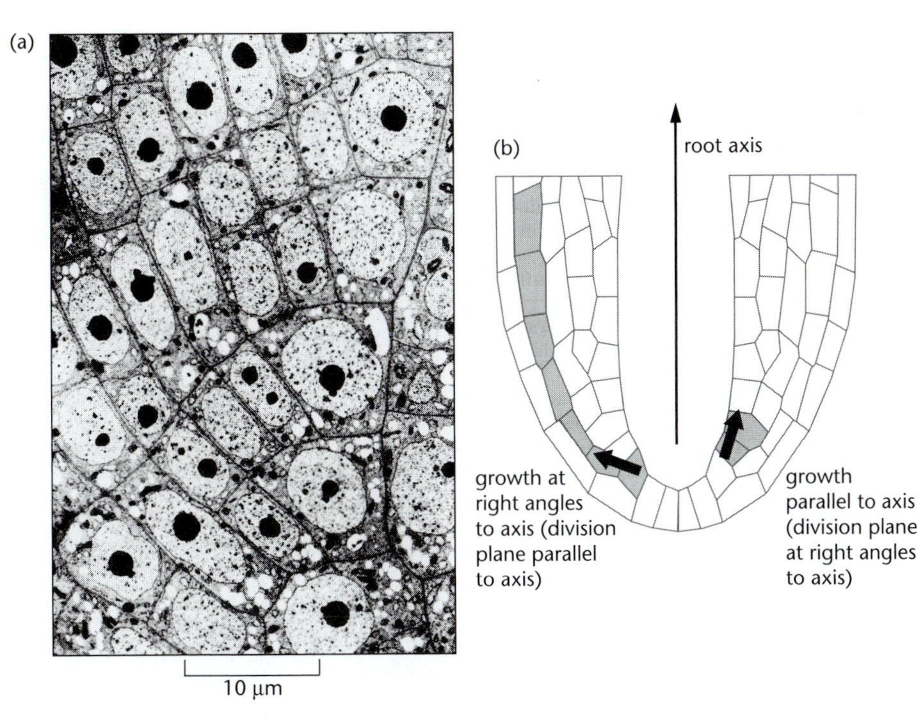


**Figure 19-2** Plant cell walls. (a) Electron micrograph of the root tip of a rush, showing the organized pattern of cells that results from an ordered sequence of cell division in cells with rigid cell walls. (b) Schematic cross-section of the tip of a root showing the two predominant orientations of cell division and growth. Selection between these two orientations plays a major part in shaping the root. (a, courtesy of C.H. Busby and B. Gunning, *Eur. J. Cell Biol.* 21: 214–223, 1980.)

plant therefore depends on orderly cell divisions followed by strictly oriented cell expansions. Most cells produced in the root tip meristem, for example, go through three distinct phases of development: division, growth (giving elongation of the root), and differentiation. These three overlapping steps, which give rise to the characteristic architecture of a root tip, are often segregated in time and space. Moreover, it has also been observed that cell growth and division in a typical root or shoot tend to be oriented with respect to the axis of elongation. Cells growing parallel to the axis, or dividing at right angles to it, contribute to the *length* of the organ. Cells that grow at right angles to the axis, or divide parallel to it, serve mainly to increase the root *thickness* (Figure 19-2).

We saw in Chapter 13 that cell division in a higher plant cell follows a different course from that in an animal cell. Rather than constricting its surface by means of a contractile ring of actin filaments, a plant cell builds up an intermediate plate of membrane and cell wall material called a phragmoplast. The position of this plate, and hence the plane of cell division, is previously marked by a ring of microtubules—the preprophase band. Growth of the phragmoplast depends on the delivery of vesicles along microtubules, which are themselves left over from the mitotic spindle.

## *Division and growth are controlled by subcortical microtubules*

Microtubules not only control the plane of cell division: they also regulate the direction of cell expansion. Increases in length or girth are driven by turgor pressure, itself caused by an influx of water into the plant vacuole due to ion imbalance. But turgor pressure is isotropic, and the *direction* of expansion depends on the arrangement of the cellulose microfibrils in the wall. Indeed, cells anticipate their future shape by controlling the orientation of microfibrils that they deposit in the wall. Cellulose molecules are spun out from the surface of the cell by an enzyme complex in the membrane using sugar nucleotide precursors supplied from the cytosol. As they grow, the nascent cellulose chains spontaneously assemble into microfibrils that form on the extracellular surface of the plasma membrane into a layer, or lamella, in which all the microfibrils have more or less the same alignment.

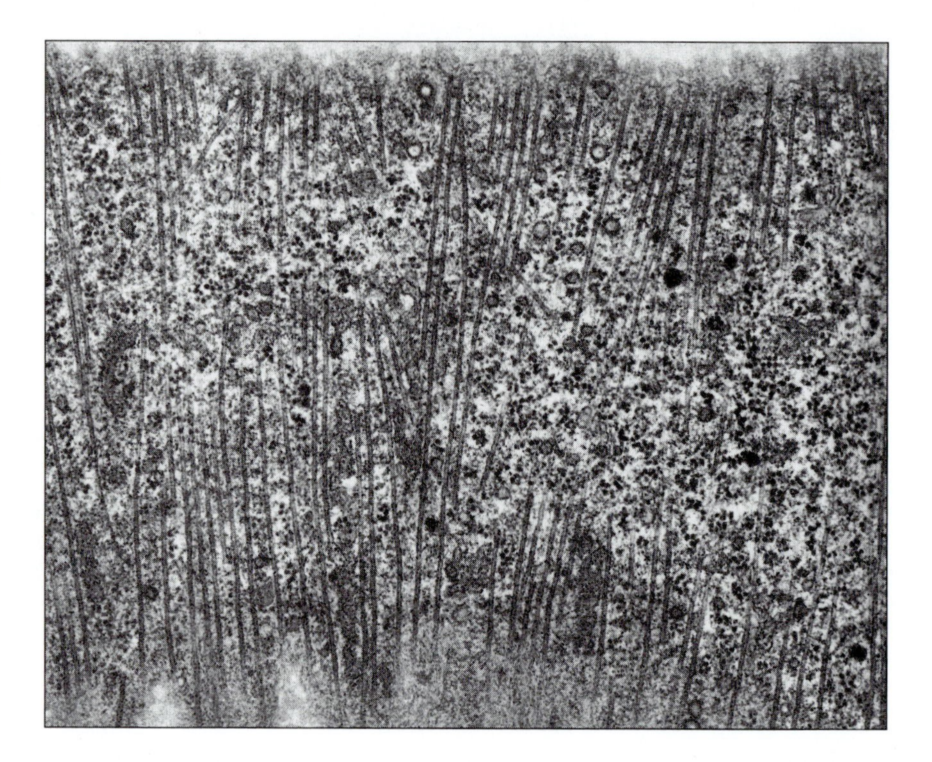

**Figure 19-3** Cortical microtubules. A grazing section of a root-tip from Timothy grass (*Phleum pratense*), showing a cortical array of microtubules lying just beneath the plasma membrane. (Courtesy of B. Gunning and M. Steer, *Plant Cell Biology: Structure and Function*. Sudbury, MA: Jones & Bartlett Publishers, 1996.)

An important clue to understanding how microfibrils are oriented came with the discovery that most cytoplasmic microtubules are arranged in the cortex of the plant cell with the same orientation as the cellulose microfibrils currently being deposited in that region. These cortical microtubules, forming what is called the *cortical array*, lie close to the cytoplasmic face of the plasma membrane, held there by protein molecules that are as yet poorly characterized (Figure 19-3). The congruent orientation of the cortical array of microtubules (just inside the plasma membrane) and cellulose microfibrils (just outside) is seen in many types and shapes of plant cells and found during both primary and secondary cell-wall deposition.

Plant cells must 'listen' to signals coming from the surroundings in order to rearrange their cortical microtubules and hence control the plane of division and the direction of cell expansion. What are these signals? They include light, temperature, water and nutrients, and mechanical stress. Because of their cell walls, plant cells are rigid structures, linked to each other by a continuous framework. Forces caused by the weight of the plant, pressures due to cell growth, movements due to wind, and other external agents, all propagate in a vectorial fashion through the framework of cell walls and could eventually influence the positioning of microtubules. Another factor that is clearly important is the presence of plant hormones. Factors such as giberellic acid diffuse from neighboring cells and are known also to influence the disposition of microtubules in the cortex. Precisely how external signals, whether chemical or physical, alter the distribution of microtubules inside a plant cell is still poorly understood.

## *Yeast budding is controlled by small G proteins*

The budding yeast *Saccharomyces cerevisiae* consists of single ellipsoidal-shaped cells that reproduce by budding. Cell shape is maintained by rigid cell walls made of polysaccharide, and growth entails a controlled assembly of the wall at specific locations, leading to the formation of a new bud (Figure 19-4). During this process, growth is initially confined to a small

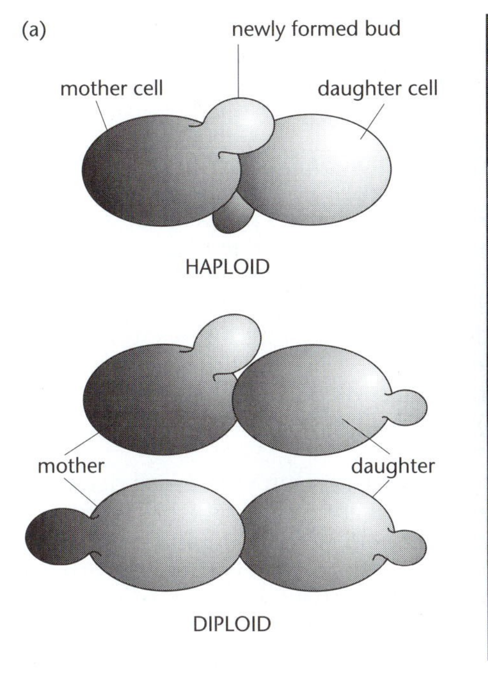


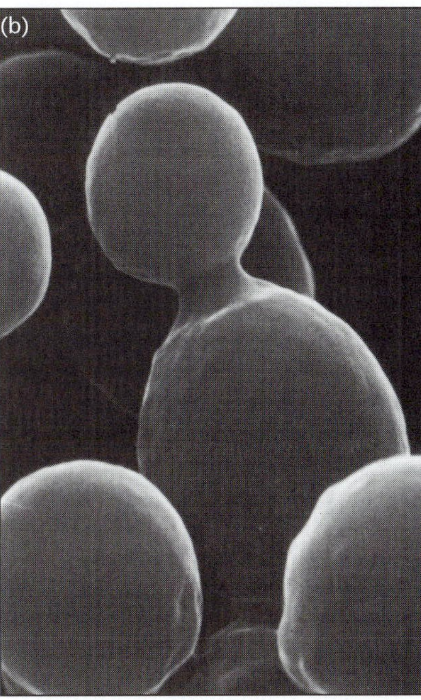


**Figure 19-4** Budding patterns of *S. cerevisiae*. (a) Haploid cells undergo axial budding in which both mother and daughter cells bud adjacent to the previous site of cytokinesis. Diploid cells show a different pattern in which mother cells bud at either pole, but daughter cells preferentially form buds opposite to the site of cytokinesis. (b) Scanning electron micrograph of budding yeast. (b, courtesy of Alan Wheals.)

region of the cell surface and then becomes more diffusely located throughout the surface as the bud enlarges. At cell division (cytokinesis) the growth machinery is directed to the neck between the mother and the bud, so that it forms a septum between the two.

*S. cerevisiae* has been enormously useful in the study of cell morphology because of many mutations that affect the budding process. Some of these define the *position* of growth sites, in which case mutants appear to be normal in size and shape but their buds are positioned erratically on the cell surface. Other mutations prevent the initial *formation* of a polarized growth site, resulting in cells that are unable to form buds and consequently become unnaturally large and rounded. Studies of these mutants show that the site of a bud depends on landmarks in the cortical cytoskeleton left behind by the preceding bud. A new cortical apparatus, based on actin filaments and incorporating secretory enzymes, is then guided to the new site by a series of Ras-related GTP-binding proteins.

For the selection of a bud site, Bud1, a small G protein is essential. If this protein, or one of the proteins that control it, is defective, the cell loses its ability to select a proper bud position and buds appear at random positions. In the organization of the bud site, Cdc42—the small G protein introduced in Chapter 6 that controls the actin cytoskeleton—is crucial, as well as its negative regulators. If Cdc42 does not work properly, the actin cytoskeleton fails to become organized and budding does not take place. Finally, another small G protein, Rho1, is required for the activity of enzymes that catalyze the wall synthesis and hence contribute directly to bud formation.

Actin in the yeast cell cortex is concentrated into discrete patches; mobile accumulations of actin and other proteins are scattered on the inside of the plasma membrane and most highly concentrated in regions of growth. Actin patches contain associated proteins, such as myosins, cofilin, and Arp2/3, and are believed to be sites at which new membrane and cell wall is deposited. The actual site of budding itself is marked by proteins called *septins*. These form a distinctive set of 'neck filaments' below the plasma membrane in the region of the mother-bud neck, appearing just before bud emergence and disappearing during cytokinesis. Note that, although actin and septins are used for bud-site selection, microtubules are not required. In fact, buds will form at the correct location in the absence of microtubules. However, cytoplasmic microtubules running from the spindle pole body into the bud *are* necessary for the correct positioning of the nucleus and spindle within the cell.

## *Microtubules maintain the polarity of fission yeast*

Fission yeast cells of the species *Schizosaccharomyces pombe* grow as cylindrical rods about 3 μm in diameter and 8–14 μm in length. During the cell cycle, growth occurs first at only one end of the cell, later at both ends, and finally is targeted to the cell center for cell division. Thus in fission yeasts (in contrast to budding yeasts) the selection of sites for cell elongation and cell division appear to take place at different times and places in the cell. The proper orientation of polarized growth in these cells requires elongation in a straight line defined by the axis of the cylindrical cell.

As for budding yeasts, a variety of mutant fission yeasts with defective shapes have been found. Mutations preventing the establishment of polarity result in cells that are rounded or ellipsoidal. Mutations affecting the position of growth sites lead to growth near the poles but with an altered orientation, producing cells that are bent or, in more extreme cases, branched (Figure 19-5).

The placement of growth sites in vegetatively growing cells is controlled by an internal program. Sites of expansion are localized to the cell poles, and their proper placement ensures that elongation of the cell occurs along the cell's long axis. Unlike in budding yeasts, microtubules have an essential function in this polarization process, as shown by mutations in tubulin genes, or treatment of cells with microtubule poisons. The fission yeast gene *tea1* (named for *t*ip *e*nd *a*berrant) links the microtubule cytoskeleton with the selection of growth sites, and mutants defective in this gene bend and branch. The product of this gene accumulates at the ends of cytoplasmic microtubules and its localization at cell poles is dependent on an intact microtubule cytoskeleton. This localization is dynamic, as the protein Tea1 is maintained at cell ends for only a few minutes in the absence of cytoplasmic microtubules.

Cell division in fission yeast follows a similar pattern to that described in Chapter 13 for vertebrate tissue cells. As the nucleus commences mitosis, the interphase array of microtubules is lost and replaced by a spindle array organized from a center close to the nucleus. The accumulation of Tea1 and associated proteins at the two ends of the cell disperses, and polarized growth ceases. At this time, a ring of actin filaments is produced around the midline waist of the cell, marking the site of the eventual septum formation.

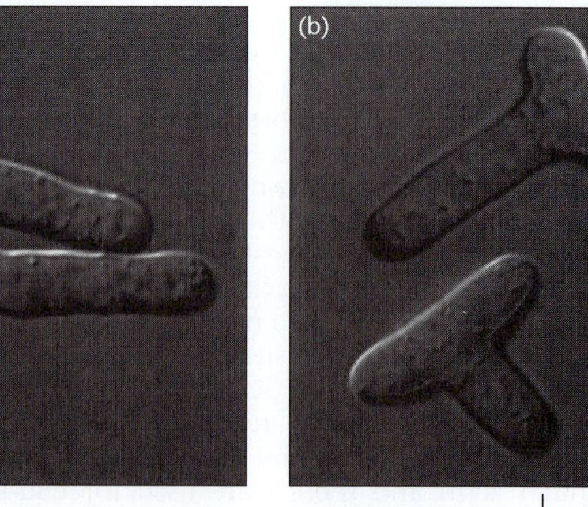


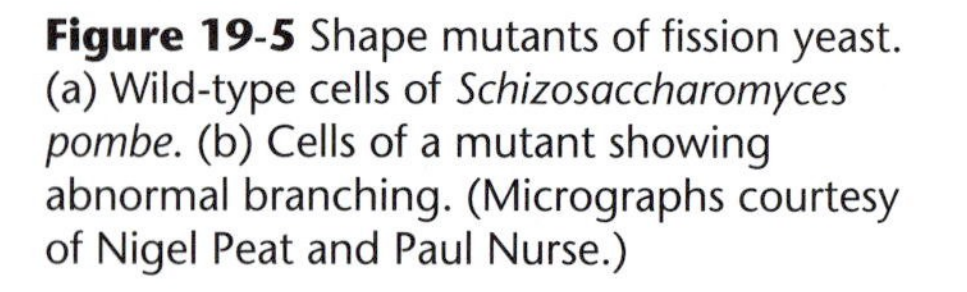
**Figure 19-5** Shape mutants of fission yeast. (a) Wild-type cells of *Schizosaccharomyces pombe.* (b) Cells of a mutant showing abnormal branching. (Micrographs courtesy of Nigel Peat and Paul Nurse.)

## *Cell wall synthesis in* Fucus *responds to external signals*

Although the cells of plants and fungi are imprisoned in rigid walls, they are still able to sense outside influences. Recall that the entire morphology of a multicellular plant depends on individual cells changing their growth and division in response to mechanical and chemical signals and light. Yeast cells also respond to external signals, most notably in the course of mating. During the mating of both budding and fission yeasts, cells form projections in the direction of the mating partner, stimulated by a mating pheromone diffusing from other cell. In both kinds of yeast this requires suppression of the intrinsic program of growth, described above, and degradation of cortical 'landmarks' that would otherwise guide wall expansion. The cell is then free to respond to the pheromone gradient.

Unfertilized eggs of the seaweed *Fucus*, have been used for the study of cell polarity for over a century. Unlike the eggs of most animal species (discussed in the following chapter), *Fucus* eggs are unpolarized and symmetrical when they are released from the adult—their nucleus is centrally located in the egg and the many organelles in their cytoplasm are uniformly distributed. Not until several hours after fertilization does the egg develop an asymmetry in the form of a rhizoid that grows out in one direction. Cell division takes place about 22 hours later, perpendicular to the long axis of the rhizoid, producing a rhizoid cell and a larger thallus cell. The rhizoid cell then divides several times to produce the holdfast, which anchors the organism to a rock, whereas the thallus cell divides continually to form a globular embryo, which is destined to become the ribbonlike leaf (frond).

Development of the rhizoid can be manipulated experimentally, and in particular by light. *Fucus* eggs are highly sensitive to light (Figure 19-6), with the rhizoid appearing on the side of the egg farthest from the light source. The light gradient is converted to a cortical asymmetry by an actin-dependent translocation of plasma membrane proteins, such as channels and receptors, to the shaded quadrant. Initially, the asymmetry is labile and can be repositioned by altering the direction of light. The colocalization of actin filaments and $Ca^{2+}$-containing vesicles then leads to targeted secretion of cell wall materials and polarized growth. At the same time, signals from the cortical domain interact with the spindle, apparently orienting the asymmetric division plane. We will see in the following chapter that the reorientation of spindles through an interaction with the actin cortex occurs also in animal development.

## *In many tissues the cells are close-packed polyhedra*

Many of the cell shapes described in this chapter, especially those that are asymmetric or highly polarized, are genetically regulated and subject to very precise molecular controls. But it is important to state that cells can

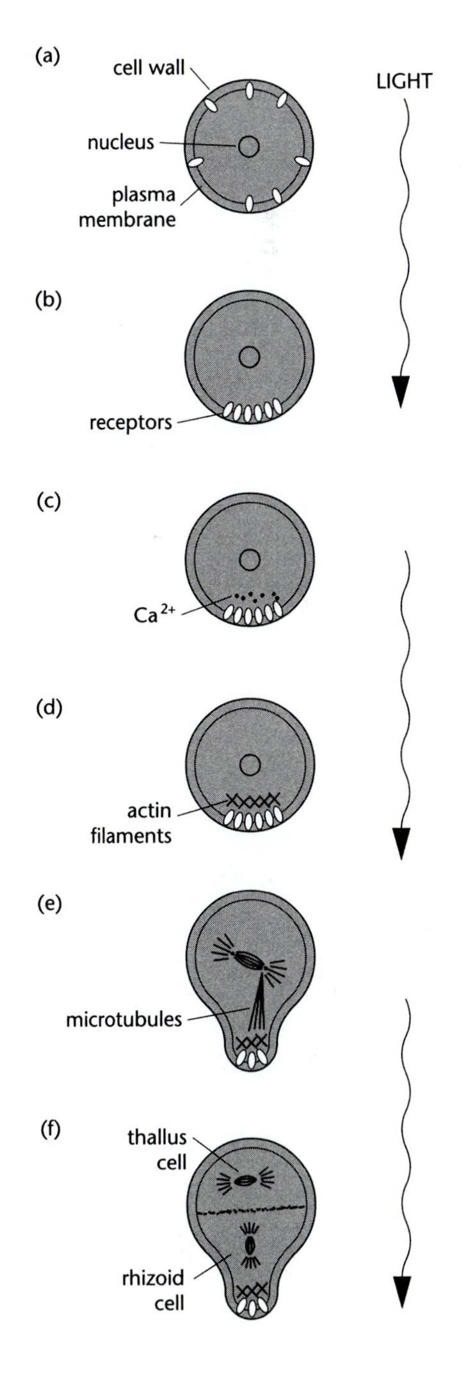


**Figure 19-6** Polarization of a *Fucus* egg by light. (a) The egg is originally spherically symmetric, with a centrally placed nucleus and uniformly distributed receptors and other molecules. (b) Exposure to light (in this case coming from the top of the page) causes specific receptors to move to the shaded quadrant of the cell. (c) A cluster of receptors triggers a local increase in concentration of $Ca^{2+}$ ions and (d) an accumulation of actin filaments and wall-synthesizing enzymes. (e) As the cell enters its first division, microtubules from one pole of the mitotic spindle attach to the patch of actin filaments and cause the spindle to rotate. (f) Division now occurs at right angles to the long axis of the cell and to the direction of light.

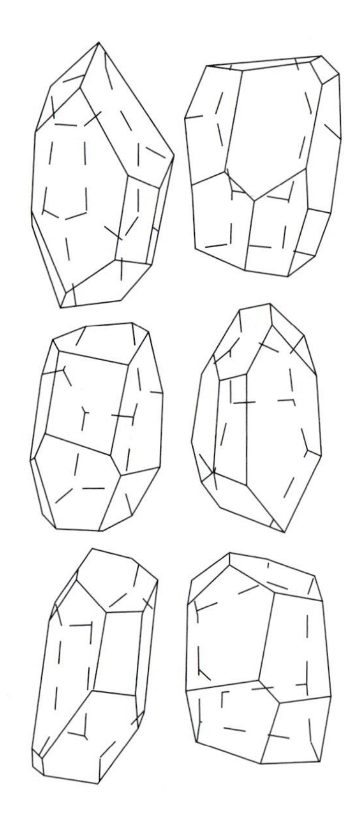

**Figure 19-7** Polyhedral cell shapes. In many plant and animal tissues, the cells are relatively uniform compact polyhedra with an average number of faces close to 14. These drawings were made of cells in the tuberous roots of *Asparagus*. (Modified from Hulbary, 1948.)

also be shaped by mechanical constraints. A typical cell from a vertebrate tissue is, in fact, roughly spherical when suspended in fluid (although its surface may be highly irregular and covered with small protrusions such as blebs and filopodia). If cells of this kind are tightly packed into a plant or animal tissue, they often take up geometrically simple polygonal forms similar to those shown by soap bubbles in foam. It seems likely that in such cases surface tension (or, its equivalent, cortical tension) has a major influence on shape.

Consider, for example, a simple two-dimensional sheet such as a sheet of plant epidermal cells or epithelial cells in tissue culture in which the cells have flattened polygonal shapes. Individual cells in such a sheet will vary in their numbers of faces, but the average will be almost exactly six as it is also for rafts of soap bubbles. Similarly, in three-dimensional tissues such as liver or adipose tissue, as well as in many undifferentiated embryonic plant or animal tissues, the cells are compact and polyhedral (Figure 19-7). No two cells have precisely the same shape but, as in the two-dimensional case, there is a *typical* form in which the average number of polyhedral faces per cell is close to 14. This rule is evidently dictated by geometrical considerations, since 14 is also the average number of faces seen in cell 'models' such as soap bubbles or compressed lead shot.

In theory, a specific 14-faced solid body, known as a Kelvin orthic tetrakaidecahedron (the Kelvin 14-hedron), made up of six 4-gons and eight 6-gons possesses the smallest surface area of any space-filling polyhedron. However, the differences in surface area (and hence energy) between this and related polyhedral forms are very small. Whenever the shapes of cells in three-dimensional tissues have been actually measured, they show a broad distribution in their numbers of polygonal faces (Table 19-1).

## Cell function follows form

The soft, deformable surface of vertebrate tissue cells allows them to take up a much wider variety of shapes than those of plants, fungi, and bacteria. Cells in the human body vary enormously in size and shape, often reflecting in their form the function they perform (Figure 19-8). A lymphocyte is tiny and irregular and equipped for rapid migration through tissues; cells in adipose tissue have bellies swollen with lipid; muscle cells are large and strong and grow thicker with exercise; cells in the ear have articulated forms that respond to vibration; nerve cells have highly branched dendrites and elongated axons to carry and integrate electrical signals.

In this connection, it is worth noting that, at least for some cells, shape itself influences biochemistry and even survival. This was demonstrated in an intriguing experiment in which endothelial cells obtained from bovine capillaries were grown on small adhesive islands of different sizes. The islands were coated with specific extracellular matrix (ECM) molecules and the cells readily took up the shape of the underlying pattern, whether square, round, or highly flattened and irregular. Significantly, cells forced to extend themselves over a wide area divided more often

**Table 19-1 Polygonal faces in three-dimensional tissues**

| polygon | 3 | 4 | 5 | 6 | 7 | 8 |
|---|---|---|---|---|---|---|
| bubbles | 0.000 | 0.105 | 0.669 | 0.221 | 0.004 | 0.000 |
| cells | 0.002 | 0.276 | 0.367 | 0.304 | 0.047 | 0.003 |

Numbers indicate the relative frequency of occurence of each polygon in a mass of either soap bubbles or cells. The average number of faces per bubble or cell is 14, so an average cell would have $0.367 \times 14 = 5.1$ pentagonal faces, and so on. (Based on Table 3 from Dormer, 1980.)

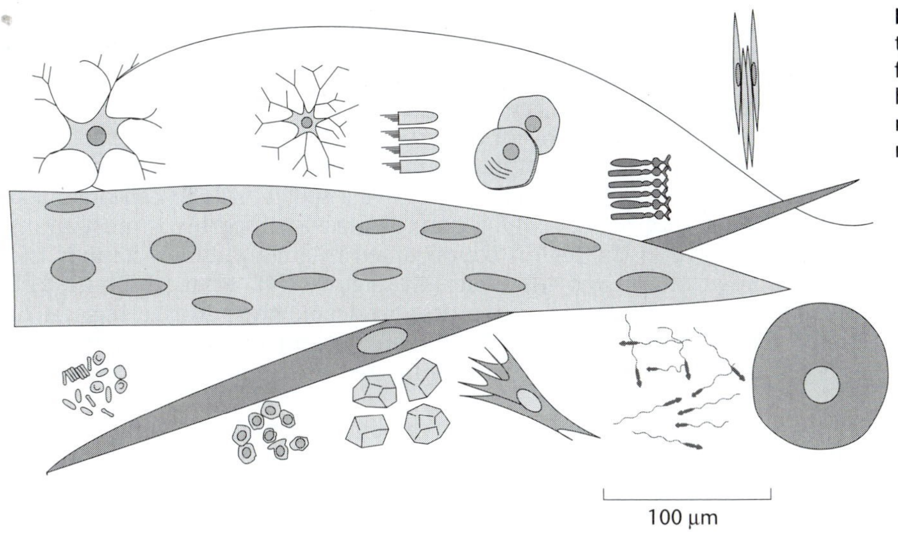


**Figure 19-8** A sample of cell shapes from the human body. Outlines include a fibroblast, red blood cells, rods and cones, hair cells, egg and sperm cells, smooth muscle, skeletal muscle, and a motor neuron—all drawn to the same scale.

than cells confined to a compact square or rounded shape. The spread-out cells also survived better and were less likely to initiate reactions leading to programmed cell death (apoptosis). The difference was not due simply to a greater area of contact with the ECM molecules but also depended on the actual shape. Thus if the surfaces were microfabricated so that different patterns were made from small dots with the same total area of ECM, the stretched cells still survived better. This makes sense in the context of tissue regeneration, since if an endothelial layer is damaged, then the remaining cells will be able to spread and this should stimulate them to grow and cover the wound. But *how* the shape of the cell regulates growth and survival is presently a mystery.

## *Localized contractions of the cortex provide a basis for changes in cell shape*

If the cortex of a cell contracts more strongly in one location than another then this will deform its contours. Indeed, as we saw in Chapter 7, this is precisely how large animal cells divide into two: following mitosis, a band of actin filaments and myosin II molecules forms around the waist of the cell, contracts, and divides the cytoplasm. It may be recalled also that the shape of crawling cells, with their irregular motile front and stubby tail, can be attributed to regional differences in the distribution of molecules such as myosin II in the actin cortex (Chapter 8).

Many types of epithelial cells have a local concentration of actin filaments in a ring situated close to their outward-facing, or 'apical' surface. The slow contraction of this apical layer deforms the cell and, if repeated over many neighboring cells, causes the epithelial sheet as a whole to roll up. We will see in the following chapter that coordinated shape changes of this kind contribute to the morphogenesis of the eye, the salivary glands, and the neural tube.

How are local differences in the actin cortex set up in the first place? One way is to impose them from the outside. Extracellular signals impinging on one region of the cell surface can induce a local restructuring of the actin cortex beneath the corresponding part of the plasma membrane. When amoeboid cells respond to gradients of chemoattractants (Chapter 8) and when tissue cells contact bacteria (Chapter 16), localized changes are produced in the cortex. In the following chapter we will see that many developmental changes in embryos are triggered by signals passing from one cell to another in a localized region of contact, often causing the recipient to become polarized.

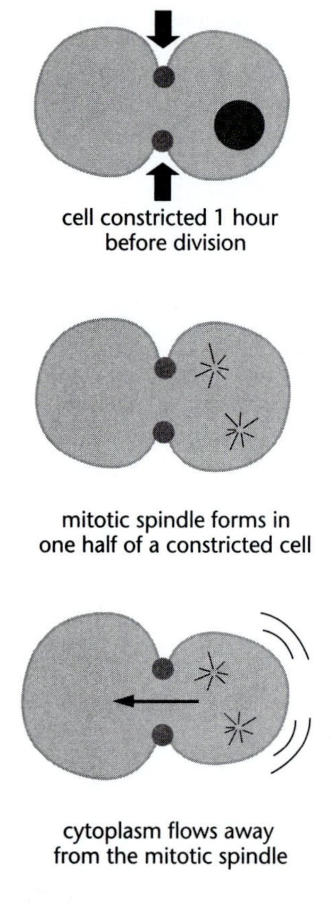


**Figure 19-9** Local contractions of the cell cortex produced by the mitotic spindle. Outline of a classic experiment performed by deforming sea urchin eggs.

## Microtubules influence the actin cortex

The other possible source of polarization is from within. Any cell with an oriented cytoskeleton in its cytosol could in principle cause an accumulation of actin filaments to occur at particular regions of the cortex. This is especially likely in the case of microtubules, which as we saw earlier in the book (especially in Chapter 11) often have a special role in cell organization. In a growing nerve cell, for example, the actin cortex is most abundant and active in the region *not* occupied by microtubules, namely the leading edge of the growth cone (see Figure 4-4). If microtubules are depolymerized with a drug such as nocodazole then motile regions of actin cortex sprout along the length of the previously quiescent axon, suggesting that they are normally inhibited in some way by the microtubules.

Another instance is the mitotic apparatus, which not only controls the movement of chromosomes but also specifies the position of the cleavage furrow. Experiments in which the mitotic spindle is pushed by a fine needle show that the position of the contractile ring of actin filaments and hence the cleavage plane of the cell is also displaced. The precise location of the plane of cleavage is fundamental to the many asymmetric cleavages that occur during development and has far-reaching consequences for the shaping of multicellular animals as well as for plants.

The mitotic apparatus appears to influence the tensile force exerted by the cell cortex. Measurements of sea urchin eggs show that the tension of the entire cell cortex rises as they enter M phase. A similar change in vertebrate fibroblasts might be responsible for their rounding-up, just before cytokinesis begins prior to division. An ingenious experiment in which the mitotic apparatus is squeezed into one half of a constricted cell shows that its presence causes a rise in cortical tension (Figure 19-9). Note, however, that the rise in cortical tension may not be uniformly distributed over the cell surface. Measurements reveal that cortical tension is high at the equator of a dividing cell and low in the polar regions. This difference could cause cortical components such as actin and myosin to move to the equatorial regions, where they could further contribute to the cortical tension.

## A marginal band of microtubules determines the shape of blood cells

Unlike mammalian erythrocytes, the erythrocytes of fish, birds, and amphibians retain both microtubules and a nucleus. In these cells a coiled loop of microtubules, like a watch spring, pushes outward against the plasma membrane and its associated spectrin-rich cortex. This *marginal band* causes the cell to flatten and enables it to spring back elastically after deformations caused by its passage in the bloodstream. The number of microtubules in the marginal band is difficult to count, since some may coil around the cell many times, but cross-sections show fewer than 10 to over 300 depending on the species.[3]

The contribution made by the marginal band to cell shape is seen in the development of chick erythrocytes. When they first form, chick erythrocytes are roughly spherical in shape but they become progressively flattened to a disc as the marginal band forms. Reversal of the process at this stage, by depolymerization of the microtubules in the band with nocodazole or colchicine, restores the spherical shape, thereby supporting the link between marginal band and flattened cell shape. As the cells mature, however, they become elliptical in profile and their shape no longer responds to microtubule depolymerization. Other components of the membrane cortex must therefore contribute to the form of the cell, having been organized by the ring of microtubules.

Figure 19-10 depicts how a marginal band may be formed and stabilized. At an early stage, microtubule cross-linking proteins attach

[3] Blood platelets also have a marginal bundle, in this case made of just one or two microtubules. Evidently these are responsible for the flattened discoid shape, since disruption of microtubules by cooling causes the marginal band to dissolve and the platelet to round up into a spherical form.

**Figure 19-10** Marginal bundle formation. Diagram based on the differentiation of chicken erythocytes. Images on the left show the flattened cell shapes (plan view); images on the right show the same cells as seen from the side. (a) In the precursor cell, microtubules emanate from a centrosome close to the nucleus. (b) As differentiation proceeds the microtubules grow at their plus ends and extend around the perimeter of the cell. Cross-linking proteins begin bundle formation while the centrosome disappears. (c) The marginal band is initially circular in profile and the cell is a circular disc in shape. (d) With later differentiation the marginal band becomes elliptical and the cell matures to an ellipsoidal form. Evidence suggests that actin filaments in the cortex are responsible for this last stage.

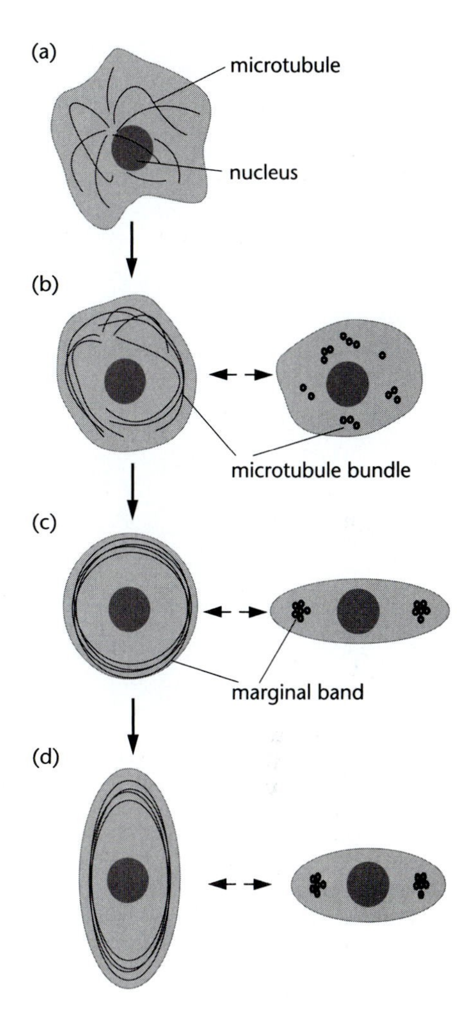


microtubules to their neighbors. Later in the differentiation process, actin filaments in the cortex align with the microtubule bundle and appear to 'mark' its location. Specific actin-binding proteins found in this region include proteins in the ERM family (ezrin, radixin, and moesin) discussed in Chapter 6. These membrane-linked proteins contribute to the stability of the marginal band, which becomes progressively more resistant to drug-induced depolymerization as the cell matures.

## *Systems of microtubules automatically find the cell center*

The array of microtubules radiating from the cell center is arguably the most fundamental determinant of eucaryotic cell shape. Whatever the final differentiated state of the cell, it usually begins, as in the chicken erythocyte just discussed, with a centrally positioned radial array of microtubules. Indeed, experiments show that an ability to find the center of a mass of cytoplasm is an inherent property of microtubule systems.

We described in Chapter 12 the large, flat pigment cells called melanophores, which can be isolated from fish scales and grown in culture. These cells contain many dark pigment granules, which are attached to microtubules and can either aggregate in the center of the cell or disperse throughout the cytoplasm, thereby allowing the fish to change its skin color. If one part of a melanophore is cut off with a needle, the cell fragment can survive for long periods even though it lacks a nucleus. The isolated fragment is evidently functional as well, since any pigment granules it contains are induced to aggregate by hormonal treatment. Interestingly, if hormone is applied immediately after surgery then the granules move toward the site of the cut. But if hormone treatment is delayed for four hours after the surgery, granules do not move to the cut site but instead move to the exact center of the cell fragment. Further investigation shows that this change results from a rearrangement of the microtubules within the fragment, so that their minus ends are now at the center of the fragment, just as they were at the center of the intact cell. In effect, the isolated cell fragment has become a mini-cell with respect to its microtubule organization (Figure 19-11).

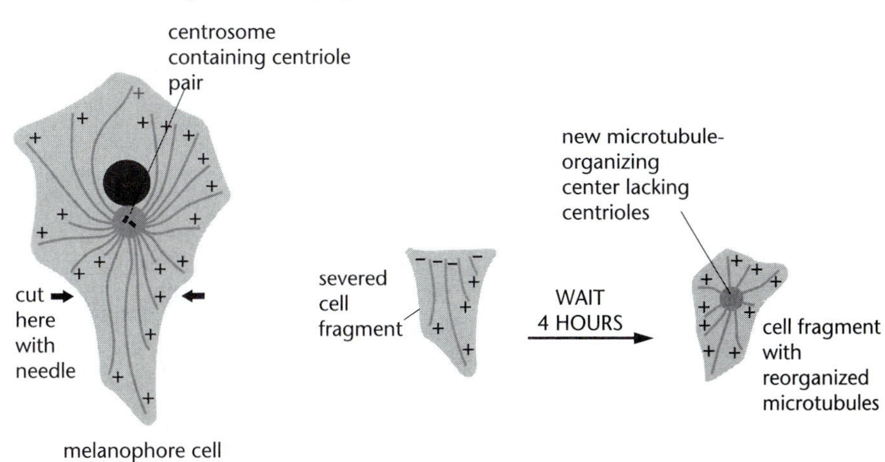


**Figure 19-11** An experiment showing that a microtubule array can find the center of a cell. After the arm of a fish pigment cell is cut off with a needle, the microtubules in the detached cell fragment reorient with their minus ends near the center of the fragment.

**Figure 19-12** Artificial asters. A mixture of unpolymerized tubulin, oligomeric kinesin, ATP, and GTP was warmed to 35°C under a microscope. The asterlike arrays of microtubules shown formed within 154 minutes. (Courtesy of Thomas Surrey and François Nédéléc.)

## *Microtubule arrays assemble spontaneously*

It is not known precisely how a new 'microtubule-organizing center' forms in a cell fragment, but it can be shown experimentally that even artificial noncellular systems of microtubules find the center of an enclosed space. Under the right conditions, a mixture of tubulin molecules and motor molecules put into an artificial chamber produce a radial array not unlike the arrays of microtubules (asters) emanating from mitotic spindles seen in living cells. The motors link together pairs of microtubules and as each set of heads moves toward its respective plus end, it causes the microtubules to telescope out from each other.[4] The combined action of many microtubules and many motors creates a center-seeking radial array (Figure 19-12).

A better spindle can be made by adding artificial chromosomes to the mixture. If polystyrene beads coated with plasmid DNA are incubated in extracts of frog egg cytoplasm together with fluorescent tubulin, the system spontaneously develops a bipolar spindle. Even though no centrosomes are present, the DNA-coated beads cluster onto a midplane analogous to the metaphase plate, while microtubules radiating out in either direction collect at foci resembling the spindle poles.

This self-assembly process is driven both by microtubule polymerization and by microtubule motors (Figure 19-13). Numerous motor proteins are present in the frog extract and, when their action is probed by antisera, specific defects appear in the assembly process. One kinesin, similar in structure to the conventional kinesin described in Chapter 12, seems to be needed to maintain the connection between chromatin and microtubules. A second kinesin, called Eg5, has a tetrameric, bipolar structure and its activity promotes antiparallel bundling of microtubules in the region of overlap of a spindle. Dynein and a minus-end directed kinesin together draw the microtubules into a polelike center.

The picture that emerges from these experiments is that in a living cell spindle formation is orchestrated by a suite of microtubule motors. How their individual activities are coordinated is a fascinating topic for future investigations.

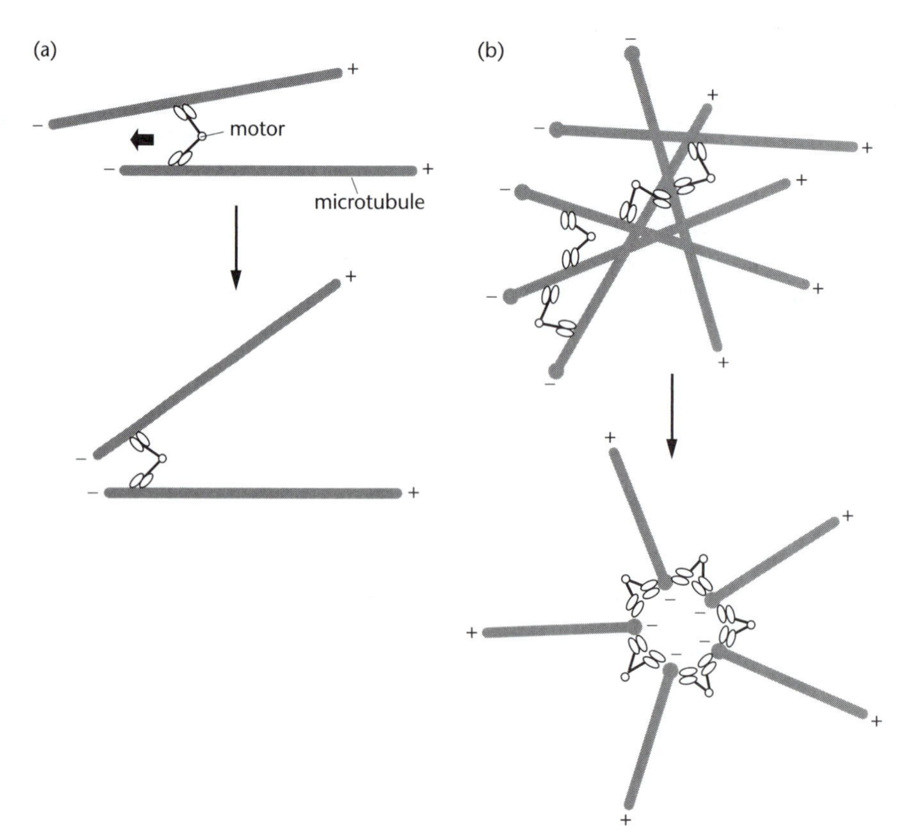


**Figure 19-13** How microtubules and motors could in principle build a spindle. (a) Multiheaded motor molecules can cross-link adjacent microtubules and move toward their ends, where they accumulate. (b) Microtubules mixed with multiple motor proteins will produce an aster of microtubules with their ends at the newly created pole. Note that in this case each microtubule is depicted with a hypothetical capping protein at its minus end so as to prevent motors from falling off when they arrive.

[4] In the original experiment the motors used were chemically linked tetramers of conventional kinesin and the microtubules congregated with their plus ends at the center. Arrays having the same polarity as those in cells (minus end inward) can, however, be made using dynein as the motor molecules.

## *Actin filaments and microtubules act together to polarize the cytoskeleton*

We see that systems of microtubules and their associated motors have a powerful capacity to self-organize. Indeed, in plants and some animal cells, microtubule arrays are produced in the absence of well-defined centrosomes, and the same is even seen in mammalian cells in which the centrosome has been artificially removed. Why then do centrosomes exist? The answer appears to be that centrosomes and other types of microtubule-organizing centers have evolved because they organize microtubules with greater efficiency. When present, they dominate the microtubule array and hence control many other parts of the cell, including the internal membranes and the actin cortex.

In a living cell, the three major types of cytoskeletal filaments are connected to one another and their functions are coordinated. The distribution of intermediate filaments in an epithelial cell in culture, for example, is radically altered if the microtubules are depolymerized by drug treatment: the intermediate filaments, which are normally arrayed throughout the cytoplasm, pull back to a region close to the nucleus. There are also many situations in which microtubules and actin filaments act in a coordinated way to polarize the whole cell—as in the growth of nerve cells or in mitosis.

Another dramatic example of an interaction between microtubules and the actin cortex is seen in the killing of specific target cells by cytotoxic T lymphocytes, an important event of a vertebrate's immune response to infection. When receptors on the surface of the T cell recognize a foreign antigen on the surface of a target cell, the receptors relay signals that alter the underlying cortex in several ways. First, proteins associated with actin filaments in the T cell reorganize under the zone of contact between the two cells. Then the centrosome reorients, moving with its microtubules to the zone of T-cell-target contact. The microtubules, in turn, position the Golgi apparatus right under the contact zone, focusing the killing machinery on the target cell (Figure 19-14).

We can perceive in this example a general strategy for polarization. First, the plasma membrane senses some difference on one side of the cell that generates a transmembrane signal. The actin cortex then reorganizes, producing an accumulation of specific proteins at that region of the cortex closest to the external stimulus. Microtubules then contact this patch of specific proteins, perhaps through a random search due to dynamic instability, and attach to them via specific protein interactions. Shortening of the microtubules, perhaps in conjunction with specific motor proteins, then moves the centrosome to that part of the cell. Subsequent changes, in which the centrosome controls the position of other organelles, then result in a cell with a strong directional focus.

We will see in the next chapter that a similar sequence of events, occurring during embryonic development, allows mitotic spindles to be oriented by external signals. Cleavage then occurs in a plane defined by neighboring cells, allowing the shape of the eventual organism to be closely controlled.

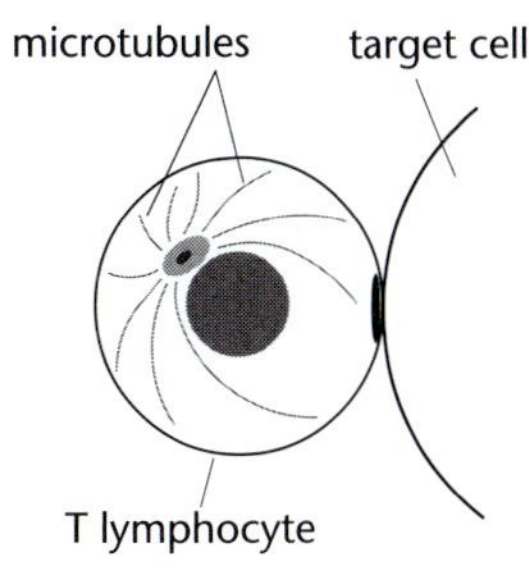

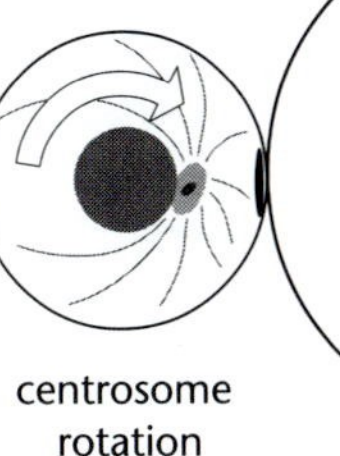

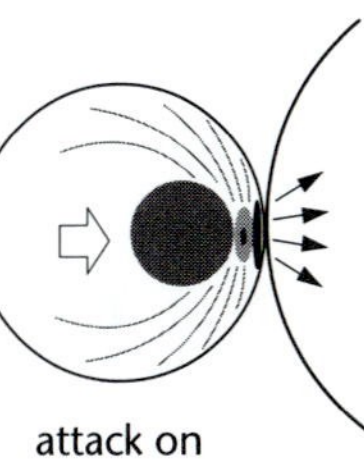


**Figure 19-14** Coordinated changes between actin and microtubules in a T lymphocyte. Contact with a target cell causes the cytoskeleton of a cytotoxic T cell to polarize and thereby direct streams of vesicles to the contact site.

## *An intricate cytoskeletal framework transmits vibrations in the ear*

It takes time to build an elaborate structure, and some of the most complex cell shapes require many days to reach their mature form. During this period, multiple episodes of selection and rearrangement of the cytoskeleton may take place, precisely tuning the shape of the cell to the function it has to perform.

The helical strip of epithelial tissue in the cochlea, known as the *organ of Corti*, detects sound vibrations entering the ear and converts them into electrical signals to be sent to the brain. The primary detectors in this process are the hair cells introduced in Chapter 6, which carry multiple rigid stereocilia on their surfaces. Sensory hair cells are arranged in rows parallel to the helical axis of the epithelial strip, held in position by a complex architecture of supporting cells. Each supporting cell has its base on a specialized basement membrane and forms a long pillarlike process that contacts hair cells at the apical surface of the epithelium. Lateral beams extend at the apices of supporting cells and substantial arrays of adherens junctions reinforce connections to hair cells. One type of supporting cell also holds the bases of hair cells in a concavity, like an egg in a cup (Figure 19-15).

The entire tissue is shaped by an intricate cytoskeletal network that fills each individual cell and makes specific contact between one cell and the next. Actin filaments forming the core of the stereocilia are linked to a highly developed fibrous meshwork of actin at the apical cortices of both the hair cells and their supporting cells. Huge bundles of microtubules, some containing several thousand microtubules, extend within the pillars and angled beams of the supporting cells. The ends of these microtubule bundles insert into compact fibrous meshworks composed of nonmuscle isoforms of actin and several types of keratins (Chapter 17). This complex organization is thought to optimize the efficiency of sound detection. During its normal function, the entire strip of epithelial tissue vibrates rapidly and the micromechanical properties conferred by its cytoskeleton will have a major influence on the sensitivity and range of response.

How the cytoskeleton of this tissue is constructed is largely unknown, but there are clues, mainly from electron microscopy. A supporting cell's microtubule bundles originally grow from a centrosome at the cell apex. This centrosome migrates, together with its pair of centrioles, into the angled beam after an additional cell-surface-associated microtubule-organizing center has been created at the top of the pillar. At the other (basal) end, microtubules appear to be captured and stabilized by a cortical meshwork of actin, and some microtubules may actually detach from the apical end and migrate down to form the bundles of pillar microtubules. The extent and detailed ultrastructure of an actin meshwork and

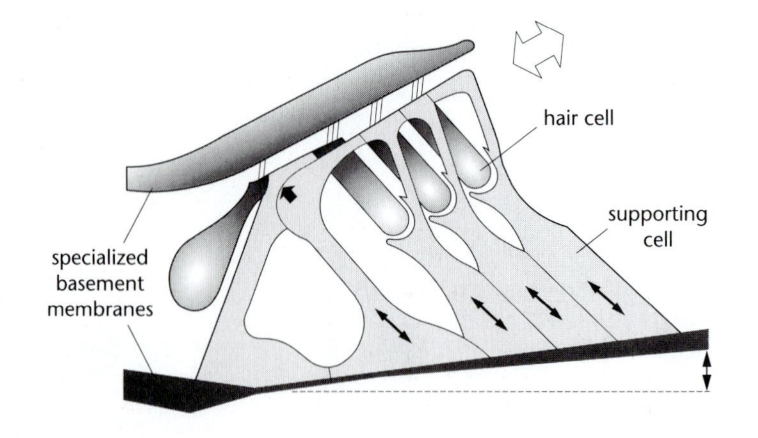


**Figure 19-15** Cells in the mouse organ of Corti. The thin sheet of epithelial tissue, shown here in diagrammatic cross-section, rests on a specialized basement membrane called a basilar membrane. As sound enters the ear it causes the basilar membrane to vibrate up and down. Stereocilia on the surfaces of hair cells deflected by this motion generate electrical signals that are relayed to the brain. Supporting cells of three kinds are shown. Their function is to hold the hair cells in the correct position and, with an appropriate stiffness, to faithfully transmit vibrations from the basilar membrane. (Based on Mogensen et al, 1998.)

its associated adherens junctions at the cell apex are regulated by the adjacent cell and if this is absent (as in certain deaf mouse mutants) the extent of the cortical meshwork increases dramatically.

## *The polarized secretion of extracellular matrix is a major source of cell asymmetry*

Although animal cells lack a rigid cell wall, they often secrete a tough 'exoskeleton' that can have a major influence on the form of the cell. In the majority of animal tissues, an *extracellular matrix* composed of tough fibrous proteins and polysaccharides occupies the spaces between cells. It is especially abundant in connective tissues, where it forms the basis of bone and cartilage, but endothelia, epithelia, nerve and muscle all rest on, or are enmeshed in, sheets of connective tissue.

The extracellular matrix influences cell morphology partly through simple physical forces. Even a planar plastic dish causes spherical cells to flatten, whereas cells growing on glass fibers become elongated in form. Similarly, fibroblasts enmeshed in a matrix of collagen fibrils tend to align with the individual fibrils, and if the matrix as a whole becomes polarized by mechanical stress then so too are the cells it includes. The interesting twist here is that in the normal development of an embryo, both collagen fibrils and the mechanical forces that distort them are produced by the action of the fibroblasts themselves. A runaway cooperative interaction between the cells and the matrix can create a large polarized structure such as tendon or bone.

Another aspect is chemical. The extracellular matrix, acting through specific transmembrane signaling mechanisms, has a major influence on the course of differentiation of most animal cells, the differentiated phenotype usually carrying with it a characteristic morphology. This influence is also seen clearly during the regeneration of damaged tissue, which is guided by any residual extracellular matrix. For example, muscle fibers that have been destroyed regain their original form, and even acquire a new neuromuscular junction at the original position on their surface, provided the basement lamina that surrounds them has remained intact.

One of the clearest examples of a signaling response between the extracellular matrix and the cytoskeleton is seen in the phenomenon of anchorage dependence of growth. Most cell types in the vertebrate body divide only when they are attached to an appropriate surface: this is presumably the way in which they prevent unrestrained proliferation within tissues. Thus most cells dissociated from a tissue are unable to divide or even to synthesize proteins when maintained in suspension: they require a culture dish coated with suitable extracellular matrix molecules before they can flatten out and commence growth. This effect depends, as we saw in Chapter 6, on transmembrane receptors such as integrins that bind selectively to specific components of the extracellular matrix on the outside of the cell, and activate cascades of biochemical signals within.

## *Outstanding Questions*

*How do external signals, whether chemical or physical, alter the distribution of microtubules inside a plant cell? How are local differences in the* Fucus *actin cortex set up in response to light? Which microtubule-binding proteins anchor them in place? What leads to the formation of the microtubule ring in a platelet? What is the 'relaxing' influence that spreads from astral microtubules to the cell cortex? When an axon grows out of a cell, what determines its initial position? Why do spread (flattened) cells survive better than rounded cells? What tells a cell that it has a specific shape? Which molecular motors maneuver microtubules to the center of a cell? How are the elaborate forms of cells in a mammalian cochlea tuned to their precise function?*

## *Essential Concepts*

- Bacteria such as *E. coli* have rigid cell walls made of cross-linked peptide glycan that are designed to withstand very high internal pressures generated by osmosis.
- Plant cells have thick, multilayered walls composed of cellulose microfibrils and associated polysaccharides and formed into compact, boxlike shapes.
- Plant morphology is determined by oriented division and expansion of its cell walls, both controlled by arrays of microtubules on the inner surface of their plasma membrane.
- Yeasts have cell walls made of polysaccharide and change their size and shape by carefully controlling enzymes that degrade and manufacture the wall.
- Budding yeasts produce local expansion of the wall at sites marked by local accumulation of septins in the actin cortex. After a bud has grown to a sufficient size, the ring of septins is also the site of septum formation leading to cell division.
- In fission yeasts the sites of wall expansion during growth are distinct from those of cell division. Mutant studies implicate both actin filaments and microtubules in the determination of polarized cell form.
- Fertilized eggs of the seaweed *Fucus* are initially symmetrical and then become polarized in response to light. Their response depends on the actin-based accumulation of wall-synthesizing enzymes in the shaded quadrant.
- Cell shape can be influenced by nongenetic, mechanical, or steric factors imposed by the surroundings. Closepacking in two or three dimensions often leads to compact polyhedral forms governed by simple geometrical rules.
- Cells in vertebrate tissues are highly diverse in their size and shape, often reflecting their particular function in the body. Cell shape itself can have a determining influence on function, as can be seen by forcing cells to adopt specific shapes in tissue culture.
- The soft and deformable surface of an animal cell is shaped by its underlying actin cortex. Regional differences in the composition and contractility of the cortex result in local constructions and expansions of the cell at that location.
- Local changes to the actin cortex are produced in response to external signals. They can also arise as a result of organizational changes in the cell, especially in their microtubules.
- It can be shown in cell-free extracts as well as in defined experiments in living cells that systems of microtubules and motors have a capacity to self-organize. When chromatin is also present they can produce bipolar spindlelike arrays.
- Once an array of microtubules has been produced, it can influence the other parts of the cell. Membrane structures are dependent upon continual shuttling of vesicles to and from the plasma membrane.
- Complex interactions between microtubules and the actin cortex occur in cells throughout the vertebrate body. These are evident, for example, in the shaping of blood platelets and nonmammalian red blood cells.
- Large cells with ornate and precise shapes, such as the cells of the mammalian cochlea, are formed by multiple episodes of selection and rearrangement of the cytoskeleton.

## *Further Reading*


Ayscough, K.R., Drubin, D.G. Actin: general principles from studies in yeast. *Annu. Rev. Cell Biol.* 12: 129–160, 1996.

Bar-Ziv, R., et al. Pearling in cells: a clue to understanding cell shape. *Proc. Natl. Acad. Sci. USA* 96: 10140–10145, 1999.

Cabib, E., et al. Role of small G proteins in yeast cell polarization and wall biosynthesis. *Annu. Rev. Biochem.* 67: 307–333, 1998.

Cayley, D.S., et al. Biophysical characterization of changes in amounts and activity of *Escherichia coli* cell and compartment water and turgor pressure in response to osmotic stress. *Biophys. J.* 78: 1748–1764, 2000.

Chen, C.S., et al. Geometric control of cell life and death. *Science* 276: 1425–1428, 1997.

Chicurel, M.E., et al. Integrin binding and mechanical tension induce movement of mRNA and ribosomes to focal adhesions. *Nature* 392: 730–733, 1998.

Clegg, J. Metabolic consequences of the extent and disposition of the aqueous intracellular environment. *J. Exp. Zool.* 215: 303–313, 1981.

Cohen, W.D., et al. Elliptical versus circular erythrocyte marginal bands: isolation, shape conversion, and mechanical properties. *Cell Motil. Cytoskeleton* 40: 238–248, 1998.

Cyr, R.J. Microtubules in plant morphogenesis: role of the cortical array. *Annu. Rev. Cell Biol.* 10: 153–180, 1994.

DeMarini, D.J., et al. A septin-based hierarchy of proteins required for localized deposition of chitin in the *Saccharomyces cerevisiae* cell wall. *J. Cell Biol.* 139: 75–93, 1997.

Dormer, K.J. Tissue Geometry for Biologists. Cambridge, UK: Cambridge University Press, 1980.

Fowler, J.E., Quatrano, R.S. Plant cell morphogenesis: plasma membrane interactions with the cytoskeleton and cell wall. *Annu. Rev. Cell Dev. Biol.* 13: 697–743, 1997.

Harold, F.M. From morphogenes to morphogenesis. *Microbiology* 141: 2765–2778, 1995.

Heald, R., et al. Self-organization of microtubules into bipolar spindles around artificial chromosomes in *Xenopus* egg extracts. *Nature* 382: 420–425, 1996.

Henderson, C.G., et al. Three microtubule-organizing centres collaborate in a mouse cochlear epithelial cell during supracellularly coordinated control of microtubule positioning. *J. Cell Sci.* 108: 37–50, 1995.

Holy, T.E., et al. Assembly and positioning of microtubule asters in microfabricated chambers. *Proc. Natl. Acad. Sci. USA* 94: 6228–6231, 1997.

Hulbary, R.L. Three-dimensional cell shape in the tuberous roots of *Asparagus* and in the leaf of *Rheo. Am. J. Bot.* 35: 558–566, 1948.

Janmey, P.A. The cytoskeleton and cell signaling: component localization and mechanical coupling. *Physiol. Rev.* 78: 763–781, 1998.

Koch, A.L. Bacterial Growth and Form. New York: Chapman and Hall, Inc., 1995.

Korinek, W.S., et al. Molecular linkage underlying microtubule orientation toward cortical sites in yeast. *Science* 287: 2257–2259, 2000.

Louvard, D., et al. The differentiating intestinal epithelial cell: establishment and maintenance of functions through interactions between cellular structures. *Annu. Rev. Cell Biol.* 8: 157–195, 1992.

Lowin-Kropf, B., et al. Cytoskeletal polarization of T cells is regulated by an immunoreceptor tyrosine-based activation motif-dependent mechanism. *J. Cell Biol.* 140: 861–871, 1998.

Madden, K., Snyder, M. Cell polarity and morphogenesis in budding yeast. *Annu. Rev. Microbiol.* 52: 687–744, 1998.

Mata, J., Nurse, P. Discovering the poles in yeast. *Trends Cell Biol.* 8: 163–167, 1998.

Mogensen, M.M., et al. Keratin filament deployment and cytoskeletal networking in a sensory epithelium that vibrates during hearing. *Cell Motil. Cytoskeleton* 41: 138–153, 1998.

Murphy, D.B., et al. Immunofluorescence examination of β-tubulin expression and marginal band formation in developing chicken erythrocytes. *J. Cell Biol.* 102: 628–635, 1986.

Nanninga, N. Morphogenesis of *Escherichia coli*. *Microbiol. Mol. Biol. Rev.* 62: 110–129, 1998.

Nédélec, F.J., et al. Self-organization of microtubules and motors. *Nature* 389: 305–308, 1997.

Pruyne, D. Bretscher, A. Polarization of cell growth in yeast. *J. Cell Sci.* 113: 571–585, 2000.

Robinson, K.R., et al. Symmetry breaking in the zygotes of the fucoid algae: controversies annd recent progress. *Curr. Top. Dev. Biol.* 44: 101–125, 1999.

Sawin, K.E., Nurse, P. Regulation of cell polarity by microtubules in fission yeast. *J. Cell Biol.* 142: 457–471, 1998.

Thompson, D.W. On Growth and Form. Cambridge: Cambridge University Press, 1963.

Tilney, L.G., Tilney, M.S. The cytoskeleton of protozoan parasites. *Curr. Opin. Cell Biol.* 8: 43–48, 1996.

Walczak, C.E., et al. A model for the proposed roles of different microtubule-based motor proteins in establishing spindle bipolarity. *Curr. Biol.* 8: 903–913, 1998.

Waterman-Storer, C., et al. Microtubule growth activates Rac1 to promote lamellipodial protrusion in fibroblasts. *Nat. Cell Biol.* 1: 45–50, 1999.

Watson, P.A. Function follows form: generation of intracellular signals by cell deformation. *FASEB J.* 5: 2013–2019, 1991.

Welch, M.D., et al. The yeast actin cytoskeleton. *Curr. Opin. Cell Biol.* 6: 110–119, 1994.

Winter, D., et al. The complex containing actin-related proteins Arp2 and Arp3 is required for the motility and integrity of yeast actin patches. *Curr. Biol.* 7: 519–529, 1997.

# Cell Movements In Embryos

This book is largely concerned with the movements of single cells and their molecular basis; but there is a contiguous but higher level of movement in which cells are themselves mechanistic elements. The movements and deformations of tissues, especially during the form-shaping episodes of embryonic development, arise from the cooperative behavior of large numbers of cells. These movements are ultimately controlled by genetic mechanisms—by the executive expression of genes in response to temporal or spatial cues—but the physical transformation of genetic blueprints into eyes, legs, and wings takes place on the factory floor of the cytoskeleton. Here an assembly line of simple cell activities based on actin filaments and microtubules gives rise to the seemingly endless diversity of animal forms.

Which cell movements underlie embryonic development? We will see in this chapter that, even before fertilization, many eggs acquire a regionally differentiated cytoplasm due to the targeted transport and transcription of mRNA molecules. Differences set up during egg formation and subsequent fertilization then take effect as the egg cytoplasm is parceled into separate cells by precisely oriented cell cleavages. Later in development, sheets of cells linked to each other through specific molecules (which extend across the plasma membrane and attach to the cytoskeletons) stretch, invaginate, and roll up into tubes to create the body plan of the future adult. Finally, and most especially in vertebrate embryos, cells migrate extensively in intermingled patterns, guided by chemical signals and selective affinities. From this intricately choreographed ballet, with a cast of billions of cells, eventually emerge organs such as the mammalian brain—the most complex biological structure known.

## The egg cytoplasm carries positional information

The cellular differences that emerge during embryonic development can frequently be traced back to regional inhomogeneities in the egg cytoplasm. In most nonmammalian species, the egg after fertilization has localized accumulations of specific proteins and mRNAs. Cleavages along precisely defined embryonic axes then segregate these components and create cells with intrinsic differences in composition. Abundant evidence shows that the controlled positioning of proteins and mRNAs is due to the cytoskeleton, arising either during egg formation or following fertilization.

Even before fertilization, the egg of the fruit fly *Drosophila* has a clearly defined polarity. If an egg is carefully punctured at its anterior end, allowing a small amount of cytoplasm to leak out, the embryo fails to develop head structures. Moreover, if cytoplasm from the posterior end of another egg is injected into the site from which the anterior cytoplasm has leaked, a second set of abdominal segments will develop, with reversed polarity, in the anterior half of the recipient egg. Key elements in this polarity determination are the products of polarity-determining genes, which become localized to specific regions of the oocyte. These include the *bicoid* mRNA, located at the anterior end of the egg, and the *oskar* and *nanos* mRNAs found at the posterior end (Figure 20-1). In another example, the *Vg1* mRNA, which encodes a growth factor thought to be involved in dorsal mesoderm induction, is localized to the vegetal pole of growing *Xenopus* oocytes.

## mRNA molecules can be precisely positioned in the cell

As these examples illustrate, positional information is commonly specified by localized mRNA molecules. Maternally derived mRNA molecules often undergo distinctive movements as development proceeds, before accumulating at one pole or the other. In cells undergoing asymmetric mitosis, such as *Drosophila* neuroblasts, specific mRNA molecules have been shown to segregate to one of the two daughter cells of a stem cell division. There are also terminally differentiated cells, such as oligodendrocytes (cells making myelin), in which mRNA localization provides a means to generate asymmetry. We saw previously that the synthesis of β-actin occurs the leading edge of a fibroblast and that synthesis of myosin II in a developing muscle cell takes place actually on the myofibrils.

Selective sorting of mRNA molecules in a cell implies that they must carry an address. Indeed, it has been shown that a number of targeted mRNAs contain specific targeting regions of sequence ('zip codes') in their 3′ untranslated regions (*UTRs*). These sequences appear to be unique to each RNA and experiments show them to be both necessary and sufficient for localization.

For zip codes to work, however, something in the cytoplasm must recognize them and deliver them to the correct location. Indeed, a growing number of proteins have been found by genetic techniques to be necessary for mRNA localization. In fibroblasts, a zipcode-binding protein (ZBP-1) binds to short RNA sequences required for localization of β-actin synthesis; a related protein binds to Vg1 in *Xenopus* oocytes. In *Drosophila*, mutations in a protein called *Staufen*, with binding sites for double-stranded RNA, disrupt the polarized distribution of several mRNAs. Intriguingly, Staufen appears to be used to target several different mRNA molecules, each to a different site.

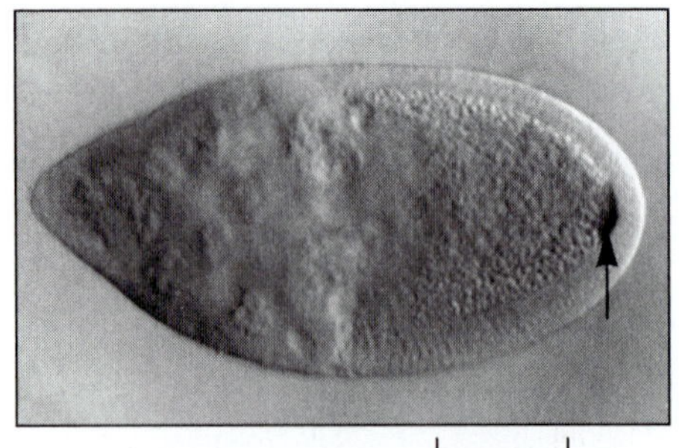


**Figure 20-1** Polar location of RNA. A stage-10 *Drosophila* oocyte has been stained for *oskar* RNA using *in situ* hybrization. The RNA is highly localized (arrow), close to the posterior pole of the oocyte. (Courtesy of Cheryl VanBuskirk.)

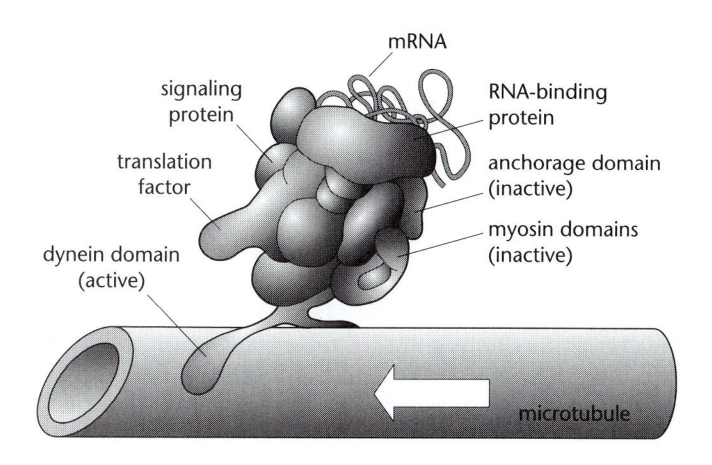


**Figure 20-2** RNA transport particle. A speculative picture showing some of the components thought to participate in the transport and eventual localization of mRNA molecules to specific locations in a eucaryotic cell.

## *Particles containing RNA and protein associate with the cytoskeleton*

We now have to ask how the mRNA/RNA-binding protein combination is actually positioned in the cytoplasm. The answer, perhaps unsurprisingly, is that it is due to the cytoskeleton.

Biochemical fractionation studies and *in situ* hybridization studies show that mRNA molecules interact with both actin filaments and microtubules. Drugs such as cytochalasin and nocadozole often inhibit the development of polarity in cells; so too does the disruption of cytoskeletal function by mutation (where it is possible to do such experiments). Unexpectedly, it seems from such experiments that actin-based and microtubule-based mechanisms are both employed—even in some instances by the same molecules. The mRNA for actin, for example, is carried in neurons in association with microtubules, whereas in a fibroblast it is attached to actin filaments. The *Drosophila* RNA-binding protein Staufen mentioned above plays a role in both the microtubule-dependent localization of mRNA in oocytes and in the actin-dependent localization of RNA in neuroblasts.

We do not understand the full significance of these observations, but they seem to point to an RNA transport mediated by multimolecular complexes, or particles (Figure 20-2). These particles contain the mRNA to be localized, together with proteins that bind specifically to that mRNA and one or more motor proteins that carry the particle along an actin filament or microtubule. The same particle may also, conceivably, carry components needed for the eventual translation of the mRNA once it arrives at a suitable location. Finally, the operation of such complicated structures in the milieu of the cytoplasm will also undoubtedly require regulation by signaling pathways, about which we presently know almost nothing.

## *The insect egg is compartmentalized by the cortical cytoskeleton*

The first nine divisions of a *Drosophila* egg, each of which takes about 8 minutes, create a cluster of nuclei in a common cytoplasm. Most nuclei then migrate to the surface to form a monolayer, by a process that depends on microtubules and cytoplasmic dynein. After another four rounds of nuclear division, plasma membranes pinch inward from the egg surface to enclose each nucleus, thereby converting the syncytial blastoderm into a cellular blastoderm comprising some 5000 separate cells (Figure 20-3). During these very rapid cycles of DNA replication there is little synthesis of RNA or protein, and development depends largely on stocks that accumulated in the egg before fertilization.

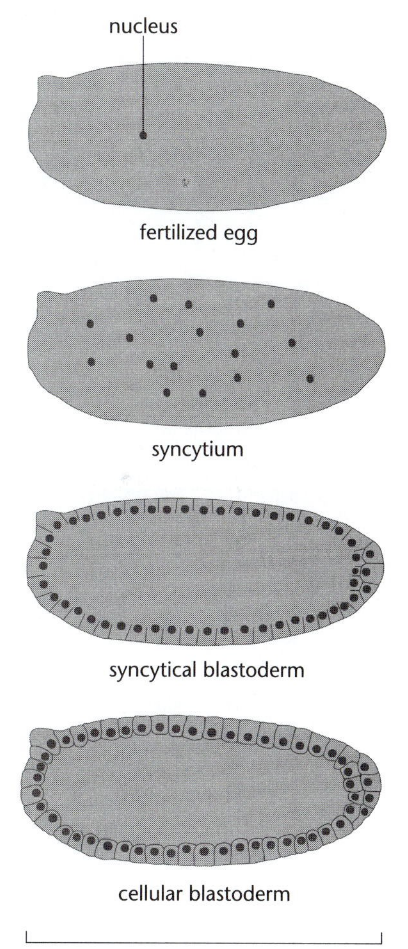


**Figure 20-3** Early development of the *Drosophila* egg. Following fertilization the egg undergoes a rapid succession of nuclear divisions without major increase in mass or in content of protein or mRNA molecules. The nuclei migrate to the periphery of the egg, where cell boundaries start to form. These are initially delineated by the cortical cytoskeleton and then by invaginations of the plasma membrane.

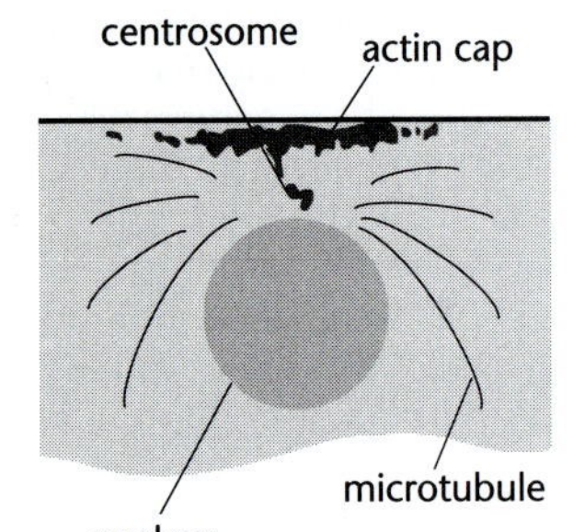


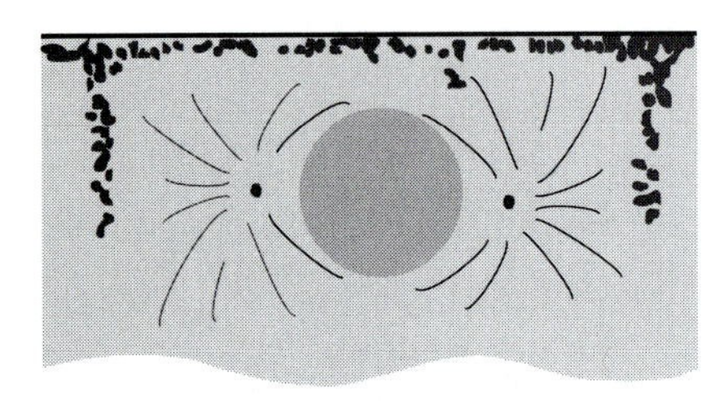


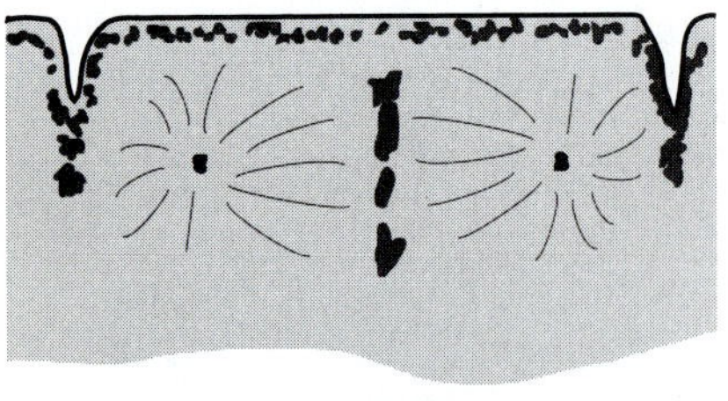


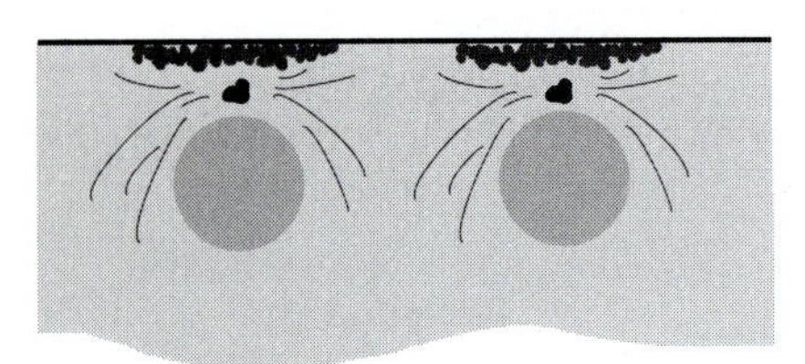


**Figure 20-4** Division cycles in the syncytial blastoderm of *Drosophila.* The arrangement of actin and microtubule-containing structures is typical of that observed at nuclear cycle 11 or 12. Membrane furrows recur in successive cycles until cellularization occurs at cycle 14.

At fertilization, the egg cortex is a layer 1–2 μm deep of actin and microtubules. As the nuclei migrate to the surface, they bring with them their centrosomes; these then nucleate microtubules and modulate the cortical actin and associated proteins, producing a local cytoskeletal environment around each nucleus.

In the interphase of each nucleus, actin filaments are concentrated in small patches just below the plasma membrane, each patch apparently holding one nucleus in position. Microtubule arrays emanate from a centrosome on the membrane side to surround each nucleus. As mitosis begins, interphase microtubules are replaced by a mitotic spindle and the actin filaments spread outward and down into the cytoplasm between adjacent nuclei. Actin walls form, leading (or pulling) transient membrane furrows in a manner reminiscent of cytokinesis (Figure 20-4).

Many different actin-binding proteins from *Drosophila* eggs have been localized by antibody staining to the actin furrows or actin caps. These proteins presumably organize actin filaments, under the coordinating influence of the centrosomes, so that they create the local nuclear compartments. The autonomous action of cortical compartmentalization is underscored by the fact that the entire sequence of events will take place in the absence of DNA synthesis. If DNA replication is inhibited by the drug aphidicolin, a specific inhibitor of DNA polymerase, the nuclei fail to divide and yet several cycles of centrosome replication and subdivision of the cortex nevertheless take place.

## *Rotation of the cortex in amphibian eggs determines the axis of the future embryo*

An amphibian egg is a large cell, about 1 mm in diameter, most of the volume of which is occupied by yolk platelets (aggregates of lipid and protein). The platelets are concentrated toward the lower end of the cell, called the *vegetal pole,* while the other end, called the *animal pole,* contains the nucleus and melanin-containing granules. Before fertilization, animal and vegetal poles of the egg contain different mRNA molecules as well as different amounts of yolk and other organelles. However, the egg is symmetrical about the animal–vegetal axis, a condition that might be expected to generate a cylindrical embryo with a head at one end and a tail at the other but no front or back.

Fertilization causes a rearrangement of the egg contents and creates a second axis of asymmetry that distinguishes the belly of the embryo (its ventral side) from its back (dorsal side). This symmetry-breaking occurs through rotation of the egg cortex relative to the core of deeper cytoplasm (Figure 20-5).

The direction of rotation is determined by the point of entry of the sperm, at a random place in the animal half. Rotation carries the cortex 30° relative to the deeper cytoplasm; the new associations set up between cortical and deeper cytoplasm then providing cues for further development. In many amphibians, the shift of cortex displaces pigment granules in the egg so that a band of pale pigmentation, known as the *gray crescent,* appears opposite the sperm entry point. The gray crescent corresponds roughly to the back and dorsal structures, whereas the sperm entry point corresponds roughly to the belly (and other ventral aspects) of the animal.

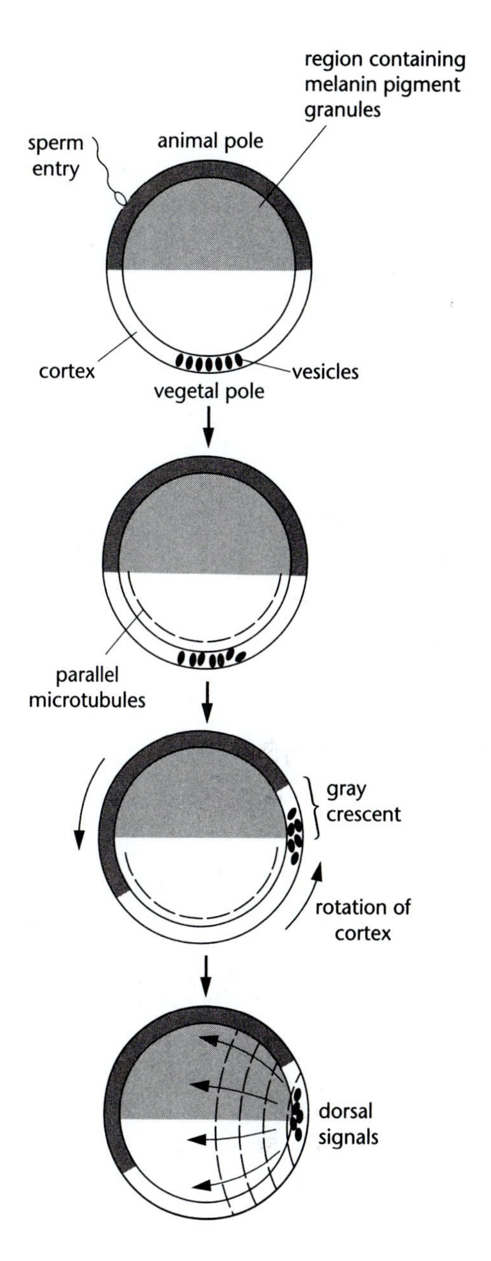


**Figure 20-5** Schematic view of cortical rotation in an amphibian egg. Before fertilization, the egg has an animal pole (distinguished by an accumulation of dark melanin-containing granules) and a vegetal pole rich in pale yolk platelets. Sperm entry triggers the assembly of a layer of a parallel microtubules beneath the egg cortex, allowing the latter to rotate (by kinesin action) around the egg as shown. Among the structures carried with this movement is a population of vesicles carrying a protein called Dishevelled that later triggers a cascade of reactions leading to the specification of dorsal structures in the embryo.

Cortical rotation appears to be due to a thin layer of parallel microtubules that forms just beneath the cortex following egg activation. Treatments that disrupt microtubules, such as exposure to colchicine, inhibit both formation of the layer and the rotation of the cortex. The presence of the sperm aster biases the orientation of this layer so that the plus ends of the microtubules point from the point of sperm entry down toward the vegetal pole. Motor proteins in the cortex (probably kinesins) then move along the microtubules, causing the entire cortex to rotate as a coherent whole. Interestingly, an egg activated in the absence of a sperm still establishes a dorsoventral axis, but with an unpredictable orientation—evidently the egg has an intrinsic capacity for symmetry-breaking that is biased in a given direction by the entry of the sperm.

The parallel array of microtubules beneath the cortex also appears to direct transport of signaling molecules to the prospective dorsal side of the embryo. One such molecule is the protein product of the *dishevelled* gene, which activates a pathway responsible for the expression of dorsal features. If this protein is labeled with a fluorescent marker, it can be seen to be in vesicles that move along the parallel microtubules during cortical rotation. We see in this example how, even before the first cleavage, the cytoskeleton can create a molecular prepattern that subsequent morphogenesis unfolds into a visible pattern.

## *Spindle growth influences the plane of cleavage*

The next step in development is to subdivide the fertilized egg into a ball of cells by successive rounds of mitosis and cytokinesis. Evidently the orientation of the mitotic spindle, and the plane of the cleavage furrow (which usually bisects the spindle, as discussed in Chapter 13), have a crucial influence on subsequent development. Cleavage not only creates the cells of the developing embryo: in those cells in which a pre-pattern exists in the cytoplasm, as it does for the eggs of most nonmammalian species, cleavage also specifies the *types* of cells that will be produced.

The pattern of cleavage at early stages of embryonic development is a highly reproducible characteristic of particular species and shows a wide diversity of geometrical relationships. Empirical rules governing these divisions were established by early embryologists, such as that cells tend to divide into two equal parts and that cleavage planes tend to be either at right angles to the surface of the embryo or parallel to it. The mechanistic bases of these crucial spatial relationships are still largely unknown.

The most common pattern seen in the first few divisions of plants and animals is that of successive orthogonal cleavages. In *Xenopus*, the first cleavage divides the egg vertically—that is, parallel to the animal–vegetal axis. The next cleavage is again vertical but perpendicular to the first, giving four cells of similar size. The third cleavage passes horizontally through these four cells but in this case slightly above the midline, thereby producing four smaller cells stacked on top of four slightly larger, more yolky cells (Figure 20-6).

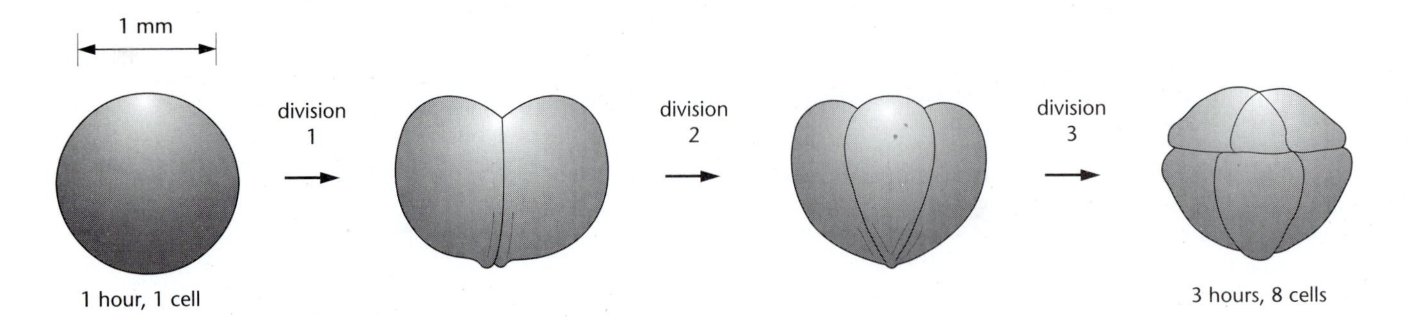


**Figure 20-6** Pattern of cleavage of a frog egg.

A simple explanation for this regular pattern of cleavage is that the spindle is forced into successive orthogonal orientations by the pressure of growth of its microtubules. During prophase, the two daughter centrosomes on the surface of the nucleus move apart—pushed, it is believed, by the growth of microtubules between them and by the action of motor proteins that cause microtubules of opposite polarity to slide against each other (Chapter 13). Assuming that each centrosome moves an equivalent distance around the nucleus, then the new spindle will naturally become oriented at right angles to the preceding spindle (Figure 20-7a).

Another simple factor that might come into play as the spindle develops is the space-filling growth of astral microtubules. If spindles grow as much as they can within the confines of the cell cortex then, by their expansion, they will tend to orient with the long axis of the cell, which is indeed the most common configuration. Operation of this rule should ensure, for example, that the third division of a *Xenopus* egg has the required orientation (Figure 20-7b). It also provides a simple explanation for the fact that nonspherical cells tend to divide their longest axis in two, and that physical barriers, such as another cell or an extraembryonic envelope, can affect cell division patterns.

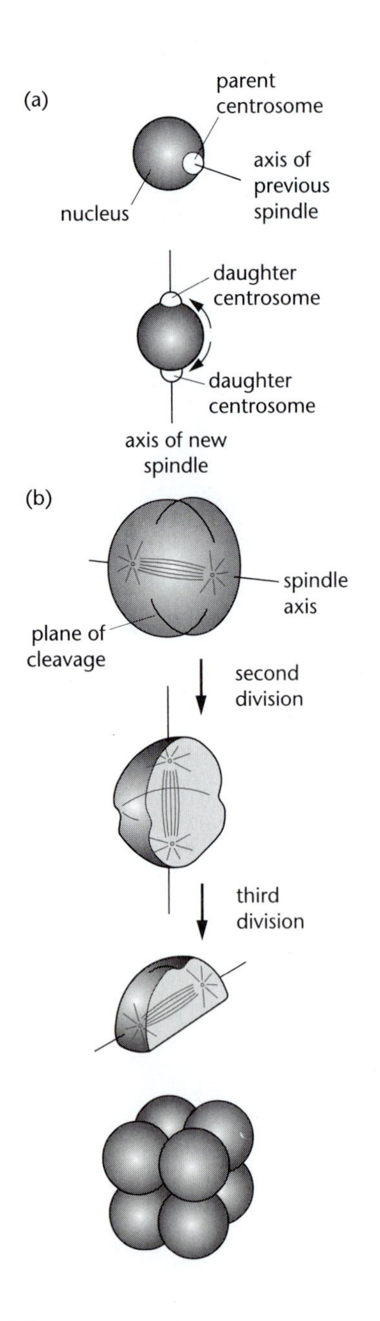


## The cell cortex can influence the plane of cell division

The space-filling properties of the mitotic spindle may provide a simple explanation for orthogonal patterns of cleavage. But there is abundant evidence that the positioning of the centrosomes and the plane of division can be controlled in a more deliberate manner. In particular, there are instances in which the cell cortex, rather than acting like a straightjacket to confine the spindle, actually provides specific cues for the positioning of the centrosome. It may be recalled from the previous chapter that a movement of this kind occurs in killer T cells when they contact a target cell.

Other examples are found in the meiotic divisions of insects and in the early mitotic divisions of sea urchin eggs and nematode worms. In the early development of the nematode *Caenorhabditis elegans*, for example, two stereotyped patterns of division can be distinguished (Figure 20-8). In one, cells proceed through a succession of orthogonal divisions as described above, giving rise to progeny cells that have equivalent developmental potentials. In the second pattern, cells divide in successive planes along the *same* axis, producing a series of cells with distinct

**Figure 20-7** Possible explanation for orthogonal cleavage. (a) When the centrosome divides, the two daughter centrosomes move apart by the action of microtubules (Chapter 13). Other things being equal, this will generate a new spindle at right angles to the first. (b) The axis of the mitotic spindle is frequently aligned with the long axis of the cell, as though tending to maximize its length. A combination of these two mechanisms can explain the sequential orthogonal cleavages seen in early amphibian oocytes.

**Figure 20-8** Nuclear rotation in the early development of the nematode. The two cells, AB and $P_1$, show a distinct pattern of centrosome movements and cleavage. In cell AB separation of the replicated centrosomes to opposite sides of the nucleus generates an orthogonal cleavage, according to the rules described above. Cell $P_1$ undergoes a similar separation but then the two centrosomes rotate together with the nucleus through 90° and the plane of subsequent cleavage is similarly rotated.

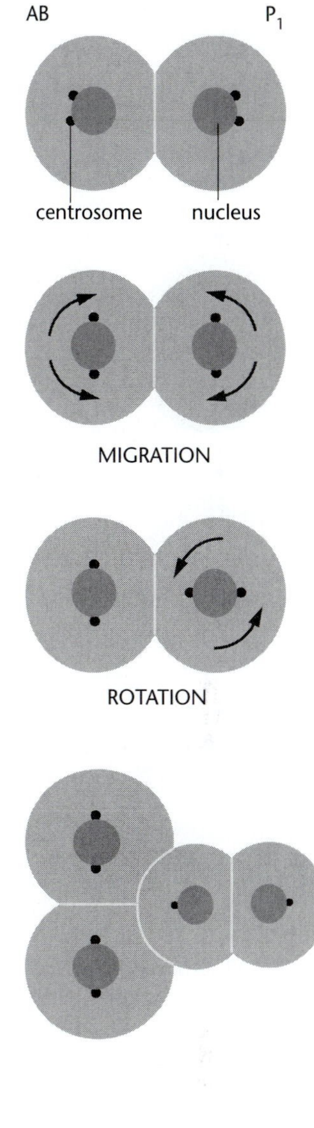


developmental potential. The difference between these patterns has been traced to changes in the positioning of centrosomes in response to their interaction with the cell cortex.

After the first division of the fertilized egg, the two centrosomes of one cell (designated AB) move apart on the surface of the nucleus, thereby generating an orthogonal cleavage, as described above. In the second cell ($P_1$), however, there is a subsequent movement in which the two centrosomes rotate as a unit together with the nucleus to lie on the longitudinal axis of the embryo—the future axis of division. Experiments in which specific regions of the cell are irradiated with a laser indicate that one of the two centrosomes, apparently selected at random, is 'pulled' by becoming attached to the cortex. This same strategy of controlled spindle rotation is also repeated in later divisions, so that the progeny of the AB cell divide in successive orthogonal axes, whereas the progeny of the $P_1$ cell divide successively along the same axis. Later in the chapter we will examine the signals responsible for one of these subsequent divisions.

## *Dynein molecules in the cell cortex may pull on microtubules*

Any interaction between microtubules growing from the centrosome and actin filaments beneath the plasma membrane is potentially of great significance. It could allow signals from neighboring cells, crossing the plasma membrane to influence the position of the spindle and cleavage furrow. But what is the molecular basis of this interaction?

In the *C. elegans* example just mentioned, the agents connecting the centrosome with the cortex are likely to be astral microtubules, since their disruption by a laser beam prevents rotation. Moreover, the rotation is inhibited by cytochalasin, so that actin filaments are also likely to play a part. In fact, a small button enriched in actin can be detected on the cortex of the P cell just prior to the attachment of astral microtubules. It has been hypothesized that motor proteins associated with this cortical button might capture microtubules emanating from one of the two spindle poles and then pull it closer.[1]

Which motor molecules? In budding yeasts the spindle pole (equivalent to the centrosome) is pulled toward the site of new bud formation. Individual microtubules can be seen to extend from the spindle pole to the cortex and mutants deficient in yeast dynein fail to show this movement. Similarly, in multinucleate *Dictyostelium* cells formed by myosin II-null mutants, centrosomes migrate through the shared cytoplasm, pulled (it seems) by unusually long microtubules. These 'guiding microtubules' connect to the cell cortex and evidence suggests that the dynein heavy chain (of which there is only one kind in *Dictyostelium*) is responsible for their gliding movements.

So dynein may be responsible for centrosomal movements—perhaps in association with dynactin, which often anchors it to the cortex. But it is not known how these proteins are themselves positioned or how they are regulated. Nor is it excluded that microtubules could be anchored to

[1] The actin button may itself be a remnant of the *previous* division, deposited on the $P_1$ cortex by the spindle midbody. If this is correct then the orientation of the $P_1$ cell division may already be determined at the first cleavage of the fertilized egg.

the cortex in other ways—the orientation of mitotic spindles in budding yeasts, for example, depends on a cortical protein, a microtubule protein, and a kinesin. There is much still to learn about this crucial interaction.

## *Cell movements are widespread in vertebrate embryos*

In a nematode or a fruit fly many features of the eventual differentiated embryo are presaged by regional changes in the cytoplasm before cell division. Accurately positioned cleavage planes often segregate volumes of cytoplasm that are already distinct in composition and potential function. However, in many other species, and especially in vertebrates, cells remain uncommitted to a particular developmental pathway for some time after they are created by cell division. A prominent feature of embryonic development in such organisms is that the cells commonly undergo large-scale movements. They typically change their location in the embryo after they are created by cell division and may move long distances before settling into their final location. Indeed, their final differentiated state often depends explicitly upon this location, being influenced by signals coming from other cells in the vicinity.

The movements of developing embryos are intricate and convoluted—a complex unfolding in time and three-dimensional space. Each species has a characteristic pattern of movements, reproducible from embryo to embryo, the detailed study of which belongs to developmental biology and embryology. But it is legitimate to ask here what the fundamental cell movements are that underpin crucial morphogenetic movements early in development. In the following pages we will examine the phases of development known as *blastula formation*, *gastrulation*, and *neurulation*, driven largely by the concerted movements of sheets of cells, and *cell migrations* in which cells move as individual units. These movements are broadly similar in all animal species and are driven by a limited repertoire of movements at the molecular level.

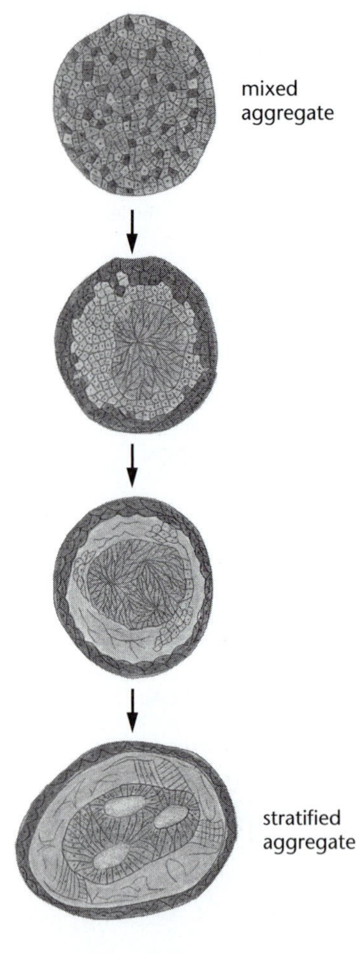


**Figure 20-9** Cell sorting. Cells from different parts of an early amphibian embryo will sort out according to their origins. In the classic experiment shown here, mesoderm cells, neural plate cells, and epidermal cells have been disaggregated and then compacted into a random aggregate. They sort out into an arrangement reminiscent of a normal embryo, with a neural tube internally, epidermis externally and mesoderm in between.

## *Morphogenetic movements are guided by cell adhesion molecules*

Before discussing the movements themselves, we ought to say a word about the factors that guide them. As they move within an embryo, individual cells must break old attachments to other cells and to components of the extracellular matrix, and make new ones. Where they finish up in the developing embryo, and what they eventually become, therefore depends on the selective expression of cell recognition molecules on cell surfaces. Groups of cells that share common surface characteristics become demarcated as development proceeds, their mutual recognition ensuring that they will remain a coherent group despite the continual random movements of individual cells.

The stabilizing effects of cell–cell recognition are sometimes so powerful that they can bring about an approximate reconstruction of the normal pattern even after the cells have been dissociated into a random mixture. Classic experiments performed in the 1930s by Holtfreter and his colleagues showed that mesoderm cells, neural plate cells, and epidermal cells will sort out from a random aggregate to form a structure with an epidermis on the outside, mesoderm immediately beneath that, and an object resembling a neural tube buried in the interior (Figure 20-9). Evidence from chick and mouse embryos suggests that this behavior depends, at least in part, on cadherins, a family of cell-adhesion glycoproteins that associate with the actin cytoskeleton (Chapter 6).

In a developing embryo, cells express different patterns of cadherins during gastrulation, neurulation, and somite formation, and antibodies against them interfere with the normal selective adhesion between cells of a similar type. In many cases it seems that cadherins act as a brake that must be removed before selective migration of particular cells can occur.

The protein activin, for example, which plays numerous roles in gastrulation, is thought to weaken the adhesion of cadherins and so allow cells to move.

## *Signals pass from one cell to the next in an embryo*

A mechanical sorting of cells on the basis of selective adhesion is of course a crude mechanism that gives little hint of the complex and sophisticated molecular changes that underpin development. Beneath the surface, so to speak, a myriad biochemical changes take place in individual cells as genes are turned on and off, specific proteins are made and degraded, and intricate structures characteristic of hundreds of differentiated cell types are assembled (including those responsible for selective adhesion). These changes are triggered by signals received by individual cells from their environment, especially those from neighboring cells in the tissue. Developmental biologists have made enormous progress in elucidating the molecular nature of these signals, and the cascades of reactions by which they act inside the cell.

For a flavor of the signaling pathways that operate during development, consider the family of *Wnt* proteins. These are cysteine-rich glycoproteins that are secreted by embryonic cells into the intercellular space and that then bind to receptors on target cells. Wnt proteins are used at multiple stages during development to influence an impressive diversity of cell fates. In mice, for example, which have at least 17 different forms of Wnt, mutations in Wnt1 cause premature death due to midbrain and cerebellum defects, mutations in Wnt2 cause placental defects, mutants defective in Wnt3a die before birth with defective notochords and somites, and Wnt4 mutants die after birth with malformed kidneys. In Wnt7a mutant females, the uterus develops abnormally so that they are sterile.

When a Wnt protein binds to a receptor on the surface of a target cell, it initiates a cascade of intracellular reactions that are complex and vary in their details from one cell to the next. A major target appears to be a large multiprotein signaling complex in the cytoplasm of target cells, which receives the signals from the Wnt receptor and relays them to the nucleus and other targets in the cell. A protein called β-catenin, especially, moves into the nucleus and activates specific genes.

It is interesting to note, in keeping with the general theme of this chapter, that Wnt proteins can also act directly on actin filaments and microtubules. An influence on actin has been found in the development of mouse kidney and in the generation of polarized cortical-based structures in fields of epithelia. Actin-rich bristles in a fruit fly wing, for example, are positioned in each hexagonal epithelial cell in a constant position with respect to the insect's body axis, and this is an effect due to Wnt signaling. With respect to microtubules, Wnt has been shown to influence the position of the mitotic spindle in both *Drosophila* sensory cells and early *C. elegans* embryos.

Following the division to the four-cell stage of a *C. elegans* embryo, described above (see Figure 20-8), secretion of Wnt by one of the progeny, known as $P_2$, induces a neighboring cell to become a precursor for endodermal structures. It does so partly by activating transcription factors and triggering expression of specific genes. But it also induces rotation of the centrosome–nuclear complex of the target cell through 90°. This rotation is essential for the subsequent determination of endoderm and is thought to depend on a Wnt-induced cortical site in the adjoining cell, to which one of the centrosomes attaches.

## *Cell junctions stabilize the blastula*

Cells at early stages of development adhere only weakly to their neighbors, their small and transient regions of contact offering little mechanical

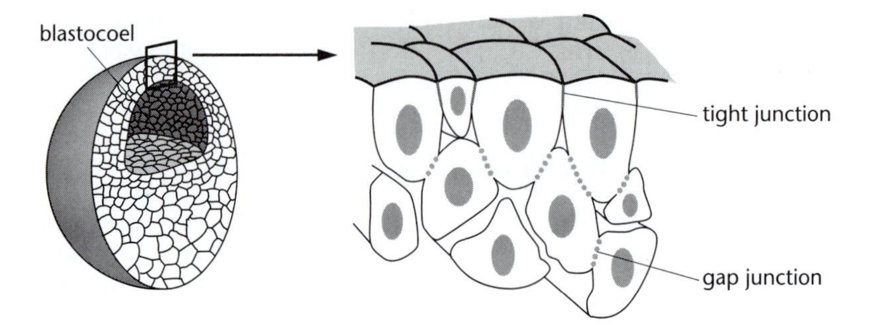


**Figure 20-10** The blastula in a frog embryo. At this stage the cells are arranged to form an epithelium surrounding a fluid-filled cavity, the blastocoel. The cells are electrically coupled via gap junctions, whereas tight junctions close to the outer surface create a seal that isolates the interior of the embryo from the external medium. Note that in *Xenopus* the wall of the blastocoel is several cells thick, and only the outermost cells are tightly bound together as an epithelium. (Adapted from B. Alberts et al, *Molecular Biology of the Cell,* 3rd ed. New York: Garland Publishing, 1994.)

resistance to cell movements. With the expression of molecules such as cadherins and extracellular matrix components, cells attach more firmly to their neighbors. Regions of cell–cell contact now become a focus for the development of mechanically strong junctions such as adherens junctions and desmosomes that link the cytoskeletons of the two cells together.

The earliest manifestation of this process is the establishment of a *blastula*, as the ball of dividing cells shapes into a hollow ball (Figure 20-10). This occurs at about the 16-cell stage in a *Xenopus* embryo, and seems to depend on expression of a specific form of cadherin known as *EP-cadherin* (*E-cadherin* in mammals). This adhesive protein becomes localized to the superficial layer of cells, thereby marking its commitment to become an epithelium. With the further differentiation of the epithelial layer, the distribution of cadherin on the cell surface becomes restricted to highly delineated regions of contact. The development of intercellular junctions is an active process, initiated by filopodia that grow into neighboring cells and produce local accumulations of cadherins. These eventually mature into the mechanically strong adherens junctions and, together with later-forming desmosomes, link one cell to its neighbor.

If the formation of specific cadherins is the first indication that a sheet of cells is about to become an epithelium, then the establishment of tight junctions makes this commitment a reality. Tight junctions between the plasma membranes of two adjoining cells prevent the movement of ions and molecules in the intercellular space between them. This creates a seal, isolating the interior of the embryo from the outside and regulating the entry and egress of ions and small molecules. Sodium ions are pumped into the spaces between the cells in the interior of the embryo, and water follows because of the resulting osmotic imbalance. As a result, the intercellular crevices deep inside the embryo enlarge and coalesce into a single cavity, the *blastocoel*. The cells not included in this epithelium become loose, *mesenchymal* cells that frequently move about beneath and between epithelial layers. Their migratory motions are relatively independent of one another, although they can aggregate to form patterned clusters of cells, such as the somites.

## Major form-shaping movements are driven by epithelia

As development proceeds, epithelial layers become more fully differentiated. Each epithelial cell acquires a definite polarity, with an apical surface facing the external medium bathing the embryo or (at later stages) a fluid-filled cavity within the body of the embryo. The apical surface may acquire secondary specializations such as secretory organelles, cilia, or microvilli. The basal surface of the epithelium rests on some other tissue—usually a connective tissue—to which it is attached. Supporting the basal surface of the epithelium is a thin, tough sheet of extracellular matrix called a *basal lamina*, composed of a specialized type of collagen (type IV collagen) and various other molecules. These include laminin, which provides adhesive sites for integrin molecules in the plasma membrane of the epithelial cells (Chapter 6).

The formation of an epithelial sheet and its function both depend

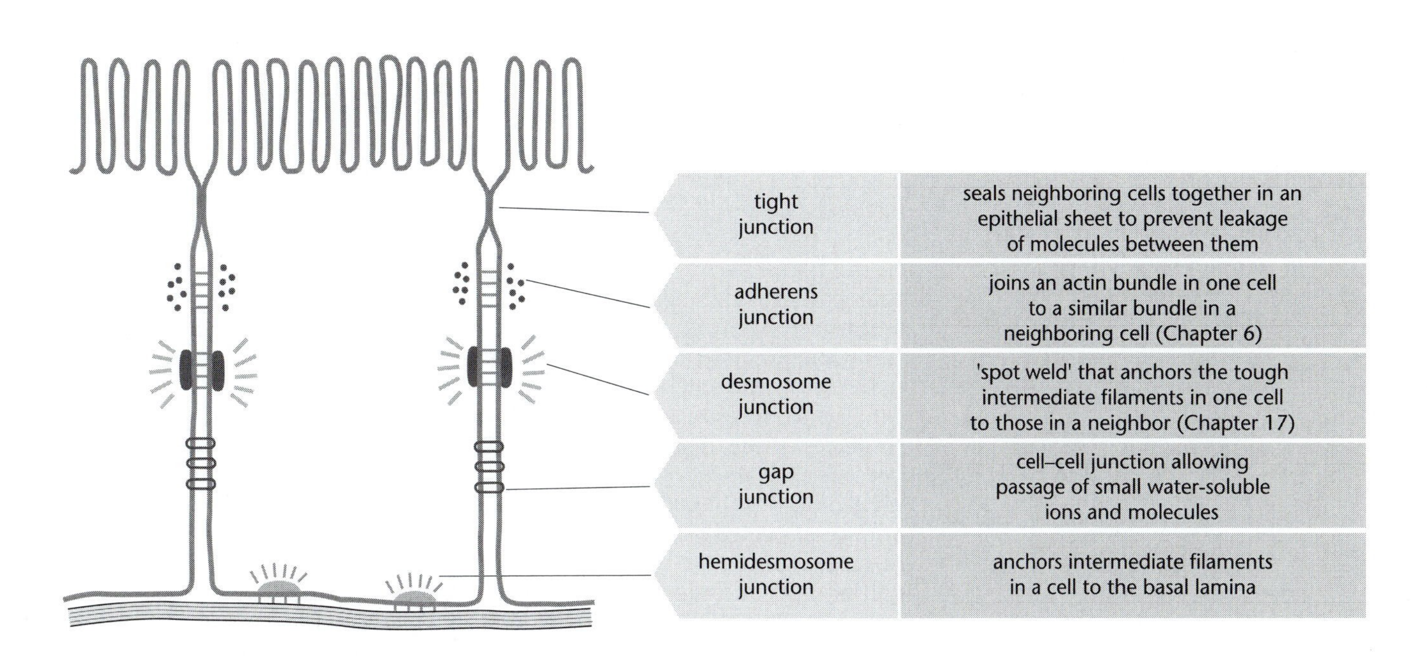


**Figure 20-11** Cell–cell junctions in animal epithelial sheets. Tight junctions are peculiar to epithelia, but the other types also occur in modified forms in various nonepithelial tissues.

crucially on the junctions linking each cell to its neighbors (Figure 20-11). Tight junctions, as mentioned above, are crucial to the sealing function of epithelia, by which they prevent leakage of molecules from the apical to the basal side. Desmosomes and hemidesmosomes, already encountered in Chapter 17, have a primary role in maintaining the sheet as a physical entity and giving it mechanical strength.

Adherens junctions, which we met in Chapter 6, are also strong but, because they are linked to the actin cytoskeleton, they have the additional feature of participating directly in active deformations. Whenever epithelial cells move or undergo coordinated changes in shape, the flat sheet of which they are part may stretch, contract, fold, invaginate, or otherwise deform while retaining its integrity as a connected tissue. As we will now relate, such changes are a major driving force for many of the form-shaping movements that an embryo undergoes.

## *Cells flatten and intercalate during epiboly*

*Gastrulation* is a dramatic process that transforms the simple hollow ball of cells of a blastula into a multilayered structure with a central gut tube and bilateral symmetry. It is a complicated transformation in which many cells on the outside of the embryo move into the interior, often in the presence of large numbers of yolk cells (Figure 20-12). Different species can vary enormously in the extent of cell division and the number of cells

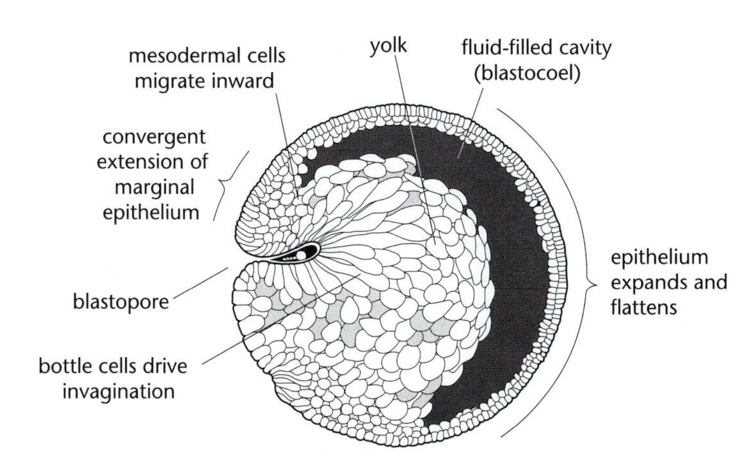


**Figure 20-12** Cell movements in gastrulation. An amphibian gastrula is shown in schematic cross-section (after Holtfreter, 1943). Major cells movements are indicated. Expansion of the future ectoderm (epiboly) occurs by cell flattening and intercalation of new cells. Wedge-shaped deformation of bottle cells drives the invagination of the epithelial layer. Migration of mesodermal cells over the fibronectin-rich roof of the blastocoel helps pull tissues into the embryo as does the migration of epithelial cells known as convergent extension.

**Figure 20-13** Epithelial spreading during epiboly. As the ectoderm expands in area, existing cells flatten while new cells intercalate from deeper layers of the embryo.

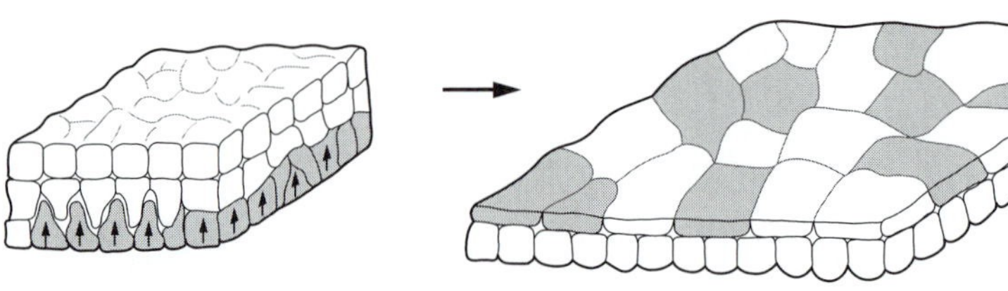

that participate in gastrulation. The gastrula of a sea urchin described below, for example, is composed of about 30 cells, whereas that of a frog has some 30,000 cells, necessitating radical differences in mechanics and patterns of movement.

The final result is, however, similar. In all animal species, gastrulation produces three distinct layers of cells. The innermost epithelial layer becomes the *endoderm*, or the tube of the primitive gut. The outermost layer becomes the *ectoderm*, the epithelium destined to form the epidermis and (unexpectedly) the nervous system. Between the two, the looser layer of tissue composed of mesenchymal cells becomes the *mesoderm*, which will produce muscle and connective tissue.

Gastrulation appears to be driven by a small number of primary mechanisms, employed in different combinations by different phylogenetic groups. One such mechanism is *epiboly*—the tendency of epithelial sheets of ectoderm to spread and surround inner sheets. During gastrulation in a frog, for example, the epithelial layer destined to become the future ectoderm increases in area by about 50%. During the development of birds and fish a sheet of blastodermal cells spreads and eventually encompasses the entire massive sphere of yolk. In these and other species, epithelial spreading can be attributed to two factors. One is the tendency of individual cells in the sheet to become flatter and more spread out; and the other is the movement of new cells into the sheet from deeper layers (Figure 20-13).

## Apical constrictions initiate blastopore formation

Gastrulation entails the active movement of cells into the interior of the embryo. A groove on the surface of an amphibian embryo, the *blastopore*, spreads and deepens to form a circular indentation and ultimately a pore. Cells are drawn through the pore and into the interior of the embryo by a number of forces, one of which is the active migration of mesodermal cells on the basal lamina beneath the presumptive ectoderm, another being the convergent migration of cells in the epithelium described below.

Formation of the blastopore, the first sign of gastrulation, illustrates a type of movement that is extremely widespread during development. This is an *invagination* in which a sheet of cells buckles inward to form a cavity. Almost a century ago it was suggested that this movement could be initiated by special cells in the epithelial layer undergoing a coordinated constriction of their apical ends. The wedge-shaped distortion of these cells (termed *bottle cells* in amphibian embryos), would then cause the epithelial layer to buckle and invaginate.

Strong evidence in support of this contention has now been obtained in experiments with sea urchin embryos. Gastrulation in this species is a much simpler process than in amphibians, involving a simple hollow ball of cells that invaginates in one region. This sheet of cells can be explanted from the embryo and its deformation studied in isolation. At the site of future invagination, a ring of future bottle cells encircles a small cluster of round central cells. Elimination of the bottle cells by laser irradiation blocks invagination, with a greater effect being shown the more the bottle cells are eliminated. The results of such experiments indicate a mechanism in which bottle cell deformations produce the primary force for the morphogenetic movement, which is coupled to the surrounding tissue through mechanically strong links (Figure 20-14).

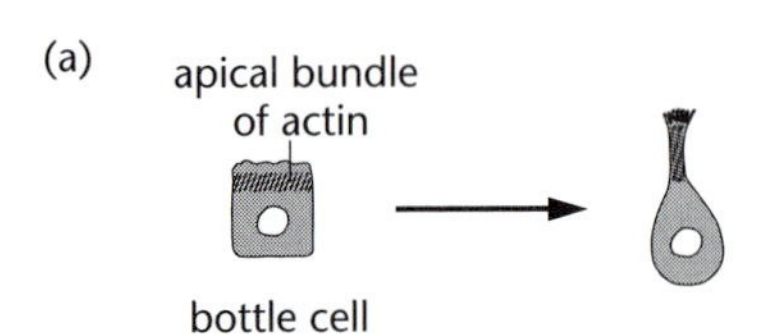


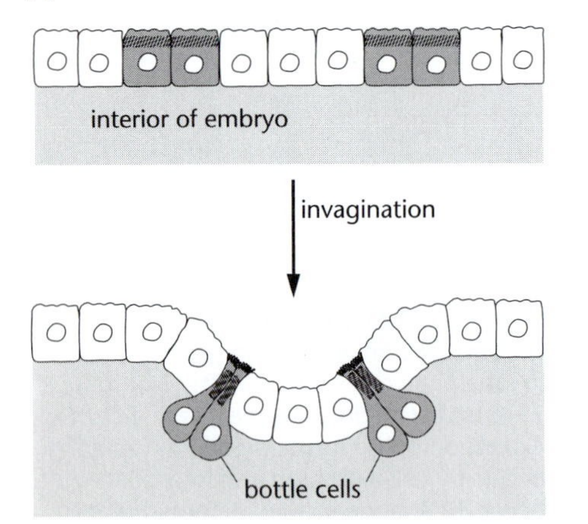


**Figure 20-14** Invagination caused by bottle cells. (a) Each bottle cell is thought to undergo contraction of its apical bundle of actin filaments, causing it to become bottle-shaped. (b) Apical contraction of a ring of bottle cells (seen here in cross-section) set into an epithelial sheet results in an invagination of the sheet.

## Embryonic invaginations are driven by actin-based contractions

Invaginations initiate developmental changes other than gastrulation. Thus, the production of most internal tubular structures, such as the gut and insect trachea as well as most vertebrate glands and sense organs, begins with the inward buckling of an epithelium. Invaginations also occur along a line rather than at a point, in this case giving rise to a groove rather than an indentation. Examples are seen in the neural groove of vertebrate embryos and the ventral furrow in insect embryos.

Although the mechanisms of these transformations have rarely been studied in such detail as that of blastopore formation in sea urchins, it seems likely that similar cellular mechanisms will be employed. Wedge-shaped cells similar to bottle cells are frequently seen, and extensive evidence supports the involvement of actin-based movement in the invagination. Thus, many embryonic invaginations are sensitive to cytochalasin treatment, and nonmuscle isoforms of myosin II are localized at the apical ends of constricting cells. Sheets of epithelia isolated from an embryo have been shown to buckle in an ATP-dependent fashion.

Another variation on this theme is the occurrence of pancellular bundles of actin and myosin in sheets of epidermis. They are seen during episodes of development such as the dorsal closure of *Drosophila*, in which an epithelial sheet closes over an underlying tissue. They can also be induced experimentally by wounding the surface of an embryo. Contraction of the actin–myosin bundle pulls opposing edges of the sheet together, rather like a suture or purse string. Its action is similar to that of the contractile ring in cytokinesis, except that the bundle in this case extends from one cell to the next, presumably being linked by adherens junctions.

Epithelial sheets can also *evaginate*. Outpocketing of an epithelium such that the apical surface lies on the convex side of the pocket occurs during the formation of feathers and scales, and during development of intestinal villi. By reasons of symmetry one might expect such deformations to be driven by a basal contraction, and bundles of actin filaments are in fact seen in the basal regions of some epithelial cells. But direct evidence is lacking.

## Cell rearrangements support the elongation of an embryo

Cells in an epithelial sheet have to stick tightly to each other so as to maintain a topologically continuous sheet. But this does not mean that they are incapable of changing their positions. Indeed, cell rearrangements occur widely within epithelia and provide an important mechanism of morphogenetic changes in embryonic tissues.

One of the most important forces driving gastrulation in an amphibian embryo is an active repacking of epithelial cells known as *convergent extension*. This takes place as cells migrate in the plane of the sheet toward a midline, thereby extending the length of this line by their increasing numbers (Figure 20-15). The phenomenon can be demonstrated outside the embryo by extirpation of a small square piece of tissue from the region close to the blastopore and watching it become thinner and longer. In an intact embryo this movement is believed to carry cells to the dorsal

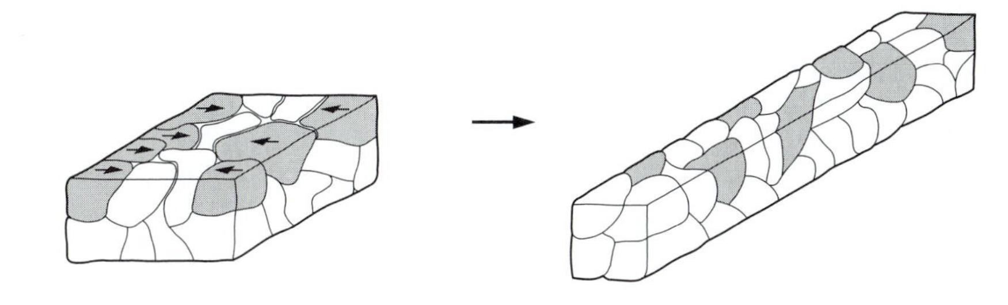

**Figure 20-15** Convergent extension. Cells in an epithelial sheet move toward the midline. As they jostle and crowd into the center they drive the extension of the tissue.

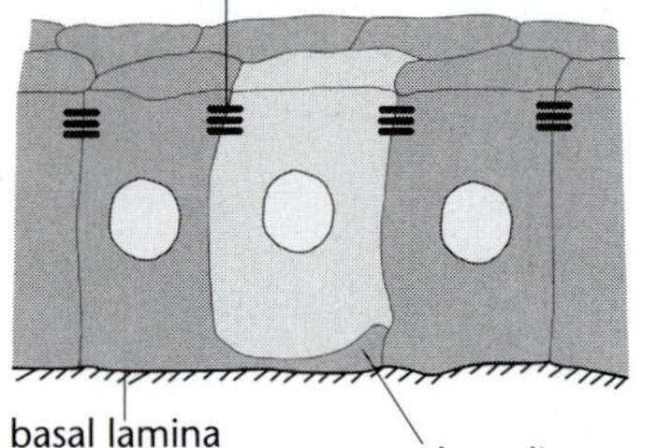


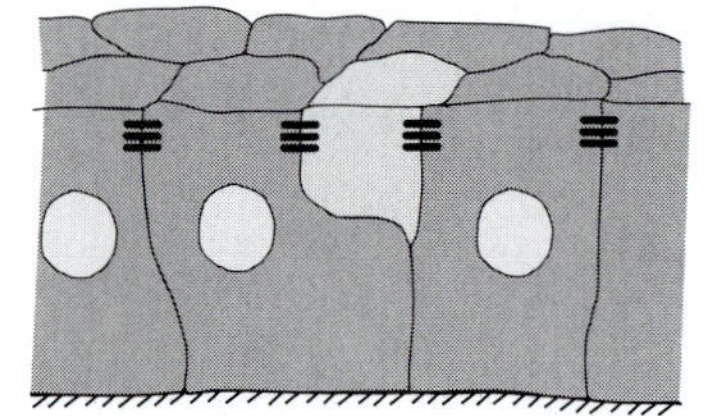

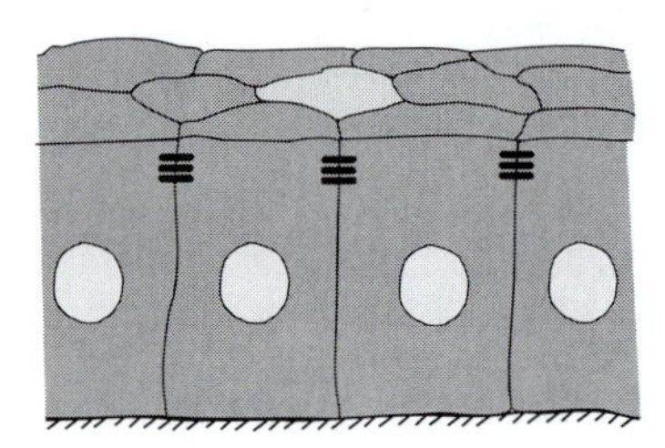

**Figure 20-16** Cell migration in a sheet. Possible mechanism by which cells move actively in an epithelial sheet. A lamellipodium extends along the basement lamina to make contact with cells in its vicinity. Selective movement of the cells results in a gradual rearrangement without a breach appearing in the epithelial sheet.

midline, into the blastopore, and hence into the interior of the gastrula, where they produce the main axis of the body.

Similar changes are seen during formation of the amphibian neural plate. Migrations of individual cells in the neural plate of a newt embryo have been tracked by continuous light-microscopic observation. There is a steady movement of cells toward the center line, where the neural plate sits over the notochord, and along the long axis of the plate, eventually changing the plate from a circular disc to a keyhole shape.

There are similarities between convergent extension and cell migration. Elongated basal extensions extending several cell diameters have been observed when active rearrangements take place, rather like the filopodia and lamellipodia of independently migrating cells (Figure 20-16). What prompts the cells to move, and how their movements are guided, are unknown.

## The nervous system begins as an epithelium rolls up

The dominant part played by epithelial sheets in development is well illustrated by *neurulation*. This is the phase in which the embryo, having formed a primitive gut, produces a thickened region of ectoderm on its dorsal surface which then rolls up to form a tube. Subsequently the tube pinches off from the rest of the ectoderm to create the *neural tube*—the precursor of the brain and spinal cord (Figure 20-17).

The cellular changes that cause neurulation are typical of those occurring throughout development. First of all cells in the dorsal ectoderm—a single sheet of epithelial cells—increase in height, changing from cuboidal cells into elongated columnar cells and thereby forming a thickened plate or *placode*. This change in cell shape, termed *columnarization*, is thought to depend on microtubules that become oriented at right angles to the plane of the cell sheet and increase in length. Treatment of embryos with inhibitors of microtubule polymerization blocks placode formation and subsequent neurulation.

Following placode formation there is an episode of convergent extension in which the cells change their position and hence cause the neural plate to become narrower and to elongate along its midline. Finally, the neural plate folds at the edges and rolls into a tube, which then separates from the surface epithelium.

## Cell migrations establish the anatomy of the brain

In the movements just described cells move as part of an epithelial sheet, still in contact with their neighbors. But vertebrate embryonic development also features episodes in which cells migrate as individuals, often over long distances. These movements, some of which were mentioned briefly in Chapter 2, include the diaspora of cells from the neural crest to multiple destinations in the developing body and the exodus of cells from the somites to give rise to skeletal muscle. Other important migrations are those of blood cell precursors and of the germ cells, which give rise to egg and sperm cells.

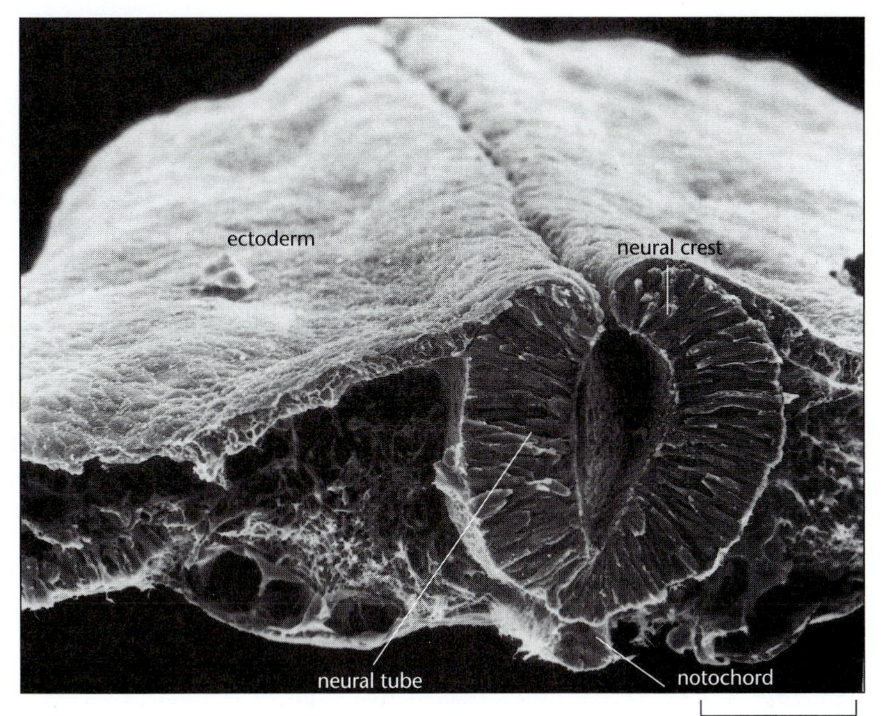


**Figure 20-17** Formation of the neural tube. Scanning electron micrograph showing a cross-section through the trunk of a 2-day chick embryo. The neural tube is about to close and pinch off from the ectoderm. At this stage it consists (in the chick) of an epithelium that is only one cell thick. (Courtesy of Jean-Paul Revel.)

The nervous system is the site of some of the most intricate and precise cell movements. During the course of the development of the vertebrate brain, billions of young neurons travel distances measured in millimeters or even centimeters. These movements establish the basic anatomical features of the developing brain, such as the eight segmental-like structures ('rhombomeres') of the embryonic hindbrain. They also define the positions of the cerebellum and the regional boundaries of the forebrain.

Reiterated cell movements are responsible for the layered structure of the brain cortex, in which sets of neurons of similar kinds (such as pyramidal cells) occupy the same stratum and form special sets of connections among themselves. These layers arise in the following manner (Figure 20-18). The first cells to develop into neurons in the neural tube are located near the inner surface of the tube. The cells divide and one of the daughters differentiates into a neuron and migrates toward the other side of the sheet. Sets of neurons that are born later migrate past those that were born earlier, and take up positions closer to the outer surface—the youngest cells therefore travel the greatest distance and end up outermost. In their migration the neurons are guided by highly elongated supporting cells—radial glia. These span the entire distance from the inner surface to the outer surface.

Migrating cells cling to radial glial cells and move along them, rather as one would climb a rope. They migrate in a distinctive fashion (termed nucleokinesis) in which a thin process first extends along the glial cell and then the cell nucleus and cell body translocates within the cytoplasmic process. The cells retract the rearward cytoplasmic remnant and move forward by repeating the sequence. Cells moving along radial glia tend to remain in line with the parent stem cells, thereby forming a column of common descent. A second form of neuronal migration also occurs widely in which cells migrate at right angles to the radial glia pillars. Even in adult mouse brain, for example, precursor cells destined to become olfactory neurons migrate in cohorts over distances up to 5 millimeters.

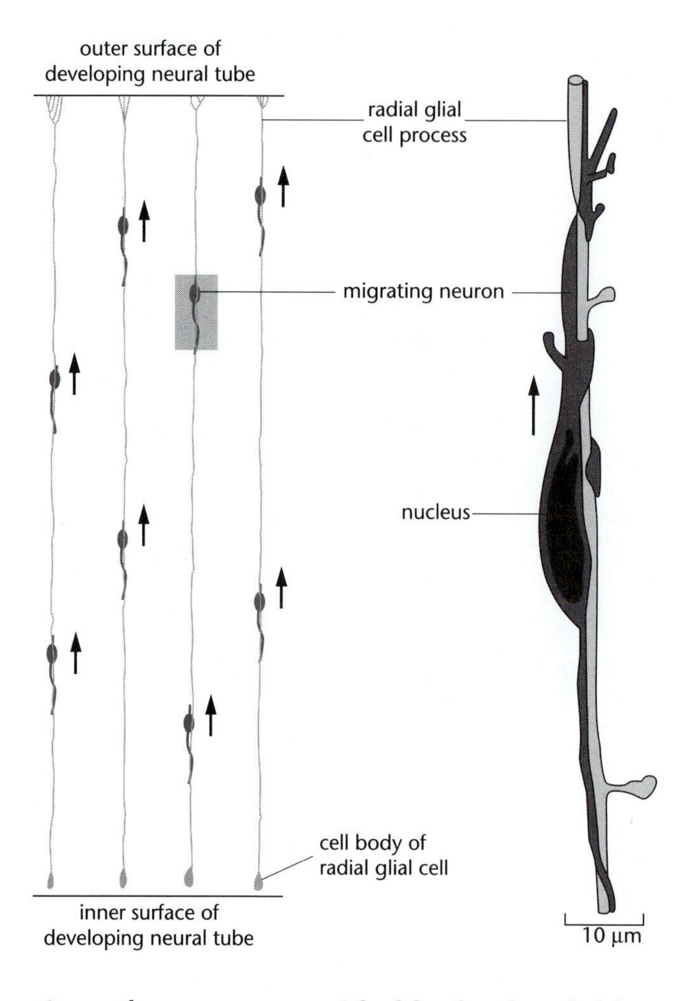


**Figure 20-18** Migration of immature neurons along radial glial cells. Neurons are born close to the inner surface of the neural tube and migrate outward. The radial glia can be considered as persisting cells of the original columnar epithelium of the neural tube that have become extraordinarily stretched as the wall of the tube thickens. (Adapted from B. Alberts et al, *Molecular Biology of the Cell*, 3rd ed. New York: Garland Publishing, 1994.)

## *Growth cones are guided by both soluble and substrate-bound factors*

Consider, finally, the detailed connections that give the brain its unparalleled computational abilities. Once a developing neuron has reached its proper location in the brain or spinal cord, it begins to grow out an axon and dendrites. These advance behind growth cones that, as we saw in Chapter 8, move by a form of actin-based migration closely related to the locomotion of fibroblasts and other tissue cells. But it is a form of migration that is exquisitely sensitive to the surroundings and capable of being guided by a multiplicity of surface-bound or diffusing chemical attractants or repellents.

For example, a group of cells in the 'floor plate' region of the developing spinal cord attracts nerve axons from the opposite side of the spinal cord, causing them to grow around the circumference. The active agent was found in tissue culture experiments to be a small protein termed *Netrin*, made and secreted by floor plate cells. The concentration gradient of Netrin set up by this secretion is detected by the growth cones of the target nerve cells and used by them to direct their migration. Multiple members of the Netrin family have now been identified, not only in vertebrates such as rat or chick, but also in species as distant as nematode worms. Interestingly, in this latter organism gradients of Netrins attract not only growth cones but also migrating mesodermal cells, thereby providing a direct link between growth cone migration and cell crawling.

Netrin is one of a growing list of proteins and other molecules (*Semaphorin* and *Slit* are two others) now known to guide growth cones. Some act as diffusible attractants, whereas others (such as Slit) act as repellents, causing responsive growth cones to 'collapse' (retract their lamellipodia and filopodia) and the axon to retreat. As well as diffusible agents, guidance molecules also occur in an 'insoluble' form, being expressed along the length of axons, for example, or woven into the extracellular matrix. Just as in the case of the diffusible guidance molecules, surface-bound guidance molecules may be attractive in their effect on growth cones or they can cause collapse and retraction.

Some diffusible guidance factors act as either attractants or repellents, depending on the type of growth cone. More specifically, the response of a growth cone to a given gradient appears to be determined by the types of receptors it carries on its surface. This was demonstrated in a clever experiment in *Drosophila* in which nerve cells were made to express one of two chimeric receptors—one mediating a repellent response to Netrin and the other a attractive response to Slit. In the first case all nerve cells were repelled by the midline of the developing central nervous system, even those normally attracted to it, whereas in the second case all nerve cells were attracted to the midline and grew across it, sometimes repeatedly (Figure 20-19). Evidently, the guidance response of each growth cone depends on the receptors it carries. Moreover, it seems from this and other experiments that the receptors function in modular fashion—their extracellular domains specifying which ligand is detected, and their cytoplasmic domain determining the motile response of the growth cone.

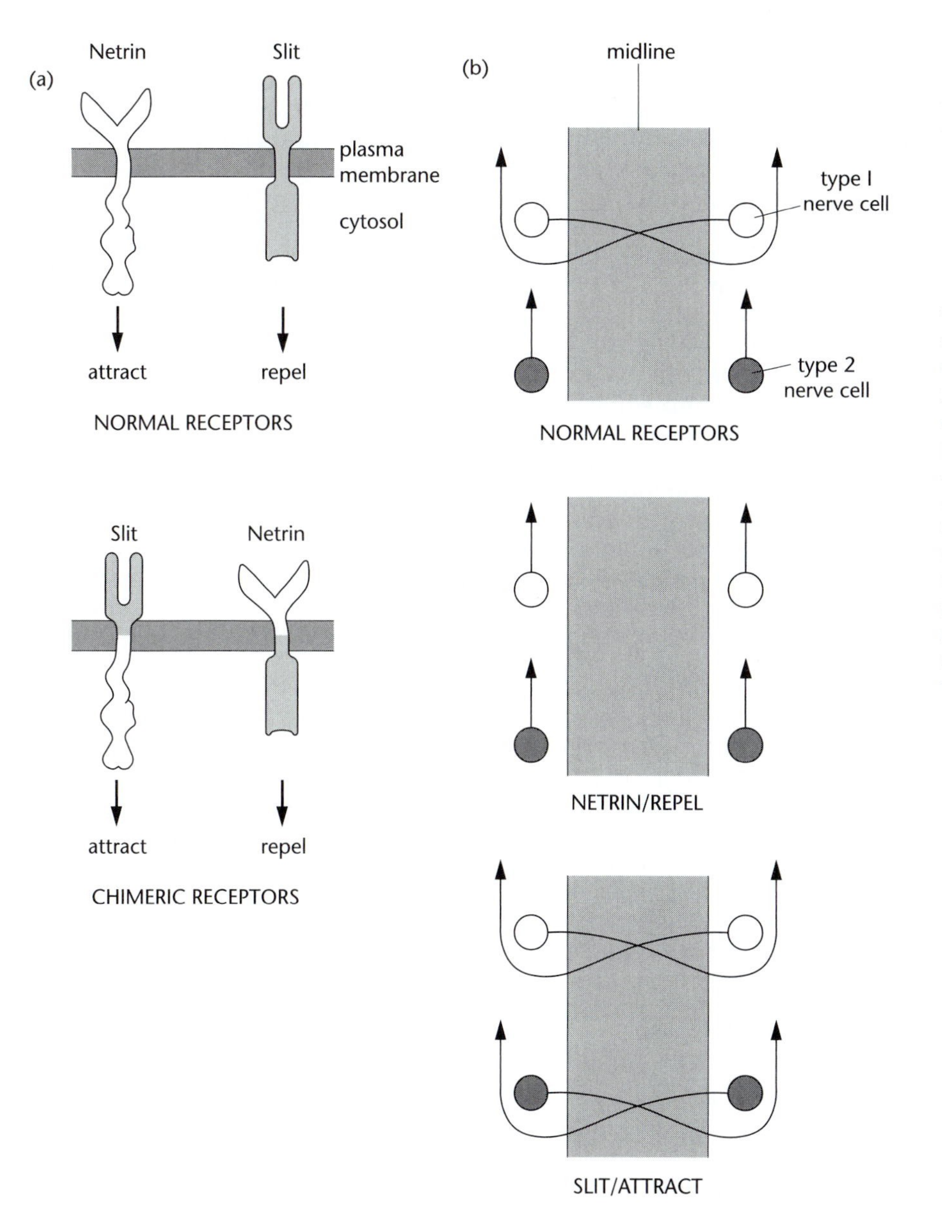


**Figure 20-19** Experiment with growth cone receptors. (a) Two kinds of receptors are normally expressed on *Drosophila* growth cones: one binds to Netrin and mediates an attractive response, the other binds to Slit and mediates a repellent response. The receptors were manipulated genetically to produce two kinds of *chimeric* receptors in which binding domains and signaling domains were swapped. (b) Patterns of axonal growth in the central nervous system of developing *Drosophila*, shown schematically. In wild type flies, two classes of nerve cell have different patterns of growth—type 1 cells are attracted to midline cells (which express both Netrin and Slit) whereas type 2 nerve cells are repelled by the midline. In flies expressing the chimeric Netrin/repel receptor at high levels, all nerve cells are repelled at the midline. Conversely, in nerve cells expressing the Slit/attract receptors at high levels, all axons cross the midline.

## *Growth cones carry a complex 'on-board' navigational system*

We see that the connections of the brain are produced by the reiterated and largely autonomous activities of growth cones. These crawl through the interstices of the surrounding tissue, searching for guidance cues that will lead them to their eventual synaptic destination. The cues are mostly markers carried on the surface of other cells in the embryo, or diffusible molecules that they secrete into the surrounding intercellular fluid. How the growth cone responds to this trail—whether it advances, retracts, or branches at a given point—is dictated by the receptors it carries on its surface, but the physical response itself is driven by the cytoskeleton and especially the actin cortex. The tips and distal portions of a growth cone are full of actin filaments and associated proteins, and these dominate its movements, as they do in any crawling cell.

How do signals from membrane receptors influence the actin cortex so as to direct growth cone movements? As we saw in Chapter 8, in connection with the chemotactic responses of crawling cells, a complex and poorly understood network of intracellular reactions interacts at multiple points with the motile machinery. Filopodial extension is driven by actin polymerization and hence is likely to engage ERM proteins, WASP proteins, and Arp2/3 complexes. In retraction we expect to find $PIP_2$ signaling and an involvement of gelsolin, cofilin, and myosins. Protein kinases will certainly be engaged, as they are in any eucaryotic signaling pathway. So too will G proteins—both the heterotrimeric kind activated by membrane receptors and the small G proteins of the Rho family, which have a special relationship with actin. Localized changes in intracellular $Ca^{2+}$ concentration can have dramatic effects on the growth cone, and initiate either attraction or repulsion depending on environmental conditions.

Lastly, we should not forget to include physical forces in the equation. Pressure and tension arise whenever a cell moves, and we saw in Chapter 18 that these can influence molecules of movement. Integrins, for example, which are carried on growth cones, establish a physically robust linkage with extracellular molecules. The tension transients experienced by integrins will modulate the biochemical signals they relay to the actin cortex.

The response of a growth cone at any specific location of the embryo will be highly individualistic, so that different growth cones moving over the same terrain can and do select different paths. Their integration of multiple environmental signals and the eventual decision to take one or another path are achieved by a complex molecular machinery built from cytoskeletal proteins and associated signaling molecules. Elucidation of the workings of this 'onboard computer' is a fascinating endeavor for the future.

## *Outstanding Questions*

*What molecular machinery carries mRNA molecules such as bicoid along microtubules in a specific direction? How is this movement regulated? What is the molecular basis of the rules that govern cleavage orientation in early embryos? How do spindle microtubules make contact with the cell cortex? Is it by dynein? If dynein is responsible for centrosomal movements, how is it positioned? What attaches it to the actin cortex? How are its movement regulated? What triggers the rearrangement of cells in an epithelial sheet (as in epiboly)? How are the cells guided—by diffusible molecules, by molecules on neighboring cell surfaces, or by the underlying basal lamina? How are the evaginations of an epithelial sheet caused? What is the molecular basis of growth cone guidance, and how do receptors on the growth cone surface mediate growth or collapse? What is the molecular logic of the guidance circuitry?*

## *Essential Concepts*

- The development of form in a multicellular animal occurs through stereotyped sequences of simple cell movements and shape changes that depend on the cytoskeleton.
- At the earliest stage, regional differences in the cytoplasm of the eggs of many species are established during oogenesis or in the course of fertilization.
- An important mechanism in the establishment of regional cytoplasmic differences is the targeted delivery of mRNA molecules by means of motor proteins and their anchorage to sites on the actin cytoskeleton.
- Precisely oriented planes of cleavage then allocate these different regions of cytoplasm to specific daughter cells.
- In specific instances, the orientation decision is controlled by specific interactions between spindle microtubules and dynactin molecules at sites on the cell cortex.
- As the ball of embryonic cells grows and divides, the expression of adhesive molecules such as cadherins on the surfaces of cells causes them to adhere together.
- A superficial layer of epithelial cells forms, establishing a centrally located cavity within which are mesenchymal cells. Many subsequent morphogenetic events depend on the separate behaviors of these two types of cells.
- Epithelial sheets are a dominant force during development. They can expand, contract, thicken, or fold through the coordinate changes in shape or movement of their constituent cells, under the ultimate direction of morphogenetic genes.
- Expansion of epithelial sheets, for example during gastrulation, is driven by the flattening of individual cells and the intercalated addition of new cells from deeper layers.
- Invagination of an epithelial sheet is initiated by the distortion of cells to a bottle shape caused by contraction of their apical actin cortex. Changes of this kind are widespread during embryonic development.
- Cells can migrate within an epithelial sheet without breaching its integrity, and concerted movements of this kind drive changes in shape of entire tissues.
- The nervous system forms from the epithelial layer of the ectoderm by an initial thickening of cells driven by microtubule elongation. Cells then roll up into a tube and elongate by mechanisms already mentioned.
- Successive generations of neuroblasts are produced by division of the basal layer of the neural ectoderm. Newly formed cells then migrate, often over large distances, to take up their final positions in the brain and spinal cord.
- In the developing vertebrate brain, successive waves of newly made neuroblasts migrate up glial pillars to form the layered structure of the different cortical layers.
- Axonal connections of the brain, on which its powers of computation depend, are established by the directed migration of growth cones, which follow multiple cues, either diffusing or in the extracellular matrix.
- The pathway followed by a growth cone depends on the receptors it carries on its surface and the complex network of biochemical and physical signals that pass from these receptors to the cytoskeleton.

## Further Reading


Adams, D.S., et al. The mechanics of notochord elongation, straightening and stiffening in the embryo of *Xenopus laevis*. *Development* 110: 115–130, 1990.

Bashaw, G.J., Goodman, C.S. Chimeric axon guidance receptors: the cytoplasmic domains of slit and netrin receptors specify attraction versus repulsion. *Cell* 97: 917–926, 1999.

Carminati, J.L., Stearns, T. Microtubules orient the mitotic spindle in yeast through dynein-dependent interactions with the cell cortex. *J. Cell Biol.* 138: 629–641, 1997.

Conklin, E.G. Effects of centrifugal force on the structure and development of the eggs of *Crepidula*. *J. Exp. Zool.* 22: 311–419, 1917.

Cooley, L.C., Theurkauf, W.E. Cytoskeletal function during *Drosophila* oogenesis. *Science* 266: 590–596, 1994.

Driever, W., Nusslein-Vollard, C. The *bicoid* protein determines position in the *Drosophila* embryo in a concentration-dependent manner. *Cell* 54: 95–104, 1988.

Goldstein, B. Cell contacts orient some cell division axes in the *Caenorhabditis elegans* embryo. *J. Cell Biol.* 120: 1071–1080, 1995.

Gumbiner, B.M. Regulation of cadherin adhesive activity. *J. Cell Biol.* 148: 399–403, 2000.

Hatten, M.E. Central nervous system neuronal migration. *Annu. Rev. Neurosci.* 22: 511–539, 1999.

Holtfreter, J. A study of the mechanics of gastrulation. Part I. *J. Exp. Zool.* 94: 261–318, 1943.

Jansen, R.P. RNA–cytoskeletal associations. *FASEB J.* 13: 455–466, 1999.

Kiehart, D.P. Wound healing: the power of the purse string. *Curr. Biol.* 9: R602–R605, 1999.

Kimberly, E.L., Hardin, J. Bottle cells are required for the initiation of primary invagination in the sea urchin embryo. *Dev. Biol.* 204: 235–250, 1998.

Kirschner, M., et al. Molecular 'vitalism.' *Cell* 100: 79–88, 2000.

Kothakota, S., et al. Caspase-3-generated fragment of gelsolin: effector of morphological change in apoptosis. *Science* 278: 294–298, 1997.

Lanier, L.M., et al. Mena is required for neurulation and commissure formation. *Neuron* 22: 313–325, 1999.

Lois, C., et al. Chain migration of neuronal precursors. *Science* 271: 978–981, 1996.

Lutz, D.A., et al. Micromanipulation studies of the asymmetric positioning of the maturation spindle in *Chaetopterus* sp. oocytes. I Anchorage of the spindle to the cortex and migration of a displaced spindle. *Cell Motil. Cytoskeleton* 11: 83–96, 1988.

Miller, J.R., et al. Establishment of the dorsal–ventral axis in *Xenopus* embryos coincides with the dorsal enrichment of dishevelled that is dependent on cortical rotation. *J. Cell Biol.* 146: 427–437, 1999.

Mittenthal, J.E., Jacobson, A.G. The Mechanics of Morphogenesis in Multicellular Embryos. Berlin, Heidelberg: Springer-Verlag, 1990.

Neujahr, R., et al. Microtubule-mediated centrosome motility and the positioning of cleavage furrows in multinucleate myosin II-null cells. *J. Cell Sci.* 111: 1227–1240, 1998.

Oleynikov, Y., Singer, R.H. RNA localization: different zipcodes, same postman? *Trends Cell Biol.* 8: 381–383, 1998.

Poo, M.M. Signal transduction underlying growth cone guidance by diffusible factors. *Curr. Opin. Neurobiol.* 9: 355–363, 1999.

Sisson, J.C., et al. *Costal2*, a novel kinesin-related protein in the Hedgehog signalling pathway. *Cell* 90: 235–245, 1997.

St. Johnston, D., Nüsslein-Volhard, C. The origin of pattern and polarity in the *Drosophila* embryo. *Cell* 68: 201–219, 1992.

Steinberg, M.S. Reconstruction of tissues by dissociated cells. *Science* 141: 401–408, 1963.

Strome, S. Determination of cleavage planes. *Cell* 72: 3–6, 1993.

Sullivan, W.P., et al. Mutations affecting the cytoskeletal organization of syncytial *Drosophila* embryos. *Development* 118: 1245–1254, 1993.

Thorpe, C.J., et al. Wnt signalling in *Caenorhabditis elegans*: regulating repressors and polarizing the cytoskeleton. *Trends Cell Biol.* 10: 10–17, 2000.

Torres, M.A., Nelson, W.J. Colocalization and redistribution of dishevelled and actin during Wnt-induced mesenchymal morphogenesis. *J. Cell Biol.* 149: 1433–1442, 2000.

Townes, P.L., Holtfreter, J. Directed movements and selective adhesion of embryonic amphibian cells. *J. Exp. Zool.* 128: 53–120, 1955.

Trinkaus, J.P., et al. On the convergent cell movements of gastrulation in *Fundulus*. *J. Exp. Zool.* 261: 40–61, 1992.

Vasioukhin, V., et al. Directed actin polymerization is the driving force for epithelial cell–cell adhesion. *Cell* 100: 200–219, 2000.

Waddle, J.A., et al. Transient localized accumulation of actin in *Caenorhabditis elegans* blastomeres with oriented asymmetric divisions. *Development* 120: 2317–2328, 1994.

Zheng, J.Q. Turning of nerve growth cones induced by localized increases in intracellular calcium ions. *Nature* 403: 89–93, 2000.

***Acanthamoeba castellani***
Species of soil amoeba that forms flexible, flattened pseudopodia. Easy to cultivate in bulk and often used as a source of cytoskeletal proteins.

**acrosomal process**
Actin-containing spike produced from the head of certain sperm when they make contact with the egg. Seen in sea urchins and other marine invertebrates, where the eggs are surrounded by a thick gelatinous coat.

**acrosome**
Head end of a sperm cell containing a sack of hydrolytic enzymes used to digest the protective coating of the egg.

**actin**
Abundant protein that forms filaments (microfilaments) in all eucaryotic cells. The monomeric form is sometimes called globular or G-actin; the polymeric form is filamentous or F-actin.

**actin-binding protein-280 (ABP-280)**—*see* **filamin**

**actin-related protein**—*see* **Arp**

**actin depolymerizing factor (ADF)**—*see* **cofilin**

**actin filament (microfilament)**
Protein filament formed by the polymerization of globular actin molecules. Major constituent of the cytoskeleton of all eucaryotic cells and part of the contractile apparatus of skeletal muscle.

**α-actinin**
An F-actin-binding and cross-linking protein originally isolated from skeletal muscle, where it is a major component of the Z-disc. Distinct isoforms of the protein are found in smooth-muscle and nonmuscle cells.

**actobindin**
Small protein from *Acanthamoeba castellani* that binds to actin and inhibits its polymerization.

**actomyosin**
Complex formed from actin filaments and myosin II.

**adaptation**
A decrease in sensitivity to a repeated or prolonged stimulus. In chemotaxis, adaptation to one concentration of attractant prepares the cell for a response to an even higher concentration.

**adducin**
Protein from erythrocytes that binds to the pointed end of an actin filament and promotes its association with spectrin. Also binds calmodulin and is regulated by phosphorylation. (Latin *adducere*, to draw together.)

**adherens junction**
Cell junction in which the cytoplasmic face is attached to actin filaments. Examples include the adhesion belts linking adjacent epithelial cells and the focal adhesions on the lower surface of cultured fibroblasts.

**adhesion belt (zonula adherens)**
Beltlike adherens junction that encircles the apical end of an epithelial cell and attaches it to adjoining cells. A contractile bundle of actin filaments runs along the cytoplasmic surface of the adhesion belt.

**adhesion plaque**—*see* **focal adhesion**

**ADP (adenosine 5′-diphosphate)**
Nucleotide produced by hydrolysis of the terminal phosphate of ATP. It regenerates ATP when phosphorylated by an energy-generating process such as oxidative phosphorylation.

**alga** (plural **algae**)
Informal term used to describe a wide range of photosynthetic organisms, either procaryotic or eucaryotic. Eucaryotic examples include *Nitella*, *Volvox*, and *Fucus*.

**amoeba**
Free-living single-celled eucaryote that crawls by changing its shape. More narrowly, a particular genus of protozoa that move in this way.

***Amoeba proteus***
Species of giant freshwater amoeba widely used in studies of cell locomotion.

**amoeboid locomotion**
Distinctive form of cell crawling typified by *Amoeba proteus*. Associated with the extension of pseudopodia and with cytoplasmic streaming.

**anaphase**
Stage of mitosis during which the two sets of chromosomes separate and move away from each other. Comprises anaphase A (chromosomes move toward the two spindle poles) and anaphase B (spindle poles move apart).

**ankyrin**
One of a family of proteins that couple cell surface proteins such as ion channels and cell adhesion molecules to the spectrin–actin skeleton on the cytoplasmic surface of the plasma membrane.

**annexins**
The annexins are a family of $Ca^{2+}$-dependent phospholipid-binding proteins. They are differentially expressed and demonstrate unique subcellular localizations. These proteins may be mediators of the intracellular calcium signal.

**anterograde transport**
Movement of organelles in a nerve axon in the direction away from the cell body. The converse of retrograde transport.

**APC (anaphase-promoting complex)**
Set of associated proteins that directs the proteolytic degradation of components of the cell cycle machinery, thereby terminating metaphase and initiating anaphase.

**apical**
Describes the tip of a cell, structure, or organ. The apical surface of an epithelial cell is the exposed free surface, opposite to the basal surface.

**Arp (actin-related protein)**
A family of proteins characterized by sequence homology with actin. Arp1 is part of the dynactin complex, and Arp2 and 3 are involved in actin filament nucleation.

**Arp2/3 complex**
Cluster of seven proteins including the actin-related proteins Arp2 and Arp3 that regulates actin filaments formation at the leading edge of cells.

**ATP (adenosine 5′-triphosphate)**
Nucleoside triphosphate composed of adenine, ribose, and three phosphate groups that is the principal carrier of chemical energy in cells. The terminal phosphate groups are highly reactive in the sense that their hydrolysis, or transfer to another molecule, takes place with release of a large amount of free energy.

**ATPase**
A large class of enzymes that catalyze a process that involves the hydrolysis of ATP. Motor proteins such as myosin and kinesin are ATPases.

**axon**
Long nerve cell process that is capable of rapidly conducting nerve impulses over long distances so as to deliver signals to other cells.

**axonal transport**
Directed transport of organelles and molecules along a nerve cell axon; can be anterograde (*from* the cell body) or retrograde (*toward* the cell body).

**axonemal dynein**—*see* **dynein**

**axoneme**
Bundle of microtubules and associated proteins that forms the core of a cilium or flagellum in a eucaryotic cell and is responsible for their bending movements.

**axoplasm**
Cytoplasm of the axon of a nerve cell, especially after it has been extruded from the axon.

**axopodium**
Type of long and rather rigid protrusion, supported by bundles of microtubules, found in radial arrays on the surface of *Heliozoa*.

**axostyle**
Bundle of microtubules found in the cytoplasm of certain protozoa that propagates vigorous bending movements along its length.

**band 4.1**
Protein that promotes complex formation between spectrin and actin and links the erythrocyte membrane skeleton to the overlying lipid bilayer. Belongs to the same family as ERM proteins.

**barbed end**
Fast-growing, 'plus' end of an actin filament—named for its appearance following decoration with myosin. *See also* pointed end.

**basal**
Situated near the base. The basal surface of a cell is opposite the apical surface.

**basal body**
Short cylindrical array of microtubules and associated proteins found at the base of a eucaryotic cell cilium or flagellum. Serves as a nucleation site for the growth of the axoneme. Closely similar in structure to a centriole.

**basal lamina** (plural **basal laminae)**
Thin mat of specialized extracellular matrix that underlies epithelial sheets, and surrounds many types of cells such as muscle cells and fat cells.

**basement membrane**—*see* **basal lamina**

**blastula**
An early embryo at the stage when the cells are typically arranged to form a hollow sphere.

**blepharoplast**
Electron-dense spherical structure found in the cytoplasm of ferns and related plant species from which basal bodies form.

**bound water**
Water that is closely associated with macromolecules: 10–20% of the total water inside cells.

**BPAG (bullosa pemphigus antigen)**
A member of the plakin family of cross-linker proteins that link intermediate filaments to other components of the cytoskeleton. Autoantibodies to BPAG cause disruption of skin and mucous membranes.

**brush border**
Dense covering of microvilli on the apical surface of epithelial cells in the intestine and kidney; the microvilli aid absorption by increasing the surface area of the cell.

**brush border myosin I**
A single-headed, membrane-associated myosin expressed in vertebrate intestinal epithelial cells. It is primarily localized within the microvilli of the apical surface where it comprises the bridges that laterally tether the microvillar actin core to the plasma membrane.

**C protein**
Protein located with the A-band of vertebrate cross-striated muscles in a series of 43 nm transverse stripes. Binds both to myosin and to titin.

**cadherin**
One of a large family of proteins that mediate $Ca^{2+}$-dependent cell–cell adhesion in animal tissues.

**caldesmon**
Actin-binding protein found in smooth-muscle and many nonmuscle cells. Shows a reversible inhibition of the ATPase activity of actomyosin, reversed by calmodulin in the presence of $Ca^{2+}$. Because of this property, caldesmon is thought to be involved in the regulation of smooth muscle.

**calmodulin**
Ubiquitous $Ca^{2+}$-binding protein whose binding to other proteins is governed by changes in intracellular $Ca^{2+}$ concentration. Its binding modifies the activity of many target enzymes and membrane transport proteins.

**calponin**
Basic, low-molecular-weight (34 kDa) actin- and calmodulin-binding protein from smooth muscle. May act as a thin filament-linked regulator of smooth-muscle contraction.

**CAM**—*see* **cell-adhesion molecule**

**capping proteins**
Ubiquitous family of proteins that bind tightly to the barbed ends of actin filaments. They are inhibited by $PIP_2$ but insensitive to $Ca^{2+}$. CapZ is a capping protein found in skeletal muscle and Cap

32/34 is from *Dictyostelium*.

**cardiac muscle**
Specialized form of striated muscle found in heart, consisting of individual heart muscle cells linked together by cell junctions.

**catenins**
Proteins that interact with the cytoplasmic region of the cell adhesion molecule uvomorulin/E-cadherin. Catenins mediate the cytoplasmic anchorage of E-cadherin to the actin filament network.

**Cdc42**—*see* **Rho**

**Cdk**—*see* **cyclin-dependent protein kinase**

**cell adhesion molecule (CAM)**
Protein on the surface of an animal cell that mediates cell–cell binding.

**cell body**
Main part of a nerve cell that contains the nucleus. The other parts are axons and dendrites.

**cell cortex**
Specialized layer of cytoplasm on the inner face of the plasma membrane. In animal cells it is an actin-rich layer responsible for cell-surface movements.

**cell fusion**
Process in which the plasma membranes of two cells break down at the point of contact between them, allowing the two cytoplasms to mingle.

**cell junction**
Specialized region of connection between two cells or between a cell and the extracellular matrix.

**cell line**
Population of cells of plant or animal origin capable of dividing indefinitely in culture.

**centractin (Arp-1)**
Member of the family of actin-related proteins. Part of dynactin.

**centrin (caltractin)**
A calcium-binding protein found in association with cytoskeletal structures such as striated rootlets and spasmonemes that undergo $Ca^{2+}$-mediated contraction.

**centriole**
Short cylindrical array of microtubules, closely similar in structure to a basal body. A pair of centrioles is usually found at the center of a centrosome in animal cells.

**centromere**
Constricted region of a mitotic chromosome that holds sister chromatids together; also the site on the DNA where the kinetochore forms and then captures microtubules from the mitotic spindle.

**centrosome (cell center)**
Centrally located organelle of animal cells that is the primary microtubule-organizing center and acts as the spindle pole during mitosis. In most animal cells it contains a pair of centrioles.

**chaperone**—*see* **molecular chaperone**

**Che protein ('key' protein)**
One of the proteins that relays chemotactic signals within a bacterium.

**chelating agent**
Organic molecule that forms multiple bonds with a metal ion such as iron, calcium, or magnesium. *See* EDTA.

**chemoattractant**
Substance that attracts a motile cell or organism.

**chemotaxis**
Motile response of a cell or organism that carries it toward or away from a diffusible chemical.

***Chlamydomonas***
Unicellular green alga with two flagella.

**chromatophore**
A pigment-containing cell found in the skin. Chromatophores of fish and amphibia provide an accessible system for the study of organelle transport.

**cilium** (plural **cilia**)
Hairlike extension of a cell containing a core bundle of microtubules and capable of performing repeated beating movements. Cilia are found in large numbers on the surface of many eucaryotic cells and are responsible for the swimming of many single-celled organisms.

**cirrus** (plural **cirri**)
A cluster of cilia, typically in a conical arrangement, that functions as a unit. Protozoa use cirri to drive water currents over the cell for example, or, in hypotrichs, as a locomotive appendage.

**clathrin**
Structural unit that forms the characteristic outer polyhedral cage on the cytoplasmic surface of coated vesicles.

**cofilin (ADF)**
Small protein, widely distributed in vertebrate cells, that binds to ADP-actin and controls depolymerization of actin filaments. Regulated by binding of phosphoinositides and phosphorylation.

**coiled-coil**
Especially stable rodlike protein structure formed by two $\alpha$ helices coiled around each other. Myosin II, tropomyosin, and intermediate filament proteins all contain extensive regions of coiled-coil.

**collagen**
Fibrous protein rich in glycine and proline that is a major component of the extracellular matrix and connective tissues. Exists in many forms: type I, the most common, is found in skin, tendon, and bone; type II is found in cartilage; type IV is present in basal laminae; and so on.

**conformation**
Spatial location of the atoms of a molecule—for example, the precise shape of a protein or other macromolecule in three dimensions.

**conformational change**
A precisely-defined, switchlike transition in the shape of a protein molecule, often triggered by an external stimulus, such as the binding of an ion or small molecule or the phosphorylation of one of its side chains.

**coronin**
An actin-binding protein from *Dictyostelium* with sequence homology to the $\beta$ subunit of trimeric G proteins. Accumulates dynamically at the leading edge of a migrating cell. Mutants lacking coronin are defective in cytokinesis and cell motility.

**cortex**—*see* **cell cortex**

**cortexillin**
Actin-binding protein from *Dictyostelium*, related to spectrin but with a parallel coiled-coil tail. Cortexillin links actin filaments to the plasma membrane and has an important function during cytokinesis.

**critical concentration**
Concentration of an unassembled protein, such as actin or tubulin, that is in equilibrium with its polymer.

**cyclins**
Proteins that activate crucial protein kinases (called cyclin-

dependent protein kinases) and thereby help control progression from one stage of the eucaryotic cell cycle to the next. The concentrations of cyclins periodically rise and fall in step with the cell cycle.

**cyclin-dependent protein kinase (Cdk)**
Protein kinase that has to be complexed with a cyclin protein in order to act; different Cdk–cyclin complexes trigger different steps in the cell division cycle by phosphorylating specific target proteins.

**cytochalasin**
One of a family of fungal metabolites that have an inhibitory effect on the polymerization of actin filaments and therefore block many actin-based cell movements such as cell crawling and phagocytosis.

**cytokeratin**—*see* **keratin**

**cytokinesis**
Division of the cytoplasm of a plant or animal cell into two, as distinct from the division of its nucleus (which is mitosis).

**cytoplasm**
Contents of a cell that are contained within its plasma membrane but, in the case of eucaryotic cells, outside the nucleus.

**cytoplasmic dynein**
A large, two-headed ATPase that moves along microtubules toward the minus end. Contributes to many intracellular movements such as the transport of membranous organelles in nerve axons and the segregation of chromosomes during mitosis.

**cytoskeleton**
Extensive system of protein filaments (and their various attachments) that enables a eucaryotic cell to organize its interior and to perform directed movements. Its most abundant components are actin filaments, microtubules, and intermediate filaments.

**cytosol**
Contents of the main compartment of the cytoplasm, excluding membrane-bounded organelles such as endoplasmic reticulum and mitochondria. Originally defined operationally as the cell fraction remaining after membranes, cytoskeletal components, and other organelles have been removed by low-speed centrifugation.

**decoration**
Technique in which myosin heads are added to actin filaments to produce a series of tangential projections, like arrowheads, thereby revealing the filament polarity.

**dendrite**
Extension of a nerve cell, typically branched and relatively short, that receives stimuli from other nerve cells.

**dense body**
Ovoid, electron-dense structure around 100 nm in diameter present in large numbers in the cytoplasm of a smooth-muscle cell. Site of attachment of myofibrils to the structural network of intermediate filaments.

**dense plaque**
Region of a smooth-muscle cell plasma membrane that forms a link between actin filaments of the contractile apparatus and the extracellular matrix. Similar to a focal adhesion but larger.

**desmin**
Intermediate filament protein abundant in muscle cells. Smooth-muscle cells in particular are full of desmin.

**desmoplakins**
Major proteins of desmosomes which anchor intermediate filaments to the desmosomal plaque. Related proteins, such as plectin and BPAG act as linking proteins for intermediate filaments in other parts of the cytoplasm.

**desmosome**
Specialized cell–cell junction, usually formed between two epithelial cells, characterized by dense plaques of protein into which intermediate filaments in the two adjoining cells insert.

***Dictyostelium discoideum***
Cellular slime mold widely used in the study of cell locomotion, chemotaxis, and differentiation.

**differentiation**
Process by which a cell undergoes a progressive change to a more specialized and usually easily recognized cell type.

**diffusion**
Net drift of molecules toward regions of lower concentrations, due to random thermal movement.

**doublet microtubule**
One of nine paired microtubules found in a ciliary axoneme. Composed of one complete microtubule with 13 protofilaments fused to an incomplete microtubule having only 10 protofilaments.

**duty ratio**
The fraction of time a motor protein spends attached to the protein filament along which it is moving. Processive motors such as conventional kinesin have large duty ratios.

**dynactin**
Protein complex, with 10 or so subunits, that links cytoplasmic dynein to actin-containing structures in the cell. Part of the molecular apparatus that drives microtubules through the cytoplasm.

**dynamic instability**
The property shown by equilibrium mixtures of tubulin and microtubules of existing in persistent states of polymerization or depolymerization and only occasionally switching between the two.

**dynein**
Member of a family of giant motor proteins that undergo ATP-dependent movement along microtubules. Axonemal dyneins, located in cilia and flagella, generate bending movements. Cytoplasmic dyneins drive organelle transport.

**dystrophin**
A high-molecular-weight member of the spectrin superfamily of cytoskeletal proteins and associated with the plasma membrane of muscle and neuronal tissues. When defective, dystrophin gives rise to the Duchenne/Becker muscular dystrophies.

***E. coli*****—*see* *Escherichia coli***

**EDTA**
Ethylenediaminetetraacetic acid, a chelating agent that binds to $Mg^{2+}$ and $Ca^{2+}$ ions.

**EF hand**
Calcium-binding motif found in troponin C, myosin light chains, and other proteins. Named for helices E and F of parvalbumin that hold $Ca^{2+}$ as if in a half-open right hand.

**efficiency**
Useful work performed by a molecular motor, such as myosin or the bacterial flagellar motor, as a proportion of the energy supplied to it (by ATP hydrolysis or a proton gradient).

**endocytosis**
Uptake of material into a cell by internalization into membrane-bounded vesicles (*see also* phagocytosis).

**endoplasmic reticulum (ER)**
Labyrinthine, membrane-bounded compartment in the cytoplasm of eucaryotic cells, where lipids are secreted and membrane-bound proteins are made.

**entropic spring**
Randomly coiled chain of amino acids that stretches reversibly under tension. It works like spring because it tends to bend and

flex as it is buffeted by thermal fluctuations. A force is required to counteract these thermal forces and to extend the chain.

**Ena/Mena**
Proline-rich adaptor proteins, related to VASP, that help nucleate actin filament assembly in filopodia and adherens junctions. Ena is produced by the *enabled* gene in *Drosophila* and Mena is its equivalent in mammals.

**epigenetic information**
Instructions that can be passed from one cell to its descendants without the participation of DNA, such as the particular orientation or configuration of a cytoskeletal structure.

**epithelium**
Sheet of one or more layers of cells covering an external surface or lining a cavity. Epithelia are usually a barrier to diffusion and often carry specialized structures on their apical surface.

**ER**—*see* **endoplasmic reticulum**

**ERM proteins**
Group of proteins, including ezrin, radixin and moesin, that link actin filaments to plasma membranes. Localized to cleavage furrows, microvilli, ruffling membranes, and cell junctions.

**erythrocyte (red blood cell)**
Small, hemoglobin-containing blood cell of vertebrates that transports oxygen and carbon dioxide to and from tissues. (From Greek *erythros*, red.)

***Escherichia coli* (*E. coli*)**
Rodlike bacterium that inhabits the colon of humans and other mammals and is widely used in biomedical research.

**exocytosis**
Process by which most molecules are secreted from a eucaryotic cell. These molecules are packaged in membrane-bounded vesicles that fuse with the plasma membrane, releasing their contents to the outside.

**explant**
Either, to take tissue from a plant or animal and put it into nutrient media suitable for growth (explantation); or, the piece of tissue thus removed.

**ezrin**—*see* **ERM proteins**

**FAK ($p125^{FAK}$, focal adhesion kinase)**
Protein tyrosine kinase found in focal adhesions. Important both for formation of focal adhesions and for the signals they send to the rest of the cell.

**fascin**
Actin-binding protein associated with actin filament bundles. Originally identified in the microvilli that form on the surface of sea urchin eggs but subsequently found in many other locations, including human cells.

**fibroblast**
Common cell type found in connective tissue that secretes an extracellular matrix rich in collagen and other extracellular matrix macromolecules. Migrates and proliferates readily in wounded tissue and in tissue culture.

**fibroblast locomotion**
Form of crawling motility shown by fibroblasts in tissue culture. Characterized by advancing lamellipodia that show ruffling movements.

**filaggrins**
A family of intermediate filament-associated, positively charged proteins expressed primarily in terminally differentiating mammalian epidermis. Filaggrins cause keratin filaments to form tight bundles.

**filamin (ABP-280)**
An abundant dimeric actin-cross-linking protein found in the cortex of vertebrate cells. Promotes orthogonal branching of actin filaments and links actin filaments to membrane glycoproteins.

**filopodium (microspike)**
Long, thin, actin-containing extension on the surface of an animal cell. Long filopodia on a neuronal growth cone have an exploratory function.

**fimbrin**
Monomeric F-actin-bundling protein found in microvilli, filopodia, stereocilia, membrane ruffles, and adhesion sites of nonmuscle cells.

**flagellin**
Protein molecule that forms the subunit of bacterial flagella.

**flagellum** (plural **flagella**)
Long, whiplike protrusion whose undulations drive a cell through a fluid medium. Eucaryotic flagella are longer versions of cilia; bacterial flagella are completely different, being smaller and simpler in construction.

**focal adhesion (focal contact, adhesion plaque)**
Small region on the surface of a fibroblast or other cell that is anchored to the extracellular matrix. The attachment is mediated by transmembrane proteins such as integrins, the cytoplasmic domains of which are linked, through other proteins, to actin filaments.

**foldback inhibition**
The strategy in which a protein molecule folds so as to inhibit or hide its active site. Specific signals, such as phosphorylation or calcium ions, then cause the molecule to unfold and become active. Seen in ERM proteins and some myosins and kinesins.

**FtsZ ('footsie')**
Bacterial protein that forms filaments in the cleavage ring of dividing bacteria. Related in structure to tubulin.

**gap junction**
Cell–cell communicating junction that allows ions and small molecules to pass from the cytoplasm of one cell to the cytoplasm of the next.

**gastrulation**
Cell movements in a developing embryo that establish three germ layers. Most importantly, the invagination and spreading of cells to form the rudiment of a gut cavity. (From Greek *gaster*, belly.)

**gelsolin**
A $Ca^{2+}$- and polyphosphoinositide-regulated vertebrate actin filament-severing protein. It is involved in the restructuring of the actin cytoskeleton in a variety of motile events.

**germ cells**
Precursor cells that give rise to gametes, which contribute to the formation of a new generation of organisms (as distinct from somatic cells, which form the body and leave no descendants).

**glial fibrillary acidic protein (GFAP)**
A member of the intermediate filament protein superfamily. Expressed almost exclusively in astrocytes and related cells and therefore used extensively as a cell type marker.

**gliding**
Form of motility in which a cell moves over a surface without locomotory organelles or obvious changes in cell shape. Shown by individual species of bacteria, algae, and protozoa.

**green fluorescent protein (GFP)**
Small, intensely fluorescent protein obtained from a jellyfish. Widely used to tag cytoskeletal proteins and follow their distribution in the cell.

**growth cone**
Migrating motile tip of a growing nerve cell axon or dendrite.

**GTP (guanosine 5′-triphosphate)**
Reactive molecule similar to ATP but with guanine in place of adenine. Often found in close association with proteins (such as tubulin and Rho) where its hydrolysis to GDP is accompanied by a conformational change.

**hair cell**
Specialized sensory epithelial cell in the ear with bundles of giant microvilli (stereocilia) protruding from its apical surface. Sound vibrations tilt the stereocilia, evoking an electrical change in the hair cell, which thus acts as a sound detector.

**haploid**
Having only one set of chromosomes, as in a sperm cell or a bacterium, as distinct from diploid (two sets of chromosomes).

**heliozoa**
Class of protozoa in which numerous long axopodia, filled with microtubules, radiate from a central cell body ('sun animalcules').

**homophilic**
Adjective used to describe a molecule that binds to others of the same kind, especially those involved in cell–cell adhesion.

**IFT particle (intraflagellar transport particle)**
Collection of preassembled axonemal parts, such as dynein arms and radial spokes, used to build a cilium or flagellum. IFT particles travel from the basal body to the axoneme tip (and back again) by means of motor proteins.

**intermediate filament**
Fibrous protein filament, about 10 nm in diameter, that forms ropelike networks in animal cells. One of the three most prominent types of cytoskeletal filaments.

**interphase**
Long period of the cell cycle between one mitosis and the next. Includes $G_1$ phase, S phase, and $G_2$ phase.

**$IP_3$ (inositol trisphosphate)**
Small water-soluble molecule produced by the cleavage of the inositol phospholipid $PIP_2$ in response to extracellular signals; causes release of $Ca^{2+}$ from the endoplasmic reticulum.

**IQ motif**
Amino acid sequences in the neck region of a myosin molecule to which light chains bind. Named for their content of isoleucine (I) and glutamine (Q).

**isoform**
One of multiple forms of the same protein that differ somewhat in their amino acid sequence. They can be produced by different genes or by alternative splicing of RNA transcripts from the same gene.

**isotype**—*see* **isoform**

**katanin**
Protein that severs microtubules. Typically associated with the centrosome, where it releases microtubules to move in the cytoplasm. Named after a *katana*—the long, single-edged sword of the Japanese samurai.

**kDa (kilodalton)**
Unit equal to 1000 daltons used in the measurement of molecular weight of protein, DNA and other macromolecules. One dalton is approximately equal to the mass of a hydrogen atom.

**keratin (cytokeratin)**
One of a large family of proteins that form keratin intermediate filaments, mainly in epithelial cells. Specialized keratins are found in hair, nails, and feathers.

**kinesin**
One of a large family of motor proteins that uses the energy of ATP hydrolysis to move along a microtubule. Conventional kinesin drives organelles along microtubules toward their plus end.

**kinesis**
The nondirectional movement of an organism or cell in response to a stimulus, the rate of movement being dependent on the strength of the stimulus. *See* taxis.

**kinetochore**
Complex structure formed from proteins on a mitotic chromosome to which microtubules attach and which plays an active part in the movement of chromosomes to the pole. The kinetochore forms on the part of the chromosome known as the centromere.

***kT***
Loosely speaking, the energy of thermal fluctuation. $k$ is the Boltzmann constant and $T$ the absolute temperature. At room temperature, $kT$ is around $4 \times 10^{-14}$ g $cm^2/sec^2$ (ergs). A particle or molecule frequently exchanges energy with its surroundings by roughly this amount.

**lamellipodium (lamellipod)**
Dynamic sheetlike extension, rich in actin filaments, found on the surface of an animal cell, especially one migrating over a surface.

**lamin**
One of a family of intermediate filament proteins that form the filamentous matrix (nuclear lamina) on the inner surface of the nuclear envelope.

**laser tweezers**
An optical device, used in conjunction with a light microscope, in which a focused beam of light from a laser is used to move small objects.

**latrunculin**
Actin-depolymerizing drug obtained from Dead Sea sponges.

**lipid bilayer**
Thin bimolecular sheet of mainly phospholipid molecules that forms the structural basis for all cell membranes. The two layers of lipid molecules are packed with their hydrophobic tails pointing inward and their hydrophilic heads outward, exposed to water.

**lobopodium (lobopod)**
Broad, thick protrusion seen in amoeboid cells. A type of pseudopodium.

**lumen**
Cavity enclosed by an epithelial sheet (in a tissue) or by a membrane (in a cell).

**lymphocyte**
White blood cell that makes an immune response when activated by a foreign molecule (an antigen).

**lysis**
Rupture of a cell's plasma membrane, leading to the release of cytoplasm and the death of the cell.

**macromolecule**
Molecule such as a protein, nucleic acid, or polysaccharide with a molecular mass greater than a few thousand daltons. (Macro from Greek *makros*, large.)

**macrophage**
White blood cell that is specialized for the uptake of particulate material by phagocytosis.

**major sperm protein (MSP)**
Small, basic polypeptide found exclusively in the sperm of nematodes. MSP forms the meshwork of fine filaments within the pseudopodia (villipodia) of these actin-deficient amoeboid cells.

**MAP (microtubule-associated protein)**
Large set of proteins, originally isolated from brain, that bind to microtubules and named MAP1, MAP2 and so on. Confusingly, the same initials are also used to designate 'M-phase-associated proteins,' only some of which bind to microtubules.

**MAP2**
Microtubule-associated protein found specifically in neuronal cells. MAP2 binds to microtubules and promotes their assembly from purified tubulin.

**MARKS (myristoylated alanine-rich C-kinase substrate)**
Substrate of protein kinase C (PKC) in neurons and leucocytes that cross-links actin filaments.

**marginal band**
Ring of microtubules that determines the shape of blood platelets and nonmammalian erythrocytes.

**meiosis**
Special type of cell division by which eggs and sperm cells are made. (From Greek *meiosis*, diminution.)

**melanophore**—*see* **chromatophore**

**membrane ruffling**—*see* **ruffling**

**metaphase**
Stage of mitosis at which chromosomes are firmly attached to the mitotic spindle at its equator but have not yet segregated toward opposite poles.

**metaphase plate**
Imaginary plane at right angles to the mitotic spindle and midway between the spindle poles; the plane in which chromosomes are positioned at metaphase.

**microfibril**
Linear bundle of cellulose molecules in the wall of a plant cell.

**microfilament**—*see* **actin filament**

**microspike**—*see* **filopodium**

**microtubule**
Long, stiff, cylindrical protein filament composed of the protein tubulin, one of the three major classes of filaments of the cytoskeleton.

**microtubule-organizing center (MTOC)**
Region in a cell, such as a centrosome or a basal body, from which microtubules grow.

**microvillus** (plural **microvilli)**
Thin cylindrical membrane projection on the surface of an animal cell containing a core bundle of actin filaments. Present in especially large numbers on the absorptive surface of intestinal epithelial cells.

**minus end**
The end of a microtubule or actin filament at which the addition of monomers occurs least readily; the 'slow-growing' end of the microtubule or actin filament. The minus end of an actin filament is also known as the pointed end.

**mitosis**
Division of the nucleus of a eucaryotic cell, involving condensation of the DNA into visible chromosomes. (From Greek *mitos*, a thread, referring to the threadlike appearance of the condensed chromosomes.)

**mitotic spindle**
Array of microtubules and associated molecules that forms between the opposite poles of a eucaryotic cell during mitosis and serves to move the duplicated chromosomes apart.

**molecular chaperone**
Protein that helps other proteins avoid misfolding pathways that produce inactive or incorrectly aggregated states. Both tubulin and actin require chaperones to fold correctly.

**morphogenesis**
The development of form and structure in a cell or organism, usually in reference to embryonic development.

**motor protein**
Protein that uses energy derived from nucleoside triphosphate hydrolysis to propel itself along a protein filament.

**MTOC**—*see* **microtubule-organizing center**

**myoblast**
Mononucleated, undifferentiated muscle precursor cell. A skeletal muscle cell is formed by the fusion of multiple myoblasts.

**myofibril**
Long, highly organized bundle of actin, myosin, and other proteins in the cytoplasm of muscle cells that contracts by a sliding filament mechanism.

**myomesin and M protein**
Related proteins that localize in the M-band region of skeletal and heart myofibrils. May be involved in the linking of the M-band to titin, which then serves to connect the myofibrils to the Z-disc.

**myosin**
Motor protein that uses ATP to drive movements along actin filaments. Conventional myosins (also known as type II myosins) are large, two-headed, filament-forming proteins similar to muscle myosin. Unconventional myosins are grouped into 12 distinct classes, each with a distinct tail region and a different function in the cell.

**myosin I**
Low-molecular-weight myosins with lipid-binding tails expressed in a wide range of organisms. Play an important role in moving membranes against actin filaments.

**myosin II**
Family of large myosins with two globular heads and a long coiled-coil $\alpha$-helical tail. A portion of the myosin rod aggregates to form the core of the thick filament found at the center of the muscle sarcomere. The globular heads contain sites for interaction with both actin filaments and ATP. This interaction generates the force of muscle cells.

**myosin V**
Two-headed motor proteins with a high affinity for actin filaments even in the presence of ATP. Appear to be responsible for the localization of vesicles to specific regions of the cytoplasm.

**myxobacteria**
Group of bacteria characterized by aggregation into multicellular 'fruiting bodies' containing resting spores.

**nebulin**
A giant actin-binding protein that serves as a length-regulating template for the thin filaments in a muscle sarcomere.

**nerve cell**—*see* **neuron**

**neural crest**
Group of embryonic cells derived from the roof of the neural tube that migrate to different locations and give rise to various types of adult cells, including nerve cells in peripheral ganglia, chromaffin cells, melanocytes, and Schwann cells.

**neurite**
Long process growing from a nerve cell in culture. A generic term that does not specify whether the process is an axon or a dendrite.

**neurofilament**
Type of intermediate filament found in nerve cells.

**neurofilament protein (NF-L, NF-M, NF-H)**
Major components of the neurofilaments (NF), the intermediate filaments found in most mature neurons. Neurofilaments are thought to be a major determinant of axonal caliber.

**neuron (nerve cell)**
Cell with long processes specialized to receive, conduct, and transmit signals in the nervous system.

**neutrophil (polymorphonuclear leucocyte)**
A white blood cell involved in the early stages of infection. An amoeboid chemotactic cell that carries out a rapid, nonspecific attack on pathogens, especially bacteria.

***Nitella***
Green algae with giant multinucleated cells. Used in studies of plant physiology and actin-based cytoplasmic streaming.

**nuclear envelope**
Double membrane surrounding the nucleus. Consists of outer and inner membranes perforated by nuclear pores.

**nuclear lamina**
Fibrous layer on the inner surface of the inner nuclear membrane made up of a network of intermediate filaments made from nuclear lamins. The lamina is thought to provide a framework for organizing the structure of the nuclear envelope and an anchoring site for chromosomes at the nuclear periphery.

**nucleoside**
Compound containing a purine or pyrimidine base linked to a sugar (ribose or deoxyribose).

**nucleotide**
A nucleoside with one or more phosphate groups, such as ATP or GDP. The two terminal phosphates are linked by phosphoanhydride bonds, and are therefore readily hydrolyzed.

**paramecium**
Freshwater protozoan of the genus *Paramecium*, having an oval body covered with cilia.

**paramyosin**
Paramyosin is a two-chain α-helical coiled-coil forming the core of myosin thick filaments in invertebrate muscles. It is present in especially large amounts in molluscan 'catch' muscles.

**parvalbumin**
Small calcium-binding protein abundantly present in fish muscle.

**paxillin**
A focal adhesion protein that interacts with vinculin and actin.

**pericentrin**
A protein that is an integral component of centrosomes and other microtubule-organizing centers. It appears to be involved in organizing the microtubule spindle during mitosis and meiosis. Unrelated to centrin.

**persistence length**
That length of a protein filament at which thermal movements completely randomize the orientation of the two ends. Persistence length increases with filament rigidity and decreases with increasing temperature.

**phagocytosis**
Process by which particulate material is endocytosed ('eaten') by a cell. Prominent in carnivorous cells, such as *Amoeba proteus*, and in vertebrate macrophages and neutrophils. (Greek *phagein*, to eat.)

**phalloidin**
One of a group of cyclic peptide toxins present in the 'death cap' fungus *Amanita phalloides*. Phalloidin binds specifically and strongly to F-actin and thereby prevents the normal disassembly to G-actin.

**phragmoplast**
Flat membrane septum that forms in the equatorial region of a dividing plant cell.

***Physarum***
A large genus of acellular slime molds. *Physarum polycephalum* has been studied for its pronounced cytoplasmic streaming.

**$PIP_2$ (phosphatidylinositol-bisphosphate)**
Minor lipid in the plasma membrane that is broken down in response to extracellular stimuli. Regulates a number of cytoskeletal proteins, such as profilin and gelsolin.

**p*K***
The pH at which an acidic group is half dissociated—a measure of its acidity. The side-chain of glutamic acid, for example, has a p*K* of 4.3, whereas that of its amide group is 9.7.

**plasma membrane**
Membrane that surrounds a living cell.

**plakin**
Family of multifunctional proteins that stabilize and cross-link cytoskeletal protein networks. Includes plectin, desmoplakin, and BPAG.

**platelet**
Cell fragment, lacking a nucleus, that breaks off from a megakaryocyte in the bone marrow. Present in large numbers in the bloodstream, platelets help initiate blood clotting when blood vessels are injured.

**plectin**
Ubiquitous and abundant member of the plakin family of cross-linking proteins, with binding sites for intermediate filaments, microtubules and actin filaments.

**plus end**
The end of a microtubule or actin filament at which addition of monomers occurs most readily; the 'fast-growing' end of a microtubule or actin filament. The plus end of an actin filament is also known as the barbed end.

**pointed end**
Slow-growing, 'minus' end of an actin filament. *See also* barbed end.

**ponticulin**
Integral membrane protein from *Dictyostelium* that binds directly to F-actin.

**primary cilium**
A rudimentary, usually nonmotile, cilium of unknown function that grows from one of the two centrioles in many vertebrate tissue cells.

**processive**
Term used of an enzyme or protein that performs multiple rounds of catalysis or conformational changes while attached to a polymer. Characteristic of motor proteins involved in transport, such as kinesin.

**profilin**
Small, globular, cytoplasmic protein that binds actin monomers, associates with plasma membranes and binds to phosphatidylinositol-4,5-bisphosphate ($PIP_2$). Important in the nucleation of actin filaments and the addition of monomers to their barbed ends.

**prophase**
First stage of mitosis during which the chromosomes are condensed but not yet attached to a mitotic spindle. Also a superficially similar stage in meiosis.

**protease (proteinase, proteolytic enzyme)**
Enzyme such as trypsin that degrades proteins by hydrolyzing some of their peptide bonds.

**protofilament**
A linear chain of protein subunits within a filament, such as a microtubule or a bacterial flagellum.

**protozoa**
Free-living, nonphotosynthetic, single-celled, motile eucaryotic organisms, especially those, such as *Paramecium* or *Amoeba*, that live by feeding on other organisms.

**pseudopodium** (plural **pseudopodia**)
Large cell-surface protrusion formed by amoeboid cells as they crawl. More generally, any dynamic extension of the surface of an animal cell, including lobopodia, filopodia, and lamellipodia (which are actin-containing structures) and axopodia (microtubule-containing).

**quorum sensing**
The capacity shown by some species of bacteria to sense the size of the population to which they belong—achieved by individually secreting small molecules and detecting their collective concentration.

**Rac**—*see* **Rho**

**radial spoke protein**
The radial spoke is a multisubunit structure found in eucaryotic cilia and flagella. Its apparent role is to regulate the bending pattern to produce an asymmetric, ciliary-type beat.

**reptation**
Snakelike movement of a long, flexible molecule in solutions of high concentration, due to the greater ease of diffusion along its length than laterally.

**retrograde transport**
Movement of organelles in a nerve axon toward the cell body. The converse of anterograde transport.

**Rho (Rho GTPase)**
Crucial component of the intracellular circuitry controlling stress fibers, focal adhesions, and other actin-based cortical structures. Related proteins are Rac, which controls lamellipodia, and Cdc42, which controls filopodia. All are monomeric GTP-binding proteins belonging to the Ras superfamily.

**ruffling**
Characteristic irregular movement of folds, filopodia, and other protrusions on the surface of many vertebrate cells. The lamellipodia of migrating fibroblasts in culture show ruffling with a predominant movement toward the rear of the cell.

**S phase**
Period of a eucaryotic cell cycle in which DNA is synthesized.

**saltatory movement**
Movement in steps or jumps interspersed by periods of rest, as shown by the movement of organelles in an animal or plant cell (Latin *saltus*, jump).

**sarcomere**
Repeating unit of a myofibril in a muscle cell, composed of an array of overlapping thick (myosin) and thin (actin) filaments between two adjacent Z-discs.

**sarcoplasmic reticulum**
Network of internal membranes in the cytoplasm of a muscle cell that contains high concentrations of sequestered $Ca^{2+}$ that is released into the cytosol during muscle excitation.

**septation**
Division of a cell by an intervening wall (septum).

**septin**
One of a family of proteins associated with septum formation in yeast cells. Also found in vertebrate cells.

**severin**
A 40 kDa protein in *Dictyostelium discoideum* that severs actin filaments, nucleates actin assembly, and caps the fast-growing end of actin filaments—all in a $Ca^{2+}$-dependent manner.

**signal transduction**
The process by which a cell converts a detected physical or chemical change into a cytoplasmic signaling response.

**signaling molecule**
Extracellular or intracellular molecule that alters the behavior of a cell in response to changes in its concentration; the concentration is in turn determined by the cell's environment.

**slime mold**
Primitive organisms that form spores, like fungi, but are phagocytic, like protozoa. Cellular slime molds such as *Dictyostelium* contain a mass of individual cells, whereas acellular slime molds such as *Physarum* are a multinucleated syncytium.

**somite**
One of a series of paired segments that forms in the mesoderm of a developing animal embryo. Each somite produces the musculature of one vertebral segment, plus associated connective tissue and vertebrae.

**spasmoneme**
System of parallel filaments capable of extremely rapid contraction found in the stalk of certain ciliates. Contains the protein centrin, which undergoes a large conformational change in response to changes in $Ca^{2+}$ concentration.

**spectrin**
Very long, flexible protein originally isolated from red blood cell membranes but subsequently found (in different isoforms) in many tissue cells. Confers mechanical stability to the membrane by being anchored to transmembrane proteins and to the actin cytoskeleton.

**spindle**—*see* **mitotic spindle**

**stathmin/Op18**
Small protein that controls the dynamics of microtubule polymerization in cells.

**stereocilium**
Large, rigid microvillus, especially one found in 'organ pipe' arrays on the apical surface of hair cells in the ear. A stereocilium contains a bundle of actin filaments, rather than microtubules, and is thus not a true cilium.

**striated rootlet**
Structure that anchors a basal body to other structures, such as the nucleus. Striated rootlets contain the calcium-binding protein centrin and are thought to determine the orientation of cilia by their $Ca^{2+}$-induced contraction.

**substrate**
Molecule on which an enzyme acts.

**substratum**
Solid surface to which a cell adheres.

**swarmer cell**
An elongated, highly flagellated form of certain bacteria that moves in groups, or swarms, over a solid surface.

**synapsin**
One of a family of closely related phosphoproteins associated with synaptic vesicle membranes that are essential for the proper assembly and release of neruotransmitter at a nerve synapse.

**syncytium**
Mass of cytoplasm containing many nuclei enclosed by a single plasma membrane. Typically the result either of cell fusion or of a series of incomplete division cycles in which the nuclei divide but the cell does not.

**syneresis**
A process in which a gel contracts and expels fluid.

**talin**
A high-molecular-weight protein concentrated at regions where bundles of actin filaments attach to and transmit tension across the plasma membrane to the extracellular matrix. Talin binds to the cytoplasmic domains of the integrin family of extracellular matrix receptors, to vinculin and to actin.

**tau**
Tau is a complex family of microtubule-associated proteins (MAPs) originally purified on the basis of their promotion of microtubule assembly from purified tubulin. May regulate microtubule dynamics during neurite outgrowth.

**taxis**
The movement of a cell or organism in a particular direction in response to an external stimulus. *See* kinesis.

**tektin**
One of a family of filamentous proteins, related to intermediate filaments, found in ciliary and flagellar axonemes. Related proteins are also present in basal bodies, centrioles, centrosomes, and mitotic spindles.

**tensegrity**
The architectural principle by which open, self-stabilizing, three-dimensional structures are built by balancing the counteracting forces of tension and compression.

**tensin**
Actin-binding protein that helps link actin filaments to membranes in focal adhesions, Z-lines of skeletal muscle, and dense bodies of smooth muscle.

**thermal ratchet**
Postulated mechanism by which a protein polymer, such as an actin filament, can produce movement through the stochastic addition of subunits to its growing end.

**thymosin $\beta_4$**
Small protein (5 kDa), present in high concentrations in blood platelets and neutrophils, that binds to and sequesters actin monomers. Homologous proteins are widely distributed in tissue cells of vertebrates and invertebrates.

**titin**
An extraordinarily long and elastic protein, abundant in the sarcomere of striated muscles. It forms an elastic filamentous matrix in the sarcomere that provides structural continuity and elastic restoring force. In developing muscles, titin molecules act as a template/scaffold for sarcomere formation.

**treadmilling**
Dynamic behavior of an actin filament or microtubule in which subunits assemble at one end and concomitantly disassemble at the other end.

**tropomyosin**
Family of rodlike $\alpha$-helical proteins that associate with actin filaments and modulate their interaction with other proteins. Tropomyosin promotes binding of troponin and myosin but inhibits access to cross-linking proteins such as $\alpha$-actinin.

**troponin**
A complex of three proteins associated with myofibrils in skeletal and cardiac muscle that, together with tropomyosin, makes contraction sensitive to $Ca^{2+}$ ions.

**tubulin**
Subunit protein of microtubules. Stable dimers of $\alpha\beta$-tubulin in solution assemble into linear protofilaments forming the wall of microtubules. $\gamma$-Tubulin is associated with microtubule-organizing centers and serves to nucleate microtubules.

**uropod**
The trailing portion of a migrating cell. (Greek, literally 'tail-foot'.)

**VASP (vasodilator-stimulated phosphoprotein)**
Proline-rich substrate of protein kinases found in most mammalian cell types, especially in blood platelets. Part of the apparatus that links actin filaments to membrane structures in focal adhesions and elsewhere.

**villin**
Major structural protein associated with microvilli in epithelial cell brush borders. Involved in the assembly of the brush border cytoskeleton and perhaps the breakdown of villi in response to calcium ions.

**villipodium**
Flattened leading edge of a migrating nematode sperm cell. *See also* major sperm protein.

**vimentin**
Vimentin is the type III intermediate filament protein characteristic of, but not restricted to, fibroblasts and other mesenchymally derived cell types *in situ*.

**vinculin**
A 117-kDa actin-binding protein located on the cytoplasmic face of adherens junctions and focal adhesions.

**WASP (Wiskott–Aldrich syndrome protein)**
Protein implicated in a human immune deficiency that participates in the polarization of T cells. The protein contains an extended polyproline sequence, like VASP, but also has a domain that interacts with the small G protein Cdc42.

**white blood cell (leucocyte)**
Nucleated blood cell lacking hemoglobin; includes lymphocytes, neutrophils, eosinophils, basophils, and monocytes.

# INDEX

Page numbers with an F refer to a figure, those with an FF refer to figures that follow consecutively; page numbers with a T refer to a table; page numbers with 'n' and a number afterward refer to a note in the page margin.


A-band, 142, 143F
A-tubule, 227, 227F
  *see also* Ciliary axonemes
*Acanthamoeba*
  myosin I discovery in, 113
  pseudopodia of, 19
Accessory proteins, 45–46, 46F
ActA protein, 74–75
Actin, 89T, 91T
  *see also* Actin filaments
  discovery of, 64
  evolutionary origin of, 65, 66F
    diversion of isoforms, 66
  F-, 66
  G-, 64, 66
  gene families for, 64–66
  globular, 64
  isoforms of mammalian, 65, 65T
  and membranes, 81
    α-actin cross-linking, 81–83
    adherens junctions, 87–89
    attachment, 84, 84F
    cell control, 97–98
    in cultured cells, 89–92
    of erythrocytes, 84–86, 85F, 86T
    filamin cross-linking, 81–83
    fragmenting proteins and reorganization, 83–84
    Rho GTPase control, 92–94
    surface structure formation and external stimuli, 94–98
  monomers of, 64, 65F, 66F
    assembly of, 66–67
  and mRNA translation, 304
  nucleation of, *see* Actin nucleation
  polymerization of, 48–49, 48FF
    asymmetrical growth, 69, 69F
    ATP cap formation, 70, 70F
    ATP hydrolysis and polymerization rate, 70, 71F
    barbed end, 69, 95
    binding proteins, 72–73, 73F
    critical concentration, 67–68
    kinetic equations, 68, 68F
    lag phase, 67
    'plus' end and 'minus' end, 69n3
    pointed end, 69
    regulators, 132–133, 133T
    reversibility, 68
    stages of, 68F
    time course, 67F
    toxin disruption, 71–72
  in *Saccharomyces* budding, 319–320
  species similarity in molecules of, 64
  thymosin sequestration of, 72
  toxins that bind to, 71–72, 71T
  tubulin similarity to, 172, 173T
  unpolymerized, sequestration of, 72
Actin arcs, 128
Actin binding proteins, 75, 75F
  *see also* Cross-linking proteins
  actin-binding domain, 76
  two binding sites, 76
  vertebrate, 75–76, 76T
Actin cortex, 81
  *see also* Cytoplasmic cortex
  α-actin cross-linking, 81–83
  actin filament regulation, 97–98
  adherens junctions, 87–89
  of cultured cells, 89–92
  cyclical flow of, 127, 128F
  of erythrocytes, 84–86, 85F, 86T
  filamin cross-linking, 81–83
  fragmenting proteins and reorganization, 83–84
  membrane attachment, 84, 84F
  microtubular interaction, 324, 327, 327F
  Rho GTPase regulation, 92–94
  surface structure formation and external stimuli, 94–98
  in T-cell response, 327, 327F
  transmembrane signaling and, 327, 327F
Actin-depolymerizing factor (ADF), 73, 73F
Actin filaments, 43, 43F
  *see also* Actin; Crawling movement; Cytoskeleton
  abundance in eucaryotic cells, 64
  in animal cell division, 110–111, 111F
  annealing, 69n2
  assembly of, 48–49, 48FF, 66–67, *see also* Actin, polymerization of
    binding proteins in, 72–73, 73F
  binding strength, 67
  cross-linking proteins of, 45–46, 46F, 81–83, 83F, *see also* Actin, and membranes
  elongation factor EF-1 binding, 77n7
  end-to-end association, 69n2
  in eucaryotic cell movement, 44–45, *see also* Cell movement
  growth rate at ends, 69–70
    ATP hydrolysis and, 70, 71F
  membrane vesicle attachment, 113–115
  microtubules and, 327, 327F
  motor proteins and, 50–51, 51F
  and myosin filaments, 142–147, *see also* Muscle contraction, molecular basis of
  myosin heads, 67, 67F, *see also* Myosin head(s)
  nucleation of
    *Listeria* motility and, 73–74, *see also* Actin nucleation
    proteins associated in, 74–75
  pointed end, 69
  polarity, 66–67
  polysome attachment, 77
  and protein synthesis sites, 77
  structure of, 66–67, 66F
  study of, 63
  tropomyosin stabilization, 109–110, 110F
Actin nucleation, 73–74
  actin-binding proteins in, 75–76
  in pathogenic bacteria, 74, 74F
  proteins associated with, 74–76
*Actinosphaerium*, axopodial movement of, 27, 238–239, 239F
Adaptation, 31–32
ADF (actin-depolymerizing factor), 73, 73F
Adherens junctions, 87, 88F, 342–343, 343FF
  and cell adhesion proteins, 87–88
  of cultured cells, *see* Focal adhesions

and embryonic cell movement, 343
location of, 88
major proteins of, 89T
structure of, 88–89
Adhesion belts, 88, 88F
Adhesion proteins, 125
at adherens junctions, 87–88, 88F
attachment and detachment, 125–126, 126F
in embryonic cell movement, 340–341
Algae
flagellar structures, 10–11
flagellar waveforms, 11–12, 11FF
gliding movement, 25–26
α-actinin, 89T, 91T
actin binding sites of, 76
cross-linking, 81–83, 83F
in myofibril development, 157
α-internexin, 279T, 287n5
α-Tropomyosin, 148
Alternate mRNA splicing, 161–162, 162F
*Amoebae*
anesthetic substances secretion, 18n2
chemotactic responses, 29, 30
complex responses, 38
crawling movement, 18, 18F
cytoplasmic streaming and hydrostatic pressure in, 128–129
cytoplasmic streaming, 19, 20F, 128–129
pseudopodial formation, 18–19, 18F, 19F
Amphibian egg
animal pole, 336
cleavage pattern, 337–338, 338F
cortical rotation and axis development, 336–337, 337F
gray crescent, 336
orthogonal cleavage, 338, 338F
vegetal pole, 336
Amphibian embryo, spindle growth and cleavage, 337–338, 338F
Anaphase, 210, 215–218
mechanisms of, 216, 216FF
Anaphase A, 216, 216F
Anaphase B, 216F, 217
Anaphase promoting complex, *see* APC
Anisotropic banding, 142
Ankyrin, 85F, 86
APC (anaphase promoting complex), 215–216
Apical constriction, 344, 344F
Appendages, surface, 4
Arps (actin-related proteins), 74–75
Arp 1, 198
Arp2/3 complex, 75
in budding yeasts, 319
in leading edge of lamellipodia, 123–124, 124F
Artificial asters, 326, 326F
ATP-cap formation, 70, 70F
ATP hydrolysis
and conformational changes of motor proteins, 51–52, 51F
by myosin heads, 104–105, 145–146, 146F
and rate of actin polymerization, 70, 71F
Attachment, to substratum, *see* Adherens junctions; Adhesion proteins
Attractants, cell, 30T
Axonal transport, 191
Axonemes, ciliary, *see* Ciliary axonemes
Axons, 192, 192F
actin filament distribution in, 45F
attractants and repellents to, 30T
cross-linking proteins of, 45–46, 46F
growth cone migration of, 25F
chemotactic, 25
filopodia in, 25, 25F
G cell and, 35
inhibition of, 36
oligodendrocyte inhibition of, 37n2
pathway selection in, 35F
growth of, 35, 348–349
microtubules in
organization of, 192, 192F, 199–200
stabilization of, 193, 193F
migration of, 25
motor, 192F
size of, 191F
organelle transport in, 191
dynein in, 197–198
kinesin in, 193–194, 194F
polarity of, 192, 199–200
shape of, 316
tensegrity of, 301, 302F
Axopodia, 27
microtubule arrangement in, 238–239, 239F
Axostyle, 237–238, 237F

B-tubule, 227, 227F
*see also* Ciliary axonemes
Bacteria
adaptive responses, 31–32, 32F
attractants and repellents, 30T
cell division of, 208, 208F
septate proteins in, 56
cell wall of, 316
chemotactic responses, 29–30, 30F, 262–263, 262T, 263F
adaption of, 265, 266F
culture patterns and, 268–269
genetic analysis of, 263–264
membrane conformation and adaptation of, 266–267, 267F
quorum sensing and, 267–268
signaling pathways in, 264–265, 264FF
variants of, 267–268
cytoplasmic organization, 56
flagella of, *see also* Bacterial flagella
compared to eucaryotic flagella, 10–11, 11F
motion, 8–9, 8F, 10F
rotation, 9, 9F
structure and motion, 9, 9F
surface motility, 26–27
flagellated forms, 9–10
gliding movement, 25–26
movements of, 257–258, *see also* Bacterial flagella
aflagellar, 272
function of, 262–263
screwing, 269–270, 270F
swarming, 10, 26, 27, 269, 269F
swimming, 8–9, 8FF, 9–10
with and without gradient of attractant, 29–30, 30F
role in meiosis, 221
Bacterial flagella, 8–9, 8F, 257–258, 257F
base connection, 259, 259F
compared to eucaryotic flagella, 10–11, 11F
flagellin structure, 258, 258F
flagellin synthesis and movement, 258–259, 259F, 259n2
hook of, 259, 259F
motion of, 8–9, 8F, 10F
motors of, 259–261, 259T, 260FF
electrostatic switching, 261
models of, 261, 261F
proton energy source, 260
reversing, 262, 262F
structure, 260–261, 260FF
rotation of, 9, 9F
rotational orientation, 262–263, 262F
chemoattractants and, 262–263, 263F
structure and motion of, 9, 9F
in surface motility, 26–27
Bacterial infection
bacterial movement and, 271
cellular uptake of bacteria, 271
signaling peptides, 31
Band 4.1, 85F, 86
Basal bodies, 243, 245, 245F
arrays, 247F
axonemal assembly, 246–247
centrioles and, 249–250
location of, 245, 245F
parent/daughter differences, 252
similarity to centrioles, 245
striated rootlets and, 248–249, 249F
structure of, 246F
Basal lamina, 141, 342
β-hemoglobin gene mutation, 46–47, 48F
β-Tropomyosin, 148
Bicoid mRNA, 334
Blastocoel, 342
Blastopore formation, 344, 344F
Blastula, 340
cell junctions of, 341–342, 342F
Blebs, 23
Blepharoplasts, 252
Boltzmann's constant, 5n1
Bottle cells, 344, 344F
Bound water, 305–306, 306F
BPAG1, 185, 185T, 288
Brownian motion, 4, 4F, 297
Brush border, of intestinal microvilli, 96
Bud1, 319
Bulk water, 305–306, 306F
Bundling proteins, 95–96, 95F

C ring, 260, 260F
Cadherins, 87–88, 88F, 89T
in embryonal cell movement, 340–341, 342
*Caenorhabditis elegans*
cortical influence on cell division, 338–339, 339F
crawling motility, 120–122
muscle mutants and genetic analysis, 167
Calcium ion concentration
and ciliary beating, 234–235
intracellular, 52, 53F, 54, 133
and muscle contraction, 147–148, 148F
and protein kinase activation, 54
Calcium release channel, 142
Calmodulin, 148–149
cAMP (cyclic AMP), 52, 53F
in cell signaling, 36–37
in ciliary beating, 234–235
Cap Z, 157
Capping, ATP, 70, 70F
Capping protein, 83–84, 124
Cardiac muscle, 163–164, 163F
Catch muscle, molluscan, 166
Catenins, 88, 89T
Caveolae, of smooth muscle, 165
Cdc42 protein, 92–93, 93F, 132, 271, 319
CdKs (cyclin-dependent kinases), 210
Cell behavior
chemotactic, 29–32, *see also* Bacteria, chemotactic responses
complex, 38–39
contact inhibitory, 36
galvanotactic, 32–33
neuronal, 35

obstacle avoidance in, 33–34
photoresponsive, 37–38
signaling, 36–37, *see also* Migration, embryonal cell; Signaling
substrate sensitive, 34
Cell contents, 19, 20F
*see also* Cytoplasm; Cytoskeleton; Nucleus
Cell cortex, *see* Actin cortex; Cytoplasm; Cytoplasmic cortex
Cell cycle, 209, 209F
anaphase, 210
checkpoints, 209, 209F
cytokinesis, 209
interphase, 209
M phase, 209
metaphase, 210
mitosis, 208–209
prophase, 210
telophase, 210
Cell division, 207
in bacteria, 208, 208F
of eucaryotic cells, 208–209, 209F, *see also* Cytokinesis; Mitosis
Cell mechanics, 293
bundled filaments
monomeric and dimeric linkages, 295–296, 296F
physicochemical process, 295–296
stored strength, 295
stress calculations, 295n2
stored energy, 296–297
contraction of cytoskeleton, 300–301, 301F
cytoplasmic
cortical tension, 302–303, 302F
hydrostatic pressure in plants and lower eucaryotes, 306–307, 307F
macroviscosity, 305
microviscosity, 305
soluble proteins, 303–304, 303FF
viscoelasticity, 299–300, 299F
water and protein interaction, 305–306, 306F
filaments
directed force of polymerization, 297–298, 297F
elasticity, 294–295
flexibility, 294, 294FF
persistence length, 294, 295T
tensegrity structure, 301–302
viscoelasticity of network, 298–299
membrane, 308–309
stimuli and response of, 309T
soluble proteins associated with cytoskeleton, 303–304, 303FF
stimuli and responses in, 309–310, 309F
Cell movement
*see also* Bacteria, movements of; Crawling movement; Gliding movement; Migration; Swimming movement
cytoskeleton in, 42–43, 42F, *see also* Cytoskeleton; Intermediate filaments
and evolution of eucaryotic cells, 55–56, *see also* Embryonal cell movement; Neural crest migration
filaments responsible for, 44T, *see also* Actin; Actin filaments; Myosin(s); Tubulin
genetic analysis of, 57, *see also* specific organism, structure or protein
signaling and, 54–55, 55F, *see also* Signaling
stimuli and responses in, 309–310, *see also* Cell mechanics; Chemotaxis
velocity comparisons, 2F
Cell shape, 42–43, 42F, 315
*see also* Cell mechanics; Cytoskeleton
apical contraction, 344, 344F
bacterial, 316
and cell size, 316
cell walls and, 316
columnarization, 346
cortical contractions, vertebrate, 323–325, 324FF
erythrocytic, nonmammalian, 324–325, 325F
eucaryotic, 325
fungal, 316, 318–321
largest, 316, 316n2
marginal bundle formation, 324–325, 325F
morphogenesis and, 315
plant, 316–318, 321
polyhedral, 321–322, 322F, 322T
smallest, 316
vertebrate, 322–323, 323F, 325
asymmetry of, 329
cortical contractions, 323–325, 324FF
cytoskeleton polarization, 327, 327F
extracellular matrix, 329
microtubule array assembly, 326, 326F
microtubule systems and cell center, 325, 325F
Cell size, 316
Cell wall, 219, 316, 321
Central apparatus, of ciliary axonemes, 227–228
Central pair, 227, 227F
Centrin, 248–249
Centrioles
and basal bodies, 249–250
centrin location, 249
centrosomal association, 250–251
and ciliary nucleation site, 252
*de novo* generation, 252–253
in multiflagellated sperm, 252
in *Naegleria*, 252
in plant sperm, 252
evolutionary origin, 253
microtubule array, 245, 251
microtubule nucleation, 251
mother and daughter roles, 251–252
origin and formation, 250–251, 251F
and primary cilia formation, 248, 248F
primary function of, 250
Centromere, 212–213, 212F
Centrosome, 181, 181F
*see also* Mitosis; Spindle
association with centrioles, 250–251
and microtubule assembly organization, 180–182, 181FF, 250
Chaperone molecules, 174–175
*che* genes, 264–265
Chemoattractants, 30T
*see also* Chemotaxis
and rotational orientation of bacterial flagella, 263F
Chemotaxis
and adaptation of sensory system, 31–32
and amoeboid responses, 29, 30
and bacterial responses, 29–30, 30F, 262–263, 262T, 263F
adaption, 265, 266F
culture patterns of, 268–269
genetic analysis of, 263–264
membrane conformation and adaptation, 266–267, 267F
quorum sensing, 267–268
signaling pathways, 264–265, 264FF
variants, 267–268
and growth cone responses, 25
and neutrophilic responses, 31
and polarization of motile mechanisms, 132, 132F
*Chlamydomonas*
ciliary beating motion, 234–235
ciliary formation and growth, 244, 245, 245F
flagellar orientation, 249
flagellar swimming, 12, 27, 229–230, 229F
parent/daughter basal body differences, 252
photoresponse, 32–33, 37–38, 38F, 234–235
Chromatids, 212
Chromatophores, organelle transport, 201–202, 202F
Chromokinesin, 214
Chromosomes, mitotic, centromere and kinetochore of, 212–213, 212F
*Chrysophytes*, flagellar swimming, 12, 12F, 12n3
Cilia, 4, 225–226
*see also* Flagella
9+0, 248
9+2, *see* Ciliary axonemes
axonemes of, *see* Ciliary axonemes
genetic analysis and biochemistry, 229–230, 230n4
growth of, 244
active transport, 247–248, 247F
cytoplasmic protein pool, 244–245, 245F
intraflagellar transport, 246–247, 246–247F
motor proteins, 247–248, 247F
nucleated, 245–246, 245FF, *see also* Basal bodies
intraflagellar transport of axonemal parts, 247
length of, 244, 244F
microtubule arrangement, 238–239, 239F
modified, 14–15, 236–239, 236FF
motion of
beating, 13–14, 13F, 14F, *see also* Ciliary beating
coordinated, 13–14
metachronal wave, 13
waveform, 226F
position and orientation of, 248–249, 249F
primary, 248, 248F
purposes of, 13–15
and similarity to flagella, 225n1
specialized, 14–15, 236–237, 236F, 238–239, 239F
in surface motility, 26–27, 27F
viscous propulsion of, 7
water movement by, 13–14
Ciliary axonemes, 226
9+2 pattern, 227, 227F
compared to myofibrils, 226
growth of, 244
active transport, 247–248, 247F
cytoplasmic protein pool, 244–245, 245F
intraflagellar transport, 246–247, 246–247F
motor proteins, 247–248, 247F
nucleated, 245–246, 245FF, *see also* Basal bodies
microtubular infrastructure, 227, 228F
central apparatus, 227–228
dynein arms, 227
nexin links, 227

protein components, 227–229, 228T
tektin protofilaments, 228–229
microtubule arrangement of, 226–227, 227F
motion of, 230–236, *see also* Ciliary beating
variants, 236–237, 236F
Ciliary beating, 13–14, 13F, 14F
calcium ion concentration and, 234–235
dynein mechanics of, 231–233, 231FF
left–right orientation and, 235–236, 236F
microtubule sliding in, 230–231
mutational defects and, 235–236, 235n5, 236F
phosphorylation regulation of, 235
radial spokes mechanics of, 233–234, 233F
and reversal of direction, 33–34, 34F
Ciliates
complex behavior, 38–39
protein conformation and movement, 50, 50F
*Ciliophora*, 13
Cirri, 15
of hypotrichs, 27
of *Stylonychia*, 27, 27F
CLIP-170, 185, 185T
and chromosomal attachment to spindle, 213
Cofilin
disassembly of actin monomers, 72–73, 73F
role in leading edge of lamellipodia, 123–124, 124F
Colcemid, 175T, 176
Colchicine, 175–176, 175T
structure of, 175F
Columnarization, 346
Contact inhibition, 36, 36F
Contractile ring, 110–111, 111F
Contractin, 198
Contraction, cell, 125
*see also* Bottle cells; Cytokinesis; Cytoplasmic cortex
Conventional mysosin, 104
Cortex, *see* Actin cortex; Cytoplasm; Cytoplasmic cortex
Cortexillin, 125
actin binding sites of, 76–77
Cortical array, 318, 318F
Cortical flow, 127, 128F
Cortical rotation, in amphibian eggs, 336–337, 337F
Cortical tension, 105–106
Costameres, 286
Crawling movement, 17, 119–120, 120F
actin-based, 122
actin filament-less, 120–122, 121F
actin polymerization in, 122–123
Arp2/3 in, 123–124, 124F
attachment to substratum, 125
cortical actin flow, 127–128, 128F
detachment process, 125–126, 126F
of differentiated tissue cells, 22–23, 23F
of embryonic cells, 21–22, 22F
of fibroblasts, 21, 21F, *see also* Fibroblasts
lamellipodia, 23–24, 24F, *see also* Lamellipodia
of mammalian blood cells, 20, 21F
membrane movement in, 130–131, 131F
microtubules in, 131–132
myosin II activity in, 126–127
pseudopodia, 18–19, 18FF
ruffling and, 23–24, 24F
steps of, 120, 120F
substrate guidance of, 34, 34F
surface drag in, 17–18
traction mechanism of, 125
unconventional myosin activity in, 127
viscous drag in, 17–18
of white blood cells, 20, 21F
Creatine phosphate shuttle, 139–140, 140F
Cross-bridge cycle, 145–146, 146F
Cross-linking proteins
*see also* Actin, and membranes; Linker proteins
α-actin, 81–83, 83F
filamin, 81–83, 83F
of nerve cells, 45–46, 46F
Cultured cells
filopodia formation, 92–93, 93F
focal adhesions, 89–90, 89F
fragmentation of, 42–43, 42F
migration of, 22–25
Cyanobacteria
aflagellar swimming, 272
gliding movement, 25–26, 270–271
Cyclic AMP (cAMP), 52, 53F
in cell signaling, 36–37
in ciliary beating, 234–235
Cyclin-dependent kinases (CdKs), 210
Cyclins, 210
Cyclosis, 103
Cytochalasins, 71F
actin binding by, 71, 71T, 122
Cytokeratin, 283n4
*see also* Keratin(s)
Cytokinesis, 209, 218–219
actin activation in, 111
actin filament activation in, 110–111, 111F
and cell migration, 128, 129F
in higher plants, 219, 220F
precision of, 111
Cytoplasm, 19, 20F
*see also* Cytoplasmic cortex
bound water in, 305–306, 306F
bulk water in, 305–306, 306F
collection of, in cell dissection, 42
composition of, 303T
cortical tension of, 302–303, 302F
hydrostatic pressure in plant and eucaryotic, 306–307, 307F
macroviscosity of, 305
microviscosity of, 305
protein and water interaction in, 305–306, 305T, 306F
soluble proteins in, 303–304, 303FF, 303T
structure of, 303, 303F, *see also* Cell mechanics
viscoelasticity of, 299–300, 299F
Cytoplasmic cortex, 81
*see also* Actin cortex; Cytoplasm
α-actin cross-linking, 81–83
actin filament regulation in, 97–98
actin filaments, 82F
actin flow, cyclical, 127–128, 128F
adherens junctions of, 87–89
contraction of
and cell shape, 323–324, 324FF
in crawling movement, 125
osmotic pressure and, 307, 307F
of cultured cells, 89–92
of erythrocytes, 84–86, 85F, 86T
external stimuli and response of, 94
filamin cross-linking, 81–83
fragmenting proteins and reorganization, 83–84
membrane attachment, 84, 84F
microtubule influence, 324
of rat livers cells, 82F
Rho GTPase regulation, 92–94
surface structure formation and external stimuli, 94–98
tension mechanics, 302–303, 302F
Cytoplasmic streaming, 19, 20F
and hydrostatic pressure in amoeboid movement, 128–129
Cytoskeleton, 41, 43F
accessory proteins, 45–46
actin and tubulin polymerization, 48–49
in cell development, 55–56
cortical, in embryonal development, 335–336, 336F
epigenetic information in, 56–58
evolution of, 56
filaments of, 43–44, 46–47, 294, 294F, *see also* Cell mechanics
polymerization of subunits, 47–48
properties, 294–297, 294FF
ionic and small molecular regulation, 52–53
lack of, in bacteria, 56
microtubules of, 43–44, *see also* Microtubules
assembly, 326, 326F
cell center and, 325, 325F
polarization of, 327, 327F
movement of, 44–45
ATP hydrolysis and motor proteins, 51–52
motor proteins, 50–51
polymerization, 49–50
protein conformation in large-scale, 50
protein kinase coordination, 54
signaling, 54–55
mRNA association, 77
organ of Corti, 328, 328F
polarization of, 327, 327F
redundancy of proteins, 58
shapes and movements of, 42–43
soluble protein association, 303–304, 303F, 303T
surface area, 44F

Daughter centriole, 251–252
Decorated filament, 67, 67F, 105
Dendrites, 192, 192F
*see also* Nerve cells
Dense bodies, 164
Dense plaque, 164–165
Desmin, 279T, 284, 286F
filaments in muscle contractions, 286, 286F
Desmoplakins, 288
Desmosome junctions, 342–343, 343FF
Desmosomes, 284, 285F, 286F
in epithelial sheets, 343, 343F
Detachment, from substratum, 125–126, 126F
Detergent-treated cells
mobility of, 43
remaining structure of, 43–44, 43F
Diatoms
gliding movement of, 25, 26
raphe of, 26
*Dictyostelium*
amoeboid signaling, 36–37, 37F
analysis of myosin genetics, 112
attractants and repellents, 30T
chemotactic responses, 30, 132, 132F
cortexillin actin binding sites, 76–77
myosin I studies in, 113

myosin II studies in, 126–127
Diffusion, cell, 4
effectiveness of, 3–6
Einstein's description of, 5
Fick's laws of, 4–5
in water, times of, 6T
Diffusion constant, 5
Dikaryon rescue, 230
*dilute* locus, 114
Dimeric links, 295–296, 296F
Dissection, of cells, 42–43
DNase I, and actin, 73n6
Doublet microtubules, 226–227, 227F
*Drosophila*
muscle mutants and genetic analysis, 167
myosin studies, 112, 115
Dynactin
coupling to dynein, 198–199, 198F
mitotic role, 214
Dynamic instability, of microtubules, 49, 178–179, 178FF
assymetric microtubule array, 181–182, 182F
catastrophe and rescue, 179, 179F
centrosomal organization of, 180–181, 181F
measurement of, 179
structural basis of, 179–180, 180F
Dynein, 185, 185T
ciliary, 197, 199, 232–233, 232F
axonemal arrangement, 227, 231–233, 231FF
beating motion mechanics, 231–233, 231FF
diversity, 233
phosphorylation regulation of, 235
cytoplasmic
mitotic role, 214
motility mechanics, 197F
organelle transport, 197–198
dynactin coupling, 198–199, 198F
microtubular interaction in embryonal development, 339–340
and microtubule sliding, 199, 231–232, 231F
Dystrophin, 162–163

E-cadherin, 342
ECM, *see* Extracellular matrix
Ectoderm, 344, 344F
Ectoplasm, 19
EF hand, 149
Elastic properties, 298–299, 298F
Electric fields
cell response to, 32–33
vertebrate cell response to, 33, 33F
Elongation factor EF-1, 77n7
Elongation factor EF-Tu, 56
Embryonic cell movement, 333
adhesion molecules in, 340–341
in blastula, 340
cell junctions, 341–342, 342F
in brain formation, 346–347, 348F
cell migration and, 340, 345–346, 346F
vertebrate, 21–22, 22F
cell rearrangement and, 345–346, 346F
cell signaling in, 341
cell sorting in, 340F
cleavage plane and spindle growth, 337–338, 338F
convergent extension, 345–346, 345F
cortical cytoskeleton, 335–336, 336F
cortical influence on cell division, 338–339, 339F
cortical rotation and embryo axis, 336–337, 337F
cytoplasm and positional information, 334, 334F
mRNA in, 334, 334F
RNA transport particle, 335, 335F
dynein interaction with microtubules, 339–340
epithelial cell folding in, 345, 345F
epithelial cells in embryonal morphogenesis, 342–343, 343F
epithelial cells in neurulation, 346, 347F
in gastrulation, 340, 343–344, 343FF
bottle cell role, 344, 344F
germ cells, 21
growth cone, 348–349, 349F, *see also* Growth cones
invagination in, 345
in neurulation, 340, 346, 347F
in vertebrates, 340
Ena, 133
Ena/VASP, 89T
Endoderm, 344, 344F
Endoplasm, 19
*Entamoeba*, pseudopodia, 19, 19F
Enzymes, insoluble, 304, 304F
Eosinophils, newt, 130n3, 131
EP-cadherin, 342
Epiboly, 344, 344F
Epidermolysis bullosa simplex, 289
Epigenetic information, cytoplasmic, 56–58, 57F
Epithelial cell folding, 345, 345F
Epithelial cells
actin cortical contraction in, 323
ciliated, 13–14, 14F
in embryonic morphogenesis, 342–343, 343F
galvanotactic response of, 33F
movement of, in sheets, 20n3
Epithelial sheets, 20n3, 342–343, 343F
cell migration of, 345–346, 345F
evagination of, 345
invagination of, 345, 345F
ERM proteins, 84, 84F, 88, 325
Erythrocyte ghosts, 85
Erythrocytes, 85F
actin and spectrin network of, 84–86, 85F
nonmammalian, 324
marginal band of, 324–325, 325F
microtubules of, 84n2
plasma membrane of, 85F, 86
proteins in cortex, 86T
spectrin network regulation, 87
*Escherichia coli*
cell shape of, 316
chemotactic responses, 29, 262, 262n5, 262T
signaling pathways, 264–265, 264FF
culture patterns of, 268–269, 268F
flagella of, 9, 9F, 257–258, 258F, *see also* Bacterial flagella
swimming motion, 8, 8FF
swimming speed, 8
*Euglena*
flagellar waveform, 11–12, 11F
soluble proteins and cytoskeleton, 304, 304F
Excitation–contraction coupling, 141–142, 141F
Exoskeleton, 329
Extensions, cell, 122–124
*see also* Crawling movement
Extracellular matrix (ECM)
molecules of, 125–126, 126F, 322–323
and vertebrate cell shape, 325
Ezrin, 84, 84F, 325

FAK (focal adhesion kinase), 91, 91T
Fascicles, 141
Fertilization, microtubules in, 220–221
FGF (fibroblast growth factor), 156
Fibroblast locomotion, 23–24, 119–120, 120F
*see also* Crawling movement; Migration of vertebrate cells, 24–25
Fibroblasts
*see also* Cultured cells
contact inhibition of, 36, 36F
cultured
focal adhesion, 89–90, 89F
movement, 22–23, 23F
microtubules of, 172F
migration of, 21, 21F, 22–23, 24F, *see also* Migration
in connective tissue, 21F
migration track of, 24F
surface protrusions, 23, 23F
Fick's first law, of diffusion, 4
Filaggrins, 288
Filaments
*see also* Actin filaments; Intermediate filaments; Microtubules; Myosin(s)
actin, 44–45
advantages of protein, 46–47, 47F, 48F
bipolar, of myosin, 107–108, 107F
directional, 51
intermediate, 43F, 44
microtubule, 43F, 44–45
polymerization of subunits of, 47–48, 48F
protein, 46–47, 47F
responsible for cell movement, 44T
Filamin
actin binding sites, 76
cross-links of, 81–83, 83F
Filopodia, 23, 23F
bundling proteins of, 95–96, 95F
in embryonic cell movement, 21–22, 22F
in fibroblast movement, 24–25
formation of
actin polymerization, 94–95, 95F
external stimulation, 94
of growth cones, 25, 25F
rigidity of, 95–96
in vertebrate cell movement, 23, 24–25
Fimbrin, 83F
actin binding site of, 76
in intestinal microvilli, 96–97, 96F
Fish pigment cells, *see* Chromatophores
Fission yeast, *see Schizosaccharomyces*
fla10 gene, 247
Flagella, 4
*see also* Cilia
bacterial, 8–9, 8F, *see also* Bacterial flagella
eucaryotic, 11–12, *see also* Cilia; Flagella
compared to bacterial, 10–11, 11F
modified, 14–15
in surface motility, 26–27
viscous propulsion of, 7
waveforms of, 11–12, 11FF
movement of, *see* Cilia; Ciliary axonemes
nonswimming uses of, 14–15
and similarity to cilia, 225n1
specialized, 236–237, 236F
Flagellin, 258, 258F
*fliD* gene, 259
Flight muscle, insect, 166
Focal adhesion kinase (FAK), 91, 91T
Focal adhesions, 89

cellular regulation, 90–91
fibroblastic, 89F
formation of, 89
in crawling movement, 125–126
integrins in formation, 89–90
lack of, 125
major proteins, 91T
molecular structure, 90F
Rho GTPase link, 92
signaling at, 90–91
signaling proteins of, 91T
tension development, 92n4
triggering formation, 91–92
visualization of, 89
Foldback inhibition
kinesin, 194
myosin, 84, 108
*Foraminifera*, 202
Formyl-methionine, 31n1
Fruiting bodies, bacterial communication and, 268
FtsZ protein, 56
in bacterial cell division, 208
similarity to tubulin, 173
*Fucus*
cell wall synthesis, 321
flagellar motion of sperm, 12n3
polarization of egg, 321, 321F
Fungi
attractants and repellents, 30T
cell wall of, 316
cytoplasmic streaming, 19
gliding motion, 25–26

G actin, 64
in crawling movement, 124
G cell, 35
G proteins, 52, 53F, 92, 132
Ras superfamily of, 92
proteins related to, 172–173, 173n2
and *Saccharomyces* budding, 318–319, 319F
Galvanotaxis, 33, 33F
γ-tubulin
centriole association with, 251
in centrosomes, 181, 250
Gap junctions, 342–343, 343FF
Gastrulation, 340, 343–344, 343FF
bottle cell role, 344, 344F
epiboly, 344
Gelsolin, 83–84
role in apoptosis, 84n1
role in crawling movement, 124
GFAP (glial fibrillary acidic proteins), 279T, 284
Gliding movement, 17, 25–26, 270–271
Globular actin, 64
Glycolysis, 139, 139F
Glycophorin, 86
Grasshopper embryo, neuronal growth behavior, 35
*Gregarines*, gliding motion, 26
Griseofulvin, 175T, 176
Growth cones
filopodia of, 25, 25F
guidance of, 348–349, 349F, 350
netrin, 348–349, 349F
migration of, 25, 25F
axonal G cell in, 35
chemotactic, 25
filopodial, 25, 25F
inhibition, 36
oligodendrocytic inhibition, 37n2
pathway selection, 35F
receptors of, 348–349, 349F
responses of, 350
GTP-tubulin, 49, 178–180, 178FF

H-zone, 142, 143F
*Halobacterium*
flagellar motion, 10, 10n2, 269–270
photoresponse, 37
HAP2 protein, 259
and mutant bacteria, 259n1
Heavy meromyosin (HMM), 104
*Heliozoa*, axopodial movement, 27
Hemidesmosome junctions, 342–343, 343FF
Hydrostatic pressure
in amoeboid movement, 128–129
in plant and eucaryotic cytoplasm, 306–307, 307F
Hypotrichs, surface motility, 27

I-band, 142, 143F
IFT (intraflagellar transport), 247
IFT particles, 247
Inertial resistance, 6
Integrins, 89–90, 91T
Intercalated discs, 163, 163F
Intermediate filaments, 43F, 44, 277
*see also* Cytoskeleton
assembly process
genetic sequences crucial in, 281
head and tail domains of protein subunits, 281
longitudinal annealing, 280
phosphorylation inhibition, 281
unit filament formation, 280
assembly process of, 280
attachment to cell, 287–288
chemistry of, 278
desmin, 279T, 284, 286F
function of, cytoplasmic, 289
GFAP proteins, 284
keratins, 278F, 279–280, 279T
of desmosomes, 284, 285F
of epithelial cells, 283–284, 284F
genetic mutations of, and disease, 289
lamins, 282–283, 282FF
linker protein attachments, 287–288
neurofilaments, 278F, 279T
of axons, 287, 287F
nonpolarity of, 282
nuclear lamina formation, 282–283, 282FF
peripherin, 279T, 284
proteins of, 279, 279F, 279T
classes, 279–280, 279T
coiled-coil subunits, 279, 279n2
compared to actin and myosin, 281
subunits, 278
smooth muscle, 164
structure of, 278
tonofibril, 278n1
vimentin, 285–286
Interphase, 209
Intraflagellar transport (IFT), 247
Invagination, epithelial sheet, 344
actin based contractions, 345
IQ motifs, 105
of myosin classes, 114F

Jet propulsion, by cells, 7, 7F

Kartagener's syndrome, 235–236
Katanin, 185, 185T
in deflagellation, 244n1
in mitosis, 210
in neuronal polarity, 199–200
Kelvin orthic tetrakaidecahedron, 322
Keratin(s), 278, 279, 279T
of desmosomes, 284
of epithelial cells, 283–284, 284F
genetic mutations of, and disease, 289
Keratocyte, crawling movement of, 122, 122F
KIF2, mouse, 195
KIF3, 247
Kin I, 195
mitotic role of, 214
Kinases, protein, 52, 53F
calcium ion activated, 54
cytoskeletal response coordination, 54, 132–133
Kinesin(s), 185, 185T
conventional, 194–195, 195T
mitotic role, 214
molecular mechanics, 196–197, 197F
motility assay of, 196–197, 197F
in neuronal organelle transport, 193–194, 194F
superfamily of, 194–196, 195T
Kinesin II, 247
Kinesis, 30
Kinetochore, 212–213, 212F
motor protein attachment, 216–217
Kinetochore microtubules, 212–213, 213F

*Labyrinthulas*, gliding motion, 25–26
Lamellipodia
actin filament organization of, 122–123, 123F
amoeboid, 19
in embryonal cell movement, 21–22, 22F
fibroblast, 23, 23F
in locomotion, 23–24, 24F, *see also* Migration
formation of
chemotactic, 31
external stimulation and, 94
Lamin A, 279T
Lamin B, 279T
Lamins, 282–283, 282FF
Laser tweezers, 106
Latrunculins, 71, 71F, 71T, 72
*Leptospira*, movement, 269–270, 270F
*Limulus* sperm, movement, 50
Linker proteins, intermediate filament, 287
filaggrins, 288
plakins, 288
Links, monomeric and dimeric, 295–296, 296F
*Listeria monocytogenes*
actin nucleation in, 74, 74F, 298
VASP protein in motility, 74–75
Lobopodia, 19, 19F

M-line, 142, 143F
M phase, 209–210, 209F
Macromolecular crowding, 304, 304F
Macrophages, attractants and repellents, 30T
MAD (mitotic arrest deficient) proteins, 215
Major sperm protein (MSP), 120–122
in motility of nematode sperm, 121F
movement based on, 121, 121F
MAPs (microtubule associated proteins), 184–185, 185T
MAP2, 184–185, 185T
binding to microtubules, 193, 193F
in spindle formation, 210
Marginal band/bundle formation, 324–325, 325F
Mechanosensitive channels, cell, 310
Meiosis
bacterial roles in, 221

chromosomal movement in, 219–220
Membrane
actin and, 81
α-actin cross-linking, 81–83
adherens junctions, 87–89
attachment, 84, 84F
cell regulation, 97–98
of cultured cells, 89–92
of erythrocytes, 84–86, 85F, 86T
filamin cross-linking, 81–83
fragmenting proteins in reorganization, 83–84
Rho GTPase regulation, 92–94
surface structure formation and external stimuli, 94–98
mechanical response, 308–309
movement and, 130–131, 131F
stimuli and, 309T
Membranelles, 15
Mena, 133
Mesenchymal cells, 342
Metachronal wave, 13
Metaphase, 210, 213–215, 214FF
Metaphase plate, 213, 214F
Methylation, protein
and adaption of chemotactic response, 265, 265F
and conformation of membrane receptors, 266
Microtubule associated proteins (MAPs), *see* MAPs
Microtubule-organizing center (MTOC), 249–250
Microtubules, 43F, 44, 172F
*see also* Cytoskeleton; Tubulin
actin filaments and, cytoskeletal, 327, 327F, *see also* Cytoskeleton
arrays of, *see also* Cell shape; Cytoskeleton; Spindle
asymmetric, 181–182, 182F
cytoskeletal, 327, 327F
in plant cortex, 318, 318F
spontaneous assembly of, 326, 326F
and cell center, 325, 325F
centrosomal organization of, 180–181, 181F
ciliary, *see* Ciliary axonemes; Ciliary beating
in crawling movement, 44–45, 131–132
cutting, 199–200
cytoplasmic movement of, 183–184, 184F
doublet, 226–227, 227F
dynamic instability of, 178–179, 178FF
structural basis, 179–180, 180F
and dynein, in embryonic cell development, 339–340
dynein in sliding action of, 199, *see* also Dynein
in fertilization, 220–221
function modification of, 184–185
GDP storage of energy in, 180n5
kinetochore, 212–213, 213F
maturation of, 182–183
detyrosination and acetylation, 183
enzyme, 183
motor proteins and, 50–51, 51F, *see also* Dynein; Kinases; Motor proteins
movement through cytoplasm, 183–184, 184F
in myoblasts and myofibril formation, 158
neuronal organization of, 192, 192F, *see also* Nerve cells
nonmammalian, 84n2, 324
marginal band, 324–325, 325F
nuclear movement along, 203
and organelle transport, 189–190, 190F, *see also* Organelle transport
membrane binding, 190–191, 191F
physical structure of, 176–177, 177F
polarization of, 177–178, 211, 211F, 327, 327F
polymerization of, 48–49, 48FF
of protozoan feeding tentacles, 202–203
ribosomal movement along, 203
sliding action of
in ciliary beating, 230–236, *see also* Ciliary axonemes; Ciliary beating
dynein and, 199, *see also* Dynein
mitotic, 217–218, 217F
in other structures, 237–238, 237F
spindle, 210–211, 211F, *see also* Spindle
chromosomal movement on, 213, 213F
structure of
composition, 171–172
cross section, 177F
formation, 174–177
lumen of, 176
protofilament, 176–178, 177F
tubulin isoforms and, 174
wall, 177
viral movement along, 203
Microvilli
bundled proteins and rigidity of, 95–96, 95F, 96F
formation of, and external stimulation, 94
intestinal
bundled proteins, 96–97
major proteins, 96F
stereocilia, 97, 97F
Midbody, 218
Migration
embryonic cell, 340, 345–346, 346F, *see also* Embryonic cell movement
in brain formation, 346–347, 348F
in epithelial sheets, 345–346, 345FF
neural crest, 21–22, 22F
vertebrate, 21–22, 22F
over surfaces, 17, *see also* Crawling movement; Gliding movement
amoeboid, 18–19, 110
animal, 119
and cell division, 128, 129F
ciliary, 26–27
crawling movement and, 17–24, *see also* Crawling movement
cultured cell, 22–25, *see also* Cultured cells
cytoplasmic streaming and, 19
differentiated cell, 22–23
electrical currents generated by, 33
fibroblastic, 21, 21F, *see also* Crawling movement; Fibroblast locomotion
filopodial, 24–25, *see also* Filopodia
flagellar, 26–27
gliding, 25–26
growth cone, 25, *see also* Growth cones
lamellipodial, 23–24, *see also* Lamellipodia
neural crest, 34
neuronal, 25, *see also* Growth cones
neutrophilic, 20, 21F
pseudopodial, 18–19
signaling and, 54–55, 55F, 132–133
stimuli of, *see* Cell behavior
substrate features influencing, 34
substrate guidance in, 34, 34F
surface drag in, 17–18
vertebrate blood cell, 20
vertebrate embryonic cell, 21–22
viscous drag in, 17–18
white blood cell, 20
Mitosis, 208–209, 209F
anaphase mechanisms, 216, 216F
anaphase-promoting complex, 215–216
chromosomal attachment to spindle, 212–213, 212F
cyclin-dependent kinases, 210
cytoplasmic cleavage in spindle plane, 218–219
early stages, 211F
kinetochore–microtubule tension, 214–215, 215F
MAD signaling, 215
microtubule assembly of spindle, 210–211, 211F, *see also* Spindle
microtubule sliding, 217–218, 217F
molecular motor roles, 214
motor roles
chromosomal movement on spindle, 216–217
nuclear membrane breakdown, 210
nuclear reformation, 218
phosphorylation in, 209–210, 210F
plant cell wall formation, 219
pole migration, 217–218, 217F
protease activation, 215–216
proteolysis in, 209–210, 210F
regulation of, 209–210
spindle formation, 210–211, 211F
spindle residues, 218
stages of, 209F, 210, *see also* specific stage
Mitotic arrest deficient (MAD) proteins, 215
Moesin, 84, 84F, 325
Molecular rulers, 150–151
Monomeric links, 295–296, 296F
Morphogenesis, 315
cytoskeletal inheritance, 57–58, 57F
MotA, 260, 260F
MotB, 260, 260F
Motility assays, 105–106, 105F
measurement of molecules in, 106–107, 106F
Motor proteins, 50–51, 51F
*see also* Dynein; Kinesin(s); Myosin(s); Organelle transport; specific kinesin
ATP hydrolysis and conformational changes, 51–52, 51F
in ciliary growth, 247–248, 247F
in cytoplasmic contraction, 300–301, 301F
duty ratio of, 196, 196F
of microtubules, 190, 191F, *see also* Kinesin(s)
mitotic role of, 214, 216–217
in neuronal polarity, 199–200, 199–200F
mRNA
actin and translation, 304
address coding of, 334
bicoid, 334
nanos, 334
oskar, 334
and positional information in embryonal development, 334, 334F
translation sites, 246
MSP, *see* Major sperm protein
MTOC (microtubule-organizing center), 249–250
Muscle contraction, 137
aerobic, 138–139, 139F
anaerobic, 138–139, 139F
creatine phosphate shuttle, 139–140, 140F
desmin filaments, 286, 286F

energy production, 138–139
excitation–contraction coupling, 141–142, 141F
Galen contributions, 137
glycolysis in, 139
isometric, 138
maximum tension of, 138, 143–144
mitochondria in energy cycle of, 139–140
molecular basis of
accessory proteins, 149–150, 149T
calcium ion sensitivity, 147–148, 148F
force production of myosin heads, 146–147, 147F
myosin head cross-bridges, 145–146, 146F
nebulin filaments, 150–151, 151F
positioning of actin and myosin filaments, 144–145, 144F
sliding actin and myosin filaments, 142–144, 143FF
titin filaments, 150–151, 151F
tropomyosin action, 147–148, 148F
troponin action, 147–148, 148F
troponin C action, 148–149
muscle fibers in, 140–141, 140F
myofibrils in, 140FF, 141
stimulation of, 141–142
oxygen metabolism, 138–139
power output, 138
sliding filament hypothesis, 143–144, 143F
stimulation of, 137–138, 141–142, 141F
tension measurement, 137–138, 138F
tetanic, 138, 138F
Muscle development, 155, 156FF
cardiac, 163–164, 163F
exercise and gene expression, 160, 160F
fibroblast growth factor in, 156
genetic analysis, 167
genetic regulation, 156, 160–161
sequence of, 160–161
innervation influence, 159–160, 159F, 159n2
insect flight, 166
molluscan catch, 166
mononucleated myoblasts in, 156, 156F
in muscular dystrophy, 162–163
myotubules in, 156
RNA splicing and protein diversification, 161–162, 162F
smooth, 164–166, 164F, 165T, *see also* Smooth muscle
somites in, 156, 156F
stem cells in, 156–157
titin and myofibril assembly, 157–158
Muscle fiber, 140
cardiac, 163–164, 163F
costameres of, 286
development of, 157–158
excitation–contraction coupling, 141–142, 141F
fascicles, 140, 140F
insect flight, 166
molluscan catch, 166
myofibrils, 140FF, 141–142
myofilament turnover, 161
organization of, 140F
RNA splicing and protein diversification, 161–162, 162F
sarcolemma, 141
sarcoplasmic reticulum, 142
smooth, 164, 164F
ATP hydrolysis, 165
caveolae, 165
contraction of, 165–166
cytoskeletal components, 165T
dense bodies of, 164
dense plaque of, 164–165
enzyme cascade, 165–166
intermediate filaments of, 164
myosin activation, 109
plasma membrane, 164
regulation of, 165
T-tubules, 142
triad junction, 142
types of, 158–159, 158F, 158T
genetic regulation, 160–161
innervation influence, 159–160, 159F, 159n2
Muscular dystrophy, molecular lesion of, 162–163
Mycoplasmas, 316
Myoblasts
microtubules and myofibril formation, 158
mononucleated, 156, 156F
myoD genes, 156
Myofibril(s), 140FF
*see also* Muscle contraction, molecular basis of
banding of, 142–144, 143FF
anisotropic, 142
basal lamina, 141
calcium release channels, 142
compared to ciliary axonemes, 226
microtubules in formation, 158
sarcomere of, 142–144, 143FF
stimulation of, 141–142
titin in assembly of, 157–158
Myofilament turnover, 161
Myosin(s), 103, 104, 104F
actin interaction with, 108
assays of, 105–106, 105F
bipolar filaments of, 107–108, 107F
in budding yeasts, 319
in cell division, 112n2
classification of, 113T
in cytokinesis, 110–111, 111F
genetic analysis of, 111–112
in hearing and vision function, 115
heavy chain development, genetic regulation, 160–161
IQ motifs and motor domains of 13 classes, 114F
light chain phosphorylation, regulation of, 109
measurement of molecule, 106–107, 106F
in membrane vesicle movement, 113–115
molecular structure of, 104, 104F
mutants, genetic analysis, 167
myosin I, 113–114
action mechanics of, 113–114
activity in crawling movement, 127
in membrane vesicle movement, 113
in sensory transduction process, 115
myosin II, 103, 104, 104F, *see also* Myosin
activity of, in crawling movement, 126–127
in cytoplasmic cortical tension, 302–303, 302F
myosin III, associated with visual defects, 115
myosin V, and membrane vesicle attachment, 114–115
myosin VI, associated with deafness, 115
myosin VII, associated with deafness, 115
phosphorylation regulation of, 108–109, 108F, 109n1
light chain, 109
recombinant DNA analysis, 112–113
sequences, 114F
tropomyosin role, 109–110, 110F
unconventional, 112–113, 113T, *see also* specific myosin class
Myosin head(s), 104–105, 104f
α-helix of, 146
actin binding action, 104–105
ATP hydrolysis, 104–105
cross-bridge cycle in muscle contraction, 145–146, 146F
force production of, 146–147, 147F
three-dimensional structure, 147, 147F
Myosin light chain kinase (MLCK), 109
catalytic domain of, 109
in smooth-muscle contraction, 165
Myotubules, 156
*Myxobacteria*, gliding movement, 26, 26F, 270–271
*Myxotricha paradoxa*, 253

*Naeglaria*, centriole formation, 244, 252–253
Nanos mRNA, 334
Ncd protein, *Drosophila*, 195
Nebulin, 150–151, 151F
Nematode sperm
crawling motility, 120–122
nonpolarity of MSP filament, 282n3
Nerve cells
actin filament distribution, 45F
attractants and repellents, 30T
axonal growth, 35, 348–349
behavior of, 35
brain, 346–350, 348FF
cross-linking proteins, 45–46, 46F
dendritic growth, 348–349
growth cone migration, 25F
axonal G cell in, 35
chemotactic, 25
filopodial, 25, 25F
inhibition of, 36
oligodendrocytic inhibition of, 37n2
pathway selection in, 35F
microtubules in
stabilization, 193, 193F
microtubules of
organization, 192, 192F, 199–200
migration of, 25
motor
axons and dendrites, 192F
size of, 191F
organelle transport, 191
dynein in, 197–198
kinesin in, 193–194, 194F
polarity of, 192, 199–200
shape of, 316
tensegrity of, 301, 302F
Net slime molds, gliding motion, 25–26
Netrin, 348–349, 349F
Neural crest migration, 21–22, 22F
substrate features influencing, 34
Neural tube formation, 346, 347F
Neurofilaments, 278, 278F, 279T
of axons, 287, 287F
Neuronal polarity, 192
motor proteins and, 199–200, 199–200F
Neurons, *see* Nerve cells
Neurulation, 340, 346, 347F
Neutrophils
attractants and repellents, 30T
chemoattractants, 31
chemotactic behavior, 31, 132
chemotactic response
lamellipodial formation in, 31
observatory chamber for, 31F
migration of, 20, 21F
Nexin links, 227

*Nitella*, studies of cytoplasmic streaming, 105–106
Nocadazole, 175T, 176
Nuclear lamina formation, 282–283, 282FF
Nucleating actin filaments
*Listeria monocytogenes* motility and, 73–74
proteins associated with, 74–75, 133, 133T
and unconventional myosins, 127
Nucleation site, centrosomal, 181, 181F
Nucleotide diphosphate (NDP), 48
Nucleotide hydrolysis, of protein filaments, 48–49, 48FF
Nucleotide triphosphate (NTP), 48
Nucleus
membrane of eucaryotic, 282–283, 282FF
movement along microtubules, 203
Nutritive tubes, 203

Oocyte
cortical response to fertilization, 94, 95
polarization
amphibian, 336–338
*Fucus*, 321, 321F
Organ of Corti, 328, 328F
Organelle transport, 189–190, 190F
in chromatophores, 201–202, 202F
dynein in, cytoplasmic, 197–198
kinesin superfamily in, 194–196, 195T
kinesins in, 193–194, 194F
motility assays, 196–197, 197F
MAP2 in, 193, 193F
membrane binding in, 190–191, 191F
and microtubule arrangement, 191–192, 192F
myosin in, 200
in nerve cells, 191, 191FF, *see also* Nerve cells
other motor proteins in, 200–201
phosphorylation regulation of, 201–202
tau protein in, 193, 193F
Oskar mRNA, 334

P-loops, 232
*Paramecium*
ciliary beating motion, 13–14, 14F, 234–235
reversal of, 33–34, 34F
ciliates of, 13
complex behavior of, 38
cytoskeletal inheritance in, 57, 57F
formation and growth of cilia, 244
obstacle avoidance, 33–34, 34F
Parvalbumin, 149
Paxillin, 91T
Pericentrin, 181
*see also* Centrioles; Centrosome
Peripherin, 279T, 284
Persistence length, filament, 294, 295T
Phagocytosis, ruffling membrane in, 94
Phalloidins, actin binding, 71, 71T
Phosphatidylinositol bisphosphate, *see* PIP2
Photoshock response, 38
Phototaxis, 37–38, 37F
*Physarum*, cytoplasmic streaming, 19
Pigment granule movement, 201, 202F
$PIP_2$ (phosphatidylinositol bisphosphate), 52, 53F, 132
Placode, 346
Plakins, 288
Plakoglobins, 88, 89T
Plant cells
centriole generation in sperm, 252
cytokinesis in higher, 219, 220F
division of, 317–318, 318F
growth of, 317–318, 318F
hydrostatic pressure in cytoplasm, 306–307, 307F
morphogenesis of, 219, 316–317, 317F
signaling in, 54n5
subcortical microtubules of, 317–318, 318F
wall of, 316
formation, 219
Plasmagel, 19, 20F
Plasmasol, 19, 20F
*Plasmodium*, gliding motion, 26
Plectin, 288
Podophyllotoxin, 175T, 176
Polar microtubules, 211, 211F
Polymerization
actin, 48–49, 48FF, *see also* Actin filaments
asymmetrical growth in, 69, 69F
ATP-cap formation in, 70, 70F
ATP hydrolysis and rate of, 70, 71F
barbed end, 69, 95
binding proteins, 72–73, 73F
critical concentration, 67–68
kinetic equations of, 68, 68F
lag phase, 67
pointed end, 69
regulators of, 132–133, 133T
reversibility of, 68
stages in, 68F
time course, 67F
toxin disruption, 71–72
movement by filament, 49–50, 49F
protein filament, 48–49, 48FF
tubulin, 48–49, 176, *see also* Dynamic instability; Microtubules
GTP-cap, 179–180, 179F
GTP-hydrolysis, 176, 179
'plus' end and 'minus' end, 177
Power stroke, 51F, 52
Primary cilia, 248, 248F
Profilactin, 73
Profilin, assembly of actin monomers, 72–73, 73F
Prophase, 210–213, 211FF
Protein(s)
conformation and movement, 50, 50F, *see also* Cell movement
in cross-links, *see* Cross-linking proteins
filaments of, 46–47, 47F, 48F, *see also* Actin filaments; Intermediate filaments; Microtubules
motor, 50–51, 51F, *see also* Dynein; Kinesin(s); Motor proteins; Myosin(s); Organelle transport; specific kinesin
signaling, *see* Signaling
three-dimensional structure and evolutionary origin, 64n1
Protein kinases, 52, 53F
calcium ion activated, 54
cytoskeletal response coordinated by, 54, 132–133
*Proteus mirabilis*
differentiation into swarmer cells, 27
flagella of, 9–10
*Protozoa*
ciliary motion, 12–14, 14F
flagellar structures, 10–11
flagellar waveforms, 11–12, 11FF
microtubules in feeding tentacles of, 202–203
parasitic, gliding motion, 25, 26
Pseudopodia
ameoboid, 18–19, 18FF
cytoplasmic formation of, 19, 20F
types of, 19F
Purkinje cell, 316F
Quorum sensing, 267–268

Rac protein, 92–93, 93F, 132, 271
Radial spokes, mechanics of, 233–234, 233F
Radixin, 84, 84F, 89T, 325
Rafts, 247
Ran1 protein, 210
Random walk, 4
mathematical description of, 5F
Ras superfamily, of G proteins, 92
proteins related to, 172–173, 173n2
Receptor–receptor coupling, 266–267, 267F
Recovery stroke, 51F, 52
Red blood cells, *see* Erythrocytes
Repellents, cell, 30T
*Reticulomyxa*, 202
Reynolds number(s), 6
in description of swimming motion, 6
fluid flow and, 6F
low, propulsion at, 7F
Rho1, 319
Rho GTPase, 92
action mechanism of, 92
link with focal adhesions, 92
regulation of surface receptors and actin cytoskeleton, 92–93, 93F, 132
Ribosomes, movement along microtubules, 203
RNA targeting, mechanism of, 77
RNA transport particle, 335, 335F
Ruffling, 23–24, 24F, 128
and cellular uptake in bacterial/viral infection, 271, 271F
external stimuli causing, 94, 271, 271F

S1 fragment, 105
S phase, 209–210, 209F
*Saccharomyces*
actin in budding, 319–320
cell shape, 316
G protein regulation of budding, 318–319, 319F
*Saccinobacculus*, 237–238
*Salmonella*, swimming speed, 8
*Sarcodina*, pseudopodia of, 18–19
Sarcolemma, 141
Sarcomere, 142–144, 143FF
development of, 157–158
filament arrangement, 143F
Sarcoplasmic reticulum, 142
Scar protein, 133
*Schizosaccharomyces*
cell shape of, 316
microtubules of, 320
polarity of, 320
Scruin, 95
Semaphorin, 348–349
Septins, 319
Shuttle streaming, 19
Signaling
in bacterial infection, 31
bacterial pathways, in chemotactic responses, 264–265, 264FF
and cell migration, 54–55, 55F, 132–133
cyclic AMP in, 36–37
in embryonal cell movement, 341
MAD, in mitosis, 215
in plant cells, 54n5
transmembrane
in extracellular matrix, 329
at focal adhesions, 89–91
of T lymphocytes, 327
Signaling molecules, 52, 53F, 54
Signaling nucleotides, of slime mold, 36–37
Signaling peptides, of bacterial infection, 31

Signaling receptors, 132–133
Situs inversus, 235–236
Sliding filament hypothesis, 143–144, 143F
Slime mold (*Dictyostelium*)
amoeboid signaling, 36–37, 37F
analysis of myosin genetics, 112
attractants and repellents, 30T
chemotactic responses, 30, 132, 132F
cortexillin actin binding sites, 76–77
myosin I studies in, 113
myosin II studies in, 126–127
Slit protein, 348–349
Smooth muscle, 164, 164F
ATP hydrolysis in, 165
caveolae of, 165
contraction of, 165–166
cytoskeletal components of, 165T
dense bodies of, 164
dense plaque of, 164–165
enzyme cascade of, 165–166
intermediate filaments of, 164
myosin activation in, 109
plasma membrane of, 164
regulation of, 165
Somites, muscle, 156, 156F
Spasmoneme contraction, 50, 50F
Spectrins, 83F
actin binding sites of, 76
in cortex of human erythrocytes, 84–86, 85F
diversity of mammalian, 86n3
gene families regulating, 86
linkage to plasma membrane of erythrocytes, 86
network regulation by erythrocytes, 87
Speed, cell, 2F
diffusion, formula for, 4
Sperm
flagellum of, 10F
*Fucus*, flagellar motion, 12n3
invertebrate
bundling proteins and rigidity, 95–96, 95F
crawling motility of, 120–122
nonpolarity of MSP filament, 282n3
*Limulus*, movement, 50
MSP protein of, 120–122
nematode sperm motility and, 121F
multiflagellated, *de novo* centriole generation, 252
plant, *de novo* centriole generation, 252
Spindle
chromosomal attachment, 212–213, 212F
chromosomal movement, 216–217
cytoplasmic cleavage in plane of, 218–219
formation of, 210–211, 211F
phosphorylation in, 209–210, 210F
proteolysis in, 209–210, 210F
kinetochore–microtubule tension, 214–215, 215F
MAD signaling, 215
microtubule formation of, 210–211, 211F
microtubule sliding in, 217–218, 217F
motor proteins and chromosomal role, 214, 216–217
pole migration on, 217–218, 217F
protease activation, 215–216
residue of, 218
Spirochetes, motility of, 10, 269–270, 270F
Src, 89T, 91T
Stathmin, in mitosis, 210
Stathmin/Op18, 185, 185T
Staufen protein, 334
Stem cells, muscle, 156–157
*Stentor*, complex behavior, 38–39, 39F
Stereocilia, 97, 97F
*Sticholonche*, axopodial movement, 27
STOPs, 185, 185T
Stress fiber, formation of, 89, 125
Striated rootlets, 248–249, 249F
*Stylonychia*, ciliary movement, 27F
Subfragment 1 (S1), 104
Surface appendages, 4
Surface drag, 17–18
Swarming
bacterial, 10, 269, 269F
of *Myxobacteria*, 26, 26F
Swimming movement, 3
bacterial and eucaryotic, comparison, 10–11
ciliary, 12–14
modified, 14–15
movement of water in, 13–14
diffusion mechanics and, 3–6
energy consumption in, 7–8
eucaryotic, 11–14
waveforms of, 11–12
flagellar, 8–11
bacterial, 9
modified, 14–15
of flagellated bacteria, 9–10
kinetics of, 3–4
protozoan ciliary, 12–13
surface drag in, 18
viscous forces and, 6–7
*Synechococcus*, motility of, 272
Syneresis, 108

T-lymphocyte immune response, 327, 327F
T-tubules, 142
Talin, 91T
Tau protein, 185, 185T
binding to microtubules, 193, 193F
Taxis, 30
Taxol, 175T, 176
structure of, 175F
teal gene, 320
Tektin, 185, 185T
Tektin protofilaments, 228–229
Telophase, 210, 218, 218F
Tensegrity, 301–302, 302F
Tensin, 89T, 91T
Thermal ratchet, 49–50, 297
Thiabendazole (TBZ), 175T, 176
Thymosin, 72
*Thyone*, isolation of actin monomer binding proteins, 72–73
Tight junction, 342–343, 343FF
Tissue culture, *see* Cultured cells
Titin, 150–151, 151F
C-terminal end, 157
and myofibril assembly, 157–158
Toxins
actin binding, 71T
disruption of actin polymerization, 71–72
Traction mechanism, 125
Treadmilling, 70
actin filament polymerization dynamics, 69–70, 69FF
lamellipodial, 123–124, 124F
microtubule, 183–184, 184F
*Treponema*, flagellar motion, 10
Triad junction, 142
*Trichomonas*, flagellar motion, 12, 13F
*Trichonympha*, specialized flagella, 15, 15F
Tropomodulin, 157
Tropomyosin, 109–110, 110F
in muscle contraction, 147–148, 148F
Troponin, in muscle contraction, 147–148, 148F
Troponin C, 148–149
Troponin I, 148
Troponin T, 148
alternate mRNA splicing and, 161–162, 162F
*Trypanosma brucei*, flagellar waveform, 12
Tubulin, 171
*see also* Microtubules; Tubulin filaments
of brain, 172
GTPase activity of, 172–173
dynamic instability of microtubules, 178–180, 179FF
isoforms of, 172–173, 181, 181F
importance of, 174
molecular structure of, 172–173, 173F
polymerization of
assembly into microtubules, 176
drugs affecting, 175–176, 175T
folding process, and chaperones, 174–175
protofilaments of, 176–177, 177F
minus end, 177
plus end, 177
polarity, 177–178
and similarity to actin, 172, 173T
vertebrate gene families of, 173–174, 174F
Tubulin acetyltransferase, 183
Tubulin filaments, polymerization of, 48–49, 48FF
Tumor cells, lamellipodia formation, 31, 32F

Unit filaments, 280

van't Hoff equation, 306
VASP (vasodilator-stimulated phosphoprotein), 74–75, 133
Velocity(ies)
comparison of cell, 2F
by diffusion, 4, 6T
*Vibrio fischerii*, 267–268
Villin, of intestinal microvilli, 96–97, 96F
Vimentin, 279T, 285–286
of *Xenopus*, 280F
Vinblastine, 175T, 176
Vincristine, 175T, 176
Vinculin, 89T, 91T
Viruses, movement along cytoplasmic microtubules, 203
Viscoelasticity, 298–299, 298F
cytoplasmic, 299–300, 299F
Viscosity, 3, 298–299, 298F
appendicular motion in, 7
close to solid surfaces, 17–18
swimming movement and, 6–7
Viscous drag, 6, 17–18
*Vorticella*, spasmoneme contraction, 50, 50F

WASP (Wiskott–Aldrich syndrome protein), 133
Waveforms, of eucaryotic flagella, 11–12, 11FF
White blood cells, migration, 20, 21F
Wnt proteins, 341
*Wolbachia*, 221, 271

Z-band, *see* Z-disc
Z-disc, 142, 143F
development of, 157–158
ZBP (zipcode binding protein), 334
*Zoothamnium*, movement, 50
Zyxin, 89T